Road and Off-Road Vehicle Dynamics

Moustafa El-Gindy · Zeinab El-Sayegh

Road and Off-Road Vehicle Dynamics

Springer

Moustafa El-Gindy
Department of Automotive
and Mechatronics Engineering
Ontario Tech University
Oshawa, ON, Canada

Zeinab El-Sayegh
Department of Automotive
and Mechatronics Engineering
Ontario Tech University
Oshawa, ON, Canada

ISBN 978-3-031-36215-6 ISBN 978-3-031-36216-3 (eBook)
https://doi.org/10.1007/978-3-031-36216-3

This Springer imprint is published by the registered company Springer Nature Switzerland AG
The registered company address is: Gewerbestrasse 11, 6330 Cham, Switzerland

To my family Iman, Nancy, Amir, and Mona El-Gindy for their unwavering support.

Moustafa El-Gindy

To the glory of Almighty; my parents Iman El-Kadi and Ali El-Sayegh; my siblings Batoul, Alyaa, Fatema, and Abbas El-Sayegh for their support.

Zeinab El-Sayegh

Preface

The need for a book to cover road and off-road vehicles including military multi-wheeled combat vehicles is high. The presented information in this book is intended for engineering students, researchers, and engineers. We have used part of the material included in this book in several undergraduate and graduate courses offered at Penn State and Ontario Tech Universities.

In writing this book, the authors have gained a lot of knowledge from the research work and projects performed with research organizations, industries, and government agencies.

The authors would like to express their appreciation to the NSERC Discovery Grant, Volvo Group Trucks Technology, General Dynamics Land Systems-Canada, US Army, and US Navy for supporting several research projects that helped us to prepare this book.

Special thanks to Prof. J. Y. Wong and Dr. Thomas Gillespie for inspiring the authors. The Books of *Theory of Ground Vehicle* by J. Y. Wong and *Fundamentals of Vehicle Dynamics* by Thomas Gillespie were great sources of information that helped us to write this book.

Also, we would like to express our gratitude to the Penn State University graduate students, Abehshik Joshi, Andrew H. Hoskins, Aric Sherwood, Brent W. Shoffner, Seokyong Chae, Jeffrey L. Slade, James Richard Allen, II, Jonathan Culp, Anstrom Joel, James D'Iorio, Yongjie Lai, Robert W. Goldman, and Meghan Henty Hoskins, who provided us with indirect support through their thesis work, advised by Prof. M. El-Gindy. Similar thanks to the Ontario Tech University graduate students: Todd MacDonald, Patrick Galipeau-Bélair, Aaron Hao Tan, Adam George Mount Cook, Adam Cameron Reid, Amr Mohamed, Hossam Ragheb, Arnold Odrigo, Fatemeh Gheshlaghi, Brett Russell, Kristian Lander, Laith Dababneh, Mehrsa Marjani, Mirwas Sharifi, and Moataz Ahmed.

Special thanks to Prof. Dr. Laszlo Palkovic and Prof. Dr. Peter Tomas of Budapest Technical University, Hungary, for using some of their work in this book. In addition to Drs. Ahmed Ouda, Ahmed Salem, AL-Hossein Sharaf, and Ahmed Khalil for their help during their Post-doctoral Fellowship at Ontario Tech University.

Special thanks to Dr. Xianke Lin for the perpetration of the Hybrid Electric Vehicle sections in Chap. 6 and the assistance received by Jonathan Tse, a research associate.

We would like also to acknowledge the references listed at the end of this chapter and express our appreciation and gratitude to several organizations and individuals for their permission to reproduce illustrations and other copyrighted materials. We would like to appreciate the effort of Mrs. Natasa Blagojevic in preparing the illustrations presented in this textbook.

Oshawa, ON, Canada

Moustafa El-Gindy
Zeinab El-Sayegh

Introduction

This book introduces advanced tire mechanics and interaction, including hard and soft soils. Emphasis will be placed on the basic mechanics governing a vehicle's directional, tractive, braking, and ride performance on both road and off-road terrains. Steering and steady-state handling characteristics of wheeled vehicles will also be introduced and analyzed. The state-of-the-art approaches to design syntheses of automotive vehicles will also be discussed, including underride crash analysis. This book will help prepare students for the research and development of modern road vehicles.

The readers of this book should be able to

- understand basic design principles and general design procedures of typical automotive components including tires, drivelines, steering systems, suspensions, and brakes;
- understand the modeling techniques of the FEA tires and soils. In addition, the use of the Smoothed-Particle Hydrodynamics (SPH) techniques for soil, water, and snow modeling;
- demonstrate a sound understanding of the fundamentals of vehicle dynamics stability and control;
- gain expertise in analyzing and evaluating road vehicle performance;
- gain experience in heavy vehicle performance evaluation and rollover dynamics;
- get familiar with the design aspects of the front, side, and rear underride protection devices of heavy vehicles.

This book includes new topics that may not be found in the available published vehicle dynamics books, for example, it includes chapters related to tire-terrain interaction, multi-wheeled combat vehicle dynamics, heavy truck performance measures, rollover dynamics as well as crashworthiness topics.

Chapter 1 starts with an introduction to the history of wheels and tires and presents the construction of various types of tires, such as radial-ply and cross-ply tires. The longitudinal and lateral forces and moments developed by the tires are discussed in detail. Analytical, empirical, and semi-empirical models of pneumatic tires are systematically developed to study the tractive, braking, ride, and cornering behavior of the vehicle, which are also presented in this chapter. Most of the materials in this chapter were prepared based on the *Road Vehicle Theory* book, by Prof. J. Y. Wong.

Chapter 2 presents tire-terrain interaction modeling and analysis. As off-road vehicles are becoming more demanding, the study of severe conditions is becoming more critical. During wintertime, the tire may be subjected to snow falling or accumulated snow layers on the ground, which can severely change the performance of the vehicle. Thus, studying tire-snow interaction becomes a critical demand as it becomes more and more important for safety and performance analysis. In order to perform any tire-terrain interaction analysis, a well-calibrated terrain model is needed. The accuracy of the terrain model drastically affects the accuracy of the tire-terrain interaction. Most of the materials in this chapter were prepared based on Dr. Z. El-Sayegh's research work under the supervision of Dr. M. El-Gindy.

Chapter 3 focuses on the performance characteristics of off-road vehicles by investigating the terrain-dependent in-plane and out-of-plane rigid ring model parameters in different conditions. The in-plane rigid ring model parameters are the longitudinal tire stiffness, vertical stiffness, and rolling resistance coefficient. The out-of-plane rigid ring model parameters include lateral stiffness, cornering stiffness, self-aligning moment stiffness, and relaxation length.

Chapter 4 discusses the evaluation of the handling characteristics of a vehicle based on its response to steering input. The control of the direction of motion of a vehicle and its ability to stabilize its direction of motion against external disturbances are the basic issues in vehicle handling.

Chapter 5 presents a comprehensive introduction to the rollover mechanics of various vehicle models. Although a technically correct definition of rollover would be the state at which the overall center of gravity (CG) of the vehicle has moved laterally past the vehicle's "balance point", researchers typically define the rollover point or rollover threshold as the state where the load from one side of the vehicle has transferred to the other side. This issue is analyzed in detail in this chapter. Most of the materials in this chapter were prepared based on the research work performed by Dr. Robert Goodman of the US Navy under the supervision of Dr. M. El-Gindy.

Chapter 6 discusses road vehicle performance characteristics. The two main characteristics are the tractive and braking efforts. Tractive effort is developed by the tires in order to overcome the resisting forces acting on a vehicle which determines the performance potential of a vehicle on flat road surfaces.

Chapter 7 presents the results of a study and focuses on examining the stability and control characteristics of multi-wheeled combat armed vehicle configuration. The vehicle is evaluated using computer simulations and field tests during step steering input (J-Turn) and lane change maneuvers. The vehicle model is validated against the measured directional responses. With increases in computational power and the accuracy of the simulation models, validated computer simulation models can be extensively used as an alternative to full-scale real tests, in particular in severe maneuvers.

Chapter 8 presents extensive discussions of multi-wheeled compact vehicles, in particular the 8×8 configurations. The off-road vehicle's behavior depends not only on the total provided power by the engine but also on the power distribution among the drive axles/wheels. In turn, this distribution is primarily regulated by the drivetrain layout and the torque distribution devices. A number of simulation studies, during longitudinal and cornering maneuvers, are conducted to investigate the contribution of typical significant parameters. In addition, the influences of different drivetrain arrangements are presented. The obtained results defined that both the traction and cornering response of multi-wheeled off-road vehicles are highly affected by the driving torque distributed between axles/wheels. This chapter is prepared with the help of Dr. Hossam Ragheb of the Egyptian Military Technical College. Most of the materials were part of Dr. Ragheb's research work under the supervision of Dr. M. El-Gindy.

Chapter 9 presents several examples of the suspension systems of road-driven cars and trucks. Vehicles use a suspension system to keep the tires on the road and to provide acceptable riding comfort. A vehicle with a solid suspension or no suspension would bounce off the ground when the tires hit a bump. Roll centers and roll axis concepts are explained, and simplified examples of road vehicle suspension systems are introduced. Most of the materials in this chapter were prepared based on the *Fundamentals of Road Vehicles* book by Dr. Thomas Gillespie.

Chapter 10, the final chapter, presents a comprehensive summary of the recent underride research work that stems from many aspects, such as the design of front, side, and rear underride protection devices for heavy trucks. The foremost motivation arises from the well-known unreasonable fatality risks to automobile occupants in the event of a car-to-heavy vehicle collision, which are attributable to (i) the high kinetic energy of the heavy vehicle with its disproportionate mass, size, and power-to-weight ratio; and (ii) the automobile under-riding the heavy vehicle structure. The under-riding, where the automobile tends to wedge under the heavy vehicle structure, is particularly associated with greater fatality risks since the impact occurs at a location other than the primary energy-absorbing structure. The severity of such collisions can be greatly reduced through the designs of safety guards that prevent intrusion of the car underneath the truck/trailer structure. This chapter was prepared based on the research work performed by Todd MacDonald and Patrick Galipeau-Belair under the supervision of Dr. M. El-Gindy.

In summary, this book summarizes some of the authors' teaching and research experience over the past years and the collaboration with industries, such as Volvo Group Trucks Technology and GDLS-Canada, and Truck manufacturers and operators in addition to the governmental research agencies that funded our research projects, such as Auto21, NSERC, BC Ministry of Transportation, US Navy, and US Army. I wish also to express my appreciation to our postdoctoral fellows, research associates, and postgraduate students, former and present, for their contributions and assistance.

Contents

Abbreviations and Nomenclature

Abbreviations

ACD	Active Center Differential
ANN	Artificial Neural Network
ASC	Active Stability Control
AWD	All Wheel Drive
AYC	Active Yaw Control
CFD	Computational Fluid Dynamics
CSV	Critical Sliding Velocity
ESP	Electronic Stability Program
FAR	Footprint Aspect Ratio
FEA	Finite Element Analysis
FSP	Front Sprung mass
FSS	Front Suspensions Roll Stiffness
FUS	Front Unsprung mass
FVM	Finite Volume Method
GA	Genetic Algorithm
GVW	Gross Vehicle Weight
GVWR	Gross Vehicle Weight Rating
HC	Hydraulic Coupling
HFU	High-Speed Utilization
HOF	High-Speed Offtracking
HTC	Hydraulic Torque Converter
LAFB	Lateral Acceleration Feedback
LFU	Lateral Friction Utilization
LOF	Low-Speed Offtracking
LQR	Linear Quadratic Regulator
LSD	Limited Slip Differential
LTR	Load Transfer Ratio
MTD	Mean Texture Depth
NATO	North Atlantic Treaty Organization

PID	Proportional-Integral-Differential
RHD	Regional Haul Drive
RMS	Root Mean Square
RSF	Roll Safety Factor
RSP	Rear Sprung mass
RSS	Rear Suspensions Roll Stiffness
RUS	Rear Unsprung mass
RWA	Rearward Amplification
SAE	Society of Automotive Engineering
S-AWC	Super All Wheel Control
SPH	Smoothed-Particle Hydrodynamics
SRT	Static Rollover Threshold
SVA	Single Vehicle Accidents
TOF	Transient High-Speed Offtracking
TWR	Track Width Ratio
UMTRI	University of Michigan's Transportation Research Institute
VDC	Vehicle Dynamics Control
VTD	Variable Torque Distribution
YDR	Yaw Damping Ratio

Nomenclature

a	Half contact length between tire and road surface
b	Plate width, m
B	Stiffness factor
c	Cohesion constant of terrain, kPa
$c_{01,}\ c_{10}$	Coefficients of Mooney-Rivlin
C	Shape factor
c_{bx}	In-plane translational damping of sidewall, kNs/m
c_{bz}	In-plane vertical damping of sidewall, kNs/m
c_{by}	Out-of-plane translational damping constant, kNs/m
$c_{b\gamma}$	Out-of-plane rotational damping constant, kNms/rad
$c_{b\theta}$	In-plane rotational damping of sidewall, kNms/rad
c_c	Critical damping constant, kNs/m
c_l	Out-of-plane slip damping constant, kNs/m
c_{vr}	Residual damping constant, kNs/m
d	Tire deflection due to loading, m
D	Peak factor
E	Young's modulus of the terrain, MPa
E_t	Tangential modulus, MPa
f^{α}	External force, N
f_n	Acceleration, m/s^2

f_r	Rolling resistance coefficient
f	Seepage force, N
F_d	Drawbar pull, N
F_k	Force on node, N
F_p	Force on particle, N
F_x	Longitudinal (tractive) force, kN
F_y	Lateral force, kN
F_{yf}	Cornering forces at the front tire, N
F_{yr}	Cornering forces at the rear tire, N
F_{ytot}	Total lateral force, kN
F_z	Vertical force, kN
G_S	Specific gravity of soil
G_L	Specific gravity of liquid
G	Shear modulus of the terrain, MPa
I_{ax}	Mass moment of inertia of tire rim about x-axis, kgm^2
I_{az}	Mass moment of inertia of tire rim about z-axis, kgm^2
I_{ay}	Mass moment of inertia of tire rim about y-axis, kgm^2
I_{bx}	Mass moment of inertia of tire about x-axis, kgm^2
I_{bz}	Mass moment of inertia of tire about z-axis, kgm^2
I_{by}	Mass moment of inertia of tire about y-axis, kgm^2
K	Bulk modulus of the terrain, MPa
C_{bx}	In-plane translational stiffness of sidewall, kN/m
C_{bz}	In-plane vertical stiffness of sidewall, kN/m
C_{by}	Out-of-plane translational stiffness, kN/m
$C_{b\gamma}$	Out-of-plane rotational stiffness, kNm/rad
C^*	Brake factor
$C_{b\theta}$	In-plane rotational stiffness of sidewall, kNm/rad
k_c	Pressure-sinkage parameter, kN/m^{n+1}
k_ϕ	Pressure-sinkage parameter, kN/m^{n+1}
k_{cx}	Longitudinal tread stiffness. kN/m
C_α	Cornering stiffness, kN/rad
C_s	Longitudinal slip stiffness, KN/slip unit
C_l	Lateral slip stiffness, kN/m
C_M	Self-aligning moment stiffness, kNm/rad
C_{tot}	Tire total vertical stiffness, kN/m
C_{vr}	Residual vertical stiffness, kN/m
C_θ	Pressure-sinkage parameter, kN/m^{n+2}
G	Gradeability
h_0	Swollen soil height, m
K_e	Engine capacity factor
L	Applied vertical load, kN
m	Mass, kg
m_a	Wheel rim mass, kg
m_b	Tire belt mass, kg
m_{tot}	Mass of the tire and rim $(m_a + m_b)$, kg

m_{tread}	Mass of the tread of the tire only, kg
M_x	Overturning moment, kNm
M_y	Rolling resistance moment, kNm
M_z	Vertical or aligning moment, kNm
n	Exponent from terrain values
P	Tire inflation pressure, kPa
p_{gcr}	Critical pressure, kPa
R	Radius of the inflated tire before loading, m
R_a	Aerodynamic resistance
R_b	Bulldozing resistance
R_e	Effective rolling radius, m
R_{drum}	Drum radius, m
s	Longitudinal slip, %
S	Shoe factor
S_h	Horizontal shift
S_v	Vertical shift
u_s	Specific fuel consumption of the engine, kg/kWh
v, v_{tire}	Tire speed, m/s
u	Lateral tire speed, m/s
v_{drum}	Drum speed, m/s
W	Strain energy function
x_n	Position of node, m
X	Slip angle/skid
X_c	Destructive angle, degree
j	Shear displacement, mm
z	Sinkage of disc in Bekker equation, m
σ	Yield stress of the soil, MPa
σ_a	Minor principal stress, MPa
σ_p	Passive earth pressure, MPa
α	Slip angle, rad
δ	Log decrement
$\delta^{\alpha\beta}$	Kronecker's delta
γ	Amplitude ratio of the yaw oscillation output
γ_m	Mass factor, kgm^2
γ_w	Sand porosity
θ_{ss}	Steady-state angle value for rotation, rad
θ_0	Angle where a point on the rim comes in contact with the terrain, rad
θ_1	First peak angle, rad
θ_2	Second peak angle, rad
θ_f	Entry angle, rad
θ_r	Exist angle, rad
η_{tr}	Transport efficiency
ρ	Density of terrain, kg/m^3
$\tau^{\alpha\beta}$	Shear stress, pa
τ	System time constant, sec

τ_d	Damped period of vibration, sec
ω	Wheel angular speed, rad/s
ω_d	Damped natural frequency, rad/s
ω_{drum}	Drum angular speed, rad/s
ω_n	Un-damped natural frequency, rad/s
ω_{path}	Path frequency, rad/s
ω_y	Yaw oscillation frequency, rad/s
ψ	Damping ratio
τ	Shear stress, MPa
τ_{max}	Maximum shear strength, MPa
ϕ	Angle of internal shearing resistance, deg
ϕ_s	Sprung mass roll angle, rad
ϕ_u	Axle roll angle, rad
$\in$	Critical damping
ξ_1	Lower gear ratio

Chapter 1
On-Road Tire Mechanics

This chapter starts with an introduction to the history of wheels and tires and presents the construction of various types of tires, such as the radial-ply and cross-ply tires. The longitudinal and lateral forces and moments developed by the tires are discussed in detail. Analytical, empirical, and semi-empirical models of pneumatic tires are systematically developed to study the tractive, braking, ride, and cornering behavior of the vehicle, which are also presented in this chapter.

1.1 History of Tires

Round logs have been used to move heavy payloads more easily since a long time ago [1]. People placed a piece of solid wood under the payload object and pulled it over one round log to the next location. The round log slowly changed to solid wheels at each end and was connected with an axle, which were first utilized for transportation in 3,500 BC. These solid wheels were primitive and later were enhanced to form the tires we have today.

It is noted that wooden solid wheels had been produced with different shapes (spokes), such as leather-tops, and iron strip-tops for a very long time. Charles Goodyear in 1839 [2] introduced the vulcanization process, which is the process of heating raw rubber with sulfur to reconstruct rubber compounds in a firm form. This vulcanization process produces ideal rubber compounds for engineering applications, including tires. After the discovery of the vulcanization process, tires were manufactured using solid rubber. These solid rubber tires were strong enough to withstand the various operating conditions and resist the abrasions. These tires were able to absorb some vibrations from the road irregularities. Nevertheless, despite these features, solid rubber tires were heavy and did not provide automobiles with a smooth ride.

Following this, in 1845, Robert Thomson [3] came up with the idea of pneumatic tires and anticipated that these tires could overcome the limitation of solid tires. However, his idea of a pneumatic tire was only a theory and was never manufactured.

M. El-Gindy and Z. El-Sayegh, *Road and Off-Road Vehicle Dynamics*,
https://doi.org/10.1007/978-3-031-36216-3_1

Meanwhile, John Dunlop [3], watched his son have difficulty riding a tricycle over a rough road. John found that the ride difficulty was due to the solid rubber tires of the tricycle. He began to find a method to improve the ride performance of the tires. John then considered that a smoother ride can be provided by an air-inflated tire that is lighter and absorbs more shocks than a solid rubber tire. Ultimately, John manufactured the first pneumatic bicycle tire in 1888 and patented it. The pneumatic bicycle tire then replaced most of the solid bicycle and tricycle tires within ten years.

Later in 1889, a bicyclist brought a punctured bicycle tire to the Michelin brothers, André and Édouard Michelin, to fix [4]. The brothers noticed a major disadvantage of the punctured tire: the tire was solidly glued to the rim. This major problem caused the punctured tire repair to become very challenging. The brothers decided to help the bicyclist and attempted to find an easier method to repair punctured tires. Eventually, the brothers manufactured a detachable pneumatic tire that could save time and effort to repair. A few years later, the Michelin brothers attempted to persuade carmakers to use inflatable tires. The brothers' effort was accepted, and, within a few years, the Michelin firm achieved extraordinary growth by serving the early stage of the automotive industry.

1.2 Construction of Pneumatic Tires

A pneumatic tire is a structure in the shape of a toroid filled with air, comprised of various flexible membranes. The primary structural element of a tire is the carcass, which consists of a number of flexible chords of a high modulus of elasticity encased in a matrix of low modulus rubber. The characteristics of the tires rely to a great extent on the design and construction of the carcass plies. The crown angle, defined as the angle between the cord and the circumferential centerline of the tire, has a significant role in the mechanics of the tire. The crown angle influences the vehicle ride and cornering characteristics considerably. Tires with low crown angles produce good cornering characteristics but a harsh ride. On the other hand, the large crown angle provides a good ride but poor cornering characteristics.

1.2.1 Bias-Ply Tires

The bias-ply tires are generally used for off-road operations. The bias-ply tires have reinforcing cords that extend diagonally across the tire from bead to bead, as shown in Fig. 1.1. The crown angle commonly ranges from 25 to 40°. The cords flow in opposite directions in each successive ply, thus the cords overlap in a diamond-shaped pattern. In a rolling bias-ply tire, the diagonal plies flex and rub, thus elongating the diamond-shaped elements and the rubber filler. This flexing action yields a wiping motion between the tread and the road, thus increasing the rolling resistance and the rate of tire wear.

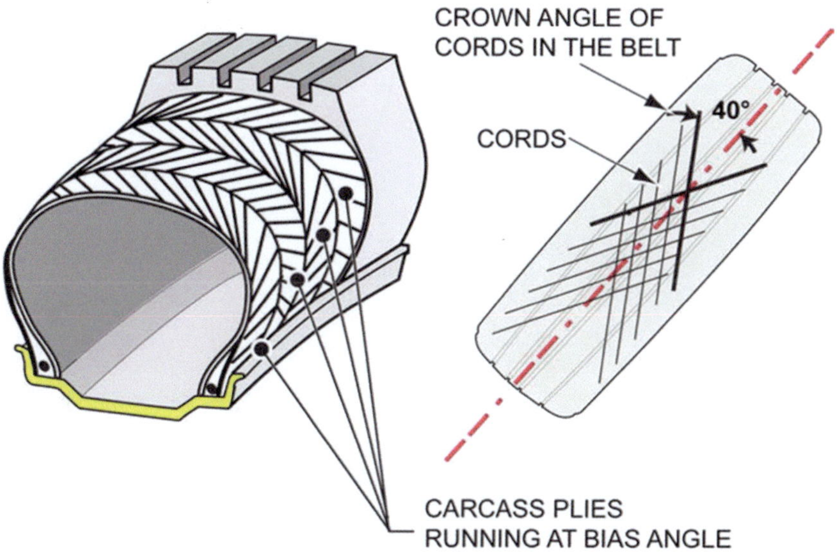

Fig. 1.1 Bias-ply tire construction [5]

Bias-ply tires are also built with belts having low crown angles. The cords in the belts are usually made of materials with a higher modulus of elasticity than those in the bias plies. The belts in such belted bias-ply tires provide high rigidity to the tread against distortion and reduce tire wear. The advantages and disadvantages of a bias-ply technology tire are listed as follows:

- In bias-ply tires, the tread and sidewalls share the same casing plies;
- All sidewall flexing is transmitted to the tread;
- Lower cost compared with radial-ply tires;
- Cushioned, smooth ride per design;
- High friction with the ground compared with radial-ply tires;
- Rapid wear causing reduced traction compared with radial-ply tires;
- Higher fuel consumption compared with radial-ply tires.

1.2.2 Radial-Ply Tires

The construction of a radial-ply tire differs considerably from that of a bias-ply tire. The cords in the carcass of a radial-ply tire are constructed in a radial direction, leading to a 90° crown angle, as presented in Fig. 1.2. The large crown angle of the cords yields a notably good ride despite poor cornering. A radial-ply tire structure is highly soft to compressive and extensive loads in comparison to a bias-ply tire. The radial-ply tire is consequently stiffened by a belt or several plies of cords of a high

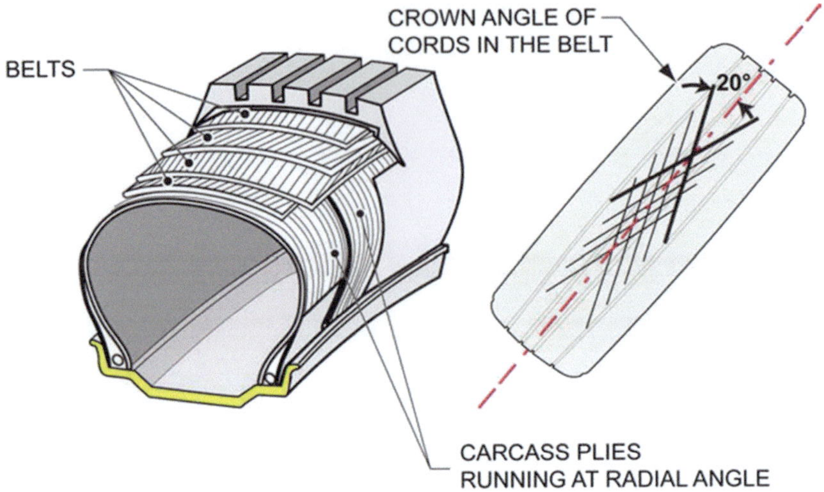

Fig. 1.2 Radial-ply tire construction [5]

modulus of elasticity laid between the carcass and the tread at a low crown angle around 20°, as shown in Fig. 1.2. The belt is essential to the proper functioning of the radial-ply tire.

In the absence of the belt, the tire periphery, when inflated, may evolve into a series of buckles due to the irregularities in cord spacing leading to an unstable tire carcass. A cross-sectional illustration of a radial tire is shown in Fig. 1.2. During operation, the flexing of the carcass involves a comparably small motion of the cords developing the belt, hence the wiping motion between the tire and the road is small. The power dissipated by the radial-ply tires could be as low as 60% of that of the bias-ply. The positives and drawbacks of the radial-ply technology tire are listed as follows:

- Outstanding traction due to flat stable crown and larger footprint;
- Better distribution of pressure in footprint;
- Reduced working time due to less tire slip, greater productivity;
- Reduced fuel consumption compared to bias-ply tires;
- Longer tread life compared to bias-ply tires;
- Better comfort and handling on the road;
- Additional cost compared to bias-ply tires.

In summary, a comparison between radial-play and bias-play tires is given in Table 1.1. It should be noted that the primary structural element of a tire is the carcass; and that the design and characteristics of the carcass plies determine the characteristics of the tire. Furthermore, the crown angle is the angle between the cord and the circumferential centerline of the tire.

The mechanical properties of the tires usually describe the tire's characteristics in response to the application of load, torque, and steering input resulting in the genera-

Table 1.1 Summary of radial-ply and bias-ply tire characteristics

	Radial-play	Bias-play
Flexing	Small	Large
Power dissipated	Small	Large
Rolling resistance	lower	higher
Crown angle	Small (about 20°)	Large (>25°)
Ride	Better	Worse
Cornering	Better	Worse
Tire wear	Lower	Higher

tion of external forces and deflections. The mechanical properties of the tires express fundamental relationships discovered via testing. These mechanical properties are interrelated; and a design decision affecting one will influence other variables.

It should be noted that

- Primary structural element of a tire is the carcass. The design and characteristics of the carcass plies determine the characteristics of the tire.
- Crown angle is the angle between the cord and the circumferential centerline of the tire.

1.2.3 Dimensions and Nomenclature

The following nomenclature is used to describe tire and rim dimensions. Many of the symbols used are established as international standards (refer to Fig. 1.3).

- **Overall Diameter (Do)**: The unloaded diameter of the new tire/rim combination.
- **Section Width (S)**: The width of the new tire section excluding any protective side ribs, lettering or decorations.
- **Section Height (H)**: The distance from the rim to the maximum height of the new tire at the centerline.
- **Aspect Ratio (AR)**: The ratio of section height to section width (HIS).
- **Static Load Radius (SLR)**: The distance from the road surface to the horizontal centerline of the rim.
- **Minimum Dual Spacing (MDS)**: The minimum dimension recommended from rim centerline to rim centerline for optimum performance of a dual wheel installation.
- **Loaded Section (LS)**: The width of the loaded tire section.
- **Footprint Length**: The length of the loaded footprint.
- **Footprint Width**: The width of the loaded footprint.
- **Gross Contact Area (A)**: The total area of the loaded footprint.

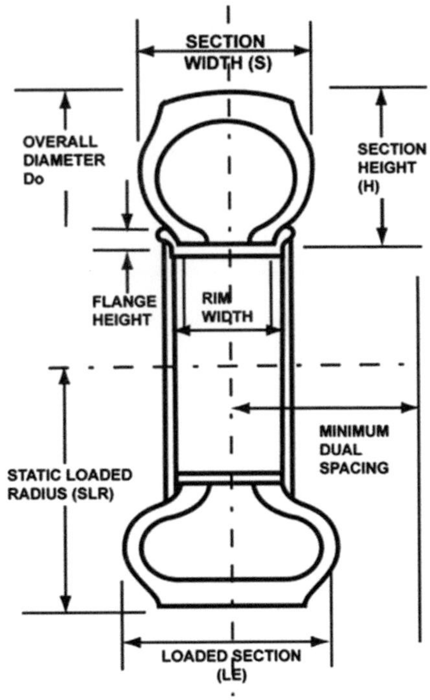

Fig. 1.3 Tire nomenclature [5]

- **Net Contact Area**: The area of the tread (excluding voids) under the loaded footprint.
- **Rim Width** (R_m): The dimension between the rim flanges.
- **Flange Height**: The height of the flange above the rim base.

1.3 Tire Forces and Moments

This section outlines the conventional system that is recommended by the Society of Automotive Engineers (SAE) as shown in Fig. 1.4. The figure shows the forces and moments that can be applied to the tire during operation.

In order to explain the tire characteristics and applied forces and moments, it is essential, first, to define an axis system that serves as a reference for the definition of various parameters. In the SAE conventional tire system, the origin of the axis system is the center of tire-ground contact. The X-axis is the interaction of the wheel plane and the ground plane with a positive direction forward. The Z-axis is perpendicular to the ground plane with a positive direction downward. The Y-axis is in the ground plane, and its direction is chosen to make the axis system orthogonal, which is on the right of the wheel plane.

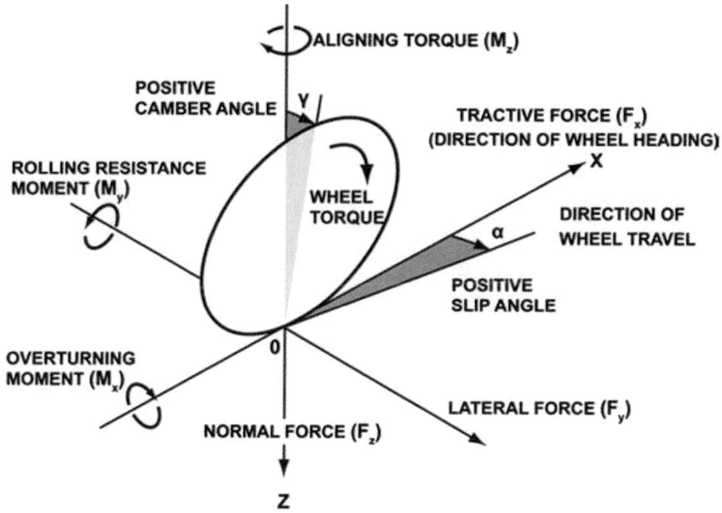

Fig. 1.4 Wheel axis, forces, and moment definitions [5]

The two major angles associated with a rolling tire are defined as follows:

- the slip angle, α, is the angle developed between the direction of wheel travel and the line of intersection of the wheel plane with the ground.
- the camber angle, γ, is the angle developed between the XZ plane and the wheel plane.

The lateral force at the tire-ground contact patch is a function of both the slip angle and the camber angle. The forces and moments are applied to the tire along all three axes during vehicle operations as shown in Fig. 1.4. Three forces and three moments are described in detail in this chapter. The three forces are, namely, the longitudinal or tractive force (F_x), the lateral force (F_y), and the vertical or normal force (F_z). The three moments are, namely, the overturning moment (M_x), the rolling resistance moment (M_y), and the vertical moment or self-aligning moment (M_z). These forces and moments are mainly due to the tire-road contact and its elastic deformation. When the tire doesn't have contact with the road surface, no forces and moments are generated on the tire.

1.3.1 Tire Normal Force

When a vehicle is running on the road, a vertical contact force exists between the tire and road surfaces due to gravity. However, when the vehicle is running on a rough road, the tire and sprung mass have vertical accelerations. Due to their vertical accelerations, the dynamic vertical force acts on the tire-road contact area that can reach up to three times higher than a static vertical force. The vertical force on the

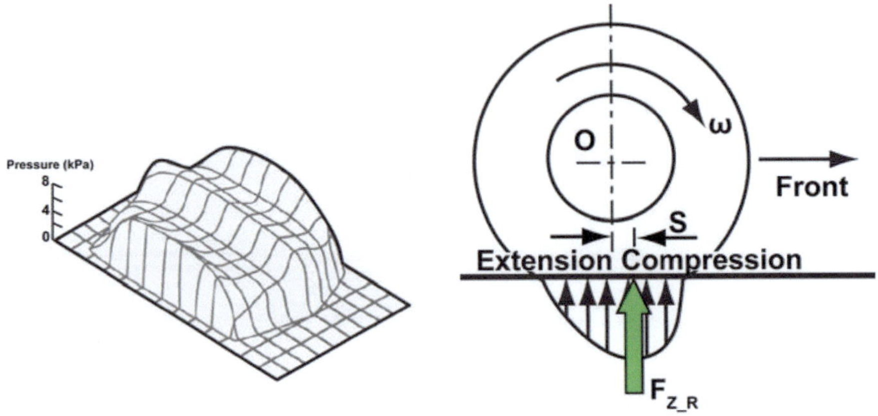

(a) Non-rolling stationary state (b) Rolling state

Fig. 1.5 Normal pressure distributions in the contact area [6]

tire is influenced mostly by the vertical acceleration of the sprung mass rather than by the vertical acceleration of the tire itself because the weight of the sprung mass is much higher than that of the tire. The vertical forces on the tire are not applied at a point but are distributed as normal pressure in the contact area as seen in Fig. 1.5.

Figure 1.5a shows the normal contact pressure distribution in the contact area for a non-rolling stationary tire. Due to the fact that the tire geometry and boundary conditions are symmetric about the center of the contact area, the pressure distribution is also symmetric. It is noted that at rated vertical load and inflation pressure, greater normal contact pressures are observed under the sidewalls and centerline of the tire due to higher vertical stiffness at those local areas.

Figure 1.5b shows the normal contact pressure distribution in the contact area for a rolling tire. In the case of a rolling tire, the boundary conditions are not symmetric about the center of the contact area. Instead, compression in the leading portion of the tire and extension in the trailing portion of the tire near the contact area are observed. Therefore, greater normal contact pressures are exerted in the leading area of contact. This results in the vertical force to the tire, F_{zR}, which can be drawn toward the leading edge with an offset, s, as seen in Fig. 1.5b. This resultant force and offset generate a moment about the tire center, O, counter to the tire rotational direction, which is called the rolling resistance moment.

The tire vertical stiffness represents the tire model's ability to resist deformation in the vertical directions from a known applied vertical force. The tire vertical stiffness, C_z, is defined as the slope of the vertical load versus deflection curve as shown in Fig. 1.6 and Eq. 1.1:

$$C_z = \left. \frac{\partial F_z}{\partial z} \right|_{z=0} \tag{1.1}$$

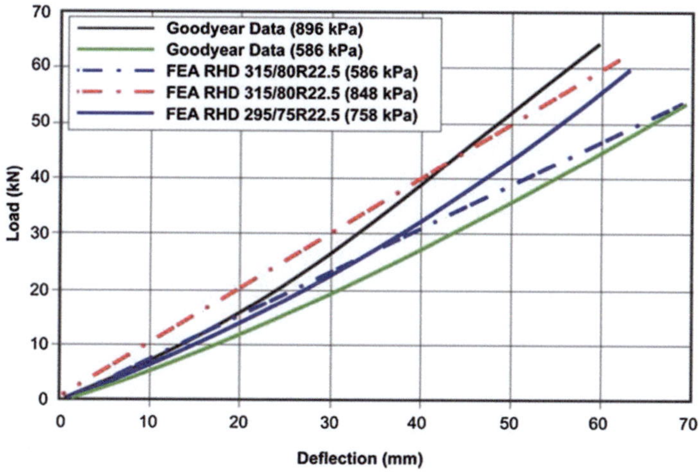

Fig. 1.6 Load as a function of deflection for RHD tire model and other similar tire measurements [7]

Figure 1.6 shows the static deflection curves from actual tire data provided by Goodyear and the simulation results using the RHD tire model over a wide range of loads and inflation pressures. It is observed that all curves have a similar trend, as the vertical load increases the deflection as well. Furthermore, the results obtained for the RHD tire at 848 kPa and those obtained from Goodyear at 896 kPa are in good agreement. It is concluded that the tire model exhibits a similar trend as that of the actual Goodyear tire.

1.3.2 Tire Rolling Resistance

The rolling resistance force is primarily produced by the hysteresis in tire materials due to the carcass deflection during the rolling motion. In addition to the hysteresis, many other factors may influence the rolling resistance of a rolling pneumatic tire. Some of the factors affecting the rolling resistance are the tire construction, materials, and its various operating conditions such as road surface condition, inflation pressure, speed, and temperature. At rated vertical load and inflation pressure on the equivalent size tires, bias-ply tires record higher rolling resistance than radial-ply tires due to greater hysteresis losses in bias-ply tires. Thicker treads, thicker sidewalls, and an increased number of carcass plies also tend to increase the rolling resistance due to greater hysteresis losses. Tires produced of synthetic rubber compounds usually have higher rolling resistance than those made of natural rubber.

Terrain conditions also influence the rolling resistance of a tire in motion. On smooth hard surfaces, the rolling resistance is significantly lower than on a rough surface. The coefficient of rolling resistance is described as the rolling resistance force divided by vertical tire load and is significant for each terrain condition. For example, for passenger car tires, the coefficient of rolling resistance is 0.013 on a concrete or asphalt road and 0.05 on an unpaved road. For common truck tires, the coefficient of rolling resistance is between 0.006 and 0.01 on a concrete or asphalt road, which is lower than that of a passenger car tire due to larger tire diameter and higher inflation pressure. On wet surfaces, a higher rolling resistance is usually observed than on dry surfaces. In addition, higher rolling resistance is also witnessed at higher speed, lower inflation pressure, and lower internal tire temperature.

1.3.2.1 Factors Affecting the Rolling Resistance

The rolling resistance of a tire on a hard surface is affected by many factors, including

- Tire construction and materials;
- Terrain roughness;
- Tire internal temperature;
- Operating conditions.

Tire Construction and Materials

Tire groove depth, diameter, and materials affect the rolling resistance of tires considerably on hard or soft grounds.

Figure 1.7a shows a finite element analysis truck tire model with different groove depth used to predict the rolling resistance coefficient on a hard surface. The rolling resistance coefficient as a function of the tire groove depth is shown in Fig. 1.7b. It is observed that as the groove depth increases, the rolling resistance coefficient slightly increases too.

Figure 1.8 presents the impact of tire diameter on the coefficient of rolling resistance at a given vertical load on various types of terrains. It is observed the coefficient

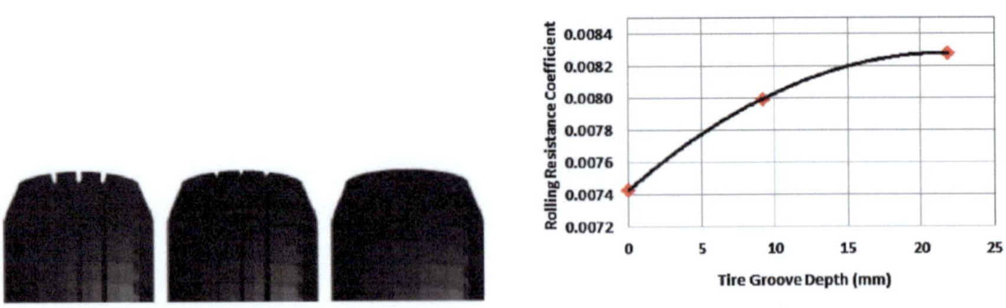

(a) Groove depth modeling (b) Effect of groove depth

Fig. 1.7 Groove depth effect on rolling resistance coefficient of a truck tire

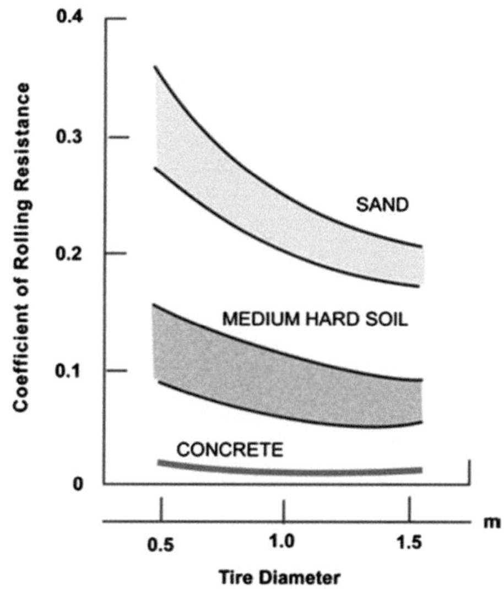

Fig. 1.8 Effect of tire diameter on rolling resistance coefficient of a truck tire [5]

of rolling resistance decreases and the tire diameter increases on all types of terrains, including soft and hard soils.

Natural rubber compounds prohibit low rolling resistance. However, factors such as tread life, wet-road grip, and tire squeal exhibit the use of natural rubber compounds, especially for treads. Synthetic rubber compounds are utilized to achieve a compromise between rolling resistance, cushioning, cornering, traction, and tire life. Butyl rubber tires are high-performance tires with superior traction, road holding, noise attenuation, and comfort characteristics. However, the hysteresis properties of these tires are very poor.

Terrain Roughness

The road roughness has a significant effect on the rolling resistance coefficient. Figure 1.9 shows roads 1, 2, and 3 which are typical road roughness and used in a truck tire-terrain model to predict the rolling resistance coefficient.

Figure 1.10 shows the rolling resistance coefficient over various road surfaces at an inflation pressure of 758 kPa and a vertical load of 26.68 kN. It should be noted that the speed for roads 1 and 2 is 70 km/h, while the speed of road 3 is 60 km/h; road 3 presents an extreme case due to large variations in surface structure. It is observed that the predicted rolling resistance coefficient is highest for road 3, which is the roughest road among the various roads used in the simulations. Also, it is observed that as the tire speed increases, the rolling resistance confidence increases due to the increased hysteresis effect.

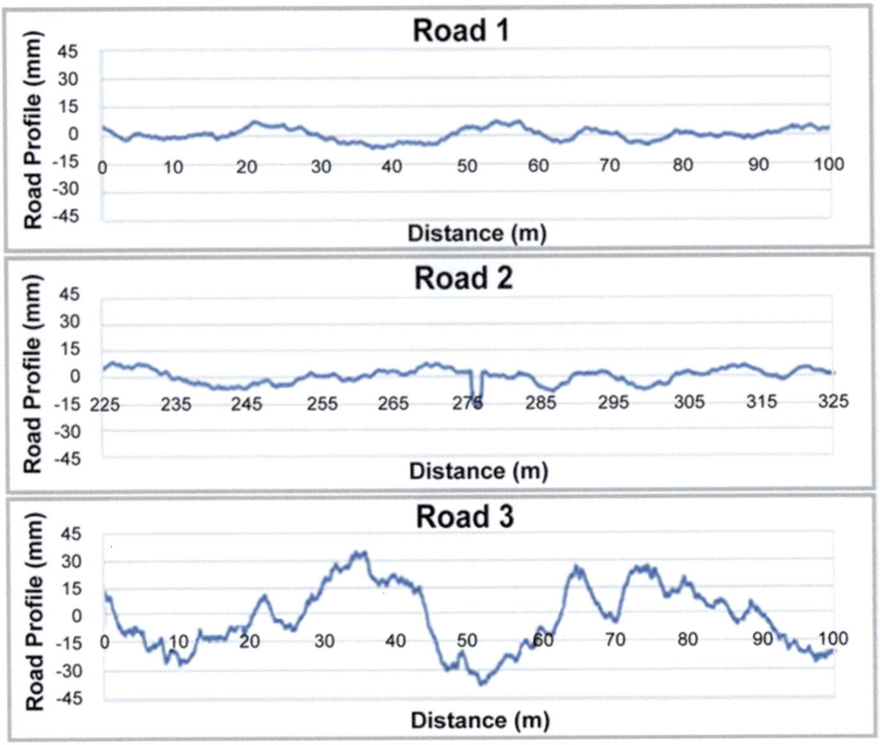

Fig. 1.9 Road profile as a function of time for various road roughness

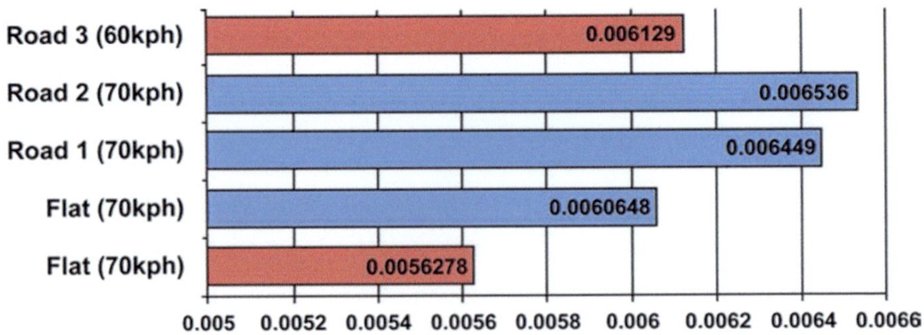

Fig. 1.10 Rolling resistance coefficient as a function of various road surfaces

In addition to the road roughness, the type of terrain has a great effect on the rolling resistance coefficient, and terrain may include on-road and off-road surfaces.

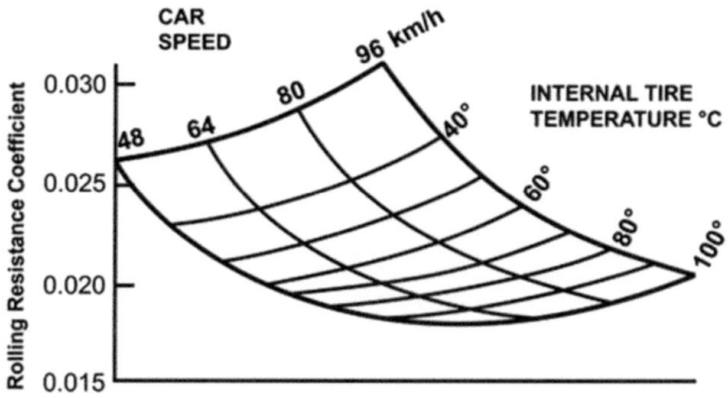

Fig. 1.11 Effect of internal temperature on rolling resistance coefficient for a car tire [5]

Tire Internal Temperature

The influence of tire internal temperature and speed on the rolling resistance coefficient is represented in Fig. 1.11. As the tire temperature increases, the rolling resistance decreases for a given speed. Furthermore, as the car speed increases, the rolling resistance coefficient increases at a given internal tire temperature.

Operating Conditions

The effect of the operating conditions on the variation of rolling resistance can be carefully selected to optimize the rolling resistance to minimize a vehicle's fuel consumption which increases as the rolling resistance increases. The flexibility of a tire is strongly related to inflation pressure. The inflation pressure, depending upon the deformability of the terrain, thus influences the rolling resistance of the tire. On hard roads, an increase in tire pressure leads to a decrease in rolling resistance. Low inflation pressure yields increased tire deflection and thus high hysteresis losses. On soft surfaces, high inflation pressure yields increased tire sinkage and thus high rolling resistance. An increase of 67 kPa in inflation pressure represents approximately a 2% reduction in rolling resistance of typical truck tires. This compares to a 0.5% fuel saving for the typical over-the-road tractor-trailer.

An increase in vehicle speed results in high rolling resistance due to the increase in work required to deform the tire tread. The influence of vehicle speed on the rolling resistance of bias-ply and radial truck tires is presented in Fig. 1.12. At a given speed, the rolling resistance coefficient of a radial-ply tire is less than that of a bias-ply tire.

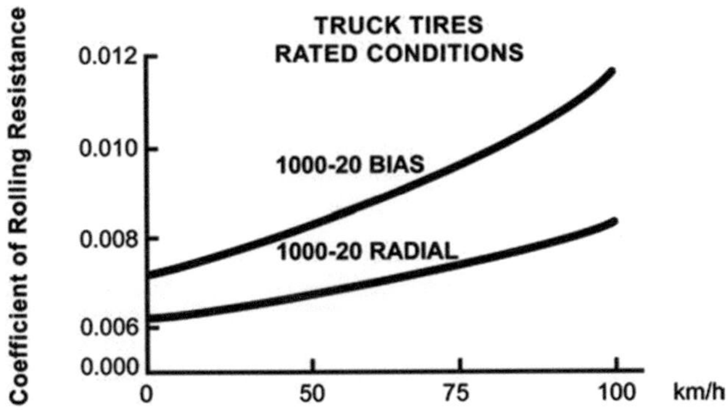

Fig. 1.12 Comparison between radial-ply and cross-ply tire coefficient of rolling resistance

1.3.2.2 Prediction of Rolling Resistance

The rolling resistance coefficients can be estimated for a radial-ply car tire at rated inflation pressure and vertical load by

$$f_r = 0.0136 + 0.40 * 10^{-7} v^2 \tag{1.2}$$

where f_r is the rolling resistance coefficient and v is the car speed for up to 150 km/h, and for a bias-ply car tire by

$$f_r = 0.0169 + 0.19 * 10^{-6} v^2 \tag{1.3}$$

For a radial-ply truck tire under rated inflation pressure and vertical load for up to 100 km/h the rolling resistance coefficient can be expressed by

$$f_r = 0.006 + 0.23 * 10^{-6} v^2 \tag{1.4}$$

and for a bias-ply truck tire by

$$f_r = 0.007 + 0.45 * 10^{-6} v^2 \tag{1.5}$$

The linear approximation of rolling resistance applies to vehicle speeds up to approximately 150 km/h; higher order terms need to be considered at higher speeds. The coefficient of rolling resistance, f_r, for passenger car tires on concrete pavement has been expressed as a function of speed:

$$f_r = f_o + f_s \left(\frac{v}{100}\right)^{2.5} \tag{1.6}$$

where f_o and f_s are the coefficients dependent upon the inflation pressure, as shown in Fig. 1.13, and v is the vehicle speed in km/h. For the most common range of

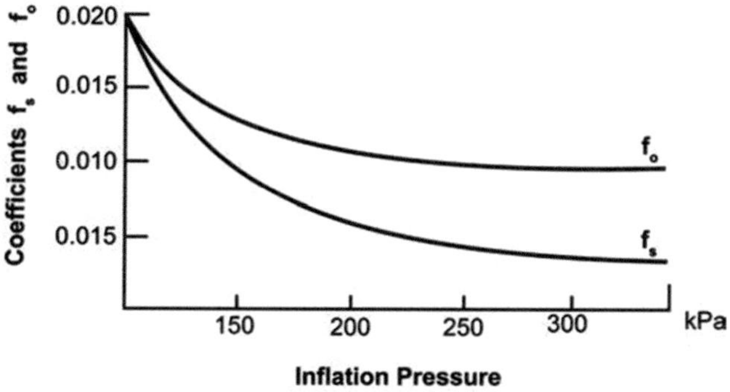

Fig. 1.13 Coefficients as a function of inflation pressure

inflation pressure (around 176 kPa or 26 psi), the rolling resistance coefficient can be approximated as a linear function of vehicle speed:

$$f_r = 0.01\left(1 + \frac{v}{160}\right) \tag{1.7}$$

Equation 1.7 predicts the rolling resistance coefficient of pneumatic tires on hard pavements with acceptable accuracy for speeds up to 128 km/h. In many cases, the coefficient of rolling resistance is expressed as a constant value neglecting the influence of vehicle speed.

1.3.2.3 Measurement of Rolling Resistance

Measurements of rolling resistance force and the coefficient is not an easy task and requires accurate and precise wheel transduces. Tests may be conducted in the field or the laboratory using flatbed testing machines or steel drums. The measured rolling resistance force is usually performed for free rolling tires.

The rolling resistance and aerodynamic drag are the primary factors given the fuel consumption of a vehicle. The rolling resistance of a vehicle is often measured on the road using a tow cable. The vehicle to be tested is towed by a second vehicle via a tow bar or tow cable. A force transducer is placed within the tow cable to measure the tractive force required to overcome the resistance. At high speeds, the aerodynamic drag and the oscillations in the cable become predominant; this test method is thus limited to low speeds only. A sample of the coefficient of rolling resistance values at low speeds for various types of tires are presented in Table 1.2.

The rolling resistance of a vehicle can be also measured using a simple coast-down test. During the test, the vehicle is driven up to a certain speed and the driveline is disconnected. The corresponding deceleration is measured as the vehicle coasts down. This test method offers the possibility to measure the entire tractive resistance and subdivides it into aerodynamic drag and rolling resistance. The deceleration of

Table 1.2 Coefficient of rolling resistance for different vehicles and surfaces

Vehicle type	Surface		
	Concrete	Medium hard soil	Sand
Passenger cars	0.015	0.08	0.30
Trucks	0.012	0.06	0.25
Tractors	0.020	0.04	0.20

the vehicle is related to total tractive resistance in the following manner:

$$ax = Av^2 + Bv + C \tag{1.8}$$

where ax is the deceleration and v is the vehicle speed. The first term in the equation represents the aerodynamic resistance. The last two terms of the equation describe that the rolling resistance is a linear function of the speed. The constants A, B, and C are determined from the experimental data, and assuming that the aerodynamic resistance is proportional to the square of the speed, the drag coefficient is computed as

$$C_D = \frac{mA}{A_f \rho/2} \tag{1.9}$$

where C_D is the drag coefficient, m is the vehicle mass, A_f is the frontal area, and ρ is the mass density of air. The aerodynamic resistance, R_a, is thus expressed as

$$R_a = \frac{1}{2}\rho v^2 C_D A \tag{1.10}$$

The rolling resistance, R_r, and the corresponding coefficient of rolling resistance, f_r, are expressed as follows, where g is the gravity in m/s^2

$$R_r = m(Bv + C) \tag{1.11}$$

$$f_r = \frac{(Bv + C)}{g} \tag{1.12}$$

1.3.3 Tractive Forces and Longitudinal Slip

The tire tractive force is classified into longitudinal frictional force and longitudinal reaction force. This section describes the longitudinal force and slip in detail.

1.3.3.1 Longitudinal Frictional Force

When a tire is undergoing acceleration or braking operations, there is a speed variation between the rolling speed of a tire and its traveling speed, which results in a particular degree of slip between the tire tread and road surface. Without acceleration or braking efforts, a slip will not occur which is the case of a free rolling tire. With a specified amount of slip, a frictional force is generated in the tire-road contact patch that allows the vehicle to be accelerated and decelerated.

In the case of a braking operation, the longitudinal slip during braking, s, can be expressed in slip ratio by

$$s(\%) = \left(1 - \frac{R_e\omega}{v}\right) \times 100 \tag{1.13}$$

During driving, the longitudinal slip, s, can be expressed in slip ratio by

$$s(\%) = \left(1 - \frac{v}{R_e\omega}\right) \times 100 \tag{1.14}$$

where R_e is the tire effective rolling radius, ω is the wheel angular velocity, and v is the tire speed.

When the braking effort is so high, it causes the tire to lock and slide on the road surface. In the case of tire lock, the slip ratio is defined as 100%. Normally, the slip ratio at which the frictional force reaches a maximum is between 10 and 30%. Accordingly, the anti-skid brake system (ABS) controls the slip ratios always between 10 and 30% to produce maximum braking efforts at full braking maneuvers.

Tire tread rubber generates friction in three major ways: adhesion, deformation, and tearing wear. Figure 1.14 presents these three components that contribute to the total friction force experienced by tread rubber on the road surface at a slip speed of v.

The surface adhesion is generated from the momentary intermolecular bonds between the tread rubber and the aggregate in the road surface. Adhesion depends on the true contact area that is determined by the road surface profile, involved material properties, and contact pressure. Normally, the adhesion component is the principal contributor to tire traction on dry and smooth roads. Nonetheless, when the road is contaminated with dust or water, the tire loses part of the contact and formation of adhesive forces. Then, the adhesion friction is diminished considerably, which results in the loss of friction.

Tread rubber in contact with a smooth surface creates friction force mainly by adhesion. Nonetheless, when rubber is in contact with a rough road surface, a different mechanism, deformation, performs an important role in friction. As the tread rubber slides on a rough road surface, the local deformations of the rubber are recognized on road surface irregularities. Friction forces due to those local deformations contribute most of the friction force between the tire and wet road surface.

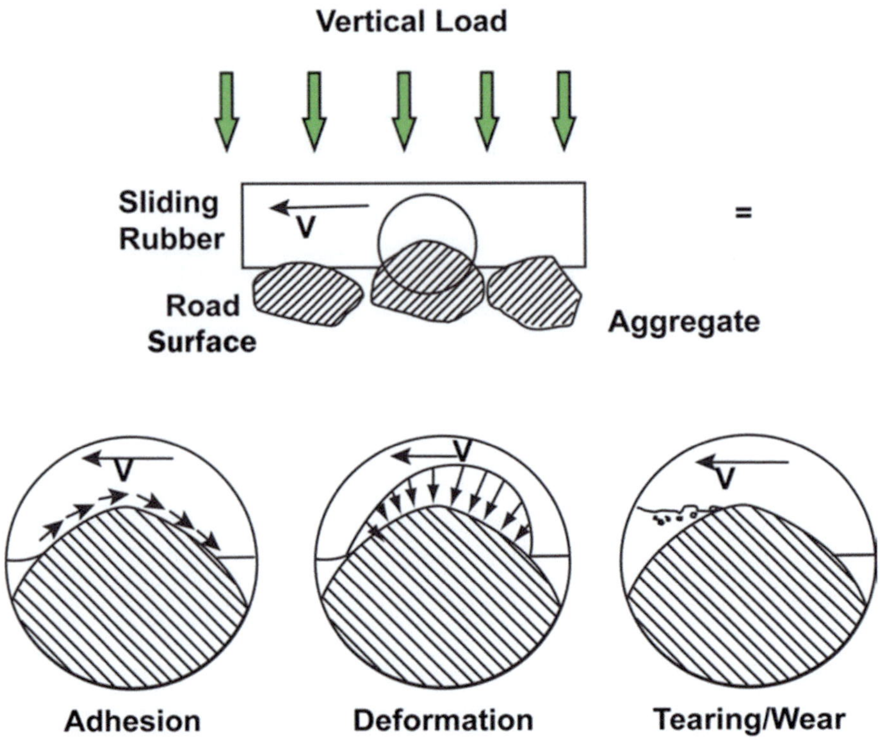

Fig. 1.14 Major components of rubber friction [8]

In addition to the adhesive friction and deformation friction, the rubber generates friction forces through the tear and wear. As the applied vertical load and sliding speeds increase, local stress can pass the tensile strength of the rubber, especially near the area of a sharp irregularity. High local stress can deform the internal structure of the rubber beyond the point of elastic recovery. When the polymer chains are stressed to failure, tearing may occur. Tearing consumes energy and results in additional friction forces at the contact surface.

It should be noted that the friction force is different from the rolling resistance force. The rolling resistance force is employed on the freely rotating tire, whereas the friction force is applied under acceleration (traction) or deceleration (braking) operations where slip exists. When the tire is under acceleration due to driving torque, the tractive force is generated in the direction of the motion. Conversely, when the tire is under deceleration due to braking, the braking force is generated on the tire against the tire moving direction. However, the rolling resistance is always generated against the tire moving direction.

Figure 1.15 shows the variation of the tractive effort as a function of longitudinal slip for a tire. The maximum tractive effort occurs at the peak and is referred to as μ_p, while the tractive effort at 100% slip is the sliding value and referred to as μ_s. The

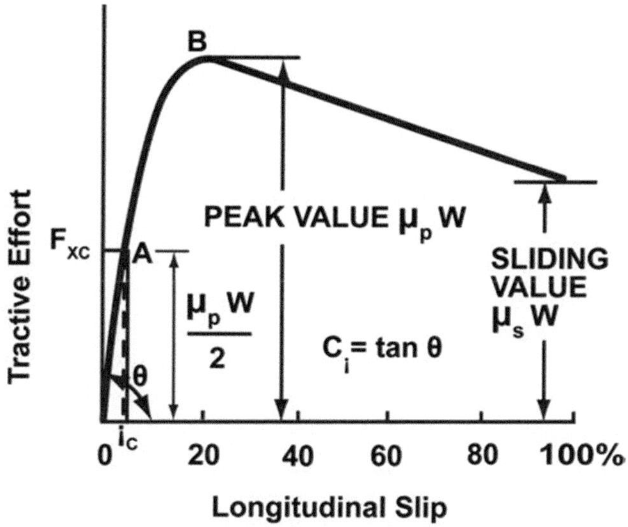

Fig. 1.15 Variation of tractive effort as a function of longitudinal slip [5]

longitudinal tire stiffness, C_s, is defined as the derivative of the longitudinal force divided by the slip at zero slip:

$$C_s = \left.\frac{\partial F_x}{\partial s}\right|_{s=0} \tag{1.15}$$

If no sliding occurs on the contact patch, the relationship between the tractive force and the slip is considered linear:

$$F_x = C_s s \tag{1.16}$$

The critical slip, s_c, and the critical tractive force, F_{xc}, at the sliding in the trailing part of the contact patch are given by

$$s_c = \frac{\mu_p F_z}{2C_s} \tag{1.17}$$

$$F_{xc} = \frac{\mu_p F_z}{2} \tag{1.18}$$

where F_z is the vertical load, the tractive force, F_{xs}, developed in the sliding region can be defined as

$$F_{xs} = \mu_p F_z \left(1 - \frac{\mu_p F_z}{2C_s s}\right) \tag{1.19}$$

and the tractive force, F_{xa}, produced in the adhesion region is given by

Table 1.3 Average values of peak and sliding coefficients [5]

Surface	μ_p	μ_s
Asphalt and concrete	0.8–0.9	0.75
Asphalt (wet)	0.5–0.7	0.45–0.6
Concrete (wet)	0.8	0.7
Gravel	0.6	0.55
Snow	0.2	0.15
Ice	0.1	0.07
Earth road (dry)	0.68	0.65
Earth road (wet)	0.55	0.4–0.5

$$F_{xa} = \frac{\mu_p^2 F_z^2}{4C_s s} \tag{1.20}$$

Thus, the total tractive force, F_x, and the slip when part of the tread is sliding is given by

$$F_x = F_{xs} + F_{xa} = \mu_p F_z \left(1 - \frac{\mu_p F_z}{4C_s s}\right) \tag{1.21}$$

The average peak and sliding values of coefficients of road adhesion for various surfaces are summarized in Table 1.3.

1.3.3.2 Longitudinal Reaction Force

When the tire runs over severe road surface irregularities or obstacles such as steps, potholes, water drainage ditches, and speed bumps, a reaction force is developed to the tire longitudinally as well as laterally and vertically. This longitudinal reaction force is generated as a shock, which can cause damage to the tire and rim. All of these longitudinal forces are usually acting opposite to the tire moving direction at the tire-road contact area. The reaction force depends significantly on the suspension characteristics of the vehicle and tire operational conditions such as vertical load, inflation pressure, and speed.

Tire longitudinal forces, developed during braking and acceleration, are affected by many factors, including road, tire, and vehicle factors.

1. Road Factors:

   - Road surface characteristics;
   - Road conditions;
   - Water depth;
   - Temperature.

2. Tire Factors:
   - Tread design;
   - Tread compound;
   - Tire construction;
   - Normal load;
   - Inflation pressure.
3. Vehicle Factors:
   - Speed;
   - Mode of operation;
   - Suspension and braking;
   - Dynamic handling properties.

Road factors can create the greatest variance in the tractive forces developed by the tire. Driving on ice, snow, mud, sand, and wet and dry pavements are just a few examples of the wide range of road conditions experienced. The performance of tires depends on wet surface texture, water depth, tread pattern, and operating mode of the tire. It is essential to maintain effective contact between the tire tread and the road in order to achieve acceptable tire performance.

1.3.4 Cornering Forces and Slip Angle

When a vehicle undertakes a cornering operation or is subjected to crosswind, lateral force is developed at the tire-road contact area. The lateral forces during a cornering maneuver and under irregular lateral wind are dynamic forces due to the lateral acceleration of the vehicle. The lateral force in reaction to a cornering maneuver is called a cornering force.

The cornering force is highly dependent on tire vertical load. As the vertical load on the tire increases under the same cornering operational condition, the cornering force also increases. Meanwhile, during cornering maneuvers, a higher vertical load is exerted on the right tires due to lateral load transfer. Therefore, higher cornering forces are also applied to the same right tires (Fig. 1.16).

The schematic contact area shapes are illustrated in Fig. 1.17 at different slip angles. Originally, the tire is rolling in the direction of the top of the page and is currently turning left. As a result, effective stationary contact (adhesive area) occurs always at the leading edge of the contact area as the tire rolls. On the other hand, slip is confined to the rear of the contact area because the resultant cornering force is applied behind the tire center. In the region of the slip, the tangential surface stresses, necessary to maintain the geometric distortion of the tread surface, exceed the local frictional stresses available. The leading edge is pointing in the steering direction, while the rearward portion lags behind on the old heading due to slip.

In addition, the cornering force depends on the slip angle of the tire. As the slip angle increases at a given vertical load on the tire, the cornering force also increases.

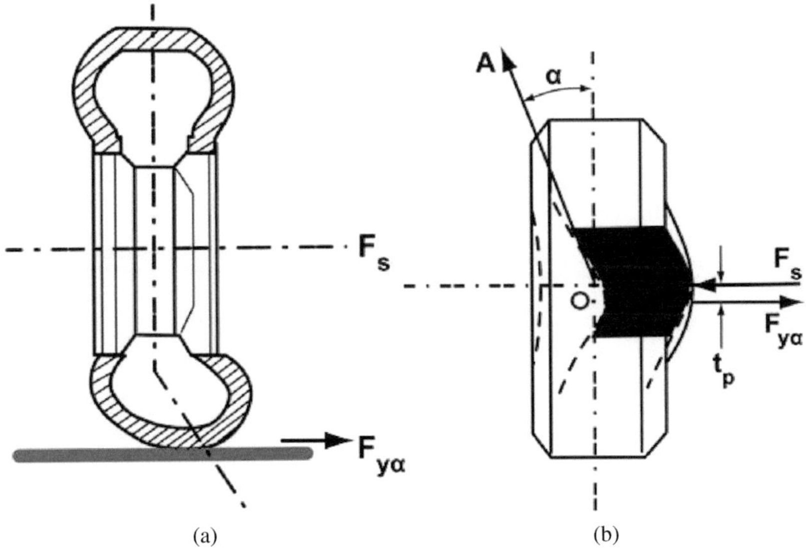

Fig. 1.16 Tire behavior when subjected to a cornering force [5]

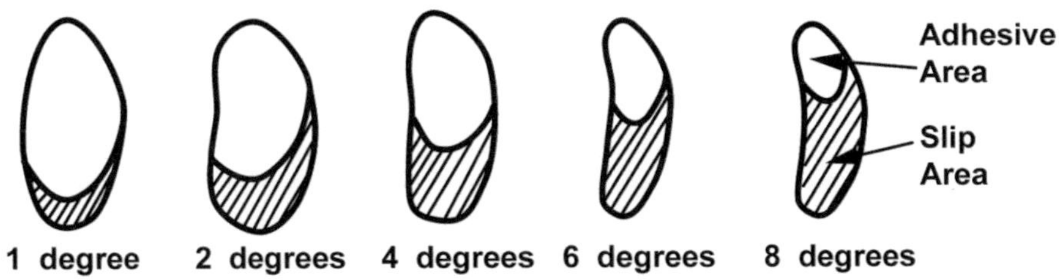

Fig. 1.17 Contact area shapes at different slip angles [8]

However, when the cornering force reaches a certain level, it does not significantly increase further. Instead, it tends to converge to an asymptote, which is a road surface adhesion limit as seen in Fig. 1.18.

The development of the slip angle is associated with the elastic nature of the tire. The tire tread grips the road surface due to friction. However, the tire also resists movement with an opposing force, yields to external force, and recovers when the external force is removed. This elastic characteristic allows the tire to have an orientation (u) different from the direction in which the vehicle is traveling (v) as shown in Fig. 1.18. The angle, α, between the two orientations is defined as slip angle and plays an important role in cornering operations. A is the front left wheel, CG is the vehicle center of gravity, O is the center of the front left wheel, PQ is the cornering curve path, u is the current wheel orientation, v is the vehicle heading direction from the CG, v' is the vehicle heading direction from the wheel A, and α is the slip angle.

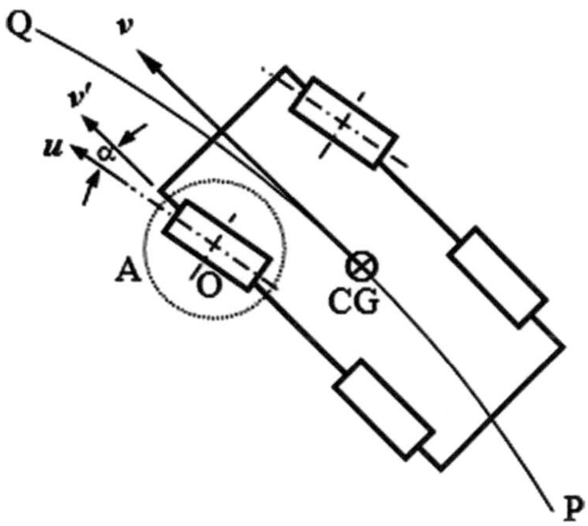

Fig. 1.18 Vehicle under cornering maneuver with slip angle

To provide a measure for comparing the cornering behavior of different tires, the “cornering stiffness”, C_α, is defined as the ratio of the derivative of the cornering force, $F_{y\alpha}$, by the slip angle at zero slip angle:

$$C_\alpha = \left.\frac{\partial F_{y\alpha}}{\partial \alpha}\right|_{\alpha=0} \tag{1.22}$$

For s small slip angle, the cornering force is written as

$$F_{y\alpha} = C_\alpha \alpha \tag{1.23}$$

Furthermore, the critical slip angle, α_c, and the critical cornering force, $F_{y\alpha c}$, at which lateral sliding in the trailing part of the contract patch begins are determined by

$$\alpha_c = \frac{\mu_p W}{2C_\alpha} \tag{1.24}$$

and the critical value of $F_{y\alpha c}$ is given by

$$F_{y\alpha c} = \frac{\mu_p W}{2} \tag{1.25}$$

When the lateral sliding occurs between the tire tread and the ground, the relationship between the cornering force and the slip angle is expressed by

$$F_{y\alpha} = \mu_p W \left(1 - \frac{\mu_p W}{4C_\alpha \alpha}\right) \tag{1.26}$$

It should be noted that various empirical equations have been proposed to represent the relationship between the cornering force $F_{y\alpha}$ and the slip angle α. The most

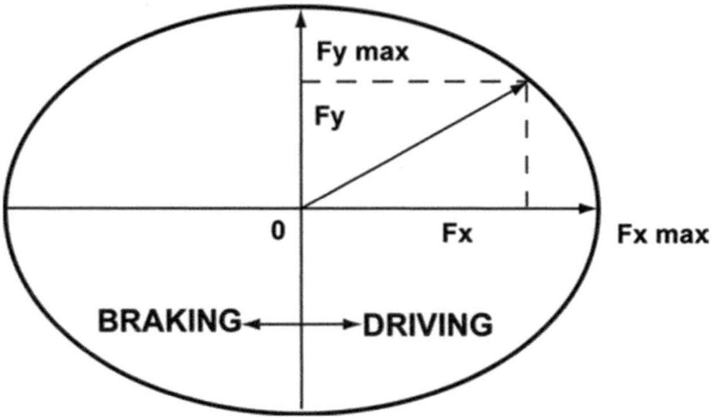

Fig. 1.19 The friction ellipse concept relating the maximum cornering force to a given longitudinal force [5]

popular relationship was proposed by Ellis [9] which was based on measured data and expressed by

$$F_{y\alpha} = c_1\alpha + c_2\alpha^2 + c_3\alpha^3 \tag{1.27}$$

where c_1, c_2, and c_3 are empirical constants derived from fitting the above equation to the measured data of a given tire.

Figure 1.19 shows the friction ellipse which is based on the assumption that the tire slides on the ground in any direction if the resultant of the longitudinal force and lateral force reaches a maximum value defined by the coefficient of road adhesion and the normal load on the tire. It should be noted that the longitudinal and lateral force components should not exceed their respective maximum values.

The friction ellipse can be used to determine the available cornering force based on the simple theory as follows:

1. Plot the relationship between the cornering force and the slip angle under free rolling conditions (i.e., in the absence of tractive or braking effort) from the measured data as shown in Fig. 1.20.
2. Plot the cornering forces at various slip angles under free rolling conditions are then marked on the vertical axis. For instance, the cornering force developed at a slip angle of 4° is identified as F_{y4} on the vertical axis, which constitutes the minor axis of an ellipse to be established.
3. Mark on the horizontal axis the maximum tractive or braking force, F_{xmax}, measured from the tire data in the absence of lateral force. This will provide the major axis of the ellipse.
4. The available cornering force F_y, at a given slip angle, such as the 4° angle shown in Fig. 1.20, for any given tractive or braking force F_x is then determined from the following equation:

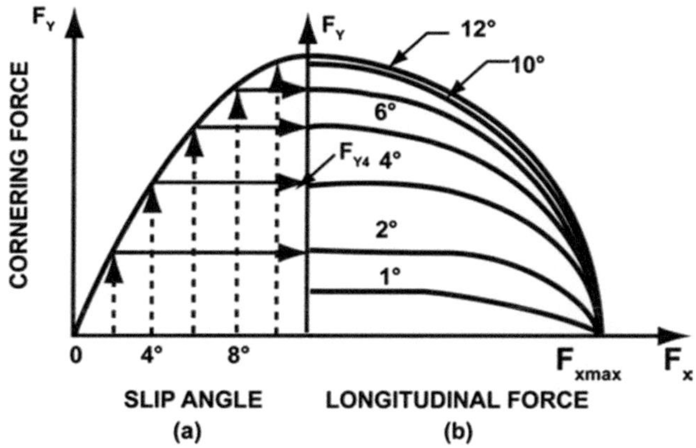

Fig. 1.20 Construction of a friction ellipse relating the cornering force to longitudinal force at a given slip angle [5]

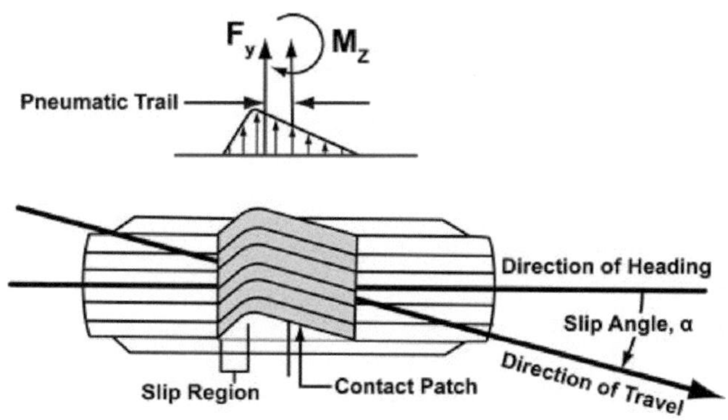

Fig. 1.21 Lateral (cornering) force at tire-road contact area during cornering maneuver

$$\left(\frac{F_y}{F_{y4}}\right)^2 + \left(\frac{F_x}{F_{xmax}}\right)^2 = 1 \quad (1.28)$$

It is noted that the above equation describes an ellipse with the measured values of F_{xmax} and F_{y4}, as the major and minor axes, respectively.

The developed cornering force at the contact area is shown in detail in Fig. 1.21. When the vehicle is turning left, a cornering force is applied on the tire to the left at the contact area. However, the cornering force is not distributed symmetrically on the contact area around the tire center. Instead, the peak cornering force moves behind the center of the tire due to the longitudinal force against the tire. Therefore,

Table 1.4 Pneumatic trail for truck tires at 1° slip angle under rated inflation pressure and vertical load

Tire type	Tire construction	Pneumatic trails (cm)
Michelin radial, 11R22.5 XZA (1/3 Tread)	Radial-ply	6.17
Goodyear unisteel II, 10R22.5 LRF	Radial-ply	6.15
Michelin radial, 11R22.5 (1/2 Tread)	Radial-ply	5.89
Goodyear unisteel G159, 11R22.5	Radial-ply	5.87
Michelin radial, 11R22.5 XZA	Radial-ply	5.51
Goodyear unisteel G159, 11R22.5 LGR	Radial-ply	5.46
Goodyear unisteel II, 10R22.5 LRF	Radial-ply	5.41
Michelin pilot, 11/80R22.5 XZA	Radial-ply	4.62
New unspecified model 10–20/F	Bias-ply	5.89
Half-worn unspecified model 10–20/F	Bias-ply	7.14
Fully worn unspecified model 10–20/F	Bias-ply	6.55

the resultant cornering force, F_y, can be drawn behind the tire center with offset. The offset is called the pneumatic trail (pt). This pneumatic trail and the resultant cornering force generate an aligning moment about the vertical axis.

The pneumatic trails for various truck tires at a 1° slip angle under rated inflation pressure and vertical load are shown in Table 1.4.

Assuming that no sliding takes place, the braking force per unit contact length at a distance x from the front contact point is expressed by

$$\frac{dF_x}{dx} = \frac{C_t x s}{1 - s} \tag{1.29}$$

where s is the longitudinal slip, and C_t is the tangential stiffness of the tire tread. If the tire develops a slip angle α, then due to the longitudinal skid, the tread in contact with the ground will be elongated at a rate of $1/(1 - s)$. As a result, the lateral deflection y' of a point on the tread in contact with the ground is given by [5]

$$y' = x \tan \frac{\alpha}{1 - s} \tag{1.30}$$

The corresponding lateral force per unit contact length is then expressed by

$$\frac{dF_{y\alpha}}{dx} = C'_y x \tan \frac{\alpha}{s} \tag{1.31}$$

where C'_y is the equivalent lateral stiffness of the tire, when no lateral sliding between the tire tread and the ground takes place. Utilizing the concept of friction ellipse described above, and noting that no sliding will take place at a point located at a distance of x from the front contact point if the resultant of the braking force and lateral force per unit contact length is less than a minimum value defined by the coefficient of road adhesion μ and the normal pressure p,

$$\frac{\mu W}{l_t} = \sqrt{\left(\frac{C_t x s}{1-s}\right)^2 + \left(C'_y x \tan \frac{\alpha}{1-s}\right)^2} \tag{1.32}$$

where W is the normal load, and l_t is the contact length of the tire. Thus, the ratio of the characteristics length, l_c, to the contact length l_t is defined by

$$\frac{l_c}{l_t} = \frac{\mu W(1-s)}{2\sqrt{\left(\frac{C_t l_t^2 s}{2}\right)^2 + \left(C'_y l_t^2 \tan \frac{\alpha}{2}\right)^2}} \tag{1.33}$$

Substituting longitudinal and cornering stiffness, C_s and C_α respectively, the above equation becomes

$$\frac{l_c}{l_t} = \frac{\mu W(1-s)}{2\sqrt{(C_s s)^2 + (C_\alpha \tan \alpha)^2}} \tag{1.34}$$

The entire contact patch is an adhesion region when the ratio of the characteristic to contact length is greater than 1 ($l_c/l_t > 1$). The braking force is then given by

$$F_x = \int_0^{l_t} \frac{C_t x s}{1-s} dx = C_s \frac{s}{1-s} \tag{1.35}$$

and the cornering force as a function of slip angle α and skid s is then given by

$$F_{y\alpha} = \int_0^{l_t} \frac{C'_y x \tan \alpha}{1-s} dx = C_\alpha \frac{\tan \alpha}{1-s} \tag{1.36}$$

If the ratio of the characteristics length, l_c, to the contact length, l_t, is defined by less than 1 ($l_c/l_t < 1$) and the tread is sliding over the ground, then, the braking force developed on the adhesion region, F_{xa}, is given by

$$F_{xa} = \int_0^{l_c} \frac{C_t x s}{1-s} dx = \frac{\mu^2 W^2 C_s s(1-s)}{4[(C_s s)^2 + (C_\alpha \tan \alpha)^2]} \tag{1.37}$$

and the braking force developed on the sliding region, F_{xs}, is given by

$$F_{xs} = \frac{\mu W C_s s}{\sqrt{(C_s s)^2 + (C_\alpha \tan \alpha)^2}} \left[1 - \frac{\mu W(1-s)}{2\sqrt{(C_s s)^2 + (C_\alpha \tan \alpha)^2}}\right] \tag{1.38}$$

The total braking force, F_x, is then defined by adding up both adhesion and sliding braking force, and it is expressed as

$$F_x = F_{xa} + F_{xs} \tag{1.39}$$

$$= \frac{\mu W C_s s}{\sqrt{(C_s s)^2 + (C_\alpha \tan\alpha)^2}} \left[1 - \frac{\mu W(1-s)}{4\sqrt{(C_s s)^2 + (C_\alpha \tan\alpha)^2}}\right] \tag{1.40}$$

In a similar approach, if the tread is sliding over the ground, then the cornering force developed on the adhesion region is defined as

$$F_{y\alpha a} = \int_0^{l_c} \frac{C'_y x \tan\alpha}{1-s} dx \tag{1.41}$$

$$= \frac{\mu^2 W^2 C_\alpha \tan\alpha (1-s)}{4[(C_s s)^2 + (C_\alpha \tan\alpha)^2]} \tag{1.42}$$

and the cornering force developed on the sliding region is given by

$$F_{y\alpha s} = \frac{\mu W C_\alpha \tan\alpha}{\sqrt{(C_s s)^2 + (C_\alpha \tan\alpha)^2}} \left[1 - \frac{\mu W(1-s)}{2\sqrt{(C_s s)^2 + (C_\alpha \tan\alpha)^2}}\right] \tag{1.43}$$

The total cornering force, $F_{y\alpha}$, is given by

$$F_y = F_{y\alpha a} + F_{y\alpha s} \tag{1.44}$$

$$= \frac{\mu W C_\alpha \tan\alpha}{\sqrt{(C_s s)^2 + (C_\alpha \tan\alpha)^2}} \left[1 - \frac{\mu W(1-s)}{4\sqrt{(C_s s)^2 + (C_\alpha \tan\alpha)^2}}\right] \tag{1.45}$$

It should be noted that the parameters, μ, W, C_s, and C_α may change with operating conditions. For instance, it has been found [10] that on a given surface, the values of μ, C_s, and C_α are functions of the normal load and operating speed of the tire. In a dynamic maneuver involving both braking and cornering, the normal load and speed of the tires on a vehicle change as the maneuver proceeds. To achieve more accurate predictions, the effects of normal load and speed on the values of μ, C_s, C_α and other tire parameters should be properly taken into account.

? Example 1.1

A truck tire 10 × 20/F with a normal load of 24.15 kN is traveling on a dry asphalt pavement with a coefficient of road adhesion 0.85. The cornering stiffness of the tire is 133.30 kN/rad and the longitudinal stiffness is 186.82 kN/unit skid. Estimate the braking force and the cornering force that the tire can develop at a slip angle of 4° and a longitudinal skid of 10%.

Solution. To determine whether sliding takes place on the tire contact patch under the given operating conditions, the ratio, l_c/l_t, is calculated as

$$\frac{l_c}{l_t} = \frac{\mu W(1-s)}{2\sqrt{\left(\frac{C_t l_t^2 s}{2}\right)^2 + \left(C'_y l_t^2 \tan\frac{\alpha}{2}\right)^2}} \tag{1.46}$$

$$= 0.442 \tag{1.47}$$

Since the ratio is less than 1, sliding takes place in part of the contact. The braking force can be calculated as

$$F_x = F_{xa} + F_{xs} \tag{1.48}$$

$$= \frac{\mu W C_s s}{\sqrt{(C_s s)^2 + (C_\alpha \tan\alpha)^2}}\left[1 - \frac{\mu W(1-s)}{4\sqrt{(C_s s)^2 + (C_\alpha \tan\alpha)^2}}\right] \tag{1.49}$$

$$= 14.3\,\text{kN} \tag{1.50}$$

while the cornering force is calculated by

$$F_y = F_{y\alpha a} + F_{y\alpha s} \tag{1.51}$$

$$= \frac{\mu W C_\alpha \tan\alpha}{\sqrt{(C_s s)^2 + (C_\alpha \tan\alpha)^2}}\left[1 - \frac{\mu W(1-s)}{4\sqrt{(C_s s)^2 + (C_\alpha \tan\alpha)^2}}\right] \tag{1.52}$$

$$= 7.14\,\text{kN} \tag{1.53}$$

1.3.5 Self-aligning Moments and Slip Angle

The moment acting on the tire spindle about the vertical axis (Z-axis) is defined as the vertical moment. Non-symmetric contact force distribution on the tire-road contact plane determines the vertical moment. In addition, during a cornering maneuver, the resultant cornering force acts on the tire with some offset behind the center of the contact area, called the pneumatic trail. The cornering force and offset create the vertical moment that tends to restore the steered tire to the original un-steered tire orientation. Therefore, this vertical moment is called the self-aligning moment or aligning moment. The self-aligning moment increases with increasing slip angle input, similar to the cornering force response. The self-aligning moment increases as the slip angle increases until a peak self-aligning moment is developed at approximately 4° to 6° of slip angle. However, once the peak is reached, the self-aligning

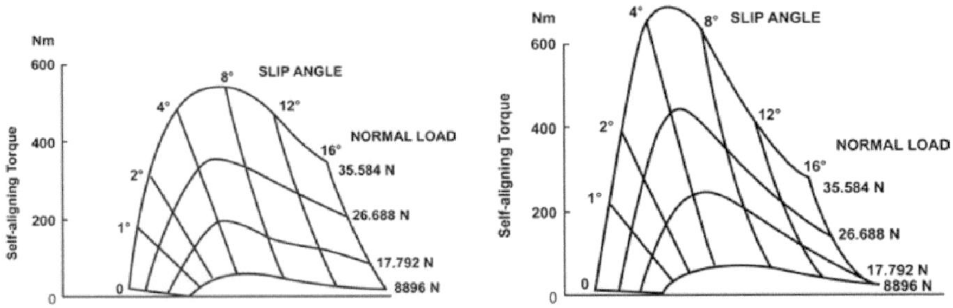

(a) Bias-ply truck tire 10-20/F (b) radial-ply truck tire 10-20/G

Fig. 1.22 Variation of the self-aligning moment as a function of vertical load and slip angle [5]

moment tends to decrease with a further increase of the slip angle because of the decrease in the moment arm (pneumatic trail).

Figure 1.22 shows the variation of the self-aligning moment as a function of vertical load and slip angle; it is observed that as the slip angle increases, the self-aligning moment increases for both radial-ply and bias-ply truck tires. The self-aligning moment reaches a maximum at a particular slip angle and then decreases with further increase of slip angle. Furthermore, the self-aligning moment increases as the vertical load increases at a given slip angle for both bias-ply and radial-ply truck tires.

The self-aligning moment, M_z, is also considered as the torque created by the tire while undergoing a cornering maneuver. The self-aligning moment stiffness, C_m, is another parameter used to determine the cornering operational performance of a tire; the relationship between the self-aligning moment, slip angle, and the self-aligning moment stiffness is shown in Eq. 1.54:

$$C_m = \left. \frac{\partial M_z}{\partial \alpha} \right|_{\alpha=0} \tag{1.54}$$

Similar to the cornering stiffness, the self-aligning moment stiffness is considered the derivative of the self-aligning moment over the slip angle at zero slip angle.

1.3.6 *Tire Camber Thrust*

The lateral force developed at the road tire interface perpendicular to the wheel plane because of the camber angle is called camber thrust. This force always acts a little ahead of the wheel center producing a moment. The camber thrust is always 1/5th of the cornering force. Figure 1.23 shows the direction of the camber thrust with respect to the direction of the camber angle.

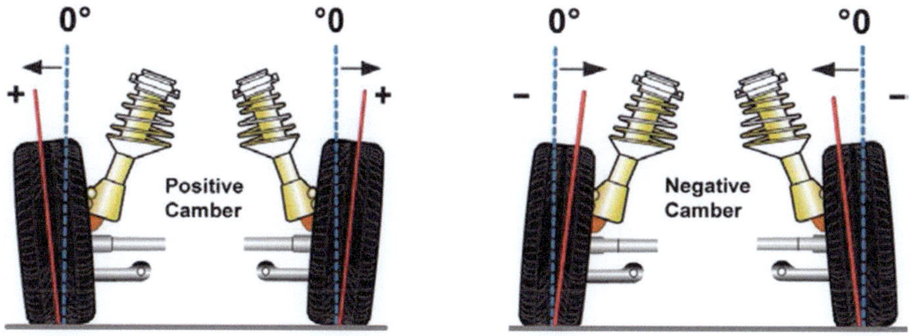

Fig. 1.23 Direction of positive and negative cambers

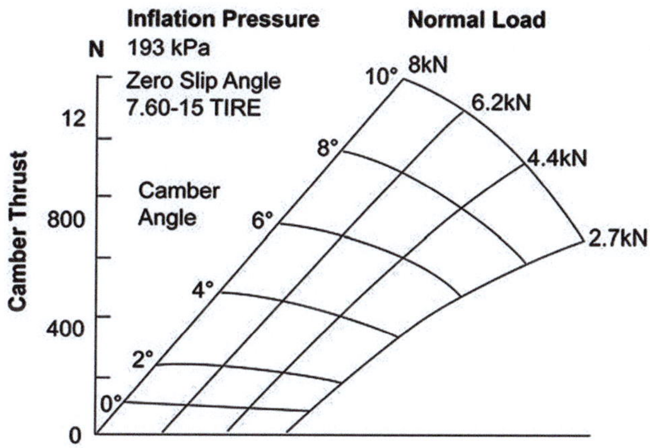

Fig. 1.24 Effect of camber angle and tire load on camber thrust [5]

The "camber stiffness" is often used to provide a measure for comparison for camber characteristics of different tires; the camber stiffness, C_γ, is defined as the derivative of the camber thrust with respect to the camber angle at zero camber angle:

$$C_\gamma = \left. \frac{\partial F_{y\gamma}}{\partial \gamma} \right|_{\gamma=0} \tag{1.55}$$

Figure 1.24 shows the effect of vertical load and camber angle on the camber thrust. It can be observed that as the vertical load increases, the camber thrust increases. Similarly, as the camber angle increases, the camber thrust increases as well.

The total lateral force, F_y, of a cambered tire operating at a slip angle is the summation of the cornering force, $F_{y\alpha}$, and the camber thrust force, $F_{y\gamma}$:

$$F_y = F_{y\alpha} \pm F_{y\gamma} \tag{1.56}$$

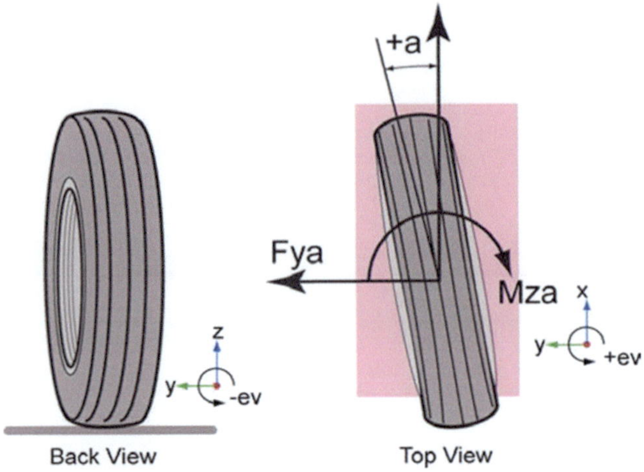

Fig. 1.25 Force and moment directions with slip angle (α)

If the camber thrust and the cornering force are in the same direction, the two positive forces are added. If the camber thrust and the cornering force are in the opposite directions, the camber force should be subtracted. In the case of a small camber and slip angle, the relationship between the lateral force and the cornering and camber forces becomes

$$F_y = C_\alpha \alpha \pm C_\gamma \gamma \tag{1.57}$$

1.3.7 Combined Cambering and Cornering Properties

The computation of the combined cornering and cambering is a challenging situation because of the changes of the camber thrust direction due to steering to the left or to the right. The camber thrust may then be added to or subtracted from the tire cornering force, but it requires advanced finite element simulations to model a cambered and steered tire as shown in Fig. 1.25.

The predicted lateral forces, aligning moment, and rolling resistance during steering to the left and right (positive and negative slip angles) are shown in Fig. 1.26a, b, and c.

Figure 1.26a shows the variation of the cornering force as a function of slip angle for different camber angles. It is observed that as the slip angle increases, the cornering force increases at a given camber angle. Furthermore, as the camber angle increases, the cornering force increases as well at a given slip angle.

Figure 1.26b shows the variation of the self-aligning moment as a function of slip angle for different camber angles. It is observed that as the slip angle increases, the self-aligning moment decreases until it reaches 4° and then the opposite is true at a

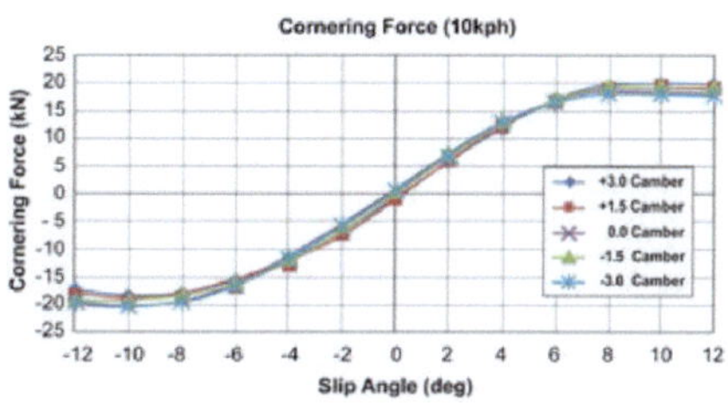

(a) Combined Cornering Force and Camber thrust

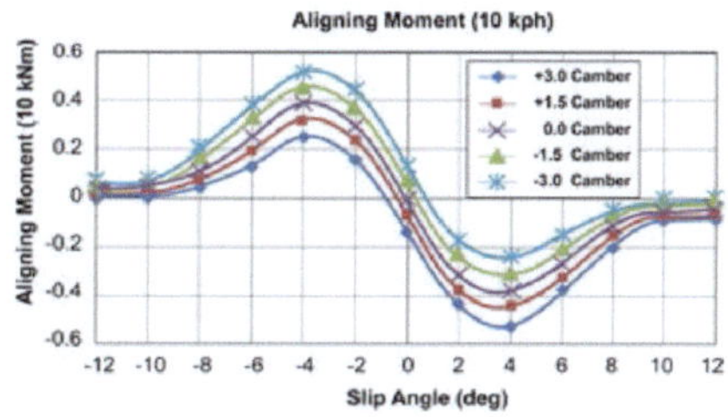

(b) Aligning moment due Combined Cornering and Cambering

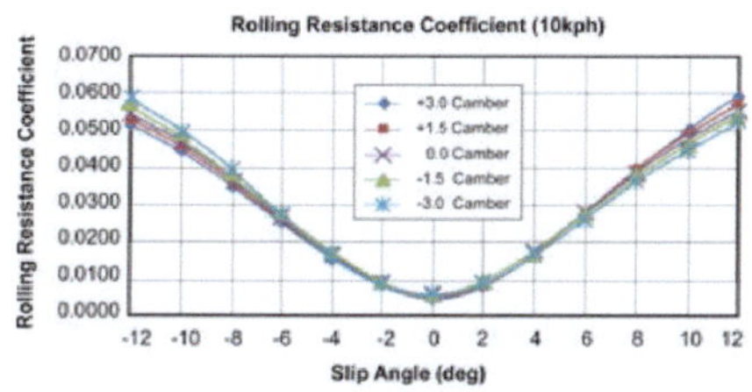

(c) Rolling Resistance Coefficients due Combined Cornering and Cambering

Fig. 1.26 Tire-road interaction characteristics for a combined steered and cambered tire

given camber angle. Furthermore, as the camber increases, the self-aligning moment increases.

Figure 1.26c shows the variation of the rolling resistance coefficient as a function of slip angle at different camber angles. It is noticed that as the slip angle increases, the rolling resistance coefficient increases as well at a given camber angle. Furthermore, as the camber angle increases, the rolling resistance coefficient also increases at a given slip angle.

1.3.8 Combined Braking and Cornering Properties

When a tire is operated under conditions of simultaneous longitudinal and lateral slip, the respective longitudinal and lateral forces are seen to differ considerably from those values derived under independent slip conditions. Driving or braking a wheel reduces the lateral force developed at the tire-road interface, as shown in Fig. 1.27.

For a given slip angle, the lateral force decreases gradually with the increase of tractive or braking force. At low values of tractive or braking effort, the cornering force decreases due to reduction in the cornering stiffness of the tire. The lateral force decreases considerably when the tractive force is further increased. This is due to the utilization of the available local friction by the tractive force, which reduces

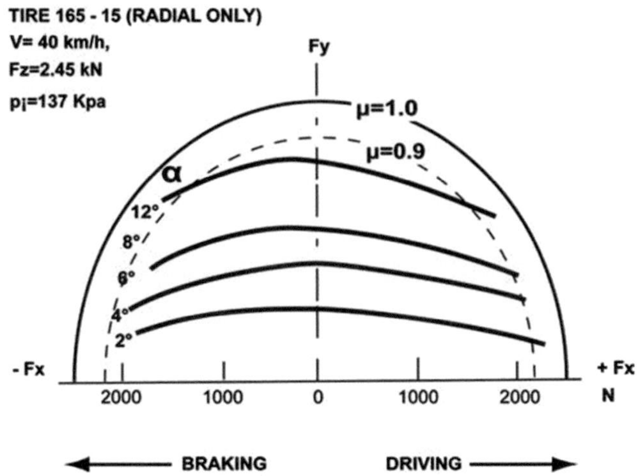

Fig. 1.27 Driving and braking lateral and longitudinal force distribution [5]

the amount available in the lateral direction. The tire approaches sliding when the resultant of tractive and lateral forces reaches the maximum value determined by the coefficient of road friction. Using the friction circle concept, the condition for sliding can be expressed as

$$\sqrt{F_y^2 + F_x^2} = \mu F_z \tag{1.58}$$

where μ is the coefficient of the road friction, and μ approaches the values of μ_p in the absence of cornering forces. The lateral force developed by the radial tires during the braking and driving is more or less symmetric, as shown in Fig. 1.27. A bias-ply tire, however, can generate larger cornering forces during braking than that during driving. Figure 1.28 shows actual measurement of the cornering and longitudinal forces.

1.4 Approach for Tire Modeling

Tire models and virtual testing have been used since the 1980s. It is important for the tire model to be able to predict the tire response from a dynamic vehicle simulation point of view.

Figure 1.29 shows the common methods available for a general field problem solution. The common methods are divided into numerical and analytical. The analytical method is determined using either an exact solution such as the separation of variables method or an approximate solution such as that implemented by the Rayleigh-Ritz. However, the numerical method involves either a numerical solution such as the numerical integration and finite differences, or the FEA technique.

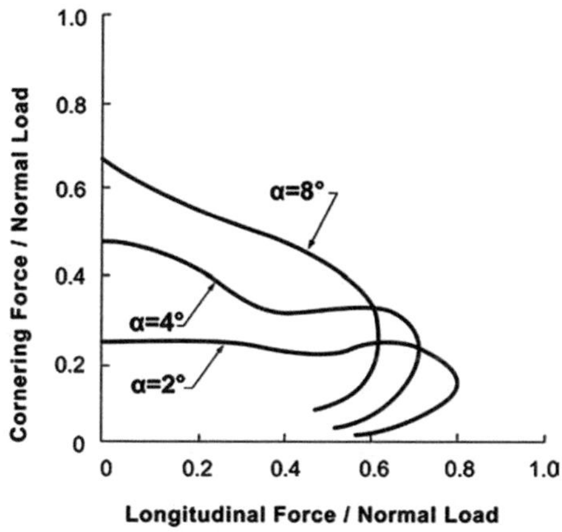

Fig. 1.28 Effect of longitudinal force on the cornering characteristics of truck tires [5]

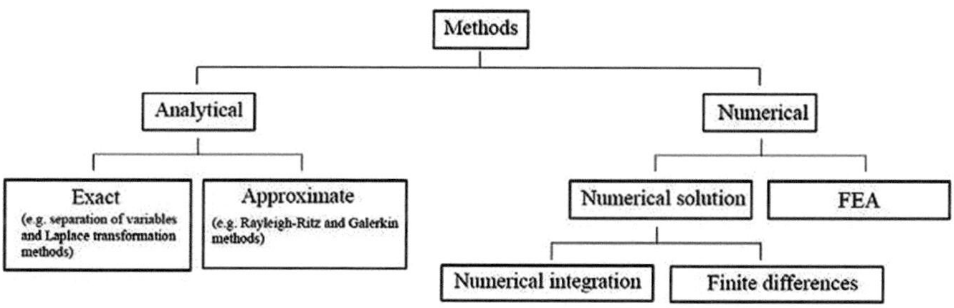

Fig. 1.29 Classification of common methods [11]

1.4.1 Lumped Parameter Modeling Technique

Ring and string models have been developed since the 1950s. These early models were based on pre-stressing the tread to string or ring. However, these early models had limitations in accessibility due to the required extensive experiments in order to determine the tire parameters' characteristics. Additionally, the validation of these tire models was restricted to a range of parameters, also the domain of validity was not always predicted in advance.

During the 1980s and 1990s, tire models mostly adopted the point contact mechanism. This mechanism assumes that the tire and road surface are in contact through a single point as shown in Fig. 1.30a. The point contact mechanism is sensitive to the road irregularities, thus it is more useful for longwave road profile inputs. Later, the effective road input model was established to overcome the limitations of the point contact model. Figure 1.30b shows the effective road input model that contributes to more realistic road input.

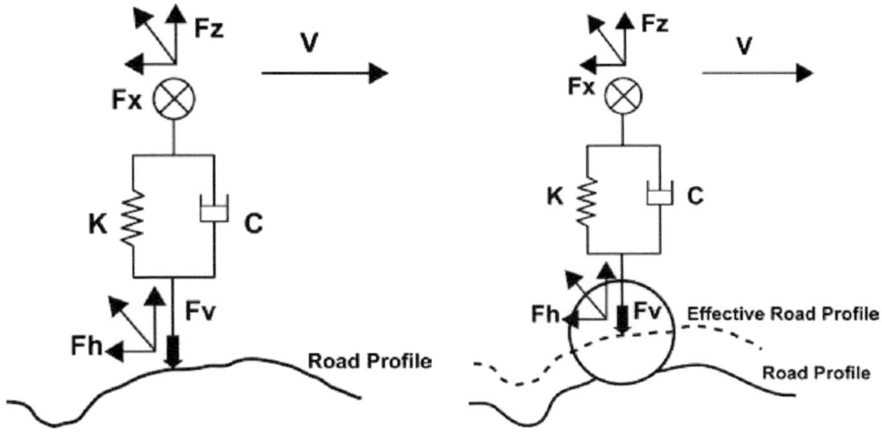

Fig. 1.30 Different tire-road contact mechanisms [12]

1.4.2 Empirical Modeling Technique

Previous studies predicted the cornering forces and self-aligning moment of a tire as early as 1987 [13]. In 1997, an empirical equation to characterize the cornering forces based on the tire measurements was developed and called the "Magic Formula" [14]. Equations 1.59 and 1.60 show the basic magic formula equations, where $Y(X)$ represents cornering force, self-aligning torque, or braking effort, and X denotes slip angle or skid. Coefficient B is called the stiffness factor, C the shape factor, D the peak factor, and E the curvature factor. S_h and S_v are the horizontal shift and vertical shift, respectively:

$$y(x) = D \sin \{C \arctan [Bx - E (Bx - \arctan Bx)]\} \tag{1.59}$$

$$\begin{aligned} Y(X) &= y(x) + S_v \\ x &= X + S_h \end{aligned} \tag{1.60}$$

Figure 1.31 shows the characteristics of the Magic Formula using the fitting tire test data. In order to complete a set of Magic Formulas, tire measurement data, such as cornering force and self-aligning moment versus slip angle, or brake-traction force versus slip angle needs to be obtained. Figure 1.31 also shows the meaning of some of the coefficients in the magic formula. D, is the peak value with respect to x and y origins. The product, BCD, is the slope of the curve at the origin (Table 1.5).

It is found that the coefficient D can be expressed as function of the vertical force, F_z, by

$$D = a_1 F_z^2 + a_2 F_z \tag{1.61}$$

a_1, a_2 are empirical coefficients. For the cornering stiffness, the product BCD is expressed as

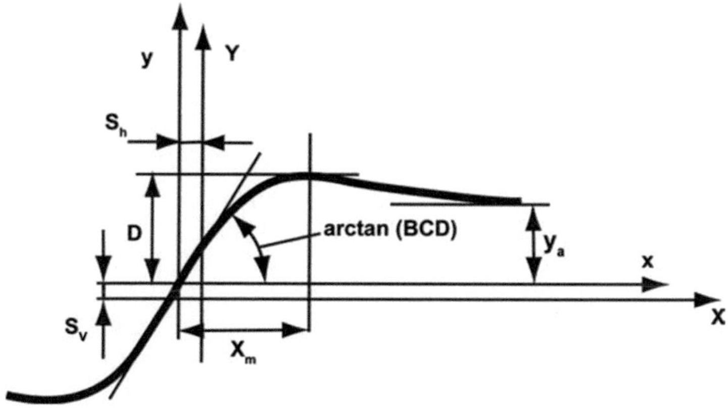

Fig. 1.31 Characteristics of the Magic Formula for fitting tire test data [14]

Table 1.5 Coefficients of the magic formula for a car tire

	F_z, kN	B	C	D	E	S_h	S_v	BCD
F_y, N	2	0.244	1.50	1936	–0.132	–0.280	–118	780.6
	4	0.239	1.19	3650	–0.678	–0.049	–156	1038
	6	0.164	1.27	5237	–1.61	–0.126	–181	1091
	8	0.112	1.36	6677	–2.16	0.125	–240	1017
M_z, N.m	2	0.247	2.56	–15.53	–3.92	–0.464	–12.5	–9.820
	4	0.234	2.68	–48.56	–0.46	–0.082	–11.7	–30.45
	6	0.164	2.46	–112.5	–2.04	–0.125	–6.00	–45.39
	8	0.127	2.41	–191.3	–3.21	–0.009	–4.22	–58.55
F_x,N	2	0.178	1.55	2193	0.432	0.000	25.0	605.0
	4	0.171	1.69	4236	0.619	0.000	70.6	1224
	6	0.210	1.67	6090	0.686	0.000	80.1	2136
	8	0.214	1.78	7711	0.783	0.000	104	2937

$$BCD = a_3 \sin\left[a_4 \arctan(a_5 F_z)\right] \tag{1.62}$$

where a_3, a_4, a_5 are empirical coefficients. For self-aligning moment stiffness or longitudinal stiffness, the product BCD is expressed as

$$BCD = \frac{a_3 F_z^2 + a_4 F_z}{e^{a_5 f_z}} \tag{1.63}$$

It should be noted that the shape factor, C, appears to be independent of the vertical load. The stiffness factor, B, is derived from

$$B = \frac{BCD}{CB} \tag{1.64}$$

Table 1.6 Coefficients a_1 to a_8 for a car tire

	a_1	a_2	a_3	a_4	a_5	a_6	a_7	a_8
F_y, N	–22.1	1011	1078	1.82	0.208	0.000	–0.354	0.707
M_z, N.m	–2.72	–2.28	–1.86	–2.73	0.110	–0.070	0.643	–4.04
F_x, N	–21.3	1144	49.6	226	0.069	–0.006	0.056	0.486

Table 1.7 Coefficients a_1 to a_8 for a car tire

	a_9	a_{10}	a_{11}	a_{12}	a_{13}
F_y, N	0.028	0.000	14.8	0.022	0.000
M_z, N.m	0.015	–0.066	0.945	0.03	0.07

The curvature factor, E, is expressed as a function of the vertical load as

$$E = a_6 F_z^2 + a_7 F_z + a_8 \tag{1.65}$$

where a_6, a_7, a_8 are empirical coefficients. Table 1.6 shows the typical coefficient values for a car tire.

The vertical and horizontal shifts, S_v, S_h, are influenced by the camber angle, γ, and the vertical load, F_z, as follows:

$$\Delta S_h = a_9 \gamma \tag{1.66}$$

$$\Delta S_v = \left(a_{10} F_z^2 + a_{11} F_z\right) \gamma \tag{1.67}$$

where a_9, a_{10}, a_{11} are empirical coefficients. Furthermore, the change in stiffness factor ΔB is obtained by

$$\Delta B = (1 - a_{12} \mid \gamma \mid) \, B \tag{1.68}$$

where a_{13} is an empirical coefficient. Table 1.7 shows the coefficient values a_9 to a_{13} for a car tire.

Vast experimental tire measurements are needed to cover a specific range of vertical tire loads. On the other side, to generalize the Magic Formula, 13 coefficients need to be calculated. These coefficients are computed from the vast and expensive experimental tire measurements at several vertical loads. The Magic Formula is often criticized due to a large number of coefficients.

? Example 1.2

Estimate the braking force developed by a tire operating at a vertical load of 8 kN that is skidding at 10% using the Magic Formula and the empirical coefficients provided in Table 1.5.

Solution. The Magic Formula represents the tractive force and thus the Y becomes F_x and X becomes s

$$F_x = D \sin\left[C \arctan\left(B(s + S_h) - E\left\{B(s + S_h) - \arctan\left[B(s + S_h)\right]\right\}\right)\right] + S_v$$

Using the empirical coefficients provided in Table 1.5 for a normal load of 8 kN and a braking effort of 10%, the tractive force becomes

$$F_x = 7711 \sin\left[1.78 \arctan\left(0.214(10) - 0.783\left\{B(10) - \arctan\left[0.214(10)\right]\right\}\right)\right] + 104$$

It should be noted that skid, s, is expressed in the Magic Formula in percentage, and the value of arctan should be calculated in radians

$$\begin{aligned} F_x &= 7711 \sin\left[1.78 \arctan\left(2.14 - 0.783\left\{2.14 - 1.13\right\}\right)\right] + 104 \\ &= 7711 \sin\left[1.78 \arctan(1.35)\right] + 104 \\ &= 7711 \sin\left[1.78 * 0.93\right] + 104 \\ &= 7783\,\text{N} \end{aligned} \tag{1.69}$$

1.4.3 Semi-empirical Modeling Technique

Another popular modeling technique is the semi-empirical technique which was developed in early 1997 [15]. A rigid ring tire model for a passenger car tire was developed based on the assumption that the tread and steel belts are modeled together as a rigid ring. Due to this assumption, new parameters were required to describe the deformation of the tire in the contact area; such parameters include the vertical residual stiffness. The tire frequency response on a 2.5 m-diameter rotating drum was measured to determine the rigid ring tire model parameters. Later, it was indicated that the in-plane vibrations are associated with the brake torque fluctuation and the irregularity of the road [16]. The rigid ring tire model was used to predict the longitudinal force and the rotational velocity at brake pressure variations.

The rigid ring model was later deployed to compute the in-plane contact problem of free rolling pneumatic tires [17]. Figure 1.32 shows the tire model constructed with an elastic ring. The flexible rigid band and belts of the tire were presented with an elastic ring. Additionally, elastic spring components were introduced to the outer surface of the elastic ring to present the radial and tangential flexibility of the tire tread rubber. In order to complete the rigid ring model, the following inputs are required: radial displacement, tangential displacement, mean radius, radial stiffness of sidewall, tangential stiffness of sidewall, normal and horizontal stiffness of tread rubber, radial and tangential damping coefficients of the sidewall, external force

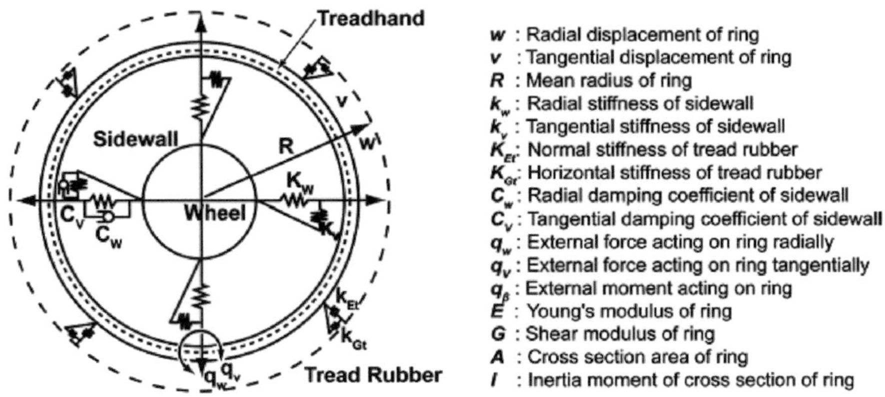

Fig. 1.32 Flexible rigid ring tire model [17]

acting radially and tangentially on the ring, external moment, Young's modulus, shear modulus, cross-sectional area, and inertial moment of cross-section.

The cornering characteristics of a rigid tire over loose sand were investigated using a driving tire model as a function of slip ratio and slip angle [18]. It was suggested that the lateral force decreases according to the increment of the slip ratio and increases according to the increment of the slip angle. Another study parameterized the in-plane rigid ring tire model using instrumented vehicle measurements during ABS braking, cleat test, and brake ramp [19]. The vibration modes of the tire using a Cleat test were extracted, which then were used to define the rigid ring model parameters. It was concluded that it is possible to derive parameters for an in-plane rigid ring tire model from vehicle measurements, however, it is vital to include the vehicle suspension into the model.

In a later attempt, a three-dimensional quasi-steady-state tire model was developed for on-road and off-road vehicle dynamics simulations [20–22]. The model implemented the brush tire model for on-road simulation and a simplified off-road tire model capable of reverting back to on-road trend. The on-road tire model is based on empirical data collected experimentally by the National Highway Traffic Safety Administration (NHTSA). Furthermore, the off-road tire model is developed based on observations of experimental data. Then, the research continued to develop an off-road flexible tire model that was later parameterized from test data to acquire static and dynamic friction coefficients, cornering and longitudinal stiffness, as well as camber stiffness.

It should be noted that in the literature there are much more analytical and numerical tire models developed by many researchers in the last decade.

A number of empirical and semi-empirical relationships between the tracking/braking force, cornering force, longitudinal force, slip force, and other design and operational parameters have been proposed. An example of this is Dugoff's tire model which is a good representative of a combined steering and braking equation as follows:

$$F_x = \frac{C_s}{1-s}.f(\lambda) \tag{1.70}$$

$$F_{y\alpha} = \frac{C_\alpha.\tan\alpha}{1-s}f(\lambda) \tag{1.71}$$

$$f(\lambda) = \lambda(2-\lambda) \text{ if } \lambda < 1 \tag{1.72}$$

$$f(\lambda) = 1 \text{ if } \lambda > 1 \tag{1.73}$$

$$\lambda = \frac{\mu_0 F_z \left[1 - \epsilon_r v\sqrt{s^2 + \tan^2\alpha}\right](1-s)}{2\sqrt{C_s^2 s^2 + C_\alpha^2 \tan^2\alpha}} \tag{1.74}$$

where F_z is the vertical load on tire (kN); μ_0 is the road coefficient of friction; ϵ_r is the speed coefficient which is equal to 0.03; v is the tire speed; s is the longitudinal slip; α is the slip angle in rad; C_s is the longitudinal stiffness (kN/unit slip) and C_α is the cornering stiffness (kN/rad).

1.4.4 Semi-analytical Modeling Technique

Finite Element Analysis is a numerical method to solve engineering and mathematical problems. It was first established as an efficient method for the approximation of problems in the field of deformable solids [23]. Later, it was proposed as a particular linear function technique and the method to solving torsion problems was applied [24]. The combination of both modifications is similar to the FEA method. However, the FEA method was later proposed in 1960 [25].

The FEA technique is widely employed in the domain of terramechanics. The FEA technique has been very useful in solving several problems in terrain mechanics. Various topics have been investigated to estimate the performance of off-road pneumatic tires using the FEA technique [26].

In 1990, Eskinazi [27] investigated the possibility of predicting the relative belt edge endurance for a car tire using the FEA technique. It was concluded that two-dimensional analysis can lead to inaccurate conclusions and thus a three-dimensional analysis under static vertical loading was performed. In 1997, Hiroma [28] implemented the FEA method to predict the tractive forces and pressure distributions beneath a rolling wheel. Hiroma compared the FEA prediction to those from measurements and found that the predictions were reasonable. Hiroma concluded that under small slip conditions, FEA methods could be used to predict traction force. In 1998, Koishi [29] computed the tire cornering characteristics using Pam-Shock, an explicit FEA software. A three-dimensional FEA tire model was built and the effect of inflation pressure, belt angle, and rubber modulus was investigated.

FEA uses the meshing methodology to establish numerical models of a physical structure with smooth and realistic discretization and representation of the boundary

(a) 295/75R22.5 Truck tire

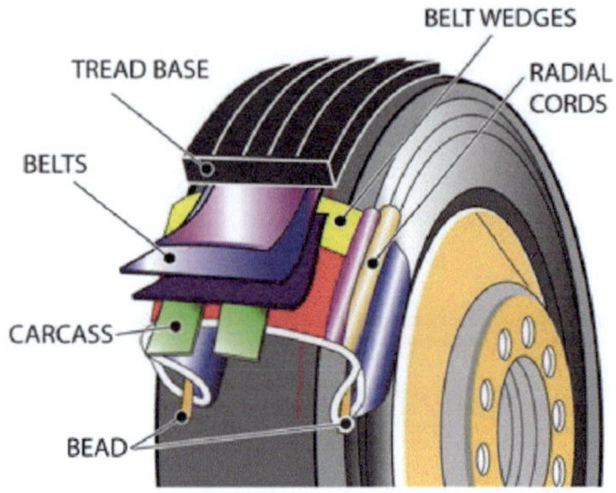

(b) Component of the tire

Fig. 1.33 Radial truck tire size and components [31]

conditions. In recent years, Pam-Crash, the virtual environment software, has been extensively adopted to build FEA tire models. Pam-Crash also implements the principle of explicit time integration which advances the solution along the time axis. The explicit solution method expresses the equilibrium equation at time t_n as shown in Eq. 1.75 [30]:

$$m\frac{d^2x_n}{dt^2} + kx_n = f_n \tag{1.75}$$

where m is the mass, x_n is the position at node n, k is the stiffness, and f_n is the acceleration. The advantage of the explicit method is that only the mass, m, appears in the denominator, however, the requirement for stability puts an upper limit on the time step.

In 2006, Chae [31] modeled Goodyear's 295/75R22.5 drive tire for tractor semi-trailers. The drive tire is a radial-ply tire with a rim diameter of 22.5 in. The truck tire-rim assembly model includes 27 different material definitions with 4200 solid elements, 1680 membrane elements, and 120 beam elements. The section width of the truck tire is 315 mm, and the aspect ratio is 75-percent. The off-road tire and components are shown in Fig. 1.33.

Later, Slade [7] modified the tire model built by Chae to represent the Goodyear off-road size 315/80R22.5 with four grooves. The cross-section was built node by node and then rotated about the tire axle axis in 6° increments to create the full tire with 60 equal pieces. This tire model is built using 9200 nodes, 1680 layered membrane elements, 120 beam elements, 27 material definitions, and one rigid body definition. The rim is defined as a rigid body for the simplicity of the model because the deformation of the rim is negligible.

The rubber material used in modeling the tire was first developed by Mooney [32] in 1940; the rubber material was isotropic and strain energy function, W, was used

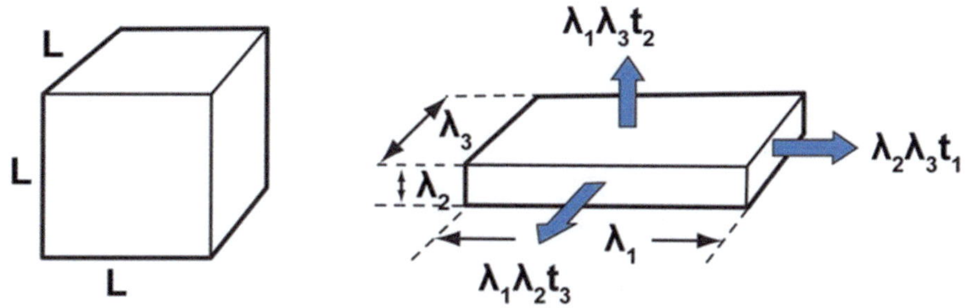

Fig. 1.34 Undeformed and deformed states of element [31]

to represent the elastic behavior. The strain energy function, W, shown in Fig. 1.34 can be written in terms of three extension ratios, λ_1, λ_2, and λ_3.

The strain energy function, W, can be written in terms of the strain invariants, I_1, I_2, and I_3, as shown in Eq. 1.76:

$$W = \sum_{i=0}^{n}\sum_{j=0}^{n}\sum_{k=0}^{n} C_{ijk}(I_1 - 3)^i(I_2 - 3)^j(I_3 - 1)^k(n = 1, 2, 3, ..., \infty) \quad (1.76)$$

Equation 1.76 can be reduced as shown in Eq. 1.77, where C_{10} and C_{01} are constants that are experimentally determined:

$$W = C_{10}(I_1 - 3) + C_{01}(I_2 - 3) \quad (1.77)$$

A typical Mooney-Rivlin tensile test can be seen in Fig. 1.35 on natural rubber vulcanizates. The coefficient C_{10} is the intercept at extension ratio of 1 within the low strain range, and the constant C_{01} is the slope of the line. The two constants, C_{10} and C_{01}, of the Mooney-Rivlin equation are determined in the low strain range because engineering applications usually fall within the low strain ranges.

In most truck models, Mooney-Rivlin material is used to model tread, under-tread, shoulder, and bead filler.

1.4.5 Artificial Neural Network Tire Model "Neuro-Tire"

One of the most difficult aspects to vehicle modeling is the accurate description of tire contact patch stress and cornering characteristics. A considerable amount of work has been done in the area of tire modeling using a range of approaches, from pure theoretical modeling to fully empirical formulae fitted to measure data.

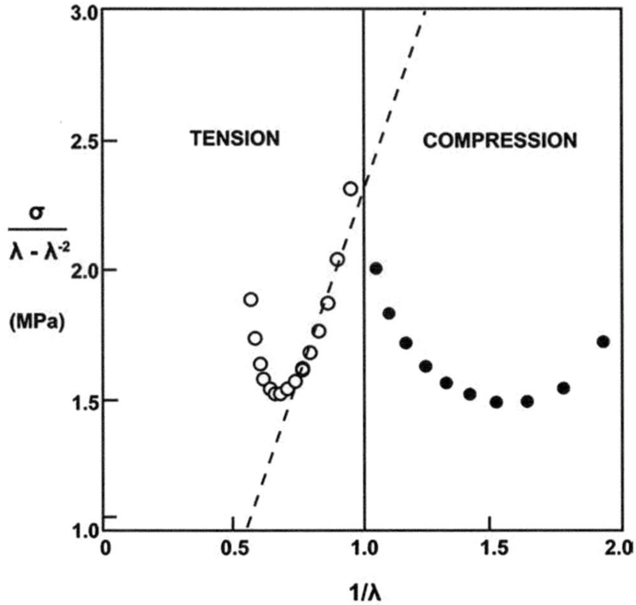

Fig. 1.35 Coefficients of Mooney-Rivlin for tensile and compression tests [33]

1.4.5.1 What is the Artificial Neural Network?

Figure 1.36 illustrates the sort of biological neuron, which has influenced the development of "artificial", or computational, neural networks. The neuron is the fundamental cell of the nervous system and, in particular, the brain. Each neuron is a simple microprocessing unit, which receives and combines signals from many other neurons through input processes (structures) called dendrites (see Fig. 1.36). Signals from the dendrites are communicated to the neuron body through synapses. If the combined signal is strong enough, it activates the firing of the neuron, which produces output signal; the path of the output signal is along a component of a cell called the axon. This simple transfer of information is chemical in nature, but has electrical side effects, which can be measured.

The human brain is the most complex computing device known to man. The brain's powerful thinking, remembering, and problem-solving capacities have inspired many scientists to attempt computer modeling of its operation. The human cerebral cortex, for instance, is comprised of approximately 100 billion (10^{11}) neurons with each having roughly 1,000 dendrites that form some 100,000 billion (10^{14}) synapses; given that this system operates at about 100 Hz, it functions at some 10,000 billion (10^{16}) interconnections per second. It weighs approximately three pounds, covers about 0.15 square meters, and is about two millimeters thick. The "soma", or the nerve cell, which is the large round central body of the neuron, is anywhere from five

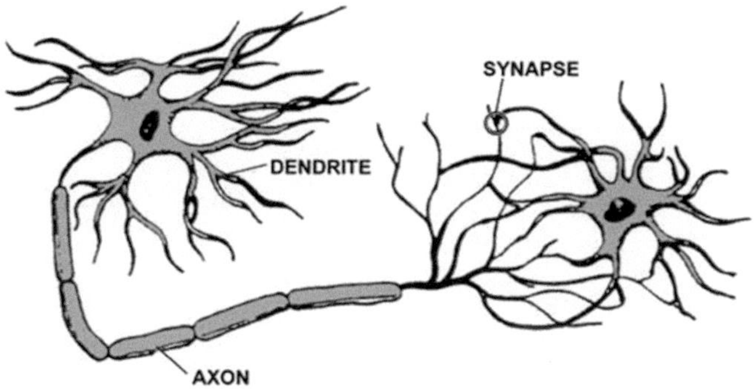

Fig. 1.36 Biological Neurons

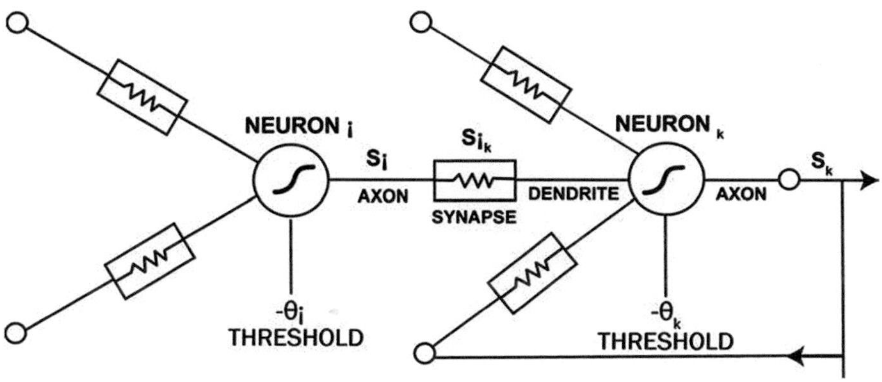

Fig. 1.37 Artificial Neuron

to 100 microns in diameter. This capability is clearly beyond anything which can be reconstructed or modeled.

In an artificial neural network (Figs. 1.37 and 1.38), the unit analogous to the biological neuron is referred to as a "processing element". A processing element (PE) has many input paths (dendrites) and combines, usually by a simple summation, the values of these input paths. The result is an internal activity level for the processing element. The combined input is then modified by a transfer function. This transfer function can be a threshold function, which only passes information if the combined activity level reaches a threshold, or it can be a continuous function of the combined input. The output value of the transfer function is generally passed directly to the output path of the processing element, which is connected to input paths of other processing elements through connection weights, which corresponds to the synaptic strength of neural connections. Since each connection has a corresponding weight, the signals on the input lines to a processing element are modified by these weights prior to being summed. Thus, the summation function is a weighted summation. In

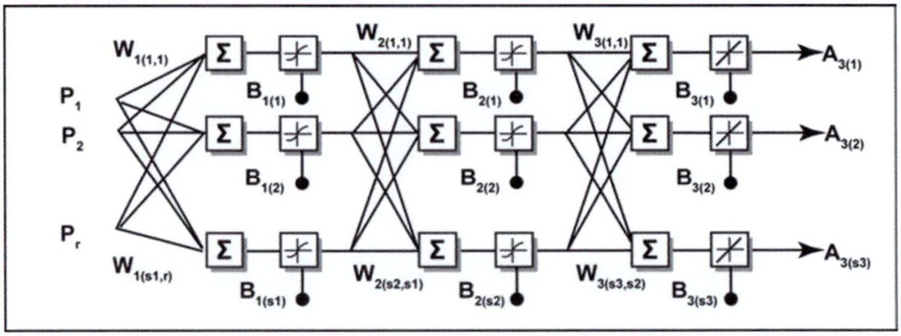

Fig. 1.38 A simple ANN architecture (obtained from ANN Matlab Toolbox)

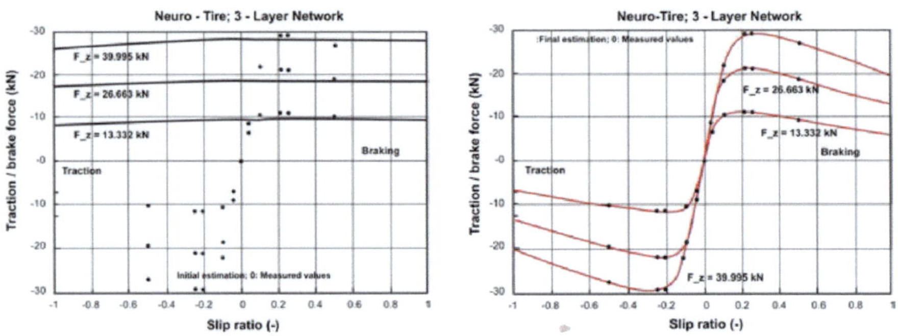

(a) Steady state (b) Measured values

Fig. 1.39 Tire traction/braking forces at different vertical loads versus longitudinal slip

itself, this simplified model of a neuron is not very interesting; the interesting effects result from the way in which neurons are interconnected.

A network consists of many processing elements joined together in the above manner. Processing elements are usually organized into groups called layers. A typical network consists of a sequence of layers with full or random connections to the outside world: An input buffer where data is presented to the network, and an output buffer which holds the response of the network to a given input. Layers distinct from the input and output layers are called hidden layers.

1.4.5.2 Example of Traction/Braking Force Prediction Model "NEURO-TIRE"

Figure 1.39 shows a comparison between the traction/braking forces at different vertical loads, developed by Neuro-Tire, and measured values.

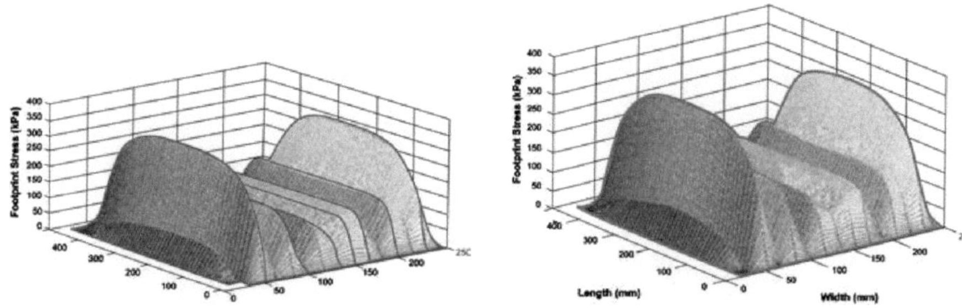

(a) Measured/ANN training set (b) Predicted/ANN output set

Fig. 1.40 Measured and predicted truck tire contact vertical stress distribution

1.4.5.3 Example of the Tire/Pavement Contact Stress Model "NEURO-PATCH"

Figure 1.40 show the training set; the neural network predicted the footprint contact stress of a truck 10×20 bias-ply at vertical Load of 56 kN and inflation pressure of 520 kPa.

1.4.6 Validation Approach

The validation of tire models is performed statically and dynamically through measurements and simulations.

1.4.6.1 Static Validation Techniques

In 1988, Ford [34] extensively investigated the heavy-duty truck tire; the study emphasized three major elements in the field of heavy-duty tires: tire design factors, performance properties, and application requirements of commercial truck tires. Ford updated the design factors of a tire to include the cross-sectional shape since radial-ply tires became more popular.

In 2002, [35] validated a passenger car tire model using the in-plane vibration modes, the standing waves, traction friction coefficient, vertical static stiffness, and contact patch. Later in 2006, Chae [31] validated the truck tire model in both static and dynamic responses. The static response is verified by vertical stiffness and static footprint tests. The dynamic drum-cleat test validates the dynamic response of the tire.

The vertical stiffness test allows for the calculation of the tire's spring rate. During the vertical stiffness test, the tire is constrained in all directions except for the vertical direction. The free motion in the vertical direction allows the tire to move on the ver-

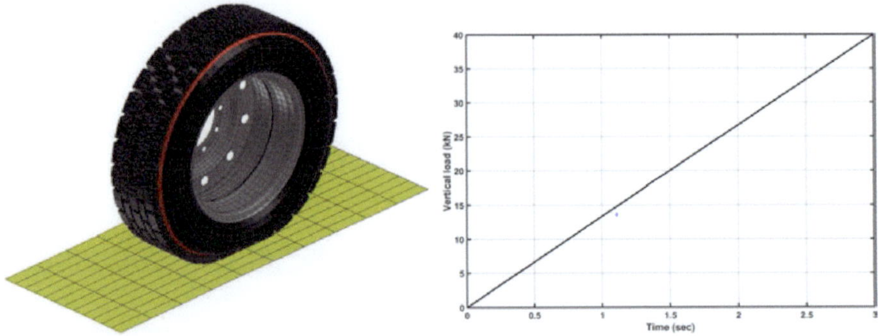

(a) Schematic of Vertical test (b) Applied ramp load

Fig. 1.41 FEA vertical stiffness test under a ramp loading up to 40 kN

tical axis as shown in Fig. 1.41. The tire is subjected to a ramp load which causes the tire to deform. The resultant deflection is then recorded for the corresponding vertical loads, and the relationship between vertical load and the deflection is considered.

The static footprint test is the second validation test applied to validate the tire. The contact patch of a tire is affected by the inflation pressure and vertical load. In the static footprint test, the same procedure of the vertical stiffness test is applied. However, in this case, the contact patch area is recorded instead of the deflection.

1.4.6.2 Dynamic Validation Techniques

Later in 1991, Yap [36] measured the cornering characteristics of a radial truck tire over the dry surface. The tire used was 11R22.5 and the tread design effect on the cornering characteristics was investigated. In 1997, Davis [37] performed physical testing to determine the mechanical properties of an aircraft tire under several conditions. The study focused on measuring the quasi-static characteristics and comparing them to the dynamic response of the tire.

In 2006, Chae [31] stated that a significant amount of the tire mass is concentrated near the tread. The rolling tire radius is not constant due to the radial deflection. However, the stiffness of the tire is affected by the inflation pressure and the material properties. During this test, a 10 mm-radius semicircular cleat test is virtually simulated with a 2.5 m diameter drum to excite the tire vertically and cyclically as the drum rotates.

Figure 1.42 shows the illustration of the vertical first mode of vibration where the entire tread band vibrates vertically without distortion about the vertically fixed rim. Due to the movement of the tread band, the force associated with the resonance is transmitted to the wheel and axle.

The tire is first inflated to the desired inflation pressure, then the tire is loaded by applying the vertical load. The drum center is constrained in all translational

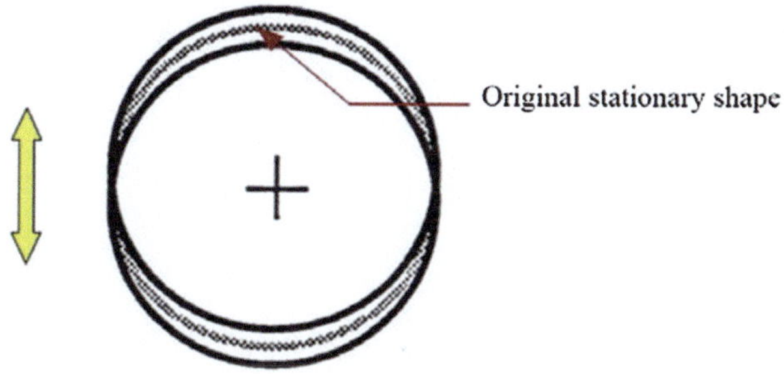

Fig. 1.42 FEA drum-cleat test and vibration mode [31]

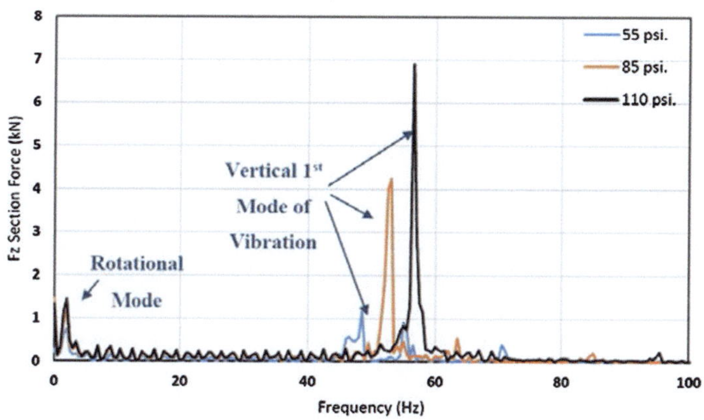

Fig. 1.43 FFT result of vertical reaction force at tire spindle [38]

directions and free in the rotational direction, while the tire center is constrained in all translational directions to detect the transmitted vertical force. An angular velocity is then applied to the center of the drum to enable free rolling of the tire at 50 km/h speed. The rotation of the drum allows the cleat to excite the tire vertically, and the vertical reaction force along with the in-plane free vibration mode is determined. The vertical and longitudinal reaction forces are measured and converted from a time domain to a frequency domain using the Fast Fourier Transformation (FFT) algorithm as demonstrated in Fig. 1.43. Tire's sidewall damping is calculated using Eq. 1.78 [31], where α is sidewall damping, $\epsilon = 5\%$ is 5-percent critical damping effect, and ω is considered as the first mode of vibration frequency:

$$\alpha = \epsilon 2\omega \tag{1.78}$$

In 2017, Lardner [38] predicted the first mode of vibration of a truck tire at different inflation pressure. Lardner concluded that as the inflation pressure increases the first mode of vibration increases as well. She concluded that the first mode of vibration

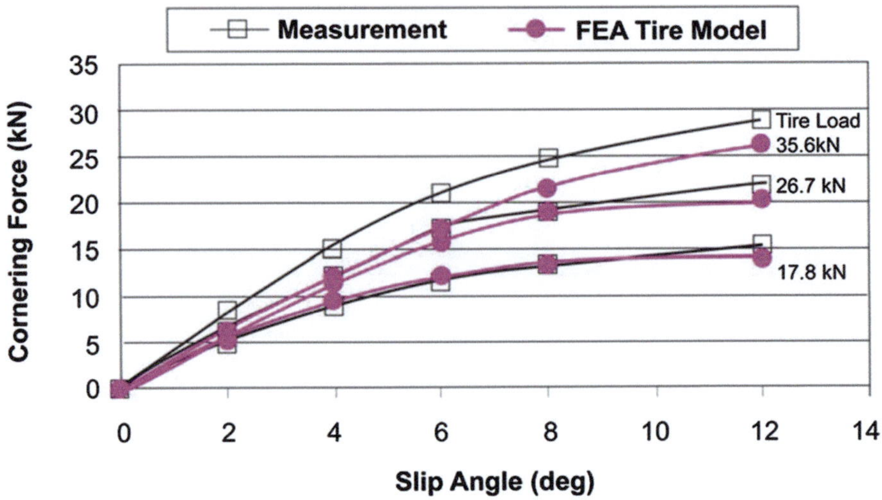

Fig. 1.44 Cornering force as a function of slip angle for FEA tire model and measurements [31]

of this specific tire is between 46 and 57 Hz depending on the inflation pressure. Lardner also predicted the sidewall damping of the tire to be 29, 33, and 36 for an inflation pressure of 380, 586, and 758 kPa, respectively.

In 2006, Chae [31] used the cornering test to validate the tire dynamic response. During this test, the tire is subjected to a lateral deflection caused by a lateral force, and the cornering force is computed at different slip angles between 0 and 12°. The cornering force versus slip angle is plotted and compared to measured data provided by the University of Michigan Transportation Research Institute (UMTRI). Additionally, the aligning moment is also computed from this test and plotted against slip angle then validated against data provided by UMTRI.

Figure 1.44 shows the variation of the cornering force as a function of slip angle for both the FEA tire model and the measurements. The applied vertical loads are 17.8, 26.7, and 35.6 kN. The predicted cornering forces and slopes at slip angle are used to compute the cornering stiffness. In general, the cornering stiffnesses are in good agreement with the measurements, especially for the two cases of lower vertical loads, 17.8 and 26.7 kN .

1.5 Experimental Tire Testing

Since Charles Goodyear invented the first pneumatic bicycle tires in 1839, a variety of measuring devices have been designed to determine the forces and moments that arise in the contact area. Although initially the interest was in load-deflection relationships in a radial direction, later, research emphasized the rolling resistance

Fig. 1.45 MTS drum tire test machines (obtained from www.mts.com)

and nonskid qualities of tires during braking and traction. The need to evaluate the cornering force and the slip angle was discussed as early as 1930 by Bradley and Allan. Later, in the 1950s, a number of simple tire tests, such as vertical displacement measurement at various loads, rolling resistance measurements, and cornering characteristics, were actively performed [39, 40]. Since then, more complicated research has been conducted in the area of viscoelastic tire material, rubber-cord composite material, and in-plane and out-of-plane transient responses of tires. In order to accomplish this complicated laboratory tire testing, large-scale experimental setups and highly experienced measurement skills and effort are usually required. Transient response measurements of a tire in particular need complex experiment facilities and data acquisition systems.

Many tire test methods have been established to measure forces and moments acting on a rolling tire. Mobile dynamometers are available which run the tire over the road using a towed trailer rig. This equipment has the advantage of providing data on actual road surfaces. However, the majority of tests have been conducted in laboratories using a large drum because the tire-drum system occupies relatively small space, and speed control of tire is easy.

Figure 1.45 shows drum tire test machines. These drum test machines are useful for comparing the characteristics of different tires. However, they do not provide exact information of the tire behavior observed on a flat surface because significant difference in the tire behavior can be induced by the curvature of the drum.

To eliminate the effects caused by drum curvature, a flat belt is used instead of a drum. The test machine consists of a steel band or conveyor belt running on two drums. The flexible belt is supported in the contact area by an extremely stiff and thin air bearing in order to minimize the air gap variation. A drawback of such an air bearing is the large air consumption [41]. Appropriately surfaced steel belts can simulate road surfaces from low to high skid resistance. Dry or wet pavement simulated conditions up to full dynamic hydroplaning from low speed to 322 km/h could be obtained [42] (Fig. 1.46).

Fig. 1.46 MTS tire test on flat belt (obtained from www.mts.com)

Using all of the available tire test machines, many researchers have performed tire tests to measure tire forces and moments in the contact area, rolling resistance, braking and traction efforts, and cornering characteristics. In 1969, Davisson mentioned three important elements to consider in tire engineering: application requirements, performance properties, and design factors. He first described design factors including tire structure, materials, stress relationships, and tread and sidewall pattern. Then, he continued with performance properties including stiffness, damping, cornering characteristics, energy loss, and durability. Finally, he completed his study with application requirements including vehicle requirements, service requirements, and economics. In his study, he included the number of tire testing data which were obtained from real experiment measurements, such as cornering forces of radial and bias-ply tires versus slip angle, contact pressure distribution, and rolling resistance versus inflation and speed.

Similarly, in 1988, Ford and Charles described heavy-duty truck tire engineering in an extensive fashion similar to Davisson in 1969. In their study, they emphasized three important elements in heavy-duty tire engineering: tire engineering factors, performance properties, and application requirements. In the section on tire engineering factors, they updated Davisson's section on design factors. For example, they updated the cross-sectional shape and structural component figure with a modern heavy-duty radial truck tire, since radial tires had popularity over bias-ply or bias-belted tires. Ford and Charles also introduced tire modeling techniques using Computer Aided Engineering (CAE) software, such as NASTRAN and CADAM. They developed an FEA truck tire model and simulated tire inflation and loading conditions. In the performance properties section, several stiffness test measurements of the latest tires in three directions (vertical, lateral, and tangential directions) were plotted. It was noted that dynamic vertical stiffness for radial heavy duty truck tires was approximately five percent less than static values. In lateral stiffness, the radial tire was somewhat less stiff than the bias tire. It was found through experimental measurements that the radial tire carcass was less stiff than the bias tire in the tangential direction. In addition, in the application requirements section, three parts were categorized, i.e., vehicle requirements, service requirements, and economics similar to Davisson [43]. However, this research provided more information about new topics, such as modal analysis results, cornering and handling, and tire selection. This large amount of information and experimental data furnished insights for tire model developments and validation.

In 1991, Yap performed cornering characteristic measurements on radial truck tires, 11R22.5, to investigate the effect of tread designs on cornering characteristics using a flat belt type of the Calspan TIRF machine. In order to limit the study to the tread design parameters, the truck tires in this study had different rib and lug tread designs, and the tread rubber compound of the tires was the same. Yap measured tire cornering forces and moments at various vertical loads, inflation pressures, slip angles, and cambers. As a result, significantly different responses between the designs could be observed with respect to cornering force and self-aligning moment, even though the tire cornering characteristics were found to be more dependent on vertical loads than on the different tread designs.

In 1997, Davis measured mechanical properties such as stiffness and damping constant of bias-ply and radial-belted 30 × 11.5–14.5/26PR aircraft tires under a variety of test conditions. The bias-ply tire was inflated at the rated inflation pressure of 1.69 MPa, while the radial-belted tire was inflated at the rated inflation pressure of 2.14 MPa. These inflation pressures correspond to a 35-percent deflection at the rated load of 111 kN. The objective of this study was to measure and compare quasi-static and dynamic response characteristics of the aircraft tire.

The vertical, lateral, and fore-and-aft deflections were recorded to determine the corresponding stiffness and damping constants. In addition, footprint area and moment of inertia were also measured and compared between the two tires. Then, the radial-belted tire was evaluated as a replacement for the bias-ply aircraft tire. Davis found that the vertical load stiffness characteristics of the two tire designs were similar. However, it was found that significantly different lateral and for-and-aft stiffness properties of the radial-belted tire might result in an increase in tire shimmy and affect the performance of an anti-skid braking system (ABS) tuned for bias-ply tires. It was also found that the smaller and more elliptical footprint shape of the radial-belted tire might be more sensitive to hydroplaning conditions. From moment of inertia testing, it was found that the radial-belted tire required less energy to spin up, which could result in reduced tire wear and heating during high-speed landings.

1.6 Tire Performance Over Flooded Surfaces

Tire performance is highly dependent on the ground conditions; one of the most important conditions is the water.

1.6.1 Hydroplaning Phenomena

Hydroplaning is the loss of tire-road surface contact due to hydrodynamic lift force of water. In other words, water layers build under the rubber tires of the vehicle and the road surface, which leads to loss of traction [44]. The contact forces between the tire and road decrease as the tire speed increases; this results in the decrease of the driving controllability of the vehicle [45]. This phenomenon was first noticed and demonstrated experimentally during a tire treadmill study in 1957 [46]. Several manifestations were accredited with the hydroplaning phenomenon; these manifestations are mentioned by [44]. Some of these manifestations are the detachment of tire footprint, hydrodynamic ground pressure, spin-down of the wheel, suppression of tire bowwear, loss of breaking, and loss of tire directional stability.

Since the 1960s, many researchers have tended to employ an analytical or numerical method to investigate the hydroplaning problem. In 1966, Martin [47] considered the total dynamic hydroplaning problem by applying the potential flow theory and conformal mapping techniques. Later in 1967, Moore [48] modeled a rubber sliding on a two-dimensional smooth sinusoidal asperity by a thin fluid film. In the same year, Eshel [49] divided the tire-water contact area into three zones based on the amount of the inertial and viscous effects and utilized a different method for each zone.

In 1972, Browne [50] studied hydroplaning phenomenon using Navier-Stokes equations and proposed a two-dimensional treatment for a three-dimensional tire deformation model. In 1972, Leland [51] reported that in shallow water around a thickness of 1 mm, tread grooves are highly effective in delaying the occurrence of hydroplaning. In 1974, Sinnamon [52] concluded that hydroplaning speed varies inversely with water depth, thus lower water depth decelerates the onset of hydroplaning phenomena.

In 1996, Gogger [53] investigated the tire velocity field and the pressure distribution of a deformable automobile tire without rotation during hydroplaning. One year later in 1998, Panagouli [54] performed an investigation on pavement textures which indicated that pavement macrotexture is a function of aggregate size, shape, spacing, and distribution of coarse aggregates.

Later in 2000, Seta [45] used an FEA to model the tire and Finite Volume Method (FVM) to model the water to simulate tire hydroplaning. One year later, Janajreh [55] used Computational Fluid Dynamics (CFD) to determine the drag force, which indicates the fluid evacuation around the tread pattern. In 2005, Fwa established a numerical simulation model for hydroplaning prediction using CFD techniques implemented by Fluent to investigate the effect of several factors such as groove width, depth, and spacing of pavement on the hydroplaning speed of smooth passage car tire [56]. In 2007, Ong [57] established a numerical simulation model for hydroplaning prediction using CFD techniques implemented by Fluent to investigate the effect of different factors such as groove width, depth, and spacing of a pavement on the hydroplaning speed of smooth passage car tire. Moreover, Oh [58] adopted two separate mathematical models to simulate hydroplaning. One year later, Jenq [44] implemented a hydroplaning model for a tire using Ls-Dyna, the model accounted for the water viscous effect. In 2012, Choi [59] estimated a wet road braking distance for vehicles equipped with Anti-lock Braking System (ABS).

How to Avoid Hydroplaning

- Make sure tires have good tread.
- Drive slowly down on flooded surface.
- Keep proper inflation pressure.
- Avoid driving through standing water.
- Avoid hard braking.

Hydroplaning Equations

To describe the hydroplaning phenomena, researchers developed a "Three-Zone" concept. This concept was first applied by Gough [60] in 1954 and then developed further to cover the rolling tire case by Moore [48] in 1967. Figure 1.47 shows a schematic of the three-zone concept.

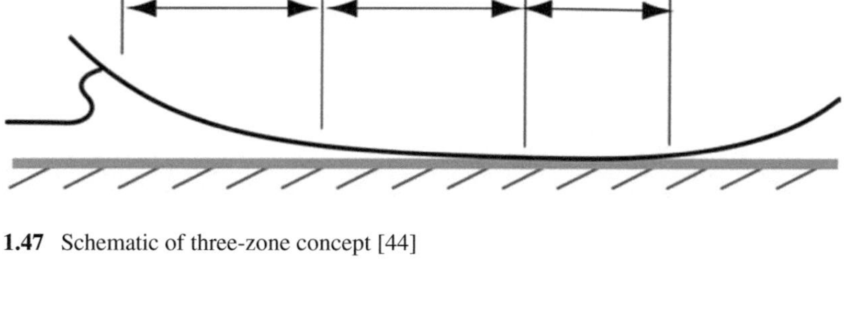

Fig. 1.47 Schematic of three-zone concept [44]

Zone A is the squeeze-film zone which is governed by the Elastohydrodynamic lubrication (EHL). In this district water the wedge penetrates in the backward direction. In this section, the frictional forces depend on the viscosity and velocity gradient in the lubricant film. Zone B is the transition zone where tire elements penetrate the squeeze-film commence to drape about the asperities of the road surface. When driving at ordinary speeds, the uplift forces produced in this zone are not great enough to produce a full dynamic hydroplaning. Zone C is the traction zone; this is the rear part of the contact area; it starts at the beginning of the end of the transition zone. In this zone, the lubricated water film is considerably removed, and the vertical equilibrium of the tread elements is attained. Depending on the tire speed, the length of this zone may vary.

In 1965, Horne [61] proposed the NASA hydroplaning equation according to aircraft tire experiments at NASA Langley Research Center. The NASA equation is shown in Eq. 1.79, where p is the tire inflation pressure in kPa and v is the minimum hydroplaning velocity km/h:

$$v = 6.36\sqrt{p} \tag{1.79}$$

Equation 1.79 can be implemented if the water depth exceeds the tire tread depth or if the tread pattern is smooth.

Later in 1986, Horne [62] developed an equation that predicts hydroplaning speed for truck tires. Horne developed Eq. 1.80; the equations relate the hydroplaning speed with the tire Footprint Aspect Ratio (FAR) and the inflation pressure:

$$v = 23.3p^{0.21}\left(\frac{1.4}{FAR}\right)^{0.5} \tag{1.80}$$

In 1979, Gallaway [63] developed another equation to predict the tire hydroplaning speed. Gallaway's equation involves the spin-down %, tire inflation pressure, tread depth, water film thickness, and mean texture depth of pavement surface. Equation 1.81 presents Gallaway equations, and Eq. 1.82 shows the parameter A, where SD is the spin-down %, t_w is the water film thickness in in, MTD is the mean texture depth in in, and TD is the tire tread depth in $1/32\,in$:

$$v = SD^{0.04}p^{0.3}(TD + 1)^{0.06}A \tag{1.81}$$

where A is

$$A = max\left[\left(\frac{10.409}{{t_w}^{0.06}} + 3.507\right), \left(\frac{28.952}{{t_w}^{0.06}} - 7.819\right)MTD^{0.04}\right] \tag{1.82}$$

In 1984, Wambold [64] developed an equation that predicts hydroplaning for low-pressure tires based one a 10% SD and 165 kPa tire inflation pressure:

$$v = 3.5k_1\left[\left(\frac{TD}{25.4}+1\right)^{k_2} MTD^{k_3}\left(\frac{k_4}{t_w{}^{k_5}}+1\right)\right] \tag{1.83}$$

Equation 1.83 shows the Wambold equation; t_w is water film thickness in mm, MTD is mean texture depth mm, TD is tire tread mm, and k_s are empirical coefficients. The empirical coefficients k_1, k_2, k_3, k_4, and k_5 are typically equal to 0.05, 0.01, 1.8798, and 0.01, respectively.

Recently in 2017, Zeinab et al. [65] developed an equation to predict the truck tire hydroplaning speed as a function of various operating conditions. The equation was based on truck tires and various hybrid FEA-SPH simulations and genetic algorithms. Equation 1.84 presents the hydroplaning speed equation as a function of the tire inflation pressure in kPa, vertical load in kN, tread depth in mm, and water depth in mm; k_s are empirical coefficients. The empirical coefficients k_1, k_2, k_3, k_4, k_5, and k_6 are typically 3.3, 9.11, 0.0167, 0.00125, 0.0623, and 0.203, respectively:

$$v = k_1 t_d + k_2 L + k_3 P + t_w{}^2\left(k_4 L - k_5 L^{-1}\right) - k_6 L t_w \tag{1.84}$$

? Example 1.3

Find the hydroplaning speed of a truck tire 315/80R22.5 running over a flooded surface at 758 kPa inflation pressure using both NASA and Horne equations.

Solution. The tire inflation pressure is given to be 758 kPa and the FAR is 80 (based on the tire size). Using the NASA equation,

$$v = 6.36\sqrt{758} = 175\,\text{km/h}$$

Using the Horne equation for truck tires,

$$v = 23.3(758000)^{0.21}\left(\frac{1.4}{80}\right)^{0.5} = 52.7\,\text{m/s}(189\,\text{km/h})$$

Figure 1.48 shows the tire-water interaction model at various speeds and at 586 kPa inflation pressure, 13 kN load, and a water depth of 50 mm. The tire-water interaction is taken using a bottom view and a pressure distribution contour to visualize the contact pressure. It is clear that at 20 km/h speed, the contact area between the tire and ground is highest; in the case of 60 km/h, the contact area is still visible as well, while for the case of 105 km/h, the contact patch is mostly blue which refers to a minimal pressure.

The simulation test of a truck tire running over flooded surface is repeated at different vertical loads of 13, 27, and 40 kN. The variation of the hydroplaning speed as a function of load is shown in Fig. 1.49 at 100 mm water depth and several inflation

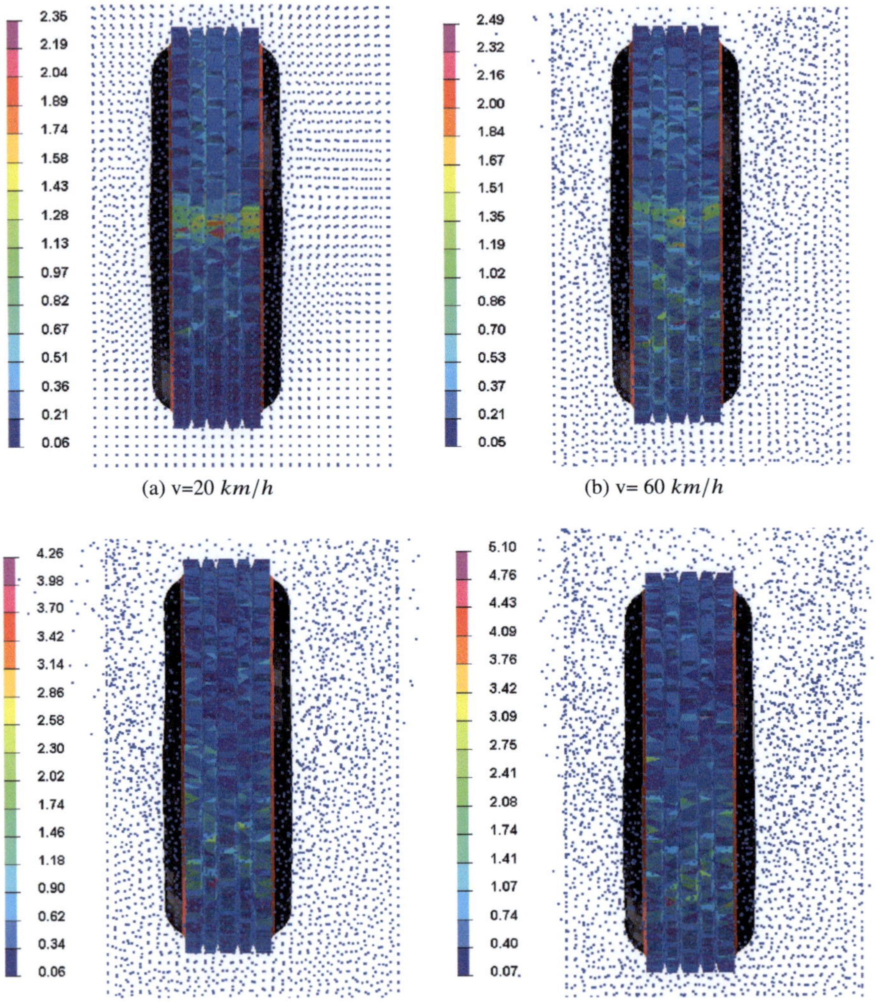

(a) v=20 km/h (b) v= 60 km/h

(c) v= 80 km/h (d) v= 105 km/h

Fig. 1.48 Truck tire at various speeds at 586 kPa inflation pressure, 13 kN load, and a water depth of 50 mm [65]

pressures. It is observed that the hydroplaning speed increases as the vertical load increases.

The increase in vertical load causes the contact area of the tire to produce more pressure, which is a result of higher contact force per unit area. The contact force growth results in greater stability control, and thus a delay to the hydroplaning phenomenon. An increase of 27 kN vertical load or in other words tripling the vertical load from 13 to 40 kN causes the hydroplaning speed to double up from 62 to 129 km/h at given inflation pressure of 379 kPa and a static water depth of 100 mm.

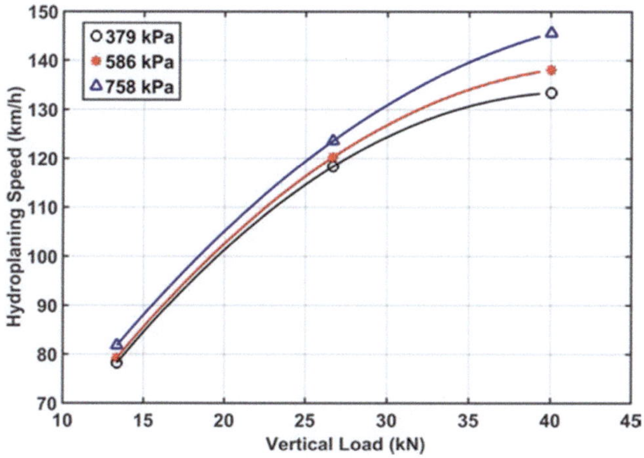

Fig. 1.49 Hydroplaning speed as a function of vertical load at 100 mm water depth and several inflation pressures [65]

Thus, the vertical loading has a significant effect on the hydroplaning speed. The tire loading acts as a stabilizer against hydroplaning.

In 2008, Oh [58] concluded that inflation pressure is a key factor affecting the hydroplaning speed. However, later in 2010, Chang [66] reported that the inflation pressure does not have a significant effect when it comes to truck tires. Nonetheless, the increase in inflation pressure causes the hydroplaning speed to increase infinitesimally. A high inflation pressure leads to a smaller contact area, therefore the contact force per unit area increases. The increase in contact force per unit area will require more hydrodynamic pressure from the water to lift the tire which leads to a higher hydroplaning speed.

As mentioned earlier, the hydroplaning speed increases as the inflation pressure increases. An increase of 300% in the inflation pressure from 379 to 758 kPa at a rated vertical load of 27 kN and 100 mm static water depth causes the hydroplaning speed to increase by 2%, which is considered significantly minimal.

Water depth has a major effect on hydroplaning speed. This section investigates the effect of water depth on the hydroplaning speed. The tire treads are covered faster when the water is deeper on the ground. It is a well-known fact that in order for a tire to hydroplane, the grooves should be filled with water. The tire treads supply stability and control during wet driving conditions, thus the filling of tire treads with water decreases the allocated stability resulting in accelerated hydroplaning. Equations relating hydroplaning speed to water depth exist only for car tires, and no previous attempt to numerically evaluate the effect of water depth on truck tire hydroplaning has been done. The water depth selected in this study is based on the literature and the tread depth of the tire used. The water depths examined are 50, 65, and 100 mm. Figure 1.50 shows the variation of the hydroplaning speed as a function of water depth at a nominal inflation pressure of 586 kPa and various vertical loads.

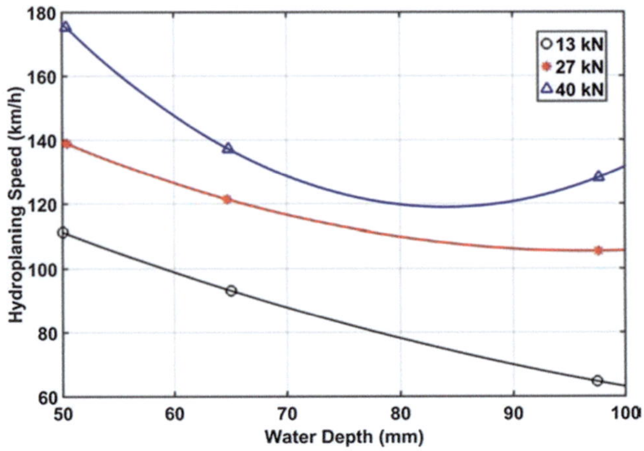

Fig. 1.50 Hydroplaning speed as a function of water depth at 586 kPa inflation pressure and several vertical loads [65]

It is observed that the increase in the water depth results in the decrease in the hydroplaning speed at all vertical loads. In other words, driving in deeper water accelerates the tire hydroplaning. For instance, if a truck is driving over a 100 mm static water depth, an inflation pressure of 586 kPa and a vertical load of 13 kN hydroplaning may happen at a speed as low as 65 km/h. Additionally, if the water depth is doubled on the ground, the hydroplaning speed may reduce by 25%. Furthermore, at higher vertical loads the effect of water depth becomes less in comparison to that of lower vertical loads.

1.6.2 Rolling Resistance Characteristics

Figure 1.51 shows a schematic of the main forces acting on a free rolling tire; the forces include the applied vertical force and the rolling resistance force.

In this simulation, the tire is first inflated to the desired inflation pressure, and then the tire is loaded on the flooded surface by applying the desired vertical load. After allowing the tire to settle, a constant linear longitudinal velocity is applied at the center of the tire. The simulation runs until the vertical and longitudinal contact forces reach a steady state. It should be noted that constant longitudinal velocity is applied at the center of the tire as shown in Fig. 1.51. However, the angular velocity of the free rolling tire may change depending on the water depth, inflation pressure, and applied vertical load. The change in angular velocity even without applying driving torque is due to the sinkage of the tire in the water, while it's towed with constant linear longitudinal velocity. Therefore, the predicted rolling resistance may not be under a pure free rolling condition as is the case on the hard surface. The variation

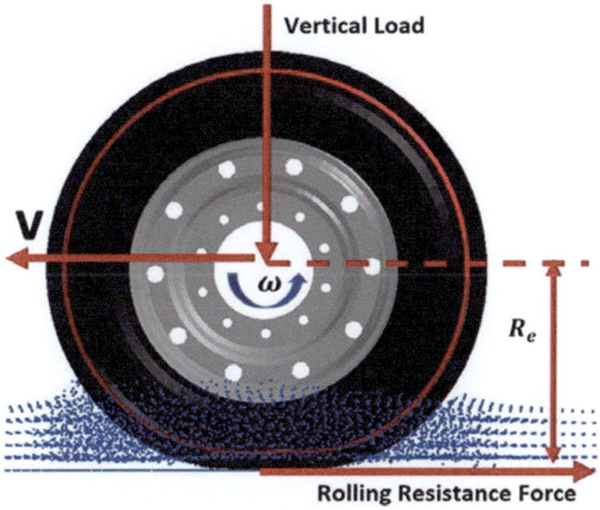

Fig. 1.51 Schematic of the main forces acting on a free rolling tire [67]

of the angular velocity at constant linear longitudinal velocity results in longitudinal slip which is found to be in the range 4–13%.

Figure 1.50 shows the variation of the rolling resistance coefficient as a function of the water depth at 758 kPa inflation pressure and different vertical loads at a given speed of 10 km/h. It is concluded that the rolling resistance coefficient decreases as the vertical load increases at a given speed and inflation pressure. In addition, the rolling resistance coefficient increases as the water depth increases. It is noted that the variation of the angular velocity at constant linear longitudinal velocity results in longitudinal slip which indicates that the rolling resistance is not calculated at a freely rolling tire. The method used to predict the rolling resistance of a tire running over a flooded surface is commonly used in physical testing.

To further understand the relationship between the rolling resistance coefficient and the tire operating conditions, Fig. 1.52b is presented. Figure 1.52b shows the variation of the rolling resistance coefficient as a function of the water depth at a given vertical load of 40 kN and different inflation pressures at a given longitudinal speed of 10 km/h. It is shown from this figure that as the inflation pressure increases, the rolling resistance coefficient decreases at a given water depth and vertical load. This result is due to the fact that the increase of the inflation pressure decreases the contact area between the tire and the surface leading to lower resistance forces. The same conclusion mentioned in the previous paragraph is also noticed here. The rolling resistance coefficient increases as the water depth increases at a given vertical load and inflation pressure. This relationship is due to the fact that the contact area between the tire and the surface increases as the water depth increases.

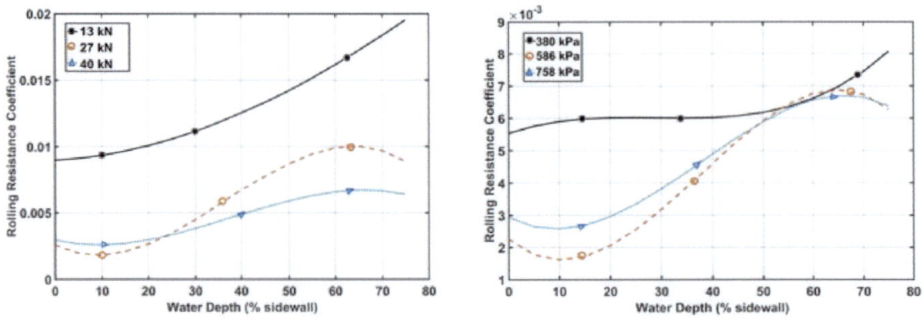

(a) 758 kPa inflation pressure (b) $40kN$ vertical load

Fig. 1.52 Rolling resistance coefficient as a function of water depth at different operating conditions

1.6.3 Traction Characteristics

The longitudinal tire stiffness of the truck tire is defined as the tire's ability to recover maximum traction after experiencing 100% slip conditions through the application of a rapid angular acceleration to the tire's center. The model setup is shown in Fig. 1.53; the tire is first inflated to the desired inflation pressure, then the desired vertical load is applied to the center of the tire. Rapid angular acceleration ($\dot{\omega}$) is then applied to the tire's center until the desired steady-state speed is reached. Due to the quick angular acceleration, the tire experiences 100% slip conditions at the beginning of the simulation. The longitudinal and angular velocities at the center of the tire, in addition to the longitudinal force at the contact patch, are computed. The simulation results are used to analyze the longitudinal force against the longitudinal slip ratio. From the relationship between the longitudinal force, F_x, and slip, s, the tire longitudinal stiffness, $C_s\big(\frac{\partial F_x}{\partial s}|_{s=0}\big)$, and the peak coefficient of adhesion, μ_p, are determined.

The longitudinal force as a function of the longitudinal slip for several applied vertical loads and inflation pressures is shown in Fig. 1.54a and 1.54b, respectively.

Figure 1.54a indicates that the longitudinal force reaches a maximum value for a longitudinal slip between 15 and 40% depending on the water depth. The longitudinal force increases as the applied vertical load increases, and the longitudinal force reach its peak earlier for a higher vertical load.

On the other side, Fig. 1.54b shows that the peak longitudinal force is slightly affected by the inflation pressure; as the inflation pressure increases, the peak longitudinal force decreases. This is due to the fact that the higher the inflation pressure, the lower the contact area is. The previously discussed results are for wet asphalt surfaces and stand true for all flooded surface cases.

Figure 1.55a shows the variation of the longitudinal stiffness, C_s, as a function of water depth at 586 kPa inflation pressure and 0 kN vertical load. The trend shows an increase in the longitudinal tire stiffness as the water depth increases. This indicates that the tire produces lower forces at higher water depth.

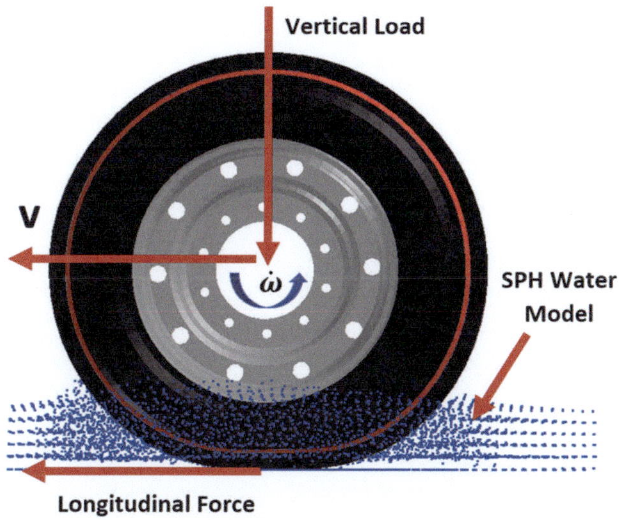

Fig. 1.53 Schematic of the longitudinal tire stiffness simulation model

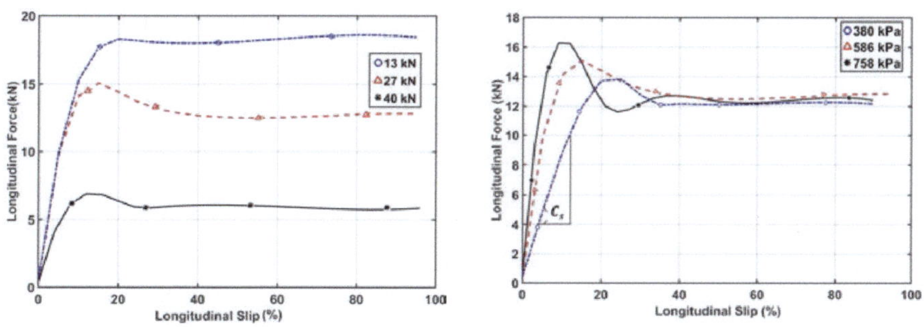

(a) 380 kPa inflation pressure (b) 40 kNapplied vertical load

Fig. 1.54 Longitudinal force as a function of longitudinal slip for several applied vertical load and inflation pressure over a wet surface [67]

The longitudinal tread stiffness, C_{cx}, is determined by dividing the tire stiffness by half of the contact length, a, of the tire which is 0.0625 m [5]. Figure 1.55b shows the longitudinal tread stiffness as a function of water depth for 586 kPa inflation pressure and different vertical loads. The longitudinal tread stiffness decreases as the water depth increases; this trend is proven for all inflation pressures and applied vertical loads.

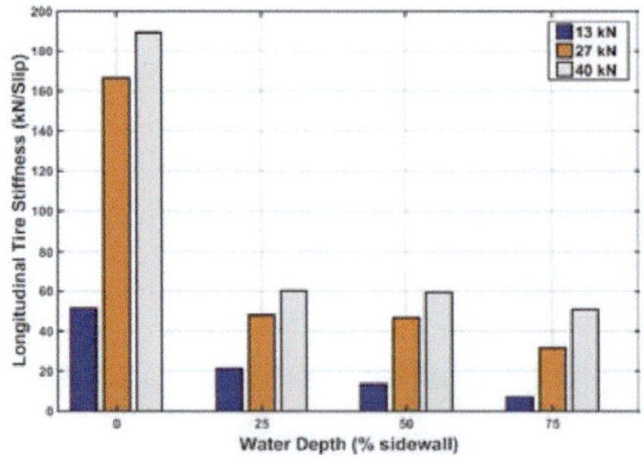

(a) Longitudinal tire stiffness

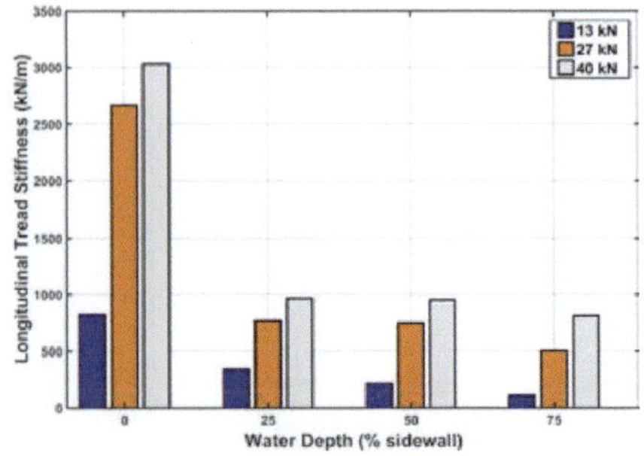

(b) Longitudinal tread stiffness

Fig. 1.55 Longitudinal tire and tread stiffness as a function of water depth at 586 kPa and different vertical loads

? Example 1.4

Using Fig. 1.54a, calculate the longitudinal tire stiffness, C_s, over a wet surface at 380 kPa inflation pressure and 13, 27, and 40 kN applied vertical loads.

Solution. Using the equation of longitudinal tire stiffness defined as

$$C_s = \left.\frac{\partial F_x}{\partial s}\right|_{s=0}$$

Thus, the slope in the linear part of the three curves is considered to be the longitudinal tire stiffness. The longitudinal tire stiffness at 13 kN based on Fig. 1.54a is:

$$C_s = \frac{10}{0.067} = 150\,\text{kN/slip}$$

In a similar manner, the longitudinal tire stiffness at 27 and 40 kN is calculated to be 150 kN/slip and 75 kN/slip, respectively.

1.6.4 Cornering Characteristics

The cornering characteristics determined in this section include the cornering stiffness, self-aligning moment, and relaxation length. Figure 1.56 shows the free body diagram of the cornering test procedure with the forces and moments acting on the tire at 25% sidewall height water depth.

The cornering stiffness, C_α, is defined as the ability of the tire to resist deformation in the shape while a vehicle is undergoing a cornering operation; furthermore, cornering stiffness can be determined from the lateral force applied on the contact area.

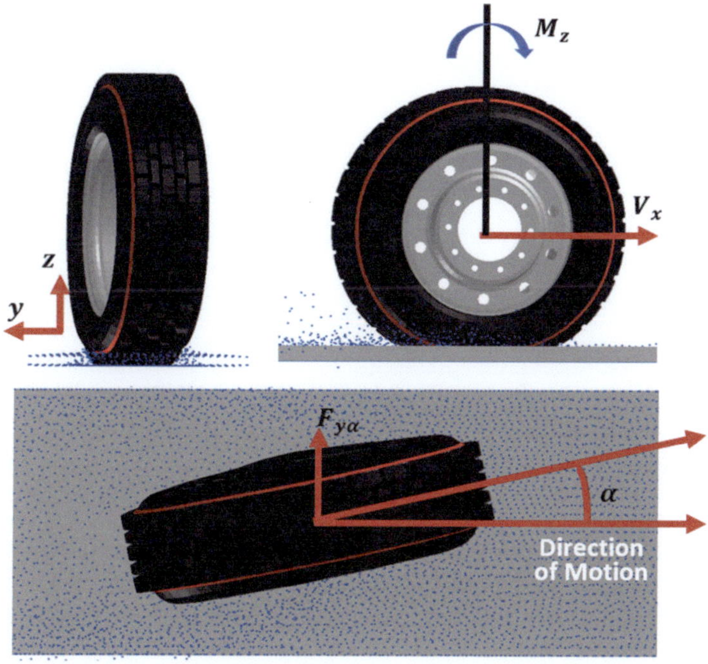

Fig. 1.56 Schematic of the cornering test of the truck tire over flooded surface simulation setup [67]

For a small slip angle, the cornering stiffness is considered linear, while for a slip angle greater than 2° the relationship can be highly nonlinear. The cornering stiffness concept is used to approximate the interaction between side and circumferential tire forces as shown in Fig. 1.57.

Figure 1.57 indicates that the cornering force appears to be approximately linear under 2° steering and increasing after that. The increase in slip angle increases the cornering force as well. Additionally, as the vertical load increases at a given surface condition, inflation pressure, and steering angle, the cornering force increases as well.

The tire is first inflated to the desired inflation pressure, then the vertical load is applied to the center of the tire. A constant longitudinal speed of 10 km/h is then applied to the center of the tire and the model is kept running for 2 s. The contact forces are computed on the contact patch between the tire-ground and tire-water. The tire is pre-steered to several slip angles between 0° and 12°; this will allow examining the influence of the slip angle on tire operational performance.

Figure 1.58a shows the variation of the cornering stiffness, C_α, as a function of vertical load for several inflation pressures over the wet surface. The cornering stiffness increases as the vertical stiffness increases at given inflation pressure. Moreover, the cornering stiffness also increases as the inflation pressure increases at a given

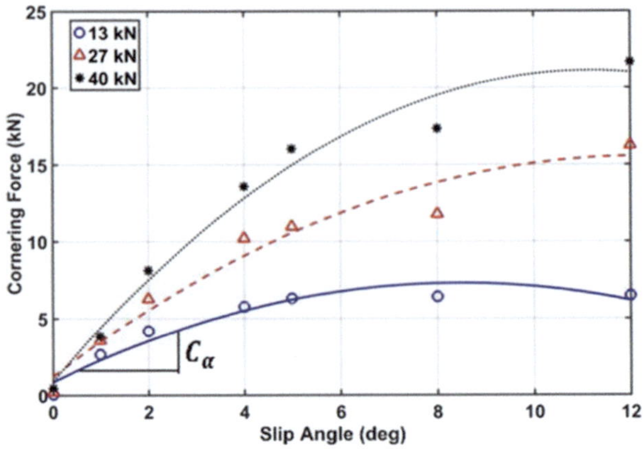

Fig. 1.57 Cornering force as a function of slip angle for 586 kPa inflation pressure and different vertical loads on wet surface

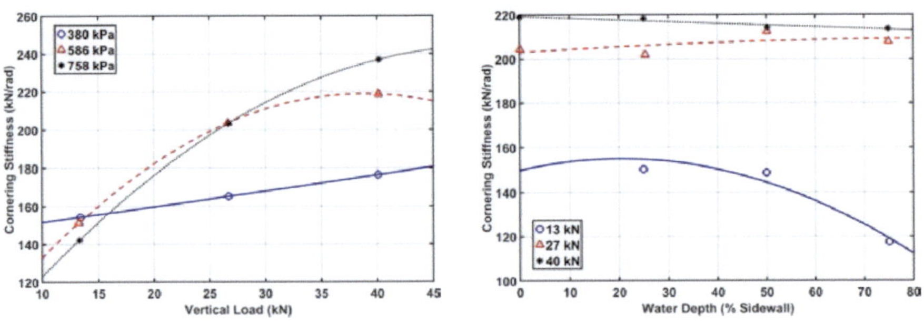

(a) Function of vertical load

(b) Function of water depth

Fig. 1.58 Cornering stiffness as a function of vertical load and water depth at different operating conditions

vertical load. The increase in loading increases the tire's ability to resist deformation while cornering which results in the increase of cornering stiffness. Additionally, when the inflation pressure increases, the contact area of the tire increases as well and thus the cornering force per unit area reduces which leads to an increase in the cornering stiffness. It is noted that the relationship between the rate of change and vertical load change is not linear and is dependent on other operating conditions at the same time; this is unlike the relationship between the lateral stiffness and vertical load rate of change.

Figure 1.58b shows the variation of the cornering stiffness, C_α, as a function of the water depth at a rated inflation pressure of 586 kPa and different vertical loads. The cornering stiffness reduces as the water depth increases at a given vertical load and inflation pressure. The rate of reduction in cornering stiffness with respect to water depth is dependent on the vertical load as well. For low tire loading, the cornering

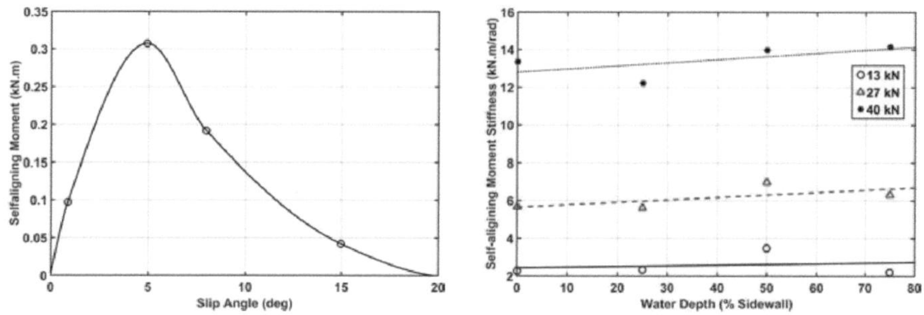

(a) Function of slip angle at 27 kN vertical load (b) Function of water depth

Fig. 1.59 Self-aligning moment and stiffness as a function of various operating conditions at 586 kPa inflation pressure

stiffness is highly sensitive to the water depth, while for higher loading (40 kN) the cornering stiffness is less affected by the water depth. Generally, for a rated inflation pressure of 586 kPa and 13 kN vertical load, the cornering stiffness is between 155 and 120 kN/rad for a water depth between 0 and 70% of sidewall height.

The self-aligning moment stiffness, C_m, as a function of the water depth for rated inflation pressure of 586 kPa and several vertical loads is presented in Fig. 1.59b. It is observed that the self-aligning moment stiffness increases as the water depth increases for a specific load and inflation pressure. The rate of increase of the self-aligning moment with respect to the water depth at a specific inflation pressure is minimally dependent on the vertical load. Additionally, the self-aligning moment stiffness increases as the vertical load increases as well at a given water depth and inflation pressure. It is noticed that the relationship between the self-aligning moment stiffness and the inflation pressure is not linear at a given water depth and vertical load.

? Example 1.5

Using Fig. 1.57, calculate the cornering stiffness, C_α, of the tire running over a wet surface at 586 kPa inflation pressure and 13, 27, and 40 kN vertical load.

Solution. Using the equation of cornering tire stiffness defined as

$$C_s = \left. \frac{\partial F_{y\alpha}}{\partial \alpha} \right|_{\alpha=0}$$

Thus, the slope in the linear part of the three curves is considered to be the cornering tire stiffness. The longitudinal tire stiffness at 13 kN based on Fig. 1.57 is

$$C_\alpha = \frac{2.5 * 180}{1 * \pi} = 143\,\text{kN/rad}$$

In a similar manner, the longitudinal tire stiffness at 27 and 40 kN is calculated to be 200 kN/rad and 205 kN/rad, respectively.

1.7 Tire Safety and General Information

1.7.1 Sidewall Information

Figure 1.60 shows the sidewall information and manufacturer data.

When the placard tire size contains a speed symbol, for example P20S/60HRIS or P20S/60RlS 90H, the replacement tire must have the same or higher speed rating symbol if the speed capability of the vehicle is to be maintained. If the replacement tire is not speed rated, the speed capability of the vehicle is limited by the speed capability of the replacement tire (Table 1.8).

The speed category is based upon indoor wheel tests conducted in accordance with the Procedure for Load/Speed Performance Tests of the Economic Commission for Europe (ECE-30). It should be noted that for tires having a maximum speed capability above 240 km/h (149 mph), a "ZR" may appear in the size designation.

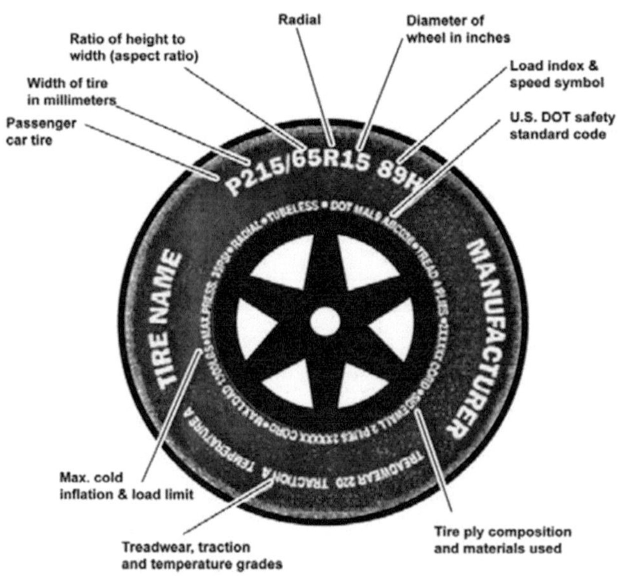

Fig. 1.60 Sidewall information and label [68]

Table 1.8 Speed category based on speed symbol [68]

Speed symbol	Speed category
S	180 km/h (112 mph)
T	190 km/h (118 mph)
U	200 km/h (124 mph)
H	210 km/h (130 mph)
V	240 km/h (149 mph)
W	270 km/h (168 mph)
Y	300 km/h (186 mph)

For example, for a tire designated with "P275/40R17 93W", the maximum speed is 270 km/h (168 mph). While a speed symbol is an indication of the speed capability of the tire, we do not endorse the operation of any vehicle in an unsafe or unlawful manner.

Determining the Correct Tire Size

1. Look in your automobile owner's manual. You'll find the size fitted on the car originally. Unless you've changed wheels, that's the recommended size.
2. The tire size is written on the tire's sidewall. Here's an example of the way tire sizing looks on the sidewall:

   a. P indicates a passenger (car-type) tire. Other options would be no P indicating metric sizing (essentially the same as P-sizing, which has its heritage in Europe) or LT for light truck. LT truck tires are designed for heavier loads and more rugged service conditions.
   b. Width of the tire, in millimeters. The higher the number, the wider the tire.
   c. Aspect ratio—The height of the sidewall section compared to the width of the tire. Example: If this number was a 50, then the tire section is half as tall as it is wide. Short sidewalls deliver crisp handling. Tall sidewalls give a smoother ride. For a specific tire width, the smaller this number, the shorter the sidewall.
   d. Construction—R indicates radial construction. Unless you specify the other option, bias (which would have a D [diagonal] or B [belted bias] in this position instead of an R), you are purchasing a radial tire.
   e. Wheel Size Designation—This indicates the distance across the "doughnut hole" of a tire, in simple terms. You must match wheel diameter and tire diameter. For example, a 15-in diameter rim; a tire with a 15-size designation will not work on a 14" wheel, and vice versa. Improperly matching of wheel and tire size can cause serious injury or death during installation.

If you are replacing the original size tires with tires of a different size, the replacement tires must have a load carrying capacity equal to or greater than the original equipment tires.

If you anticipate towing a trailer, you should see your tire dealer for advice concerning the correct size of tire and inflation pressures. Tire size and air pressure depend upon the type and size of the trailer and hitch utilized, but never exceed the maximum cold inflation pressure or the maximum tire load rating. The only sure way to prevent overload is to weigh, axle by axle, the fully loaded vehicle on a reliable platform scale. Check the tire placard on the vehicle and the owner's manual supplied by the manufacturer of your vehicle for further recommendations on trailer towing.

Wheel alignment and balancing are important for safety and maximum mileage from your tires.

Inspect Your Tires Regularly

At least once a month inspect your tires closely for signs of uneven wear. Uneven wear patterns may be caused by improper inflation pressure, misalignment, improper balance or suspension neglect. If not corrected, further tire damage will occur. These conditions shorten the life of your tires and may result in loss of vehicle control and serious personal injury.

If any of these conditions exist, the cause may often be corrected at your tire dealer or other service facility. Your tires will then last longer.

Proper Tire Rotation is Important

If you notice irregular or uneven tread wear, the tires should be rotated to alleviate the problem. Remember: it is important to check your tires and wheels for signs of possible damage (as previously discussed); also check your vehicle for any mechanical problems and correct if necessary. You should follow the rotation pattern or procedure indicated in your limited warranty and the vehicle owner's manual. We recommend you rotate tires on front wheel drive vehicles and/or all season-tires on any vehicle every 8,000 miles to equalize the rate of wear; however, rotate your tires earlier if signs of irregular or uneven tire wear arise, and have the vehicle checked by a qualified technician to determine the cause of the wear condition. The first rotation is the most important.

Sometimes, front and rear tires on a vehicle use different inflation pressures. After rotation, adjust individual tire air pressure to the figures recommended by the vehicle manufacturer for the new locations—front or rear—as shown on the tire placard on the vehicle.

Tire Mixing Can be Dangerous

Most passenger tires today are radial tires. For best performance, we recommend the same size and type of tire be used on all four wheel positions unless the vehicle manufacturer specified different sizes, front and rear, as original equipment. Check the vehicle placard. If only two radials are mounted with two non-radials, the radials should be mounted on the rear. If tires of different types are mixed on a vehicle in any configuration, they should not be used for long periods and speeds should be kept to a minimum.

Mixing or matching of tires on four-wheel drive vehicles requires special precautions. Always check the vehicle manufacturer's manual for their recommendations.

Tire Alterations are Dangerous

Do not perform any alteration on your tires. Alterations may prevent proper performance, leading to tire damage, which can result in sudden tire destruction. Tires, which have been altered, are excluded from warranty coverage.

For repairs, see any tire dealer at once. Never use a plug-only or patch-only repair. If any tire has sustained a puncture, have the tire dismounted and inspected internally by any tire dealer for possible damage that may have occurred.

Punctures in the tread of passenger tires which do not exceed 1/4-in (6 mm) in diameter can be repaired by following the Rubber Manufacturers' Association (RMA) recommended repair procedures. A plug by itself or a patch by itself is an unacceptable repair. The repair material used—for example, a "combination patch and plug" repair—must seal the inner liner and fill the injury to be considered a permanent repair. Never use a tube in a tubeless tire as a substitute for a proper repair.

If the tire has a puncture in the tread which exceeds 1/4-in (6 mm), any puncture in the sidewall, or if more than one radial cable per casing ply is damaged, the tire must be replaced.

When a speed rated tire has been repaired, the speed rating no longer applies. The speed capability of the vehicle is limited by the speed capability of the repaired tire.

Exercise Care When Storing Tires

When tires are stored, they should be stored in a cool, dry place away from sources of sunlight, heat, and ozone such as hot pipes and electric motors. Tires should be stored so that there is no danger of water collecting inside them. Be sure that surfaces on which tires are stored are clean and free from grease, gasoline, or other substances, which could deteriorate the rubber. Tires exposed to these materials during storage or driving may be weakened and subject to sudden failure. Also, be sure to allow air to circulate around all sides of the tires, including underneath, to prevent moisture damage.

When storing tires flat (one on top of the other), stack them so that tires on the bottom retain their shape.

If storing tires outdoors, protect them with an opaque waterproof covering and elevate them from the ground. Do not store tires on black asphalt, other heat-absorbent surfaces, snow-covered ground, or sand.

Driving on Studded Snow Tires

If studded tires are installed on the front of any vehicle, they must also be installed on the rear.

The beginning of movement and acceleration of any vehicle in snow, ice, and other adverse cold weather conditions is highly dependent on the traction qualities of the tires on the driving axle. The controlled handling and braking of a vehicle after it is in motion in adverse weather conditions, however, is highly dependent on the traction of the rear tires. Consequently, the rear tires of any vehicle must have equal or higher traction capabilities than the front tires for safe vehicle operation.

Because of the higher traction qualities of studded snow tires under most winter weather conditions, installation of only two studded snow tires on the front of any vehicle (especially front wheel drive vehicles) without two studded snow tires on the rear can cause adverse (unsafe) handling characteristics. Consult the tire manufacturer for the correct stud size. If you sell and/or install studded snow tires on vehicles, you must follow the procedures listed as follows:

- Only new tires should be fitted with studs. Never insert studs in a used tire (even if only slightly used).
- Without studded snow tires on the rear, which have the same traction qualities as the studded front tires, adverse (unsafe) handling and braking characteristics are introduced into the vehicle. This may result in loss of vehicle control, which could cause serious injury or death.
- If studded snow tires are installed on only the rear of any vehicle, it is recommended (but not required) that they be installed on the front. Only if studded tires are installed on all wheel positions of a vehicle will optimum handling characteristics be achieved.

Tire Spinning is Dangerous

Avoid tire spinning. The centrifugal forces created by a rapidly spinning tire can cause an explosion by tearing the tire apart. These forces act on the complete tire structure, and can be of such magnitude as to break beads as well as rupturing the entire carcass.

When stuck on ice, snow, mud, wet grass, etc., the vehicle should be rocked gently (alternately using forward and reverse gears) with the least amount of wheel spinning. DO NOT exceed 35 mph as indicated on the speedometer. Never allow anyone to stand near or directly ahead of or behind the spinning tire.

Do not spin if a drive wheel is off the ground. Serious personal injury or death can result from the explosion of a spinning tire.

Tire mounting can be dangerous and should be done only by trained persons using proper tools and procedures.

Serious injury or death may result from explosion of tire/rim assembly due to improper mounting. Always have your dealer mount your tires on rims. If you are not thoroughly familiar with the Rubber Manufacturers Association mounting procedures, never attempt to mount tires.

For speed-rated tires, if the replacement tire is not speed rated, the speed capability of the vehicle is limited by the speed capability of the replacement tire.

When replacing tires, consult the placard (normally located on a door frame, door edge, or glove box door) or the owner's manual for correct size. If the tires shown on the vehicle placard do not have speed ratings, the appropriate size tire with any speed rating may be applied.

Problems

1. A passenger vehicle travels over a flooded pavement. The inflation pressure of the tires is changed from 140 to 125 kPa. If the initial speed of the car is 100 km/h and brakes are then applied, determine which inflation pressure will be critical for possible hydroplaning.
2. A passenger vehicle travels over a flooded pavement. The inflation pressure of the tires is 179.27 kPa. If the initial speed of the car is 100 km/h (62 mph) and brakes are then applied, determine whether or not the vehicle will be hydroplaning.
3. Using the Magic Formula described in Sect. 1.4.2, estimate the cornering force of a car tire with a slip angle ranging from 0 to 16° and longitudinal skid ranging from 0 to 100% at vertical loads of 2, and 4 kN. The values of the empirical coefficients in the Magic Formula for the tire are given in Table 1.5.
4. A tire with vertical load of 24.78 kN travels on a dry concrete pavement at speed, v, of 30 m/s with a peak value of the coefficient of road adhesion J.1P = 0.80. The longitudinal stiffness of the tire during braking, C, is 224.64 kN/unit slip and Ca is 132.53 kN/rad. Use Dugoff's tire model to plot the relationship between the braking force, Fx, and the cornering force, Fy, of the tire versus slip angle in the range 0–16° of the tire at slip ratios s = 5 and 60%. Assume that no steering is applied (steering angle is zero) and the adhesion reduction coefficient Er = 0.015 (make any necessary assumptions).
5. Compare the power required to overcome the rolling resistance of a passenger car weighing 15.57 kN and having radial-ply tires with that of the same vehicle but having bias-ply tires in the speed range 40–100 km/h (25–62 mph). The variations of the coefficient of rolling resistance of the radial-ply and bias-ply passenger car tires with speed are described as follows: For radial-ply car tires: $f_r = 0.0136 + 0.40x10^{-6}v^2$. And for bias-ply car tires: $f_r = 0.0169 + 0.19x10^{-6}v^2$, where v is speed in km/h.

References

1. The evolution of the wheel. https://www.historyanswers.co.uk/inventions/how-was-stained-glass-made/. Accessed 12 June 2017
2. Goodyear C (1853) Gum-elastic and its varieties: with a detailed account of its applications and uses, and of the discovery of vulcanization, vol 2. Published for the author
3. The history of pneumatic devices; pneumatic devices - pneumatic tube. http://inventors.about.com/library/inventors/blpneumatic.htm. Accessed 12 June 2017
4. Trade Europe (2001) The automotive news Europe book of lists 2001. Automot News Eur 6(1):6–19
5. Wong JY (2008) Theory of ground vehicles. Wiley
6. Gillespie TD (1992) Fundamentals of vehicle dynamics. Technical report, SAE technical paper
7. Slade JL (2009) Development of a new off-road rigid ring model for truck tires using finite element analysis techniques. Master's thesis, The Pennsylvania State University
8. Haney P (2004) Rubber friction. The racing and high-performance tire, sports car magazine January
9. Ellis JR (1994) Vehicle handling dynamics
10. Seminar Notes (1996) The mechanics of heavy-duty trucks and truck combinations. University of Michigan Transportation Research Institute
11. Barkanov E (2001) Introduction to the finite element method. Institute of Materials and Structures Faculty of Civil Engineering Riga Technical University
12. Sui J, Hirshey J (1999) Evaluation on analytical tire models for vehicle vertical vibration simulation using virtual tire testing method. Technical report, SAE technical paper
13. Bakker E, Nyborg L, Pacejka HB (1987) Tyre modelling for use in vehicle dynamics studies. Technical report, SAE technical paper
14. Pacejka HB, Besselink IJM (1997) Magic formula tyre model with transient properties. Veh Syst Dyn 27(S1):234–249
15. Zegelaar PWA, Pacejka HB (1997) Dynamic tyre responses to brake torque variations. Veh Syst Dyn 27(S1):65–79
16. Zegelaar PWA, Pacejka HB (1996) The in-plane dynamics of tyres on uneven roads. Veh Syst Dyn 25(S1):714–730
17. Kim S-J, Savkoor AR (1997) The contact problem of in-plane rolling of tires on a flat road. Veh Syst Dyn 27(S1):189–206
18. Yoshida K, Ishigami G (2004) Steering characteristics of a rigid wheel for exploration on loose soil. In: 2004 IEEE/RSJ international conference on intelligent robots and systems (IROS) (IEEE Cat. No. 04CH37566), vol 4. IEEE, pp 3995–4000
19. Tuononen AJ, Hartikainen L, Petry F, Westermann S (2012) Parameterization of in-plane rigid ring tire model from instrumented vehicle measurements. In: Proceedings of the 11th international symposium on advanced vehicle control (AVEC '12), pp 9–12
20. Chan BJ, Sandu C (2014) Development of a 3-d quasi-static tyre model for on-road and off-road vehicle dynamics simulations: Part i-on-road flexible tyre model. Int J Veh Syst Model Test 9(1):77–105
21. Chan BJ, Sandu C (2014) Development of a 3-d quasi-static tyre model for on-road and off-road vehicle dynamics simulations: Part ii-off-road rigid wheel model. Int J Veh Syst Model Test 9(2):107–136
22. Chan BJ, Sandu C (2014) Development of a 3-d quasi-static tyre model for on-road and off-road vehicle dynamics simulations: Part iii-off-road flexible wheel model. Int J Veh Syst Model Test 9(2):151–176
23. Ritz W (1909) Über eine neue methode zur lösung gewisser variationsprobleme der mathematischen physik. Journal für die reine und angewandte Mathematik 135:1–61
24. Courant R (1943) Variational methods for the solution of problems of equilibrium and vibrations. Lect Notes Pure Appl Math 1
25. Dugdale DS (1960) Yielding of steel sheets containing slits. J Mech Phys Solids 8(2):100–104

26. Yong RN, Fattah EA, Boonsinsuk P (1978) Analysis and prediction of tyre-soil interaction and performance using finite elements. J Terrramech 15(1):43–63
27. De Eskinazi J, Ishihara K, Volk H, Warholic TC (1990) Towards predicting relative belt edge endurance with the finite element method. Tire Sci Technol 18(4):216–235
28. Hiroma T, Wanjii S, Kataoka T, Ota Y (1997) Stress analysis using fem on stress distribution under a wheel considering friction with adhesion between a wheel and soil. J Terrramech 34(4):225–233
29. Koishi M, Kabe K, Shiratori M (1998) Tire cornering simulation using an explicit finite element analysis code. Tire Sci Technol 26(2):109–119
30. PAM System International (2014) Pam-crash user manual version 2014. ESI Group
31. Chae S (2006) Nonlinear finite element modeling and analysis of a truck tire. PhD thesis, The Pennsylvania State University
32. Mooney M (1940) A theory of large elastic deformation. J Appl Phys 11(9):582–592
33. Yeoh OH (1990) Characterization of elastic properties of carbon-black-filled rubber vulcanizates. Rubber Chem Technol 63(5):792–805
34. Ford TL, Charles FS (1988) Heavy duty truck tire engineering. Technical report, SAE technical paper
35. Chang Y-P (2002) Nonlinear FEA rotating tire modeling for transient response simulations. PhD thesis, Pennsylvania State University
36. Yap P (1991) Measurement of radial truck tire dry cornering characteristics. Technical report, SAE technical paper
37. Davis P, Martinson V, Yager T, Stubbs S (1997) 26x6. 6 radial-belted aircraft tire performance. Progr Technol 66:251–260
38. Lardner KL (2017) Prediction of the off-road rigid-ring model parameters for truck tire and soft soil interactions. Master's thesis, University of Ontario Institute of Technology, Canada
39. Gardner ER, Worswick T (1951) Behaviour of tyres at high speed. Trans Inst Rubber Ind 27:127–146
40. Powell HT (1957) Studies in the genus fucus li fucus distichus l. emend. powell. J Marine Biol Assoc UK 36(2):407–432
41. Bruinsma F (1968) Design and measurements of an air bearing for a running belt wit an electro-hydraulic vibrator system. Technical report, VRLD report
42. Bird KD, Martin JF (1973) The calspan tire research facility: design, development, and initial test results. SAE Trans 2012–2027
43. Davisson JA (1969) Design and application of commercial type tires. SAE Trans 1–32
44. Jenq S-T, Chiu Y-S et al (2009) Hydroplaning analysis for tire rolling over water film with various thicknesses using the ls-dyna fluid-structure interactive scheme. Comput, Mater Contin (CMC) 11(1):33
45. Seta E, Nakajima Y, Kamegawa T, Ogawa H (2000) Hydroplaning analysis by fem and fvm: effect of tire rolling and tire pattern on hydroplaning. Tire Sci Technol 28(3):140–156
46. Horne WB, Dreher RC (1963) Phenomena of pneumatic tire hydroplaning
47. Martin CS (1966) Hydroplaning of tire hydroplaning final report. Project B-608, Georgia Institute of Technology
48. Moore CG, Porter M (1967) Structural characterization of vulcanizates. Part vi. the 2-mercaptobenzothiazole-accelerated natural rubber–sulfur system. J Appl Polymer Sci 11(11):2227–2253
49. Eshel A (1967) A study of tires on a wet runway. Ampex Corporation, RR, pp 24–67
50. Browne A, Cheng H, Kistler A (1972) Dynamic hydroplaning of pneumatic tires. Wear 20(1):1–28
51. Leland TJW (1972) An evaluation of some unbraked tire cornering force characteristics
52. Sinnamon JF (1974) Hydroplaning and tread pattern hydrodynamics
53. Grogger H, Weiss M (1996) Calculation of the three-dimensional free surface flow around an automobile tire. Tire Sci Technol 24(1):39–49
54. Panagouli OK, Kokkalis AG (1998) Skid resistance and fractal structure of pavement surface. Chaos, Solitons Fractals 9(3):493–505

55. Janajreh IM (2001) Tire having a groove wall lining for reducing formation of anomalies causing subjective user dissatisfaction, April 10 2001. US Patent 6,213,181
56. Chu L, Fwa TF (2016) Incorporating pavement skid resistance and hydroplaning risk considerations in asphalt mix design. J Transp Eng 142(10):04016039
57. Ong GP, Fwa TF (2007) Wet-pavement hydroplaning risk and skid resistance: modeling. J Transp Eng 133(10):590–598
58. Oh C-W, Kim T-W, Jeong H-Y, Park K-S, Kim S-N (2008) Hydroplaning simulation for a straight-grooved tire by using fdm, fem and an asymptotic method. J Mech Sci Technol 22(1):34–40
59. Choi JH, Cho JR, Woo JS, Kim KW (2012) Numerical investigation of snow traction characteristics of 3-d patterned tire. J Terrramech 49(2):81–93
60. Gough VE (1954) Cornering characteristics of tyres. Automob Eng 44:137
61. Horne WB, Joyner UT (1965) Pneumatic tire hydroplaning and some effects on vehicle performance. Technical report, SAE technical paper
62. Horne WB, Yager TJ, Ivey DL (1986) Recent studies to investigate effects of tire footprint aspect ratio on dynamic hydroplaning speed. In: The tire pavement interface. ASTM international
63. Gallaway BM, Ivey DL, Hayes G, Ledbetter WB, Olson RM, Woods DL, Schiller RF Jr (1979) Pavement and geometric design criteria for minimizing hydroplaning. Technical report, Federal Highway Administration
64. Wambold JC, Yeh EC, Henry JJ (1984) Methodology for analyzing pavement condition data (mapcon). Technical report, Maintenance and operations manuals for the IBM 370 version, volume II
65. El-Sayegh Z, El-Gindy M (2017) Sensitivity analysis of truck tyre hydroplaning speed using fea-sph model. Int J Veh Syst Model Test 12(1–2):143–161
66. Changyong C (2010) Skid resistance and hydroplaning analysis of rib truck tire. PhD thesis, National University of Singapore
67. El-Sayegh Z (2020) Modeling and analysis of truck tire-terrain interaction. PhD thesis
68. Hankook USA: Tires for ev, passenger cars, suvs and more

Chapter 2
Off-Road Terrain Characterization and Modeling

As off-road vehicles are becoming more demanding, the study of severe conditions is becoming more critical. During wintertime, the tire may be subjected to snow falling or accumulated snow layers on the ground, which can severely change the performance of the vehicle. Thus, studying tire-snow interaction becomes a critical demand as it becomes more and more important for safety and performance analysis. In order to perform any tire-terrain interaction analysis, a well-calibrated terrain model is needed. The accuracy of the terrain model drastically affects the accuracy of the tire-terrain interaction.

2.1 Stress Distribution Under Load

Boussinesq developed the first method to calculate the stress distribution in a semi-infinite, homogeneous, isotropic, elastic medium subject to vertical load. Equation 2.1 presents the expression of the vertical stress, σ_z, at a point in the elastic medium [1]. Figure 2.1 shows the coordinates used to develop the expression.

$$\begin{aligned}\sigma_z &= \frac{3W}{2\pi} z^3 \left(x^2 + y^2 + z^2\right)^{-5/2} \\ &= \frac{3}{2\pi} \frac{1}{\left[1 + (r/z)^2\right]^{5/2}} \frac{W}{z^2} \\ \sigma_z &= \frac{3W}{2\pi R^2} \left(\frac{z}{R}\right)^3 \end{aligned} \tag{2.1}$$

Assuming $r = \sqrt{x^2 + y^2}$ and $R = \sqrt{z^2 + r^2}$. Equation 2.1 can be expressed in polar coordinates to determine the radial stress, σ_r, as shown in Eq. 2.2.

$$\sigma_r = \frac{3W}{2\pi R^2} \cos\theta \tag{2.2}$$

M. El-Gindy and Z. El-Sayegh, *Road and Off-Road Vehicle Dynamics*,
https://doi.org/10.1007/978-3-031-36216-3_2

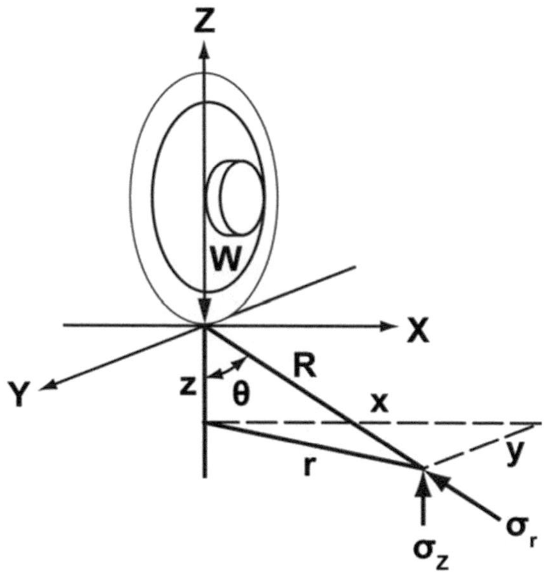

Fig. 2.1 Stresses at point in a semi-infinite elastic medium subject to a point load [1]

It is noticed that the radial and vertical stress are independent of the modulus of elasticity of the material and are solely a function of the vertical load, W, and the distance from the point of application of the load to the point. It should be noted that Eqs. 2.1 and 2.2 are only applicable for a point not too close to the point of application of the load.

Applying the principle of superposition for a vertical stress in a semi-infinite elastic medium below the center of a circular loading area is shown in Fig. 2.2.

The vertical stress at a depth, z, below the center of the contact area is determined using Eq. 2.3 [1]. The load acting upon the contact area can be presented as $\partial W = p_0 \partial A = p_0 r \partial r \partial \theta$.

$$\partial \sigma_z = \frac{3}{2\pi} \frac{p_0 r \partial r \partial \theta}{\left[1 + (r/z)^2\right]^{5/2} z^2} \tag{2.3}$$

Double integrating equation 2.3 for θ and r.

$$\sigma_z = \frac{3}{2\pi} p_0 \int_0^{r_0/z} \int_0^{2\pi} \frac{r \partial r \partial \theta}{\left[1 + (r/z)^2\right]^{5/2} z^2} = 3 p_0 \int_0^{r_0} \frac{r \partial r}{\left[1 + (r/z)^2\right]^{5/2} z^2}$$

Assuming that $u = r/z$, vertical stress, σ_z, is written as

$$\sigma_z = 3 p_0 \int_0^{r_0/z} \frac{u \partial u}{\left[1 + u^2\right]^{5/2}} = p_0 \left[1 - \frac{z^3}{\left(z^2 + r_0^2\right)^{3/2}}\right] \tag{2.4}$$

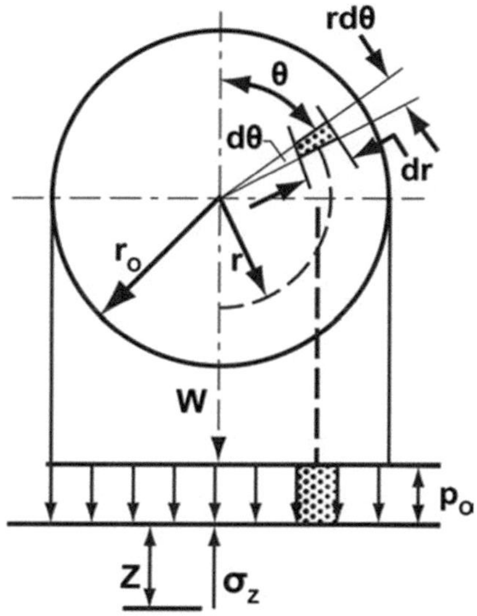

Fig. 2.2 Vertical stress in a semi-infinite elastic medium below the center of a circular loading area [1]

In a similar manner, the stresses in an elastic medium due to uniform pressure, p_0, can be computed as follows [1]. Research has revealed that there is a tendency for the compressive stress in the soil to concentrate around the loading axis. This tendency becomes greater when the soil becomes more plastic due to increased moisture content or when the soil is less cohesive, such as sand. Various semi-empirical equations have been developed to account for the different behavior of various types of soil. Frohlich introduced a concentration factor v to Boussinesq's equations. The factor v reflects the behavior of various types of terrain in different conditions. The vertical and radial stress in the soil can be expressed in the following forms:

$$\sigma_z = \frac{vW}{2\pi R^2}\cos^v\theta = \frac{vW}{2\pi z^2}\cos^{v+2}\theta \tag{2.5}$$

$$\sigma_r = \frac{vW}{2\pi R^2}\cos^{v-2}\theta = \frac{vW}{2\pi z^2}\cos^v\theta \tag{2.6}$$

It is noticed that if v equal to 3, Eq. 2.5 and v-r then becomes identical to Eqs. 2.1 and 2.2, respectively.

? Example 2.1

The contact patch of an un-treaded tire on a hard and dry soil may be approximated by a circular area of radius of 28 cm. The contact pressure is assumed to be a uniform 750 kPa. For this type of soil, the concentration factor v is assumed to be 4. Calculate the vertical stress, σ_z, in the soil at a depth of 20 cm directly below the center of the contact area.

Solution. When the concentration factor v is 4, the vertical stress, a, at a point in the soil due to a point load W applied on the soil surface is expressed by

$$\begin{aligned}\sigma_z &= \frac{4W}{2\pi R^2}\cos^4\theta \\ &= \frac{4W}{2\pi}\frac{z^4}{(z^2+r^2)^3} \\ &= \frac{4W}{2\pi}\frac{1}{z^2[1+(r/z)^2]^3}\end{aligned}$$

The vertical stress σ_z at a depth z directly below the center of a circular contact area of radius r_0 and with a uniform contact pressure p_0 is given by

$$\begin{aligned}\sigma_z &= \frac{4p_0}{2\pi}\int_0^{r_0}\int_0^{2\pi}\frac{r\partial r\partial\theta}{z^2[1+(r/z)^2]^3} \\ &= 4p_0\int_0^{r_0/z}\frac{u\partial u}{[1+u^2]^3} \\ &= p_0\left[1-\frac{1}{[1+(r_0/z)^2]^2}\right]\end{aligned}$$

It is noted that $u^2 = r^2/z^2$. Using $p_0 = 750\,\text{kPa}$, $r_0 = 28\,\text{cm}$, and $z = 20\,\text{cm}$,

$$\begin{aligned}\sigma_z &= 750\left[1-\frac{1}{[1+(28/20)^2]^2}\right] \\ &= 664.4\,\text{kPa}\end{aligned}$$

2.2 Off-Road Terrain Characterization

The bevameter, the cone penetrometer, triaxial apparatus, and the traditional civil engineering techniques are a few of the types of equipment that are used to measure the soil properties. The penetrometer and the bevameter are oftentimes utilized for vehicle applications. In 1964, Onafeko and Reece [2, 3] measured the stresses beneath a tire in radial and tangential directions over a span of longitudinal slip/skid ratios for both driven and free-rolling tires. Later in 1967, Wong and Reece [4, 5] developed the radial and tangential stress distributions underneath a rolling tire as functions of the tire sinkage and slip/skid ratio (Fig. 2.3).

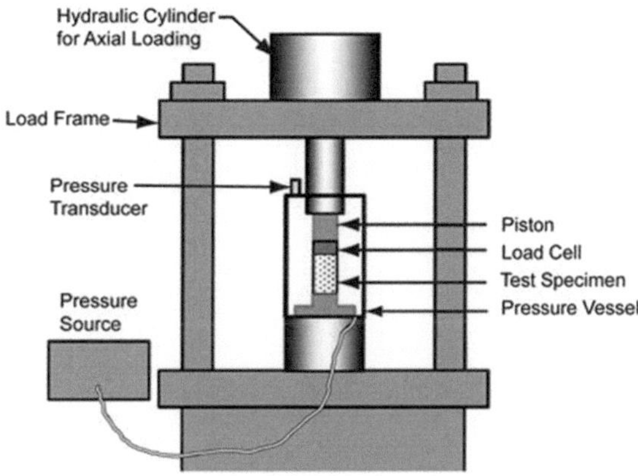

Fig. 2.3 Schematic of the cone penetrometer [6]

Triaxial apparatus is equipment to measure the mechanical properties of deformable solids and soils. During the test, a cylindrical specimen of soil is subjected to hydrostatic pressure and axial load. The triaxial apparatus features a measurement feedback control system that can simulate idealized states such as hydrostatic and triaxial compression as well as uniaxial strain loading/unloading. This feature is essential for characterizing compressibility, shear-strength, and unloading behavior of soil [7].

2.2.1 Pressure-Sinkage Test

The pressure-sinkage relationship is determined using a bevameter. Bekker designed the bevameter machine illustrated in Fig. 2.4 in the 1950s. The bevameter measures soil characteristics by applying pressure-sinkage and shear-strength tests. The pressure-sinkage test also known as the plate penetration test is conducted with a plate on top of the soil. During a pressure-sinkage test, pressure is applied to the top plate and the normal displacement is measured. The shear-strength test is conducted with a finned plate being twisted within the soil, and the shear stress is measured [8].

If a terrain is considered homogeneous within the depth of interest, it may be characterized with the following equation proposed by Bekker [9]:

$$p = \left(\frac{k_c}{b} + k_\phi\right) z^n \tag{2.7}$$

where p is the applied pressure, b is the smaller dimension of the contact patch, which is the radius of a circle or the width of a rectangle, and n, k_c, k_ϕ are pressure-sinkage parameters. The values of n, k_c, and k_ϕ are derived from the results of at least two pressure-sinkage tests with two different plate sizes. The two tests result in

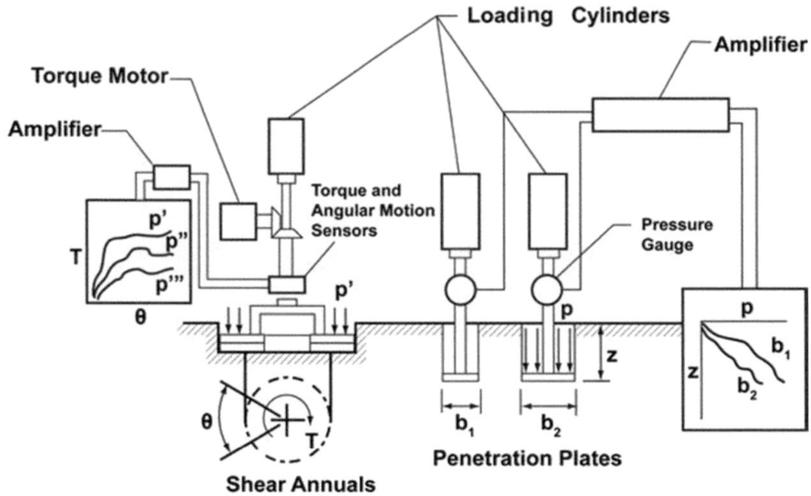

Fig. 2.4 Schematic view of a bevameter for measuring terrain properties [9]

two different curves shown in Eqs. 2.8 and 2.9:

$$p_1 = \left(\frac{k_c}{b_1} + k_\phi\right) z^n \tag{2.8}$$

$$p_2 = \left(\frac{k_c}{b_2} + k_\phi\right) z^n \tag{2.9}$$

Figure 2.5 shows the classical method to determine the values of the pressure-sinkage parameters. The method requires to convert Eqs. 2.8 and 2.9 into the logarithmic domain as shown in Eqs. 2.10 and 2.11.

$$\log p_1 = \log\left(\frac{k_c}{b_1} + k_\phi\right) + n \log z \tag{2.10}$$

$$\log p_2 = \log\left(\frac{k_c}{b_2} + k_\phi\right) + n \log z \tag{2.11}$$

where a_1 and a_2 are defined as follows:

$$a_1 = \frac{k_c}{b_1} + k_\phi \tag{2.12}$$

$$a_2 = \frac{k_c}{b_2} + k_\phi \tag{2.13}$$

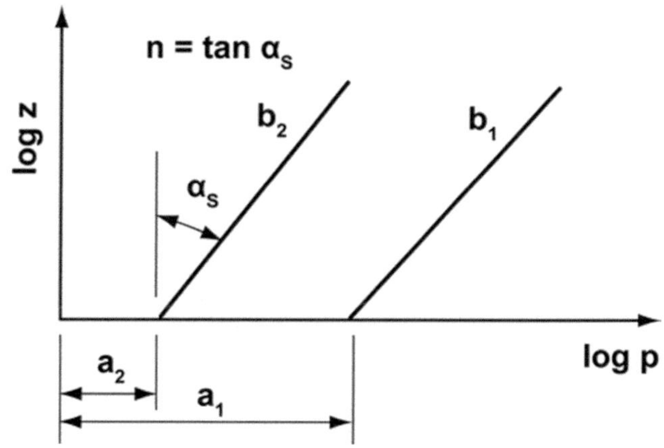

Fig. 2.5 Pressure-sinkage relationship in logarithmic domain [9]

Figure 2.5 can be utilized to determine the terrain parameters k_c and k_ϕ using the following equations:

$$k_\phi = \frac{a_2 b_2 - a_1 b_1}{b_2 - b_1} \tag{2.14}$$

$$k_c = \frac{a_1 - a_2}{b_2 - b_1} b_1 b_2 \tag{2.15}$$

? Example 2.2

Given a circular plate with a radius of 5 cm placed on top of sandy terrain with terrain values of $k_c = 5.27\,\text{kN/m}^{n+1}$, $k_\phi = 1515\,\text{kN/m}^{n+2}$, and $n = 0.7$. Find the plate sinkage at a pressure of 0, 50, and 100 kPa.

Solution. Using the pressure-sinkage relationship and the terrain values,

$$p = \left(\frac{k_c}{b} + k_\phi\right) z^n$$

$$p = \left(\frac{5.27}{0.05} + 1515\right) z^{0.7}$$

$$p = 1620.4 z^{0.7}$$

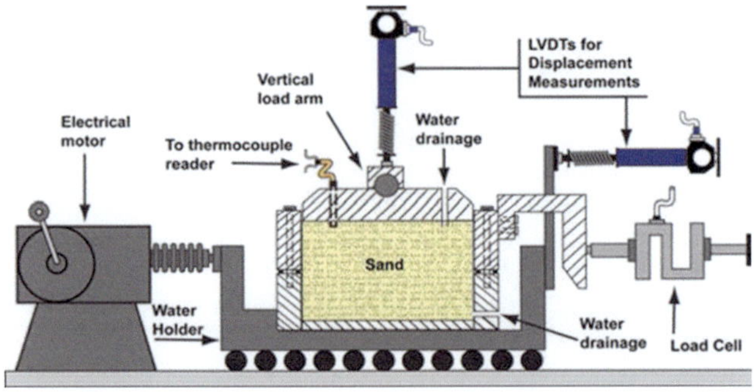

Fig. 2.6 Schematic of the direct shear-strength test with soft terrain [9]

substituting for a pressure of 0 kPa which gives a sinkage of 0 mm. Similarly, for $p = 50\,\text{kPa}$, the sinkage is $z = 6.9\,\text{mm}$, and at a $p = 100\,\text{kPa}$, the sinkage is $z = 18.7\,\text{mm}$.

2.2.2 *Direct Shear-Strength Test*

The direct shear-strength test device consists of an electrical motor that yields a constant displacement rate to the lower part of the shear box, as illustrated in Fig. 2.6. A digital load cell is connected to the upper part of the shear-strength box to constrain its movement parallel to the shear plane.

The direct shear-strength test is performed under several pressures to determine the shear-strength versus pressure curve. The curve indicates the soil cohesion and angle of shear resistance (also known as friction angle).

Figure 2.7 shows the Mohr-Coulomb criterion on the Mohr circle of stress. If a sample of the soil is subject to different states of stress, for each mode of failure, a Mohr circle can be constructed. If a straight-line envelope is drawn to the set of Mohr circles so obtained, the straight line will have an equation as that presented below:

$$\tau_{max} = c + \sigma \tan \phi \tag{2.16}$$

? Example 2.3

Find the shear-strength equation for a sandy terrain with a cohesion of 1.04 kPa and an angle of shear resistance of 28°. What is the shear stress, τ, when a square box of sandy terrain with a side of 600 mm is subjected to a pressure force of 30 kN?

Solution. The shear-strength equation is presented in Eq. 2.16. Using the cohesion and angle of shear resistance,

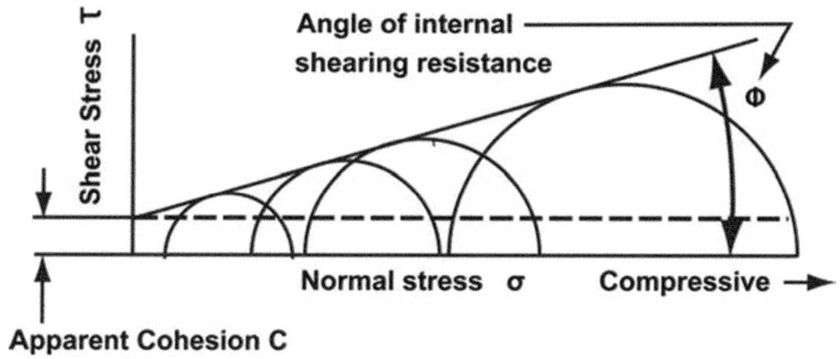

Fig. 2.7 Mohr-Coulomb failure criterion [9]

$$\tau = c + \sigma \tan \phi$$
$$\tau = 1.04 + \sigma \tan 28$$
$$\tau = 1.04 + 0.532\sigma$$

where τ and σ are in kPa. Given the force and side length,

$$\sigma = p = \frac{F}{A}$$
$$\sigma = \frac{30}{0.6^2}$$
$$\sigma = 83.3\,\text{kPa}$$

Using the calculated value of σ and the shear-strength equation to calculate τ,

$$\tau = 1.04 + 0.532 \times 83.3 = 45.37\,\text{kPa}$$

2.3 The Finite Element Analysis Approach for Terrain Modeling

In 1997, Heroma [10] embraced the FEM modeling technique to describe soil. The soil model was assumed to be viscoelastic with distinct moisture content. The tractive forces acting on the contact area between the tire and soil were analyzed at various slip angles.

Subsequently, in 2006, Shoop [11] utilized the FEM technique to model snow and compressed sand using steady-state plasticity. The model was then validated using the pressure-sinkage laboratory and field testing. In 2008, Hambleton and Drescher [12] explored soil rutting using FEM elastic-plastic soil models implemented in ABAQUS. The study deduced that the effects of indentation are negligible for clayey soils and significant for sands. Figure 2.8 illustrates an example of soil rutting. The

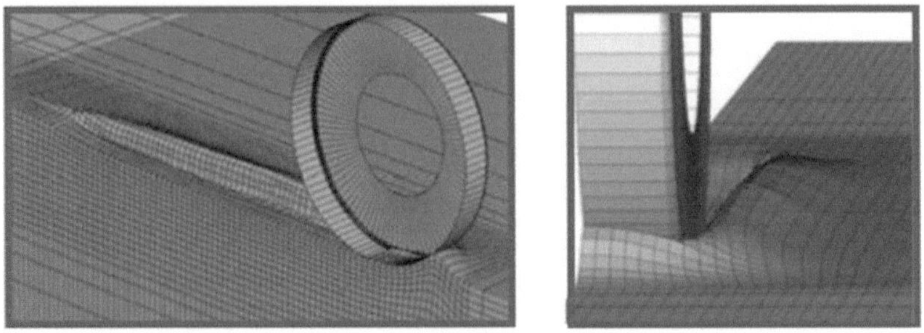

Fig. 2.8 Side and front illustration of rut formation in FEA soil [12]

investigation further concluded that the rutting process of a rolling wheel is steady, indicating that the analytical model can predict sinkage under steady-state conditions.

In 2009, Slade [13] modeled sandy loam soil using FEA technique. Nevertheless, the hysteresis and damping effects were not taken into consideration in the elastic-plastic soil model. The elastic model behaves like springs at stresses lower than that of the yield stress and deforms at stresses higher than that of the yield stress. Due to software restrictions using FEA techniques, the Mohr-Coulomb failure criterion for soil during shearing was not enforced. Slade suggested investigating SPH techniques to enhance the soil model's precision.

In 2010, Lescoe [14] modeled soil using FEA and SPH techniques in Pam-Crash for dense sand and solved the equation of state to resolve the pressure-volume relationship for elastic materials. Furthermore, Lescoe classified terrain materials according to the Idaho Association of Soil Conservation Districts [15] as illustrated in Fig. 2.9. Consequently, in 2013, Dhillon [16] validated additional FEA and SPH soil models through Pam-Crash by utilizing pressure-sinkage and shear-strength tests for diverse soils including dry sand and clayey.

In 2016, Marjani [17] optimized soil models using FEA and SPH techniques and compared FEA and SPH outcomes for the pressure-sinkage test in Pam-Crash. Furthermore, Marjani created a new modeling combination that eases the computational time, combining hybrid FEA-SPH soil models for an optimized tire-soil interaction approach. In 2017, Shahram [7] designed soil models in LS-Dyna utilizing the material for soil and foam. Shahram developed and validated several soil models including high-density clayey sand, low-density dry sand, high-density wet sand, and high-density flooded sand.

2.3.1 Terrain Calibration

The terrain calibration is performed using only the pressure-sinkage test, due to the fact that the FEA technique cannot predict the shearing characteristics between the

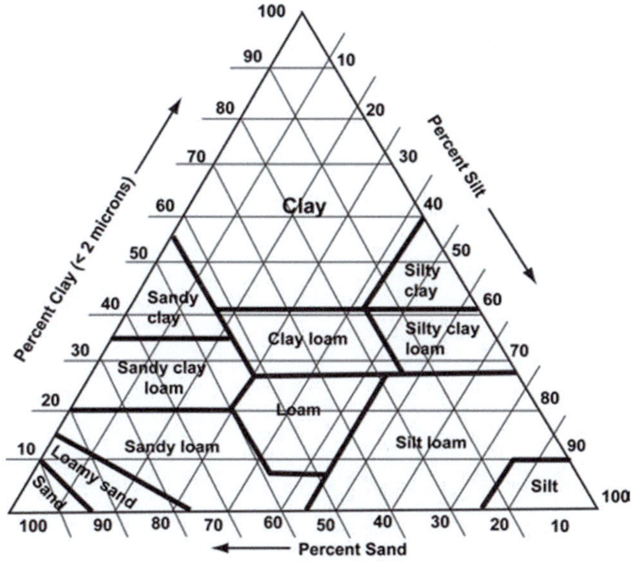

Fig. 2.9 Soil composition ratios [15]

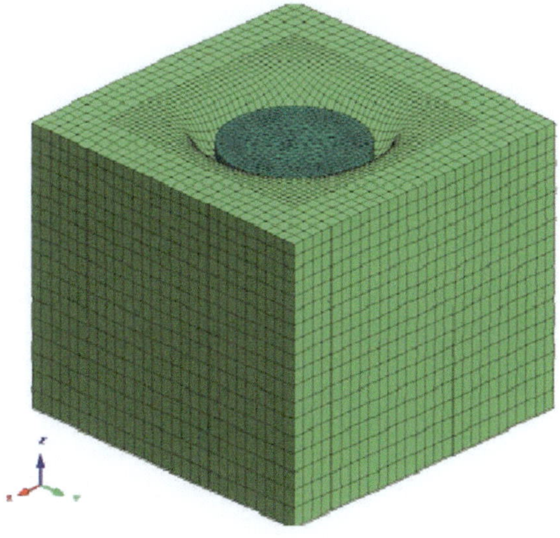

Fig. 2.10 Virtual pressure-sinkage test using a 15 cm circulate plate

soil particles. Figure 2.10 shows an example of a pressure-sinkage simulation of the soil with a rigid 15 cm circular plate.

Another method used to validate the soil models was to qualitatively compare the pressure distributions and soil flows in the soil for four standard test cases including

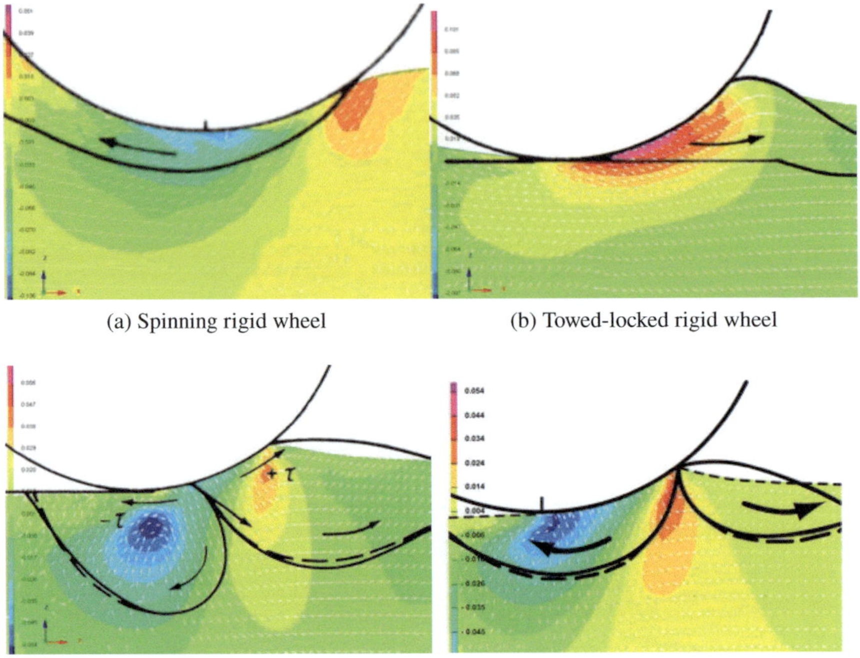

(a) Spinning rigid wheel (b) Towed-locked rigid wheel

(c) Towed-rolling rigid wheel (d) Driven rigid wheel

Fig. 2.11 Pressure distribution and soil flow for a rigid wheel running over an FEA soil [13]

the spinning, towed-locked, towed-rolling, and driven. Figure 2.11 shows the pressure distribution and soil flow for all four cases.

The towed-locked rigid wheel does not rotate, so it is considered to have 100% slip. The spinning rigid wheel rotates but is not allowed to translate in any direction and therefore it is also considered to have 100% slip. The pressure distributions and flows were simulated for speeds of 25 km/h and a friction coefficient of 0.8. The white arrows in the soil represent the velocity of soil elements and the black lines represent the rigid wheel and the approximate shape of soil flow. The black lines and arrows on the figures are superimposed diagrams of soil flow under the four cases described above. The colors in the soil indicate the pressure at that location, with the cooler colors representing lower pressures and the warmer colors representing higher pressures.

2.4 The Smoothed-Particle Hydrodynamics Approach for Terrain Modeling

Smoothed-Particle Hydrodynamic (SPH) is a new method incorporated into the simulation software Pam-Crash. SPH was originally used to simulate astrophysical fluid dynamics. When the deformation of materials, modeled by FEA, becomes very high, element tangling may occur [18]. In this case, the FEA technique is no longer reliable. SPH is usually used in combination with FEA to model explosions, bird strike analysis, sloshing of liquids, and wave impact in marine structures.

2.4.1 Fundamentals of SPH Modeling

The SPH approach is established on the interpolation method, which permits any function to be exposed regarding its values at a set of disordered points. In 1977, Gingold and Monaghan presented the SPH idea [19]. The integral interpolant also known as the "kernel estimate" of any function $A(r)$ is described in Eq. 2.17, where r and r' are contained in the integration domain, and h is the smoothing length defining the influence domain $A(r)$ [20].

$$A(r) = \int A(r')W(r - r', h)dr' \tag{2.17}$$

W is an interpolating kernel also known as the smoothing function and should satisfy the three properties in Eqs. 2.18, 2.19, and 2.20 [21]. Equation 2.18 is the normalization condition of the function W; Eq. 2.19 is the Delta function property when the normalization length, h, approaches zero. Equation 2.20 is the compact condition; κ defines the domain of $A(r)$ and is a constant related to the smoothing function for point at r as shown in Fig. 2.12.

$$\int W(r - r', h)dr' = 1 \tag{2.18}$$

$$\lim_{h=0} W(r - r', h) = \delta(r - r') \tag{2.19}$$

$$W(r - r', h) = 0 \text{ when } | r - r' | > \kappa h \tag{2.20}$$

Initially in 1977, Monaghan and Gingold [21] used the Gaussian kernel which is described in one-dimensional space. A kernel based on splines has been revealed to be computationally efficient. Nonetheless, to discover the physical interpolation of an SPH equation, it is consistently best to suppose the kernel is a Gaussian. The cubic spline interpolation function presented by Monaghan [23] is implemented to model soil behavior. The cubic spline function is described in Eq. 2.21 for three different x conditions, where x is the relative distance between r and r'.

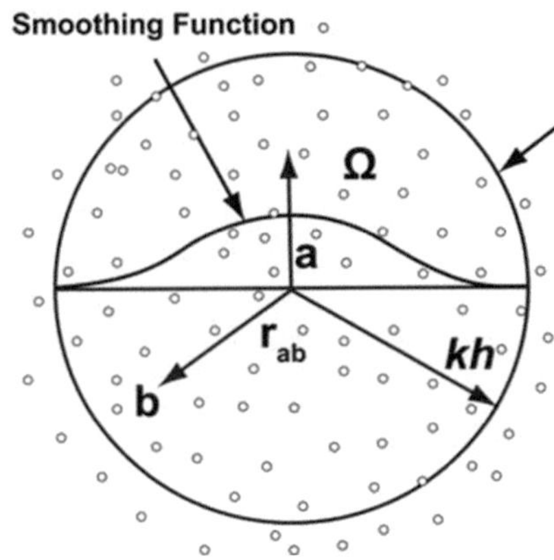

Fig. 2.12 Particle interpolation using particles within the influence domain of W for particle a [22]

$$W(x,h) = \alpha_d \times \begin{cases} 1.5 - x^2 + 0.5x^3 & 0 \le x < 1 \\ \frac{(2-x)^3}{6} & 1 \le x < 2 \\ 0 & x \ge 2 \end{cases} \tag{2.21}$$

α_d is a function defined in one-dimensional space as $\frac{1}{h}$, in two-dimensional space as $\frac{15}{7\pi h^2}$, and in three-dimensional space as $\frac{3}{2\pi h^3}$. The spatial derivative of a vector quantity $\nabla A(r)$ is derived by substituting $A(r)$ with $\nabla A(r)$ in Eq. 2.17, and Eq. 2.22 shows the spatial derivative after applying the compact condition.

$$\nabla A(r) = -\int A(r')\nabla W(r - r', h)dr' \tag{2.22}$$

For numerical analysis, the integral interpolant is approximated by a summation interpolant, and thus $A(r)$ is written in summation interpolant, where b denotes a particle label, and the summation is over all the particles. Particle b has a position r_b, mass m_b,and density ρ_b. A_b denotes the value of any quantity A at r_b. Consecutively, the particle approximation of the spatial derivative of the function at point a can be written as shown in Eq. 2.23 and the function $\nabla_a W_{ab}$ is the gradient of $W(r_a - r_b, h)$ taken with respect to the coordinates of particle a [24].

$$\nabla A_a(r) = \sum_b m_b \frac{A_b}{\rho_b} \nabla_a W_{ab} \tag{2.23}$$

2.4.1.1 SPH Water Model

Navier-Stokes equations are known as the governing equations for fluid flow. The conservation of mass and moment stated by the Lagrangian description are presented in Eqs. 2.24 and 2.25, respectively [24], where α, β are used to express the coordinate directions, $\sigma^{\alpha\beta}$ is the total stress tensor, and f^{α} is the component of acceleration caused by an external force.

$$\frac{D\rho}{Dt} = -\rho\frac{\partial v^{\alpha}}{\partial x^{\alpha}} \tag{2.24}$$

$$\frac{Dv^{\alpha}}{Dt} = \frac{1}{\rho}\frac{\partial \sigma^{\alpha\beta}}{\partial x^{\beta}} + f^{\alpha} \tag{2.25}$$

The stress tensor $\sigma^{\alpha\beta}$ is defined in Eq. 2.26 and it consists of two parts; the isotropic pressure p and the viscous shear stress $\tau^{\alpha\beta}$, where $\delta^{\alpha\beta}$ is Kronecker's delta, which is equal to 1 if $\alpha = \beta$ and zero otherwise.

$$\sigma^{\alpha\beta} = -p\delta^{\alpha\beta} + \tau^{\alpha\beta} \tag{2.26}$$

Water is considered a Newtonian flow and thus the viscous shear stress is proportional to ϵ the shear rate through the viscosity μ as shown in Eq. 2.27.

$$\tau^{\alpha\beta} = \mu\left(\frac{\partial v^{\beta}}{\partial x^{\alpha}} + \frac{\partial v^{\alpha}}{\partial x^{\beta}} - \frac{2}{3}\left(\frac{\partial v^{\gamma}}{\partial x^{\gamma}}\right)\delta^{\alpha\beta}\right) \tag{2.27}$$

The equation of state is also utilized to estimate the above Navier-Stokes relationships. Furthermore, the equation is embraced to evaluate the change in pressure of SPH water particles. Murnaghan equation of state [21] is utilized for liquids with artificially expanded compressibility. Equation 2.28 is utilized to execute a particular class of hydrodynamic simulations where the flow velocity stays entirely below the physical speed of sound [25].

$$p = p_o + B\left(\left(\frac{\rho}{\rho_0}\right)^{\gamma} - 1\right) \tag{2.28}$$

$$B \geq 100\rho_0\frac{{v_{max}}^2}{\gamma} \tag{2.29}$$

Equation 2.28 demonstrates the relationship between the pressure and the density ratios, where γ is a constant parameter that is equal to seven in most cases, ρ_0 is the initial density, and B is a parameter that depends on the problem conditions and designates a boundary to the maximum variation in density. The parameter B can be determined using Eq. 2.29, where v_{max} is the maximum water speed.

2.4.1.2 SPH Soil Model

The conservation equation of mass and moment is nevertheless utilized to predict the density and motion of soil particles. The main distinction between the soil and water model is the stress tensor. As mentioned in Sect. 2.4.1.1, the water adheres to the Murnaghan equation of state, while in the case of soil, the pressure and the stress-strain behavior are deemed to be hydrodynamic elastic-plastic. Unlike water, the stress tensor of the modeled soil displayed in Eq. 2.30 consists of two parts; the deviatoric shear stress, s, and the isotropic pressure, p [25, 26].

$$\sigma^{\alpha\beta} = -p\delta^{\alpha\beta} + s^{\alpha\beta} \tag{2.30}$$

The pressure is typically estimated using the "equation of state", which is a function of the density change and internal energy. The equation of state deployed is that of hydrodynamic elastic-plastic material as indicated in Eq. 2.31, where c_0 to c_6 are material constants, $\mu = \rho/\rho_0 - 1$ is the ratio of current over initial mass density, and E_i denotes the internal energy [25].

$$p = c_0 + c_1\mu + c_2\mu^2 + c_3\mu^3 + (c_4 + c_5\mu + c_6\mu^2)E_i \tag{2.31}$$

The solution of equation 2.31 depends on the soil being modeled and a calibration procedure is required to obtain the proper material behavior; further research on different material properties can be found in literature [27].

$$\tau_{max} = c + \sigma \tan\phi \tag{2.32}$$

The shear stress components should be limited to the failure of the surface, which happens when the plastic flow commences. The Mohr-Coulomb criterion is implemented to determine the plastic flow region indicated in Eq. 2.32, where c is the cohesion constant, σ is the applied pressure (p), and ϕ is the angle of internal shearing resistance [9]. It is noted that c and ϕ are terrain properties determined experimentally through the direct shear test.

? Example 2.4

Assuming an elastic-plastic sandy terrain, where c_2 to c_6 are neglected. Calculate the equation of state parameters c_0 and c_1, knowing that at an initial pressure of zero the density ratio is 1 and at a pressure of 100 kPa the density is increased by 50%.

Solution. Simplifying Eq. 2.31 for elastic-plastic sandy material,

$$p = c_0 + c_1\mu$$

knowing that at $p = 0$ the density ratio is 1, thus $\mu = 0$;

$$0 = c_0 + c_1 \times 0$$
$$c_0 = 0$$

Furthermore, at $p = 100\,\text{kPa}$, the density ratio is 1.5, thus $\mu = 0.5$;

$$100 = 0 + c_1 \times 0.5$$
$$c_1 = 200$$

Thus, the final equation of state is $p = 200\mu$.

2.4.1.3 Two-Phase Interaction

The conservation equations developed in Sects. 2.4.1.1 and 2.4.1.2 are solely for one flow phase. The seepage force is introduced to model the interaction between two phases (in this case sand and water). The gravitational and seepage forces emerge on the sand particles and are added to the momentum equations for both sand and water as external forces. Equation 2.33 is founded on Darcy's law [28], where the unit weight of water is represented by γ_w, n is the porosity, and k is the dry sand permeability which is selected from literature, in this case, to be 5 $cm/hour$ for dry sand [29].

$$f = \frac{\gamma_w n}{k} (v_{water} - v_{soil}) \tag{2.33}$$

The movement of SPH particles is computed independently using each material SPH governing equations. The governing equations are superimposed, and the interaction connecting the two phases is reflected through the seepage force. The pressure exerted on the sand due to water is allowed to contribute to the sand pressure throughout the overlapping method.

2.4.2 Terrain Calibration Through Virtual Testing

The SPH terrain calibration is based on physical testing data results. The two tests mainly used for calibration are the pressure-sinkage and shear-strength tests; these two tests are also performed experimentally using bevameter and direct shear-strength machines.

2.4.2.1 Pressure-Sinkage Test

A rectangular soil domain of 800 × 800 × 800 mm dimensions is filled with SPH terrain particles. A rigid circular plate of a 150 mm radius is created on top of the box. The plate is then subjected to a range of pressures between 0 and 200 kPa. The pressure is applied in a step change going directly into the desired pressure, and the duration of the test is 0.4 s, as it only takes a part of a second for the sinkage to achieve a steady state. The sinkage of the plate is then measured as an output for

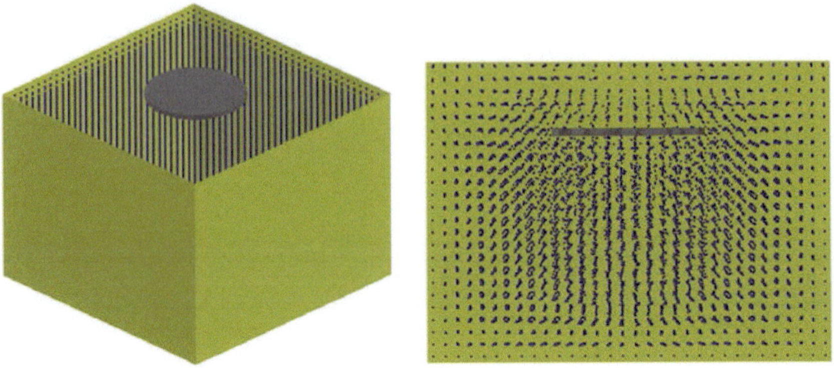

(a) Initial state (b) Final state

Fig. 2.13 Pressure-sinkage test with SPH terrain [27]

various pressures, and a curve is fitted to present the pressure-sinkage relationship. Figure 2.13 illustrates the SPH box initial and final states

The pressure-sinkage experimental results can be found in published terramechanics data [9] for terrain materials, and the equations for pressure-sinkage relationships are also provided. Since soil is homogeneous terrain, it may be characterized by the following equation proposed by Bekker [30]:

$$p = \left(\frac{k_c}{b} + k_\phi\right) z^n \tag{2.34}$$

The pressure-sinkage simulation results are compared with the results obtained from Eq. 2.34 that contains measured terrain parameters illustrated in Table 2.1 [9], where p is the pressure in kPa, b is the smaller dimension of the contact patch, that is the width of a rectangular contact area or the radius of a circular contact area in m, z is the sinkage of the plate in m, and n, k_c, and k_ϕ are pressure-sinkage terrain parameters. Finally, the process is repeated several times for different SPH material parameters until the best fitting pressure-sinkage result is obtained.

? Example 2.5

Determine the sinkage of a plate filled with dry sand when a 100 kPa pressure is applied to it. Compare the dry sand to dense sand plate sinkages under the same condition. Note that the plate is circular with a 150 mm radius.

Solution. Using Table 2.1 to obtain the terrain properties for dry sand and inputting the properties into Eq. 2.34,

Table 2.1 Terrain properties of various soils [9]

Soil name	n	k_c (kN/m^{n+1})	k_ϕ (kN/m^{n+1})	c (kPa)	ϕ (deg)
Dry sand	1.1	0.99	1528.43	1.04	28
Loose sand	0.2	2.56	43.12	1.38	38
Sandy loam	0.7	5.27	1515.04	1.72	29
Clayey soil	0.5	16.03	1262.53	2.07	10
Lean clayey	0.2	16.64	1724.69	68.95	20
Snow Sweden	1.44	10.55	66.08	6	20.7
Snow U.S.	1.6	4.37	196.72	1.03	19.7

$$
\begin{aligned}
p &= \left(\frac{k_c}{b} + k_\phi\right) z^n \\
100 &= \left(\frac{0.99}{0.15} + 1528.43\right) 150^1.1 \\
z &= 85\,\text{mm}
\end{aligned}
$$

Similarly, for dense sand, the plate sinkage at 100 kPa pressure is 19.8 mm.

2.4.2.2 Shear-Strength Test

The shear-strength test is conducted by constructing a rectangular box of 400 × 200 × 240 mm size filled with SPH soil particles, illustrated in Fig. 2.14 [9]. The box is made up of three parts; the top plate where pressure is applied, the sliding plate, and the lower plate which is constrained from moving in all directions. During the test, a constant pressure ranging between 0 and 200 kPa with an increment of 50 kPa is applied to the top plate of the box. Then, a small ramp displacement is applied to the upper and the top plates at a rate of 10 mm/s. The simulation runs until the top box displacement reaches 100 mm. The shear stress is calculated and plotted against the shear stress displacement relationship defined by the exponential function proposed by Janosi and Hanamoto in Eq. 2.35 [30]. The results are compared with Bekker's shear-strength relationship illustrated in Eq. 2.16 [9]. The test is repeated for different SPH material properties until optimal results are achieved.

$$
\begin{aligned}
\tau &= \tau_{max}\left(1 - e^{(-j/k)}\right) \\
&= (c + \sigma \tan\phi)\left(1 - e^{(-j/k)}\right)
\end{aligned}
\tag{2.35}
$$

where τ is the shear stress, j is the shear displacement, c and ϕ are the cohesion and the angle of internal shearing resistance of the terrain, respectively, and k is referred

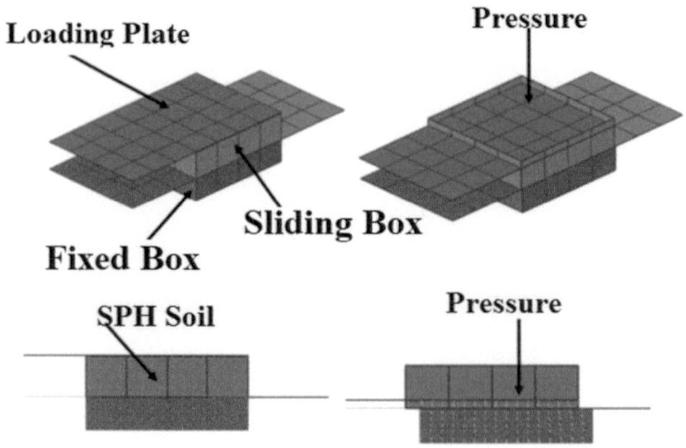

Fig. 2.14 Shear-strength test with SPH soil

to as the shear deformation modulus. The maximum shear τ_{max} can be determined through the Mohr-Coulomb failure criterion in Eq. 2.16.

$$\tau_{max} = c + p \tan \phi \tag{2.36}$$

It should be noted that both the pressure-sinkage and shear-strength tests must be simulated using identical SPH material properties to get optimal behavior for both tests. During the calibration process, the SPH material parameters are continuously adjusted, and both tests are continuously repeated until the best agreement is reached between the simulation and experimental results.

? Example 2.6

Similar to Example 2.5, calculate the maximum shear for dry sand and dense sand at 100 kPa applied pressure.

Solution. Using Table 2.1 to obtain the terrain properties for dry sand and inputting the properties into Eq. 2.36,

$$\tau_{max} = c + p \tan \phi$$
$$\tau_{max} = 1.04 + 100 \tan 28$$
$$\tau_{max} = 54.2\,\text{kPa}$$

Similarly, for dense sand, the maximum shear at 100 kPa pressure is 57.15 kPa.

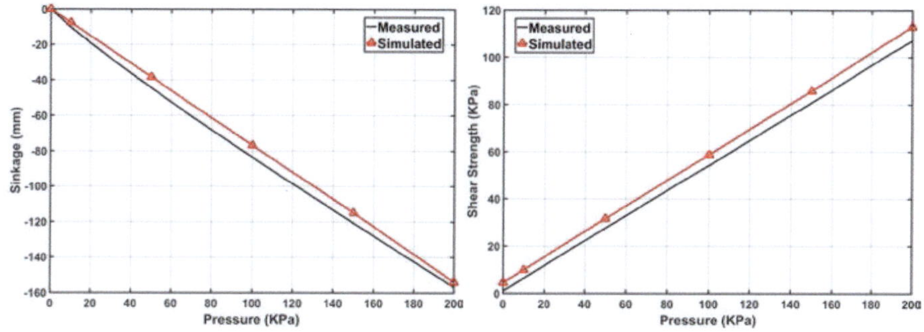

(a) Pressure-sinkage (b) Shear-strength

Fig. 2.15 Pressure-sinkage and shear-strength relationship for dry sand simulation versus measurement

2.4.2.3 SPH Calibration Sample

Dry sand can be found in most countries and consists of non-homogeneous granular particles with different material properties. Dry sand behaves like a fluidic flow when disturbed since it has no moisture. Figure 2.15a illustrates the pressure-sinkage relationship results for both simulated and measured cases. It is shown that the two lines corresponding to simulation and measured exhibit similar values.

The simulated sinkage at 200 kPa is computed to be 157 mm, while the measured one is 156.8 mm. Figure 2.15b illustrates the shear-strength relationship results for dry sand. The slope of the shear-strength curve is known as the $\tan\phi$ and the y-intercept is known as the cohesion of the curves in Fig. 2.15b. The cohesion simulated is computed to be 4.5 kPa, while the one measured is 1.04 kPa. The simulated internal friction angle is computed to be 28.4°, while the measured one is 28°. These results are in good agreement with the physical measurements and within a minimal error range.

Dense sand or sandy loam is another type of soil; it is mostly composed of sand particles and a small amount of clay. The moisture content is around 15% which makes it suitable for gardening and drainage. The same analysis method is performed on the dense sand. The two tests are implemented, and the results are compared with the measurements. Figure 2.16a and b illustrates the results of the pressure-sinkage and the shear-strength for both simulated and measured cases.

The pressure-sinkage relationship for dense sand is fitted as a second-degree quadratic equation that takes the parabolic shape. The simulated sinkage at 200 kPa is 56.66 mm, while the measured one is 53.64 mm. The shear-strength parameters are determined from the curve fitting. The cohesion coefficient simulated is 9.8 kPa, and the measurement is 1.72 kPa. The internal friction simulated is 29.4°, and the measurement is 29°. The pressure-sinkage test and the shear-strength test indicate similar results.

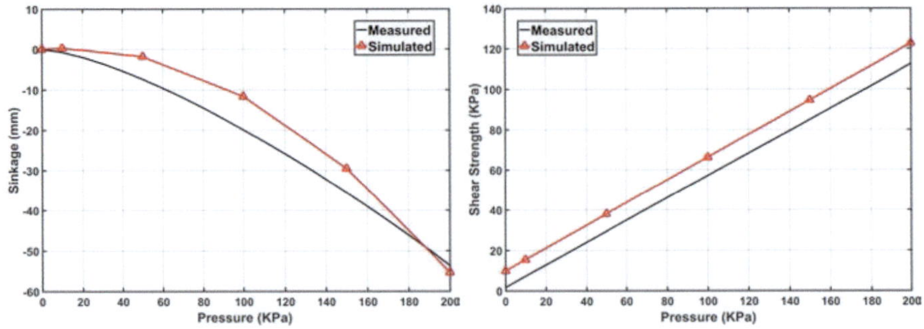

(a) Pressure-sinkage (b) Shear-strength

Fig. 2.16 Pressure-sinkage and shear-strength relationship for dense sand simulation versus measurement

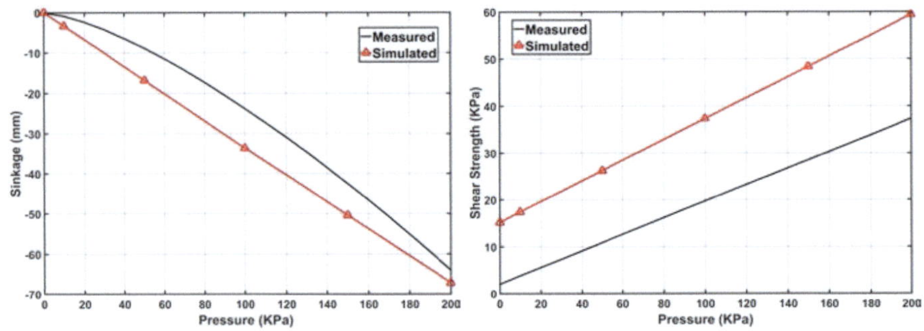

(a) Pressure-sinkage (b) Shear-strength

Fig. 2.17 Pressure-sinkage and shear-strength relationship for clayey soil simulation versus measurement

Clayey soil and its counterpart from Thailand consist of clay minerals and a relatively high moisture content, resulting in a more cohesive type of soil than sand. The same analysis applied for sand is also applied for clayey soil from Thailand. Figure 2.17a and b illustrates the pressure-sinkage and shear-strength results.

The simulated cohesion is 15.7 kPa, while the measured one is 2.07 kPa. The simulated internal friction is 12°, while the measured one is 10°. The results indicate a difference between the simulations and the measured, as clayey properties may vary depending on the sample collection and the material properties.

The snow model presented is considered the first attempt to understand the behavior and modeling of tires over snow using the Smoothed-Particle Hydrodynamics method. The snow model is founded on SPH particles similar to those of the soil and is subject to soil validation technique since literature noted snow within terrain materials [9] and identified its terrain properties including cohesion and internal friction angle. Thus, the soil validation approach is performed to validate the snow model. In other words, the two tests pressure-sinkage and shear-strength are applied to the

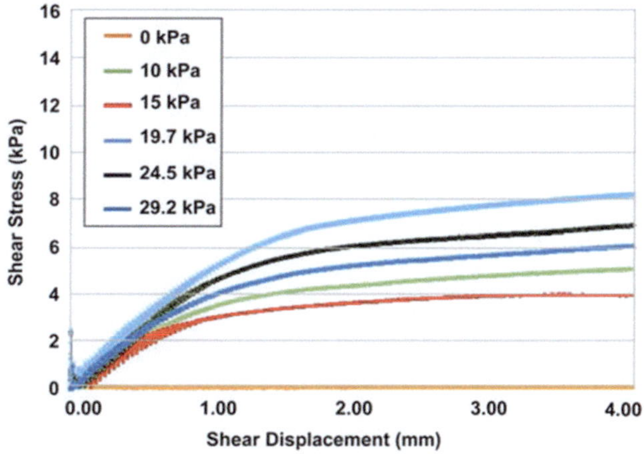

Fig. 2.18 Shear stress as a function of shear displacement obtained from shear-strength test for snow at different applied pressures [32]

snow model. The snow properties highly depend on the location of sample collection and can range between U.S snow and snow from Sweden. Swedish snow has considerably greater cohesion than sand but less than clayey soil. In a comparison of the Swedish snow to snow from the U.S., Swedish snow has the lowest index n and the highest cohesion which is 6 kPa. The snow pressure-sinkage test is performed for pressure between 0 and 50 kPa since the sinkage reaches about 500 mm at 50 kPa pressure. Note that the pressure-sinkage box utilized has a depth of only 800 mm [31], and thus the depth should be respected while running the simulations. As for the shear-strength test, the same procedure is applied to a displacement speed of 5 mm/s for a 2-s duration.

Figure 2.18 illustrates the shear stress in kPa versus the shear displacement in mm for various pressures between 0 and 29.2 kPa. The figure indicates a rapid increase and some oscillations at the beginning of the displacement curve. The curves are then continuous in an approximately steady-state motion. The curves can be compared with those from physical measurements [9]. The results show similarities in shape and slope.

Figure 2.19a and b illustrates the pressure-sinkage and shear-strength relationships. As shown in Fig. 2.19a, the snow model simulated falls within the range of the snow measurements. Note that this test is only done for a pressure of up to 50 kPa. The snow model may vary significantly depending on the desired terrain properties. The snow model proposed is obtained from various calibrations and simulations. The most sensitive parameter noticed was the tangent modulus. Calibration of this parameter was accomplished by a compensation between the pressure-sinkage and shear-strength. The results from the two tests varied inversely with the variation of the tangent modulus. The snow model validated and calibrated is valid for various snow environments, as it is fitted between different snow ranges. It is noted that

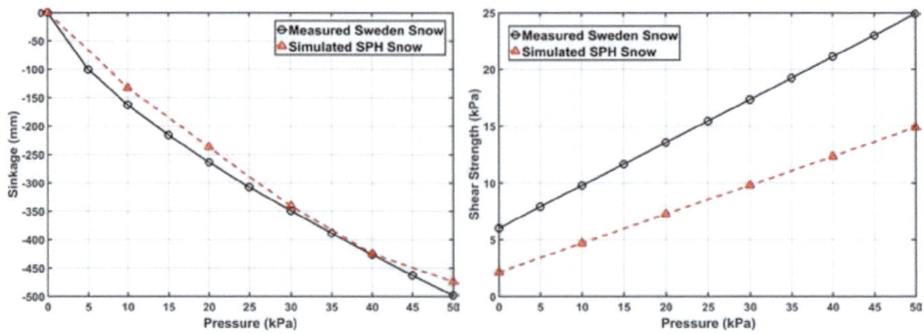

(a) Pressure-sinkage (b) Shear-strength

Fig. 2.19 Pressure-sinkage and shear-strength relationship for simulated snow versus measurement [32]

other snow models such as snow from different countries may not adhere to the same behavior.

It is observed that the snow calibrated is neither snow from Sweden nor that from the U.S. However, it is an attempt to capture a snow behavior using the SPH technique. The SPH snow sinks about 473 mm at a pressure of 50 kPa while that of Sweden sinks 500 mm. The cohesion of the modeled snow is around 1 kPa while that from Sweden is 6 kPa; in addition, the friction angle of the snow modeled is 16.2 degrees and that from Sweden is 20.7 degrees. It should be noted that the SPH snow parameters still need tuning and further work is required.

2.4.3 Sensitivity Analysis of SPH Material

The SPH material is sensitive to any change in parameters which can result in changes in the pressure-sinkage and shear-strength relationships. In this section, sensitivity analysis for various parameters is executed. The effect of these parameters on the material behavior is explained. The parameters examined include displacement speed in the shear-strength test, tangent modulus, yield stress, and the coefficient of the pressure-volume equation of state.

2.4.3.1 Displacement Speed

The displacement speed is the speed at which the box is being dragged. This parameter is relevant when performing the direct shear box test. It determines how fast the SPH particles are moving, and thus how quickly the particles will react to the change. Three different speeds are tested under constant material conditions. All material and SPH control parameters are kept constant while varying the speed of shear displacement.

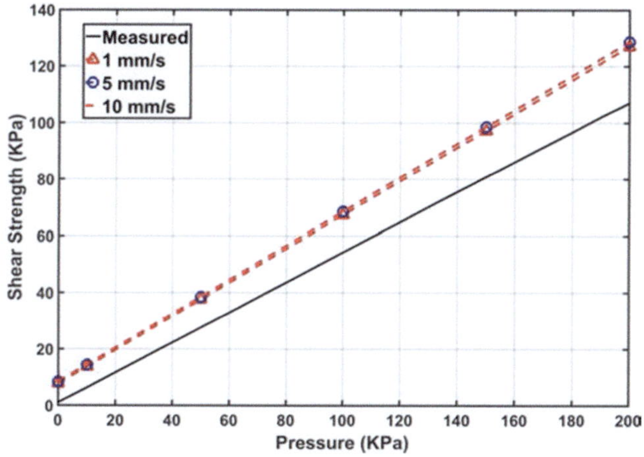

Fig. 2.20 Shear-strength as a function of pressure for several shear displacement speeds

Table 2.2 Shear-strength terrain properties for several simulated displacement speeds

Terrain properties	Measurement	10 (mm/s)	5 (mm/s)	1 (mm/s)
c (kPa)	1.04	8.48	8.41	7.77
ϕ (deg)	28	30.99	30.95	30.8

A total displacement of 10 mm is required to complete this test. Therefore, the box with speed of 10 mm/s runs for 1 s, the box with speed of 5 mm/s runs for 2 s, and the box with speed 1 mm/s runs for 10 s.

Figure 2.20 illustrates the results of the three different boxes from the simulation and the measurements. Table 2.2 outlines the parameters required to compare the results. As concluded, the speed of the box has a minimal effect on the shear-strength relationship. As the speed becomes 10 times slower between 10 and 1 mm/s, the cohesion changes between 8.48 and 7.77 kPa which is only 8%. Note that a speed difference of 10 times also means approximately 10 times more computational time to reach the same displacement.

2.4.3.2 Equation of State Coefficients

The second parameter examined is the coefficients of the pressure-volume equation of state c_1, and it is estimated by solving the pressure-volume equation of state. This parameter affects both pressure-sinkage and shear-strength relationships. Similar to the previous test, all parameters are kept unchanging, and only c_1 is being changed. The box is moving at a constant speed of 5 mm/s for 10 s. Three different values for c_1 have been selected 2.5, 5, and 11 MPa. Figure 2.21a and b illustrates the results for the pressure-sinkage and shear-strength, respectively.

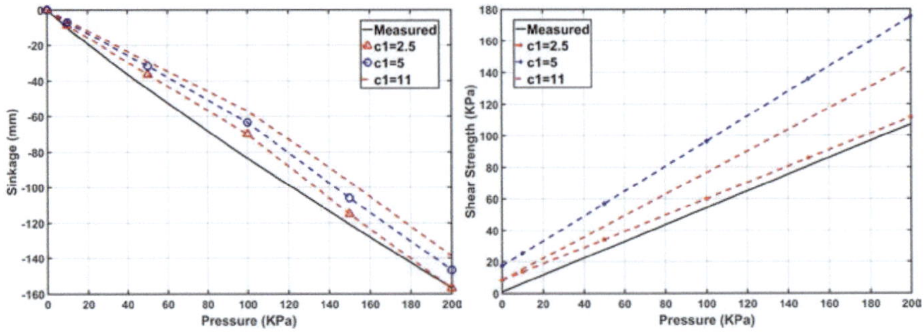

(a) Pressure-sinkage (b) Shear-strength

Fig. 2.21 Calibration results for various c_1 coefficients of 2.5, 5, and 11 MPa versus measurement

Table 2.3 Terrain properties summary for various simulated c_1 coefficients and measurements

Terrain properties	Measurement	2.5 (MPa)	5 (MPa)	11 (MPa)
Sinkage (mm)	156.8	157	147	139
c (kPa)	1.04	8.31	8.41	8.47
ϕ (deg)	28	27.29	30.95	34.2

Table 2.3 exemplifies a summary of the results of both tests, and the sinkage is estimated at 200 kPa applied pressure. The sinkage of the soil is highly affected by the value of c_1. As c_1 increases from 2.5 and 11 MPa, the sinkage decreases between 157 and 139 mm. For the shear-strength test, as c_1 increases between 2.5 and 11 MPa, the cohesion increases between 8.31 and 8.47 MPa. Similarly, the internal friction angle also increases between 27.29° and 34.2°. c_1 affects the stiffness of the material, for example, having a smaller value will give a softer material resulting in more sinkage. The best result is obtained when c_1 is equal to 2.5 which is close to the measured sinkage. The effect over the shear-strength is similar to that over sinkage as c_1 increases so does the internal friction that increases the shear force. As c_1 increases, the material becomes denser and thus the plate sinks less. In this case, the optimal value of c_1 is taken to be 2.5 MPa.

2.4.3.3 Yield Stress

The third parameter examined is yield stress (SIGMAy). The yield stress presents the value of the stress at the yield strength. Stress higher than the yield stress will lead to a permanent deformation also known as plastic deformation. After performing several simulations on the shear box test, it was concluded that the yield stress has a minimal effect on the shear-strength relationship. Nevertheless, it has a significant

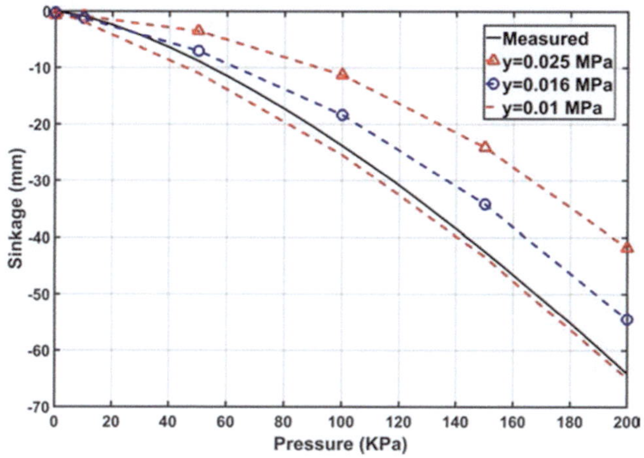

Fig. 2.22 Pressure-sinkage relationship for various yield stresses of 0.01, 0.016, and 0.025MPa versus measurement

effect on the pressure-sinkage relationship. During this test, all parameters were kept constant, and only yield stress is changed to 0.01, 0.016, and 0.025 MPa.

Figure 2.22 illustrates the pressure-sinkage relationship for various yield stresses. The effect of yield stress is similar to that of the coefficient of the equation of state. As the yield stress increases between 0.01 and 0.025 MPa, the sinkage decreases between 64.03 and 39.7 MPa. Therefore, as the yield stress increases, it results in a denser material. In this case, the optimal value of the yield stress is the nearest one to the measurement which is 0.01 MPa.

2.4.3.4 Tangent Modulus

The last parameter considered is the Tangent Modulus (E_t). Tangent Modulus is defined as the slope of the stress-strain curve at a point of interest and is known as Young's Modulus when the point of tangency falls within the linear range of the stress-strain curve. The tangent modulus is valuable in defining material behavior since it quantifies the softening of the material when determining the modulus in the plastic range of the material stress-strain curve.

In this test, all parameters are kept constant, and the tangent modulus is varied. This test was done for dense sand at a constant displacement speed of 5 mm/s for 10 s. Different tangent modulus of 0.4, 0.35, 0.3, and 0.2 MPa magnitude are tested. Figure 2.23a and b illustrates the results of the pressure-sinkage and shear-strength tests, respectively. As the tangent modulus increases between 0.2 and 0.4 MPa, the sinkage decreases between 115.13 and 53.64 mm. The increase in the tangent modulus results in stiffer soil. In this case, the optimal modulus based only on the sinkage test is 0.4 MPa. On the other side, as the tangent modulus increases between

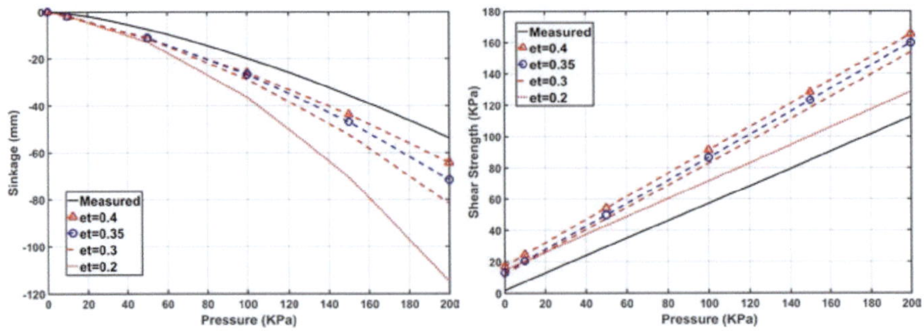

(a) Pressure-sinkage (b) Shear-strength

Fig. 2.23 Calibration results for various tangent modulus

0.2 and 0.4 MPa, the cohesion increases between 9 and 14.1 kPa. The increases result in a greater cohesion than the measured value. Similarly, the internal friction increases between 30° and 37.3°. The optimal tangent modulus according to the shear-strength relationship only is 0.2 MPa.

The tangent modulus affects both pressure-sinkage and shear-strength tests inversely. However, the optimal soil properties should include only one tangent modulus regardless of the test performed. Thus, all parameters should be consistent for both tests. In this case, a compromise should be made while choosing the tangent modulus. The compromises could include choosing a mean value for E_t and then making soil stiffer by changing the equation of state constant c_1.

Note that the tangent modulus affects the slope of the shear force versus shear displacement. As the tangent modulus increases, the slope of the curve increases, resulting in a non-steady curve. It is recommended to keep the tangent modulus low to have a stable shear test.

2.4.4 Moisture Terrain Modeling

In this section, the technique employed to model moist soil with different moisture contents is examined. The technique adopts the theory of pressurizing the water particles into the sand particles in a soil box. Further, the calibration techniques utilized to calibrate soils including the pressure-sinkage and shear-strength tests are explained.

The calibration procedures including pressure-sinkage and shear-strength tests were performed. The type of sand used in this study is considered to be dry sand with no moisture content and is modeled and calibrated in a previous publication [27]. The sand is modeled by several layers of an individual type of particles created in the bottom of a soil box, and the water is modeled by several layers of an individual type of particles created on top of the sand in the soil box. The effect of pressure and

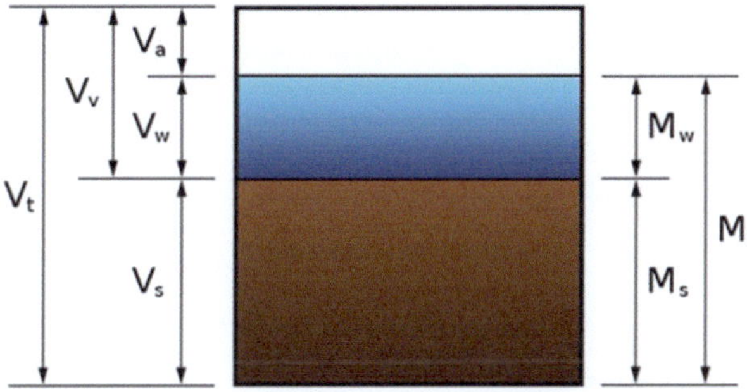

Fig. 2.24 Soil composition by volume and phase of sand, water, and total [33]

displacement on the moist sand is computed by permitting the pressure of the water to contribute to the pressure of the sand.

The depth of sand and water defined by a ratio of soil volume and the volumetric water content percentage, w, is computed as shown in Eq. 2.37 and Fig. 2.24 [33].

$$w = \frac{V_w}{V_{wet}} * 100 = \frac{V_w}{V_s + V_w} * 100 \tag{2.37}$$

where V_w is the volume of water and V_{wet} is the volume of the sand and water added. For the purpose of this research, the water content percentage will be used and referred to as the moisture content and the sand-water model will be referred to as moist sand. Another way to define the water content also using Fig. 2.24 is using the gravimetric water content percentage, u, described in Eq. 2.38.

$$u = \frac{m_w}{m} * 100 \tag{2.38}$$

where m_w is the mass of the water and m is the mass of the sample that is the mass of the dry sand and water combined.

? Example 2.7

Given that the total wet volume of a moist sand soil is 250 mm^3, find the water volume to obtain a water content percentage of 10, 20, and 50%.

Solution. Using Eq. 2.37 and the 10% water content,

$$\begin{aligned} w &= \frac{V_w}{V_{wet}} * 100 \\ 10 &= \frac{V_w}{250} * 100 \\ V_w &= 25\,\text{mm}^3 \end{aligned}$$

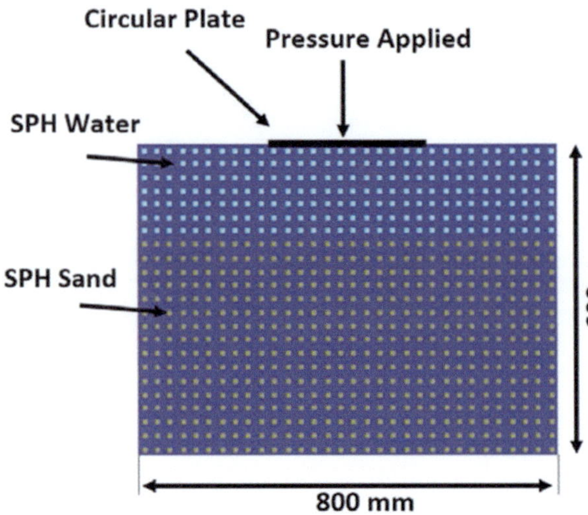

Fig. 2.25 Pressure-sinkage test condition for 30% moist sand

Similarly, the water volume to obtain 20% water content is 50 mm^3, and the water volume to obtain 50% water content is 125 mm^3.

2.4.4.1 Pressure-Sinkage Test

As previously mentioned, the soil moisturizing technique is examined using the pressure-sinkage and the shear-strength tests. Figure 2.25 illustrates the pressure-sinkage procedure for moist sand with 30% moisture content. The pressure-sinkage test is performed by applying a known pressure to a circular plate with a 150 mm radius placed on top of a box (600 × 800 × 600 mm) filled with the SPH soil particles [9]. The SPH particles are subjected to a range of pressure between 0 and 200 kPa, and the sinkage of the plate is measured. The relationship between the pressure applied to the circular plate and the circular plate sinkage is computed.

It is noted that in the case of moist sand, the water particles are pressurized into the sand particles and allowed to settle. This measure is performed before the simulation procedure starts and is considered a pre-simulation before the pressure-sinkage is applied.

The pressure-sinkage simulation results are fitted in Eq. 2.39 known as Bekker's equation [9], where p is the pressure in kPa, b is the radius of the circular plate in mm, z is the sinkage of the plate in mm, and n, k_c, and k_θ are terrain values.

$$p = \left(\frac{k_c}{b} + k_\theta \right) z^n \tag{2.39}$$

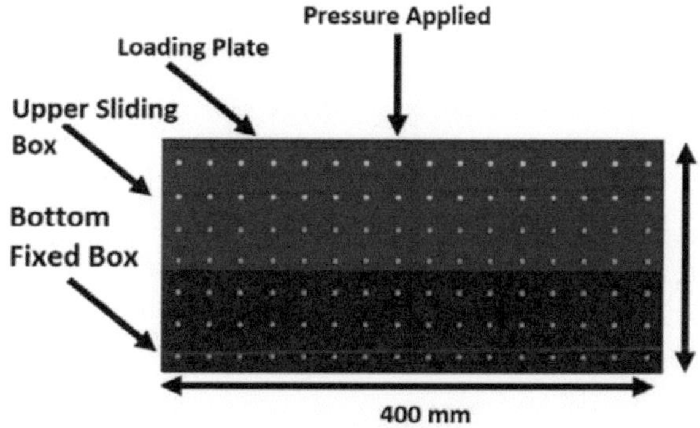

Fig. 2.26 Shear-strength test condition for sand with 30% moisture

2.4.4.2 Shear-Strength Test

The shear-strength test illustrated in Fig. 2.26 is performed by constructing a rectangular domain of $400 \times 200 \times 240$ mm size filled with SPH soil materials [9]. Similar to the pressure-sinkage, the soil consists of water layered on top of the sand in the soil box and pressurized into the sand. The domain is composed of three parts: the top plate in which pressure is applied, the upper box which is the sliding box, and the bottom box which is constrained from moving in all directions.

A known pressure ranging between 0 and 200 kPa is applied to the top plate of the domain, then a small ramp displacement is applied to the upper box and the top plate at a rate of 10 mm/s. The shear force is computed until the upper sliding box displacement reaches 100 mm. The results are fitted to Mohr-Coulomb failure criteria displayed in Eq. 2.40 [9], where c and ϕ are previously terrain values in Sect. 2.4.1.2.

$$\tau_{max} = c + p \tan \phi \tag{2.40}$$

2.4.4.3 Laboratory Testing of Moist Soil

The test sand used in the experiment is classified as poorly graded and non-plastic sand in accordance with the Unified Soil Classification System (USCS) [34]. The maximum dry density, $\rho_{d,max}$, and optimum moisture content of the soil, w_{opt}, were measured at 1850 kg/m^3 and 10%, respectively, according to Standard Proctor test [35]. The ASTM standard [35] was used to determine the particle size distribution for the sand; furthermore, the soil contained approximately 5% fines passing through a 0.075 mm sieve.

The shear-strength testing was carried out in the Minto Centre for Advanced Studies in Engineering at Carleton University, Ottawa, Canada, in accordance with ASTM D3080/D3080M-11. Figure 2.27 shows the equipment used to perform the shear test for moist sand at different moisture contents; the equipment consists of

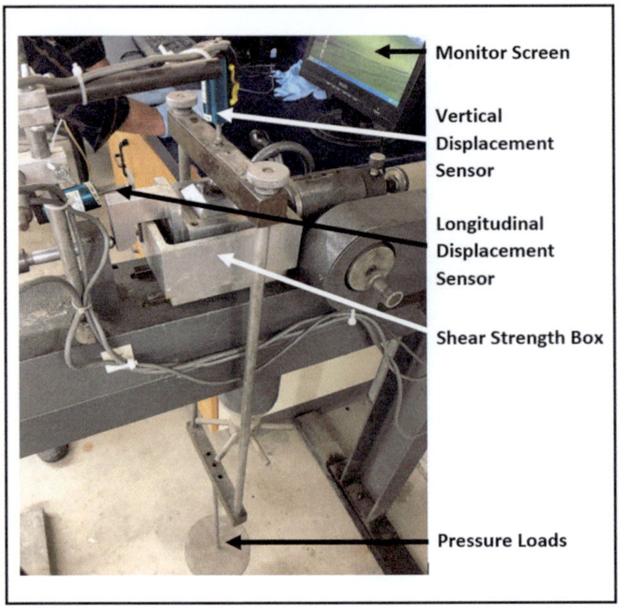

Fig. 2.27 Shear-strength test equipment used during testing (courtesy of Carleton University, Ottawa, Canada)

two boxes and a pressure plate in addition to two sensors to measure the lateral and vertical displacement of the soil.

The apparatus features a gearbox that controls the motion of the electrical motor and enables adjustment of the shear velocity. The horizontal and vertical displacements were measured through Linear Variable Differential Transducers (LVDT). The LVDT was connected to a digital logging station using LabView software. The apparatus frame facilitates applying normal stress to the top of the test specimen by incorporating a steel bearing arm. For soil shear testing, a regular coupling shear box was used. This shear box had inner plan-view dimensions of 60 mm by 60 mm and a depth of 40 mm.

Each test is repeated 3 times at different applied pressures to obtain one angle of shear resistance and one soil cohesion at a specific moisture content; thus to obtain soil characteristics at six different moisture contents, at least 18 samples were required.

Table 2.4 illustrates the dry and bulk density for each sample in addition to the mass and the volume of the shear box. The test was repeated for 6 additional samples with different moisture contents, and the bulk density of each condition was estimated using Eq. 2.41 and controlled throughout the test.

$$\rho = \rho_d \left(1 + \frac{w}{100}\right) \tag{2.41}$$

Table 2.4 Soil sampling and properties specification

	Dry density (kg/m^3)	Moisture content (%)	Bulk density (kg/m^3)	Mass (g)
1	1300	0	1300	196.56
2	1300	5	1365	206.39
3	1300	15	1430	216.216
4	1300	25	1625	245.7
5	1300	35	1755	265.36
6	1300	50	1950	294.84

Table 2.5 Water content and mass distribution

Moisture content (%)	Sample mass (g)	Water mass (g)	Sand mass (g)
0	196.56	0	196.56
5	206.39	10.32	196.07
10	216.22	21.62	194.59
25	245.7	61.42	184.27
35	265.35	92.87	172.48
50	294.84	147.42	147.42

where ρ is the bulk density, ρ_d is the dry density, and w is the water content percentage. It should be noted that the sample mass calculated in Table 2.4 is the mass of the water and sand together, and thus, using Eq. 2.38, the mass of each of the water and sand can be calculated. Table 2.5 summarizes the mass of the water and sand in addition to the water content percentages.

To prepare the test samples, the respective masses of the dry sand and water were weighed and then mixed thoroughly to achieve a homogenous mixture. The mixture was then packed into the shear box to reach the corresponding bulk densities. The shear box was then scaled on the direct shear test apparatus and the desired normal stress was, respectively, applied on top of the sample using a predefined dead load. The shear tests were performed under a controlled strain rate where the loading was applied at a constant linear velocity of 0.01 mm/s. For each water content, shear-strength properties including the cohesion and angle of shear resistance were determined under several normal loads ranging between 16 and 100 kPa as indicated in Table 2.6.

Figure 2.28a illustrates the variation of the shear stress as a function of the shear strain at three different pressures for sand with 50% moisture content. It can be noticed that the shear stress increases as the shear strain increases for a shear strain ranging between 0 and 2%. Afterward, the shear stress reaches a steady state as the shear strain keeps on increasing until 10%.

Table 2.6 Normal stress and load during the shear-strength test

Normal stress (kPa)	Normal mass (kg)	Steel holder mass (kg)	Normal load needed (kN)
16	5.87	5.7	0.073
50	018.35	5.8	12.55
100	36.7	5.8	30.91

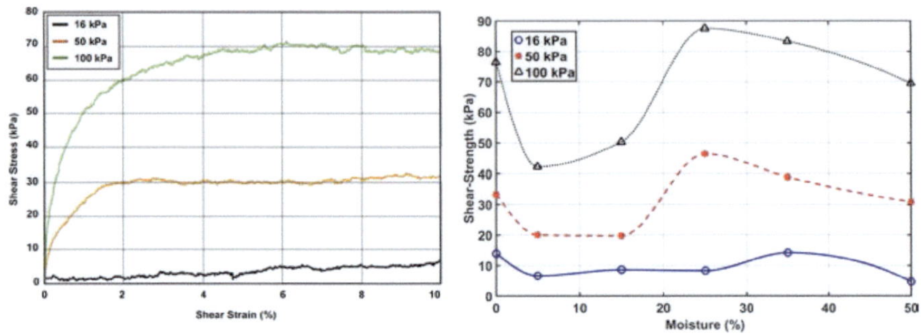

(a) Sample of the shear stress measurements (b) Shear-strength as a function of moisture content

Fig. 2.28 Sample of the shear stress experimental results

It is also noticed that the maximum shear-strength is recorded at the highest pressure; for a 50% moist sand at an applied pressure of 100 kPa, the maximum shear-strength is recorded to be around 70 kPa.

On the other side dry soil, for instance, experienced strain-softening after showing well-defined peak strength, especially under the high normal stress of 100 kPa. This could be attributed to the initial volume contraction which densified the soil and resulted in higher strength; however, as the loading progresses, the soil sample dilated and peak strength slightly decreased.

Figure 2.28b illustrates the variation of the shear-strength as a function of the moisture content for the different applied loads. The shear-strength is calculated from the shear-strength versus the shear-strain curve at a steady state. The relationship between the shear-strength and moisture content is not linear and reaches a peak value at around 25% moisture content. Additionally, the shear-strength increases as the applied pressure increases. For a low applied pressure such as 16 kPa, the shear-strength is less affected by the moisture content in comparison to the 100 kPa pressure which is highly affected by the change in soil moisture.

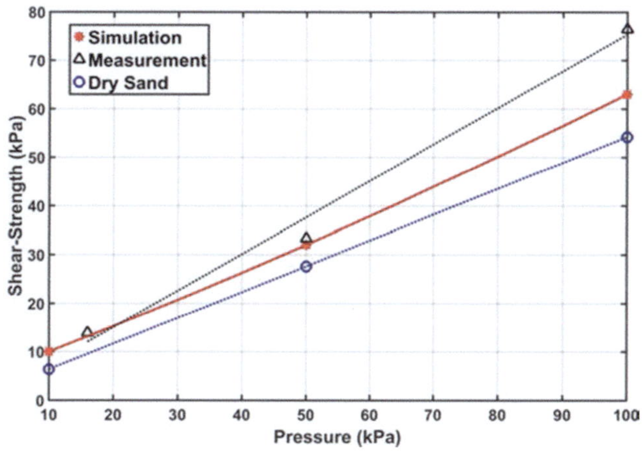

Fig. 2.29 Shear-strength as a function of pressure for dry sand

2.4.4.4 Moist Soil Calibration Results

In order to validate the results acquired using the novel moisturizing technique, the shear-strength results at different moisture contents are discussed. It should be noted that the sand tested has similar characteristics to that simulated; nonetheless, it is slightly different, due to the limitations in the available sand for testing, this sand type was selected. The simulated sand was calibrated based on published terramechanics research [6].

Figure 2.29 illustrates the shear-strength results of the dry sand for the simulated, measured, and previously published results [36]. The SPH soil was calibrated based on published terramechanics research due to the limitation of results and the wide change of soil characteristics from one sand to another. The angle of shear resistance and the cohesion are determined using the shear-strength line fitting. The angle of shear resistance is computed using the slope of the line, while the cohesion is computed as the y-intercept, which is in this case, the intersection of the shear-strength line with the shear axis.

In the case of sand with 0% moisture content, the measured cohesion is about 0.45 kPa while the simulated one is around 1 kPa. The measured angle of shear resistance is recorded to be 36° while the simulated one is 33°. In comparison between the simulated and published terramechanics results, the published cohesion is 1.04 kPa and the angle of shear resistance is around 28°.

In a similar manner, sand with 50% moisture content in both measured and simulated curves exhibits similar trends. The measured cohesion is recorded to be 7.6 kPa while the simulated one is around 4.5 kPa, and the measured angle of shear resistance is recorded to be 32° while the simulated one is 25°.

Figure 2.30a illustrates the variation of the sinkage as a function of the pressure applied to the plate for different moisture contents. At 0% moisture content, the sand

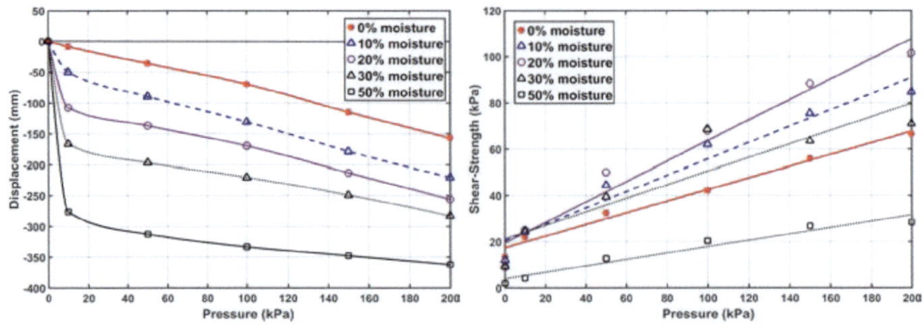

(a) Pressure-sinkage (b) Shear-strength

Fig. 2.30 Pressure-sinkage and shear-strength relationship for moist sand at different moisture contents

is fully dry and the box is filled with several layers of sand particles, and the maximum sinkage recorded is 157 mm at 200 kPa plate pressure. It can be deduced that as the moisture content increases, the sinkage of the plate increases as well causing the soil to be diluted. Nevertheless, the increase in sinkage is not linearly dependent on the increase in moisture content.

Figure 2.30a illustrates the variation of the shear-strength as a function of pressure for different moisture contents. The soil cohesion comprehended as the component of shear-strength of soil that is independent of inter-particle friction is regarded as the y-intercept in the shear-pressure line, and the slope of the shear-pressure line is known to be the internal shear friction of the soil. It is observed that as the moisture content increases, the cohesion of the soil reduces, as does the internal shear friction. In a similar methodology, different moist sands are modeled depending on the desired moisture content. These results are in good agreement with the results obtained from shear-strength physical testing.

Figure 2.30b illustrates the variation of the soil sinkage obtained from simulation at 100 and 200 kPa pressure as a function of soil moisture content, and these results are in good agreement with previously measured data [37]. It is observed that as the moisture content increases, the sinkage increases for a constant pressure (Fig. 2.31).

It can also be seen that the rate of increase of sinkage as a function of that of moisture is almost linear for constant pressure. It should be noted that the sand at 0% moisture content is the dry sand that has been previously modeled and calibrated [27] and has a plate sinkage of 157 mm at 200 kPa pressure.

Figure 2.32a illustrates the angle of shear resistance as a function of moisture content for both measurement and simulation. It is observed that although the measurement and simulation results are not equal, they exhibit a similar trend. As the moisture content increases, the angle of shear resistance increases until a peak is achieved at around 25% moisture. After the peak, the angle of shear resistance reduces as the moisture content continues to increase. It should be noted that the sand tested has similar characteristics to that simulated; however, it is slightly different, due to the limitations in the available sand for testing, this sand type was selected.

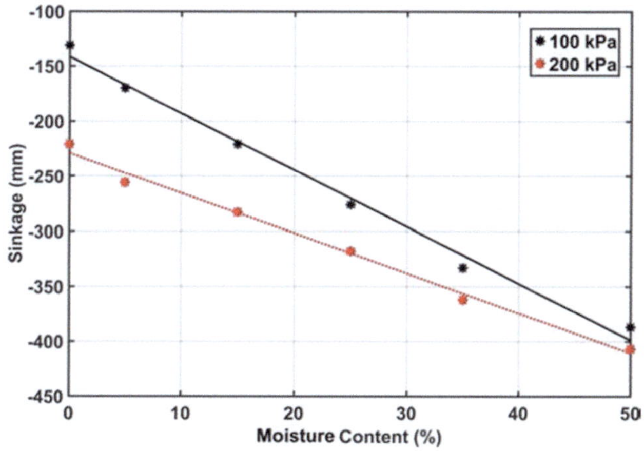

Fig. 2.31 Sample of simulation results for pressure-sinkage characteristics at different moisture content and applied pressure

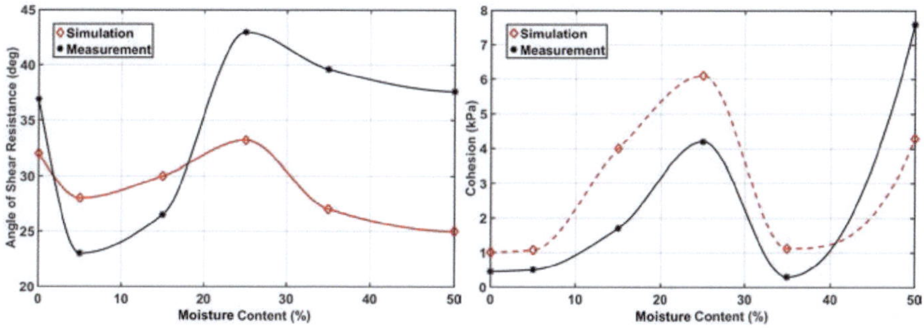

(a) Angle of shear resistance

(b) Cohesion

Fig. 2.32 Shearing characteristics at different moisture contents for simulation and measurement

Both the measurement and simulated results indicate a similar trend in variation concerning moisture content and show a peak at the same moisture content. Further, the angle of shear resistance varies between 42° and 23° for measurements, while for angle varies between 34° and 25° for simulations. It is noticed however that the rate at which the angle of shear resistance varies with respect to the moisture content is almost the same for both measurement and simulation.

Figure 2.32b illustrates the variation of the soil cohesion as a function of moisture content for both measurements and simulations. It is noticed that as the moisture content increases, the soil cohesion increases until a peak is reached at a moisture content of 25%. Afterward, as the moisture content increases, the cohesion exhibits a parabolic relationship. The cohesion ranges between 0.45 and 7.6 kPa for a moisture content ranging between 0 and 50%.

The sand behavior is attributed to the role of water in soil particle arrangement and the soil fabric. As water content increases from 5 to 25%, water may have acted as a lubricant between the particles leading to denser soil and more compact fabric. For higher water contents, the water-filled pore voids of the soil lead to the slightly loose fabric.

Problems

1. The contact patch of an un-treaded tire on hard and dry soil may be approximated by a circular area of radius of 28 cm. The contact pressure is assumed to be a uniform 380 kPa. For this type of soil, the concentration factor v is assumed to be 5. Calculate the vertical stress a, in the soil at a depth of 20 cm directly below the center of the contact area.
2. The contact area of a truck tire on a fairly hard and dry soil may be approximated by a circle having a radius of 30 cm. The contact pressure is assumed to be a uniform 750 kPa. For this type of soil, the concentration factor v is assumed to be 3. Calculate the resultant vertical stress a, in the soil at depths of 20 and 40 cm below the center of the contact area. At what depth below the center is the vertical stress one-tenth of the contact pressure?
3. Given a circular plate with radius of 5 cm placed on top of sandy terrain with terrain values of $k_c = 16.4\,\mathrm{kN/m}^{n+1}$, $k_\phi = 1725\,\mathrm{kN/m}^{n+2}$, and $n = 0.2$. Find the plate sinkage at a pressure of 300 kPa.
4. Find the shear-strength equation for a sandy terrain with the cohesion of 13.8 kPa and an angle of shear resistance of 22°. What is the shear stress, τ, when a square box of sandy terrain with a side of 600 mm is subjected to a pressure force of 50 kN?
5. Assuming an elastic-plastic sandy terrain, where c_2 to c_6 are neglected. Calculate the equation of state parameters c_0 and c_1, knowing that at an initial pressure of 10 kPa the density ratio is 1 and at a pressure of 100 kPa the density is increased by 10%.
6. Given that the total wet volume of moist sandy soil is 250 mm^3, find the water volume to obtain a water content percentage of 10, 20, and 50%.

References

1. Bekker MG (1956) Theory of land locomotion
2. Onafeko O (1964) Instrumentation for measuring radial and tangential stresses beneath rigid wheels. J Terrramech 1(3):61–68
3. Onafeko O, Reece AR (1967) Soil stresses and deformations beneath rigid wheels. J Terrramech 4(1):59–80
4. Wong J-Y, Reece AR (1967) Prediction of rigid wheel performance based on the analysis of soil-wheel stresses part i. Performance of driven rigid wheels. J Terramech 4(1):81–98

5. Wong J-Y, Reece AR (1967) Prediction of rigid wheel performance based on the analysis of soil-wheel stresses: part ii. Performance of towed rigid wheels. J Terramech 4(2):7–25
6. Wong JY (2009) Terramechanics and off-road vehicle engineering: terrain behaviour, off-road vehicle performance and design. Butterworth-Heinemann
7. Shokouhfar S (2017) A virtual test platform for analyses of rolling tyres on rigid and deformable terrains. PhD thesis, Concordia University
8. Chu L-M, Yin J-H (2005) Comparison of interface shear strength of soil nails measured by both direct shear box tests and pullout tests. J Geotech Geoenviron Eng 131(9):1097–1107
9. Wong JY (2008) Theory of ground vehicles. Wiley
10. Hiroma T, Wanjii S, Kataoka T, Ota Y (1997) Stress analysis using fem on stress distribution under a wheel considering friction with adhesion between a wheel and soil. J Terrramech 34(4):225–233
11. Shoop S, Kestler K, Haehnel R (2006) Finite element modeling of tires on snow. Tire Sci Technol 34(1):2–37
12. Hambleton J, Drescher A (2008) Development of improved test rolling methods for roadway
13. Slade JL (2009) Development of a new off-road rigid ring model for truck tires using finite element analysis techniques. Master's thesis, The Pennsylvania State University
14. Lescoe R (2010) Improvement of soil modeling in a tire-soil interaction using finite element analysis and smooth particle hydrodynamics. Master's thesis, The Pennsylvania State University
15. Soil traingle, the idaho association of soil conservation districts. http://www.oneplan.org/Water/soil-triangle.asp. Accessed: 11 June 2017
16. Dhillon RS (2013) Development of truck tire-terrain finite element analysis models. Master's thesis, University of Ontario Institute of Technology, Canada
17. Marjani M (2016) Development of fea wide-base truck tire and soil interaction models. Master's thesis, University of Ontario Institute of Technology, Canada
18. Bui HH, Fukagawa R, Sako K, Ohno S (2008) Lagrangian meshfree particles method (sph) for large deformation and failure flows of geomaterial using elastic-plastic soil constitutive model. Int J Numer Anal Methods Geomech 32(12):1537–1570
19. Gingold RA, Monaghan JJ (1977) Smoothed particle hydrodynamics: theory and application to non-spherical stars. Mon Notices R Astron Soc 181(3):375–389
20. Libersky LD, Petschek AG, Carney TC, Hipp JR, Allahdadi FA (1993) High strain lagrangian hydrodynamics: a three-dimensional sph code for dynamic material response. J Comput Phys 109(1):67–75
21. Monaghan JJ (1994) Simulating free surface flows with sph. J Comput Phys 110(2):399–406
22. Bui HH, Sako K, Fukagawa R (2007) Numerical simulation of soil-water interaction using smoothed particle hydrodynamics (sph) method. J Terramech 44(5):339–346
23. Monaghan JJ, Lattanzio JC (1985) A refined particle method for astrophysical problems. Astron Astrophys 149:135–143
24. Monaghan JJ (1992) Smoothed particle hydrodynamics. Ann Rev Astron Astrophys 30(1):543–574
25. El-Sayegh Z, El-Gindy M, Johansson I, Öijer F (2017) Truck tire-terrain interaction modelling and testing: literature survey. Int J Veh Syst Model Testing 12(3/4):163–216
26. PAM System International (2014) Pam-crash user manual version 2014. ESI Group
27. El-Sayegh Z, El-Gindy M, Johansson I, Öijer F (2018) Improved tire-soil interaction model using fea-sph simulation. J Terrramech 78C:53–62
28. Biot MA (1956) Theory of propagation of elastic waves in a fluid-saturated porous solid. ii. Higher frequency range. J Acoust Soc Am 28(2):179–191
29. Peck RB, Hanson WE, Thornburn TH (1974) Foundation engineering, vol 10. Wiley, New York
30. Bekker MG (1969) Introduction to terrain-vehicle systems. part i: The terrain. part ii: the vehicle. Technical report, DTIC document
31. Hettiaratchi DRP, Reece AR (1974) The calculation of passive soil resistance. Geotechnique 24(3):289–310

32. El-Sayegh Z, El-Gindy M (2019) Modelling and prediction of tyre-snow interaction using finite element analysis-smoothed particle hydrodynamics techniques. Proc Inst Mech Eng, Part D: J Automob Eng 233(7):1783–1792
33. Robinson DA (2009) Field estimation of soil water content: a practical guide to methods, instrumentation and sensor technology. Soil Sci Soc Am J 73(4):1437
34. AS ASTM (2011) Standard practice for classification of soils for engineering purposes
35. ASTM Committee D-18 on Soil (2007) Plasticity rock. Subcommittee D18. 03 on texture, and density characteristics of soils. Standard test method for particle-size analysis of soils. ASTM International
36. El-Sayegh Z, El-Gindy M, Johansson I, Öijer F (2018) Off-road soft terrain modeling using smoothed particle hydrodynamics technique. 15th international conference on design education, international design engineering technical conferences, IDETC. Quebec City, Quebec, Canada, pp 2018–85005
37. Abu-Farsakh M, Coronel J, Tao M (2007) Effect of soil moisture content and dry density on cohesive soil-geosynthetic interactions using large direct shear tests. J Mater Civ Eng 19(7):540–549

Chapter 3
Performance Characteristics of Off-Road Vehicles

This chapter focuses on the performance characteristics of off-road vehicles by investigating the terrain-dependent in-plane and out-of-plane rigid ring model parameters at different conditions. The in-plane rigid ring model parameters are the longitudinal tire stiffness, vertical stiffness, and rolling resistance coefficient. The out-of-plane rigid-ring model parameters include the lateral stiffness, cornering stiffness, self-aligning moment stiffness, and relaxation length.

The deformation in the soil after the tire passes is permanent, due to compression. Depending on the soil pressure-sinkage and shear-strength characteristics, the tire may sink deeply in the soil and more bulldozing effect is created; this is usually the case of clayey soil.

Furthermore, the flow patterns and soil wave under a truck tire in different conditions are demonstrated in Fig. 3.1. Generally, there exist two zones of soil under a rolling tire: in the first zone the soil flows forward and in the second zone, the soil flows backward. It should be noted that the flow patterns under a tire in the longitudinal plane depend on multiple factors including the tire slip, vertical load, and inflation pressure.

Figure 3.1a describes the case when the tire is spinning, and the longitudinal velocity is zero (the tire is not moving in the longitudinal direction). In this case, the longitudinal slip is 100% and the soil flows in a single direction backward. Moreover, if the tire is locked as shown in Fig. 3.1b, here the tire does not rotate but rather moves with a constant longitudinal speed. The soil, in this case, moves in a forward zone only and forms a shape similar to the wedge shape. The tire-terrain interaction in this case behaves like a bulldozing blade.

Figure 3.1c shows a driven tire over clayey soil at 586 kPa inflation pressure and 27 kN vertical load. The logarithmic spiral AB is clearly observed in the zone where the soil moves forward, and the logarithmic spiral AC is clear in the backward zone. Moreover, Fig. 3.1d shows a towed tire over clayey soil at 586 kPa inflation pressure and 27 kN vertical load. The shear direction under the tire and in the soil is noticed to have two flow zones and intersects at point A.

M. El-Gindy and Z. El-Sayegh, *Road and Off-Road Vehicle Dynamics*,
https://doi.org/10.1007/978-3-031-36216-3_3

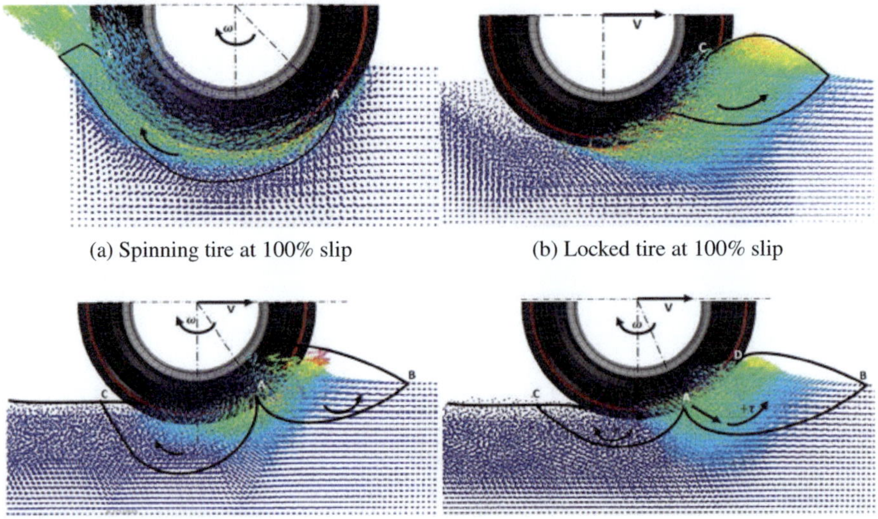

(a) Spinning tire at 100% slip (b) Locked tire at 100% slip

(c) Driven tire (d) Towed tire

Fig. 3.1 Flow patterns and soil wave for different driving conditions under a tire at 586 kPa inflation pressure and 27 kN vertical load [1]

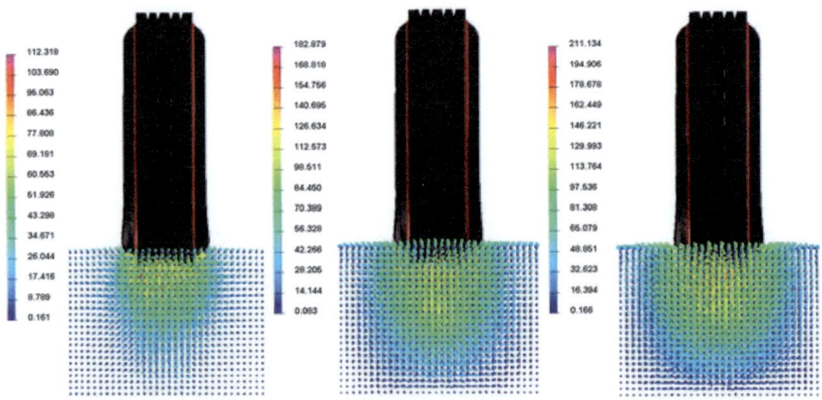

(a) 25% moisture content (b) 50% moisture content (c) 62% moisture content

Fig. 3.2 Tire sandy loam interaction at different moisture contents and 586 kPa inflation pressure and 27 kN vertical load

Figure 3.2 demonstrates the tire-sandy loam interaction at different moisture contents. The three simulations are performed at 586 kPa inflation pressure and 27 kN vertical load. The contour shown in this figure is for the normal displacement; it is observed that the sandy loam with 62% moisture content had the highest recorded displacement which is 211 mm. This indicates that the soil with higher moisture content has a higher sinkage as well. The soil in the bottom of the box of 25% moisture content shows a very low displacement and is almost not compacted which is the opposite of what is seen in the case of 62% moisture where almost all the soil region is compacted.

It should be noted that for a tire with lugs, such as tractor tires, the pressure distribution and vertical displacement over the contract area differ from that of a passenger tire. When the tire with lugs runs over hard, dry soil, the lugs carry the entire load. The pressure over the contact area of the lugs is three to four times higher than that of an equivalent tire without lugs.

3.1 Mechanics of Tire-Terrain Interaction

Figure 3.3 shows the possible active and passive failures for terrains. Both active and passive terrain failures can be quantified using Mohr circles shown in Fig. 3.3.

For the active failure, the normal stress, σ, on the base of the element at depth z can be found by

$$\sigma = \gamma_s z \tag{3.1}$$

where γ_s is the unit weight of the terrain, and σ is considered the major principal stress. The circle C_a represents the state of stress at that point. It is shown from Fig. 3.3 that the lines OM and OM_1 are tangent to the circle and represent the Mohr-Coulomb failure criterion. The minor principal stress can be determined as the point of intersection between the circle and the horizontal axis. The minor principal stress, σ_a, is referred to as the normal stress required to bring the mass at that point into active failure. It should be noted that σ_a is sometimes referred to as the active earth pressure which is given by [2]

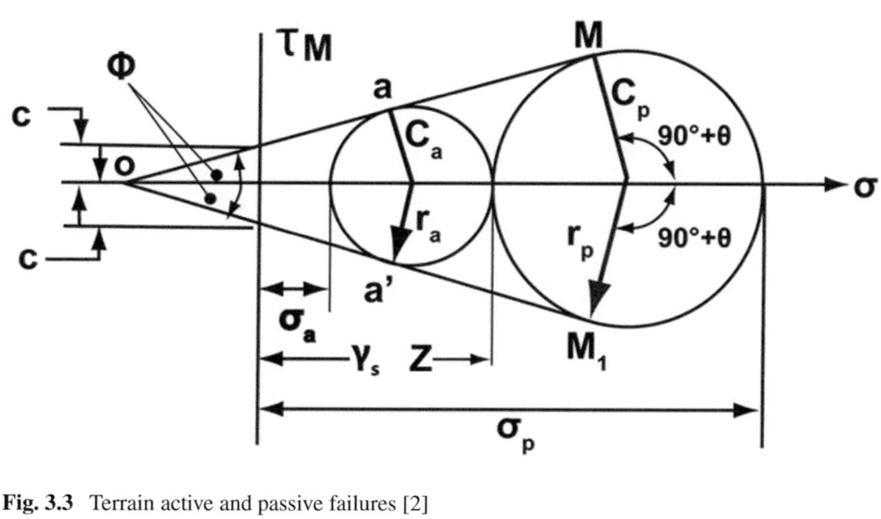

Fig. 3.3 Terrain active and passive failures [2]

$$\sigma_a = \gamma_s z - 2r_a \tag{3.2}$$

where r_a is the radius of circle C_a, and it can be expressed as

$$r_a = \frac{1}{1 - \sin\theta}(c\cos\phi + \sigma_a \sin\phi) \tag{3.3}$$

Substituting r_a into σ_a gives

$$\sigma_a = \gamma_s z - \frac{2}{1 - \sin\theta}(c\cos\phi + \sigma_a \sin\phi) \tag{3.4}$$

Equation 3.4 can be simplified as

$$\begin{aligned}
\sigma_a &= \frac{\gamma_s z}{(1 + \sin\theta)/(1 - \sin\theta)} - \frac{2c\cos\phi/(1 - \sin\theta)}{(1 + \sin\theta)/(1 - \sin\theta)} && (3.5)\\
&= \frac{\gamma_s z}{\tan^2 45^\circ + \phi/2} - \frac{2c\tan 45^\circ + \phi/2}{\tan^2 45^\circ + \phi/2} && (3.6)\\
&= \gamma_s z\frac{1}{N_\phi} - 2c\frac{1}{\sqrt{N_\phi}} && (3.7)
\end{aligned}$$

where $N_\phi = \tan^2 45^\circ + \phi/2$ and is referred to as the flow value.

On the other side, the passive failure is caused by the lateral compression; the normal stress, σ, acting on the bottom of the element is defined as the minor principal stress. The circle C_p shown in Fig. 3.3 thus represents the stress condition of a point in the state of incipient passive failure. The intersection between circle C_b and the horizontal axis is referred to as the passive earth pressure, σ_p, and it can be written as

$$\sigma_p = \gamma_s z + 2r_p \tag{3.8}$$

where r_p is the radius of circle C_p and can be expressed by

$$r_p = \frac{1}{\sin\phi}(c\cos\phi + \gamma_s z \sin\phi) \tag{3.9}$$

The passive earth pressure, σ_p, can then be written as

$$\begin{aligned}
\sigma_p &= \gamma_s z\frac{1 + \sin\phi}{1 - \sin\phi} + 2c\frac{\cos\phi}{1 - \sin\phi} && (3.10)\\
&= \gamma_s z\tan^2 45^\circ + \phi/2 + 2c\tan 45^\circ + \phi/2 && (3.11)\\
&= \gamma_s z N_\phi + 2c\sqrt{N_\phi} && (3.12)
\end{aligned}$$

Furthermore, if pressure q is applied to the terrain surface, then the normal stress at the base element at a depth z is defined as

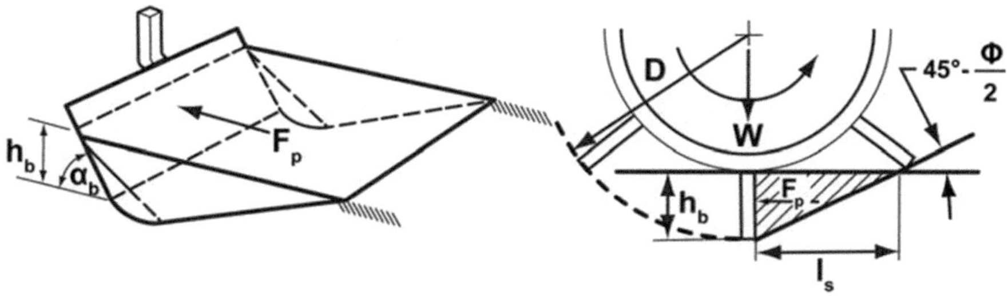

Fig. 3.4 Terrain active and passive failures [2]

$$\sigma = \gamma_s z + q \tag{3.13}$$

The pressure q is referred to as the surcharge, and the active pressure is given by

$$\sigma_a = \gamma_s z \frac{1}{N_\phi} - 2c \frac{1}{\sqrt{N_\phi}} + q \frac{1}{N_\phi} \tag{3.14}$$

while the passive pressure can be written as

$$\sigma_p = \gamma_s z N_\phi + 2c\sqrt{N_\phi} + qN_\phi \tag{3.15}$$

The passive failure of the terrain may be caused by the vehicle running the gear and other terrain engaging devices. The bulldozer effect can also be modeled using the passive failure theory. Consider a vertical cutting bulldozer blade that is being pushed against the terrain as shown in Fig. 3.4.

The terrain in the front will experience a passive failure, and the following should be noted:

- If the ratio of the blade width to the cutting depth is large, then the problem may be assumed as two-dimensional.
- If the blade is vertical and smooth, then the normal pressure exerted by the blade on the terrain will be the major principal stress and will be equal to the passive earth pressure.

If there is no surcharge ($q = 0$), then the resultant force acting on the blade per unit width can be calculated as [2]

$$F_p = \int_0^{h_b} \sigma_p dz \tag{3.16}$$

$$= \frac{1}{2}\gamma_s h_b^2 N_\phi + 2ch_b\sqrt{N_\phi} \tag{3.17}$$

In case there is a charge q, then the resultant force acting on the blade per unit width can be calculated by

$$F_p = \int_0^{h_b} \sigma_p dz \tag{3.18}$$

$$= \frac{1}{2}\gamma_s h_b^2 N_\phi + 2ch_b\sqrt{N_\phi} + ah_b N_\phi \tag{3.19}$$

Various limitations exist on the application of the earth pressure theory to the solution of practical problems. For example, the surfaces of bulldozer blades or lugs are usually not smooth, as assumed in the simple theory. It has been found that the angle of soil-metal friction δ may vary from 11° for a highly polished, chromium-plated steel with dry sand to almost equal to the angle of internal shearing resistance of the soil for very rough steel surfaces [3].

Shear stresses on the terrain-blade interface will form due to the existence of friction and/or adhesion between the soil and the blade surface. This phenomenon is noticed when the terrain adjoining the blade is brought into a state of plastic equilibrium. Therefore, the normal pressure on the contact surface will no longer be the principal stress, and the failure pattern of the soil mass will be as shown in Fig. 3.5.

The soil mass in zone ABC is in the Rankine passive state, which is characterized by straight slip lines inclined to the horizontal at an angle of $45° - \phi/2$. Zone ABD adjacent to the blade is characterized by curved and radial slip lines, and is usually called the radial shear zone. The shape of the curved slip lines, such as DB in

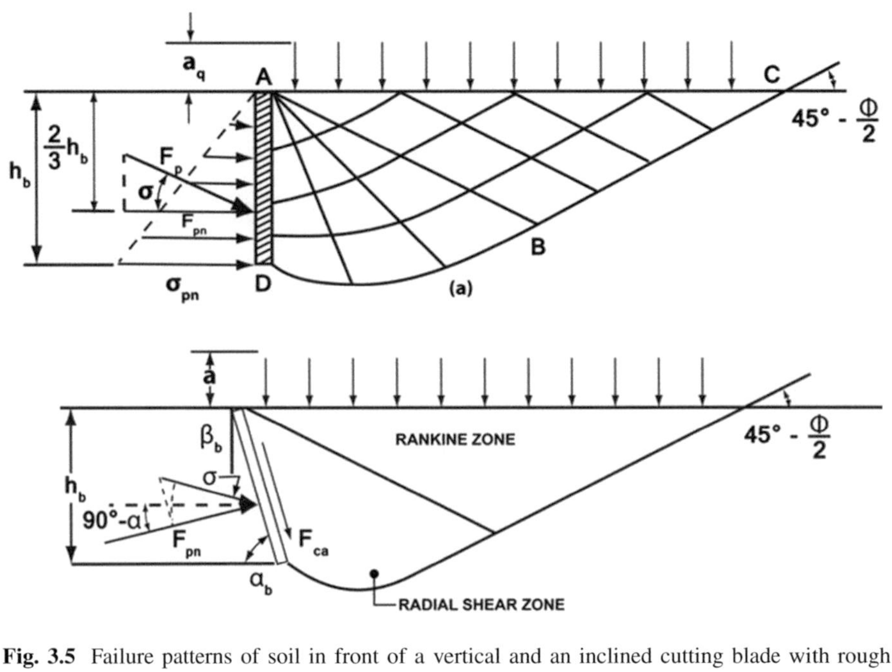

Fig. 3.5 Failure patterns of soil in front of a vertical and an inclined cutting blade with rough surface [2]

Fig. 3.5, can be considered, with sufficient accuracy, as being either a logarithmic spiral (for materials with frictional property) or an arc of a circle (for cohesive materials).

Using Fig. 3.5, the normal component of the passive earth pressure, σ_{pn}, acting on the vertical rough blade at a depth z below A is estimated by the linear equation

$$\sigma_{pn} = \gamma_s z K_{p\gamma} + q K_{pq} + c K_{pc} \tag{3.20}$$

where q is the previously defined surcharge, and $K_{p\gamma}$, K_{pq}, and K_{pc} are constant parameters (function of the terrain internal friction angle, and friction between the soil and the blade). These parameters can be estimated by various methods, including the logarithmic spiral method and the friction circle method [4, 5].

The resultant force F_p will be at an angle δ to the normal of the blade, which is equal to the terrain-metal friction angle, as shown in Fig. 3.5. If the blade is inclined with an angle, α_b, with the horizontal as shown in Fig. 3.5, then the resultant force F_{pn} can be expressed by

$$F_{pn} = \frac{1}{\sin\alpha_b}\int_0^{h_b} \left(\gamma_s z K_{p\gamma} + q K_{pq} + c K_{pc}\right) dz \tag{3.21}$$

$$= \frac{1}{2}\gamma_s h_b^2 \frac{K_{p\gamma}}{\sin\alpha_b} + \frac{h_b}{\sin\alpha_b}\left(q K_{pq} + c K_{pc}\right) \tag{3.22}$$

Combining the normal force with the frictional component ($F_{pn} \tan\delta$), the resultant force becomes

$$F_p = \frac{F_{pn}}{\cos\delta} = \frac{1}{2}\gamma_s h_b^2 \frac{K_{p\gamma}}{\sin\alpha_b \cos\delta} + \frac{h_b}{\sin\alpha_b \cos\delta}\left(q K_{pq} + c K_{pc}\right) \tag{3.23}$$

In case of adhesion, c_a, between the terrain and the surface of the blade, the adhesion force, F_{ca}, is expressed by

$$F_{ca} = \frac{h_b}{\sin\alpha_b} c_a \tag{3.24}$$

? Example 3.1

In order to increase the traction of a wheeled vehicle, a device with 10 lugs has been attached to a narrow rim. The outside diameter of the device measured from the lug tip is 1.72 m. The lugs are 30 cm wide and penetrate vertically 20 cm into the ground. Estimate the tractive effort per lug developed in the vertical position in clayey soil with a cohesion of 20 kPa and angle of shear resistance of 10°. Note that $\gamma_s = 15.7$ kN/m^3, assume that the lug is smooth, and neglect the friction and adhesion between the lug and clayey soil.

Solution. The distance between two lugs can be calculated using the perimeter of the tire and it is equal to 54 cm. The rapture distance l_s with a penetration of 20 cm is calculated by

$$l_s = \frac{h_b}{\tan(45^\circ - \phi/2)} \tag{3.25}$$

$$= \frac{20}{\tan(45 - 10)} = 28.56 \text{ cm} \tag{3.26}$$

The horizontal force F_p acting on the lug in the vertical position is calculated using Eq. 3.19, and the width of the lug, b, as

$$F_p = b\left(\frac{1}{2}\gamma_s h_b^2 N_\phi + 2ch_b\sqrt{N_\phi} + ah_b N_\phi\right) \tag{3.27}$$

$$= 0.3\left(0.5 \times 15.7 \times .2^2 \times 2.04 + 2 * 20 \times 0.2 \times 1.43\right) \tag{3.28}$$

$$= 3.6 \text{ kN} \tag{3.29}$$

It should be noted that the penetration of the lug as well as the inclination changes when the wheel is rotating. Thus, the tractive effort developed by the lug varies with its angular position. Since more than one lug is in contact with the terrain, the total tractive effort that the device can develop is the sum of the horizontal forces acting on all of the lugs in contact with the ground.

3.2 Free Rolling Performance

The rolling resistance of a pneumatic tire depends on its mode of operation. The rigid mode of operation is referred to when the tire remains round like a rigid rim. This happens when the sum of the inflation pressure, p_i, and the pressure produced by the stiffness of the carcass, p_c, are greater than the maximum pressure that the terrain can support at the lowest point of the tire circumference. If the terrain is hard and the inflation pressure is low, a portion of the circumference of the tire will be flattened; this is referred to as elastic mode of operation.

Figure 3.6 shows the behavior of a pneumatic tire in different operating modes. It is vital to start by determining if the tire is operating in rigid rim or elastic wheel mode under a given operating condition. It the tire operates like a rigid rim, then Bekker's pressure-sinkage relationship should be utilized to determine the normal pressure at the lowest contact point, p_g, as follows:

$$p_g = \left(\frac{k_c}{b} + k_\phi\right) z_0^n \tag{3.30}$$

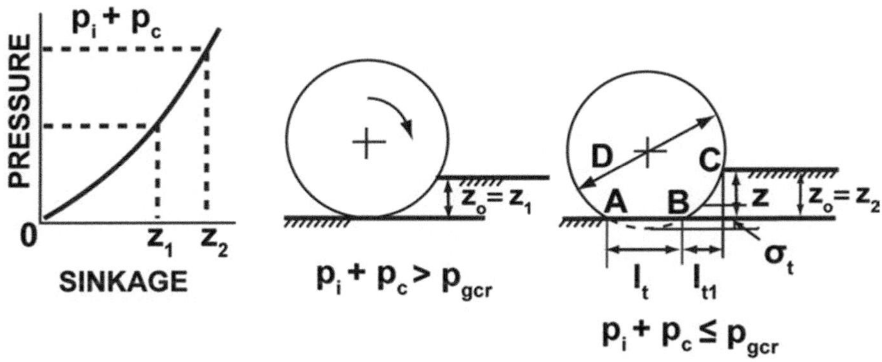

Fig. 3.6 Behavior of a pneumatic tire in different operating modes [2]

where b is the width of the equivalent plate, and z_0 is the depth as shown in Fig. 3.6. It should be noted that the lowest contact point is defined at the bottom dead center, and the pressure p_g can be expressed as

$$p_g = \left(\frac{k_c}{b} + k_\phi\right)^{1/(2n+1)} \left[\frac{3W}{(3-n)b\sqrt{D}}\right]^{2n/(2n+1)} \tag{3.31}$$

where W is the vertical load, and D is the tire diameter. The pressure, p_g, may also be defined as the critical pressure, p_{gcr}. If the sum of the inflation pressure and the pressure due to the carcass is greater than the critical pressure, then the tire remains in the rigid rim mode, and the tire sinkage can be calculated by

$$z_0 = \left(\frac{3W}{b(3-n)\left(\frac{k_c}{b} + k_\phi\right)\sqrt{D}}\right)^{\frac{2}{2n+1}} \tag{3.32}$$

And the rolling resistance due to terrain compaction can be determined using

$$R_c = \frac{1}{A(n+1)b^{1/(2n+1)}} \left[\frac{3W}{\sqrt{D}}\right]^{(2n+2)/(2n+1)} \tag{3.33}$$

where A is defined as

$$A = (3-n)^{(2n+2)/(2n+1)} \left(\frac{k_c}{b} + k_\phi\right)^{1/(2n+1)} \tag{3.34}$$

If the critical pressure is greater than the sum of the inflation pressure and the pressure due to the carcass, then the tire is operating in the elastic mode. In this case, the contact pressure will be equal to $p_i + p_c$, since the tire is flattened. The tire sinkage, z_0, can be determined by modifying Bekker's pressure-sinkage equation:

$$z_0 = \left(\frac{p_i + p_c}{\frac{k_c}{b} + k_\phi} \right)^{1/n} \tag{3.35}$$

The rolling resistance of an elastic tire due to terrain compaction can then be expressed as

$$R_c = \frac{b(p_i + p_c)^{(n+1)/n}}{(n+1)(\frac{k_c}{b} + k_\phi)^{1/n}} \tag{3.36}$$

The pressure due to the carcass stiffness depends on the inflation pressure and normal load, thus it is difficult to determine. It is proposed to use the average ground pressure, p_{gr}, of the tire on the hard ground to represent $p_i + p_c$. Implementing the average ground pressure in Eqs. 3.35 and 3.36 leads to

$$z_0 = \left(\frac{p_{gr}}{\frac{k_c}{b} + k_\phi} \right)^{1/n} \tag{3.37}$$

$$R_c = \frac{b p_{gr}^{(n+1)/n}}{(n+1)(\frac{k_c}{b} + k_\phi)^{1/n}} \tag{3.38}$$

Using Fig. 3.6, the contact length, l_t, can be expressed as a function of the tire deflection, δ_t,

$$l_t = 2\sqrt{D\delta_t - \delta_t^2} \tag{3.39}$$

The contact length should be used to replace b, when its value is less than the width of the tire:

$$z_0 = \left(\frac{p_{gr}}{\frac{k_c}{l_t} + k_\phi} \right)^{1/n} \tag{3.40}$$

The normal load on portion AB, W_{AB}, can be determined using the group pressure, p_{gr}, on the flat portion. On the other side, the normal load on portion BC, W_{BC}, can be determined by assuming a circular approximation with radius $r = D/2$:

$$W_{BC} = -b \int_0^{z_0} p dx = -b \left(\frac{k_c}{l_t} + k_\phi \right) \int_0^{z_0} \frac{z^n \sqrt{D} dz}{2\sqrt{z_0 + \delta_t - z}} \tag{3.41}$$

Using the u-substitution method to solve the integral, let $t^2 = z_0 + \delta_t - z$ then $2tdt = -dz$ and

$$W_{BC} = -b \left(\frac{k_c}{l_t} + k_\phi \right) \sqrt{D} \int_{\sqrt{\delta_t}}^{\sqrt{z_0 + \delta_t}} \left(z_0 + \delta_t - t^2 \right)^n dt \tag{3.42}$$

Using series expansion and taking the first two terms only, the equation can be rewritten and solved:

$$W_{BC} = \frac{b}{3}\left(\frac{k_c}{l_t} + k_\phi\right)\sqrt{D}(z_0 + \delta_t)^{n-1}\left[(3-n)(z_0+\delta_t)^{3/2} - (3-n)\delta_t^{3/2} - 3z_0\sqrt{\delta_t}\right] \tag{3.43}$$

The normal force acting on the tire can then be expressed as

$$W = bp_{gr}l_t + W_{BC} \tag{3.44}$$

The normal force, W, for a given tire is a function of the deflection, contact length, and tire sinkage. It should be noted that that it is recommended to utilize an iterative procedure to determine the value of the tire deflection. The iterative procedure requires the assumption of a first value for δ_t and substitution in Eq. 3.39 to calculate the contact length, l_t. Then, utilize Eq. 3.40 to calculate the tire sinkage, z_0, and Eq. 3.44 to calculate the normal force on the tire. If the assumed value for the tire deflection, δ_t, is correct, then the calculated normal force and the given normal load should be equal. Otherwise, a new value for δ_t should be assumed and the whole process should be reported until convergence is achieved.

After the correct value of δ_t is obtained and the appropriate contact length and tire sinkage are calculated, the compaction resistance can be calculated by

$$R_c = b\left(\frac{k_c}{l_t} + k_\phi\right)\left(\frac{z_0^{n+1}}{n+1}\right) \tag{3.45}$$

In addition to the compaction resistance, the tire also experiences energy dissipation in the force of hysteresis and other internal losses. These losses appear as a resisting force acting on the tire. The resistance force is due to the tire deformation which depends on the tire design, construction, and material, in addition to the operating conditions. The resistance force is generally determined experimentally, however, Bekker proposed an equation to predict the rolling resistance due to the tire deformation [6]:

$$R_h = \frac{3.581bD^3 p_{gr}\epsilon(0.0349\alpha - \sin 2\alpha)}{\alpha(D - 2\delta_t)} \tag{3.46}$$

The parameter α is defined as the contact angle in degrees and is calculated as follows:

$$\alpha = \cos^{-1}\frac{D - 2\delta_t}{D} \tag{3.47}$$

The parameter ϵ is calculated using the tire section height, h, and a parameter related to the tire construction, k_e:

$$\epsilon = 1 - e^{\frac{-k_e\delta_t}{h}} \tag{3.48}$$

The value of k_e is 15 for bias-ply tires and 7 for radial-ply tires. It should be noted that if the tire experiences significant sinkage, the bulldozing resistance has to be taken into consideration when calculating the total rolling resistance of the tire.

The compaction resistance can be computed using various simulation models. The simulation procedure includes inflating the tire to the desired inflation pressure, and then the tire is loaded on the terrain by applying the desired normal load. After allowing the tire to settle, a constant linear longitudinal velocity is applied at the center of the tire. The simulation runs until the vertical and longitudinal contact forces reach a steady state. It should be noted that constant longitudinal velocity is applied at the center of the tire, however, the angular velocity of the free rolling tire may change depending on the terramechanics terrain properties, inflation pressure, and applied vertical load. The change in angular velocity even without applying driving torque is due to the sinkage of the tire in the water while it's towed with constant linear longitudinal velocity.

Figure 3.7a shows the variation of rolling resistance coefficient as a function of applied vertical load for different soils and hard surface at a rated inflation pressure of 586 kPa. The clayey soil has the highest rolling resistance coefficient which ranges from 0.30 to 0.64 for a vertical load of 13–40 kN, respectively, while the hard surface had the lowest rolling resistance coefficient which varies between 0.008 and 0.004 for the given vertical load range.

The rolling resistance coefficient over soft terrain including dry sand, dense sand, and clayey soil adhere to the same behavior, as the vertical load increases the rolling resistance increases. This is due to the fact that as the vertical load increases, the tire tends to sink more into the soil which leads to a larger contact area and thus more forces which increase the rolling resistance coefficient. The rolling resistance coefficient over dry and dense sand is relatively in agreement regarding variation as a function of the vertical load as both adhere to almost a parallel behavior.

Figure 3.7a also shows that the rolling resistance coefficient over a hard surface is remarkably lower than that of soft terrains. The rolling resistance coefficient over a

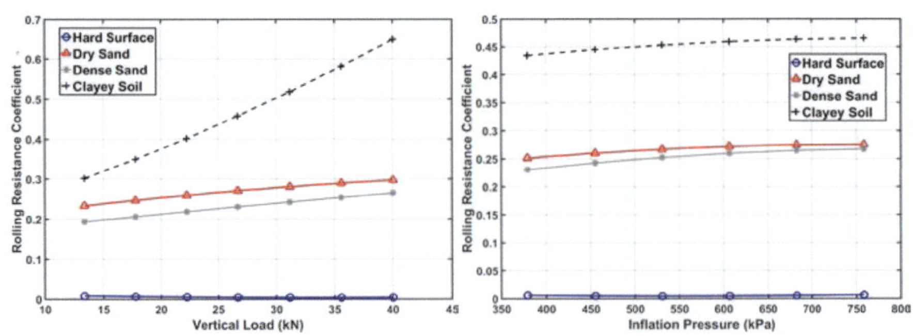

(a) Effect of vertical load (b) Effect of inflation pressure

Fig. 3.7 Rolling resistance coefficient as a function of vertical load and inflation pressure over different soils [7]

hard surface increases as the vertical load increases, which is similar to the case of soft terrains. On a hard surface, as the vertical load increases, the contact area between the tire and surface increases which leads to an increase in the rolling resistance.

It is concluded that when the truck is driving over clayey soil, the rolling resistance is highest and thus fuel consumption is higher than on sand. Moreover, for the case of dry and dense sand, dry sand has a higher rolling resistance than the dense one. This is due to the fact that the dry sand has a higher sinkage at the same applied pressure.

Figure 3.7b shows the variation of the rolling resistance coefficient as a function of tire inflation pressure for different soils and hard surfaces at a vertical load of 27 kN. Similar to the previous results, the clayey soil has the highest rolling resistance coefficient varying between 0.43 and 0.46 for an inflation pressure range of 379–758 kPa.

In addition, dry and dense sand adhere to similar behavior, and hard surfaces record the lowest rolling resistance coefficient. Despite the fact that the clayey soil has less sinkage than that of dry sand, the resistance force is higher due to the higher density of the clayey soil.

It is noted that in this case, unlike the soft terrains, the hard surface has an opposite behavior. On a hard surface, as the inflation pressure increases, the rolling resistance coefficient reduces. The increase in inflation pressure reduces the contact area between the tire and hard surface which leads to a reduction in the rolling resistance coefficient.

? Example 3.2

A pneumatic truck tire 315/80R22.5 is to be installed on an off-road truck. The tire has a radius of 546 mm, a section height of 266 mm, and a width of 315 mm. The tire is to carry a load of 41 kN, and the vehicle is to operate on a soil with pressure-sinkage parameters $n = 1$ and $k_\phi = 680$ kN/m^3. The vehicle analyst is proposing two inflation pressures 300 and 500 kPa. The average ground pressure at the given load and 300 kPa inflation pressure is 200 kPa, while that of 500 kPa inflation pressure is 300 kPa. Compare the sinkage and compaction resistance of the truck tire at the proposed inflation pressures.

Solution. For the given soil conditions, the critical pressure, p_{gcr}, for the truck tire is

$$
\begin{aligned}
p_{gcr} &= 680^{1/3}\left(\frac{3 \times 41}{(3-1)0.315\sqrt{1.092}}\right)^{2/3} \\
&= 288 \text{ kPa}
\end{aligned}
$$

For an inflation pressure of 300 kPa, the average ground pressure is 200 kPa which is less than the critical ground pressure, thus the tire operates in the elastic mode. It is noted that in an elastic mode the lower part of the tire in contact with the terrain is flattened, and the tire sinkage can be calculated as

$$
\begin{aligned}
z_0 &= \left(\frac{p_{gr}}{k_\phi}\right)^{1/n} \\
&= \frac{200}{680} = 0.29 \text{ m}
\end{aligned}
\tag{3.49}
$$

The compaction resistance at 300 kPa inflation pressure can then be calculated as

$$
\begin{aligned}
&= bk_\phi \left(\frac{z_0^{n+1}}{n+1}\right) \\
&= .315(680)\frac{0.26^2}{2} = 9.01 \text{ kN}
\end{aligned}
$$

For an inflation pressure of 500 kPa, the average ground pressure is 300 kPa which is greater than the critical ground pressure, thus the tire operates in the rigid mode. The tire sinkage can then be calculated as

$$
\begin{aligned}
z_0 &= \left(\frac{3W}{b(3-n)k_\phi\sqrt{D}}\right)^{\frac{2}{2n+1}} \\
&= \left(\frac{3 \times 41}{0.315 \times 2 \times 680 \times 1.045}\right)^{2/3} = 0.42 \text{ m}
\end{aligned}
$$

And the compaction resistance is calculated by

$$
\begin{aligned}
R_c &= bk_\phi \frac{z_0^{n+1}}{n+1} \\
&= 0.315 \times 680 \frac{0.42^2}{2} = 18.89 \text{ kN}
\end{aligned}
$$

It can be concluded that the compaction resistance of a truck tire at an inflation pressure of 500 kPa is approximately double that at an inflation pressure of 300 kPa.

3.3 Traction-Braking Performance

In order to evaluate the traction effort and the slip relationship for a tire over soil, the shear displacement along the tire-terrain interface has to be developed. The shear displacement can be determined based on the analysis of the slip and velocity, v, relationship. For a rigid tire, the slip velocity, v, of a point on the rim relative to the

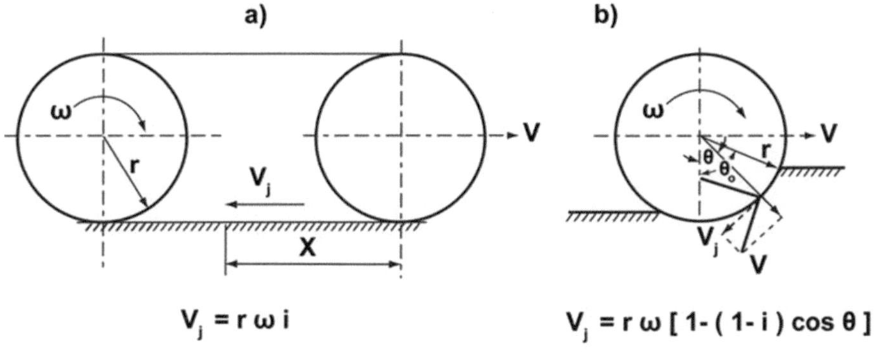

Fig. 3.8 Development of shear displacement under a rigid tire and a track [2]

terrain is the tangential component of the absolute velocity at the same point. Figure 3.8 shows the development of the shear displacement under a tire compared to that under a track.

The magnitude of the slip velocity, v, of a point on the rim is defined by

$$v = r\omega\left[1 - (1 - s)\cos\theta\right] \tag{3.50}$$

where θ is the angle shown in Fig. 3.8; it can be seen that the slip velocity is a function of both the angle θ and the slip, s. The shear displacement, j, along the tire-terrain contact patch is defined as the integral of the slip velocity as a function of time, and it can expressed by

$$j = r\left[(\theta_0 - \theta) - (1 - s)(\sin\theta_0 - \sin\theta)\right] \tag{3.51}$$

where θ_0 is the angle where a point on the rim comes in contact with the terrain; it is also referred to as the entry angle. Furthermore, the shear stress distribution along the contact area of a rigid tire can be defined by

$$\tau(\theta) = \left[c + p(\theta)\tan\phi\right]\left(1 - e^{-j/k}\right) \tag{3.52}$$

where $p(\theta)$ represents the normal pressure along the rigid tire, which can be estimated using the simplified method proposed by literature [8, 9].

The traction force is determined by integrating the horizontal component of the tangential stress over the entire contact area; it can be expressed by

$$F = \int_0^{\theta_0} \tau(\theta)\cos\theta d\theta \tag{3.53}$$

While the horizontal component of the tangential stress is used to determine the traction force, the vertical component is used to support the part of the vertical load

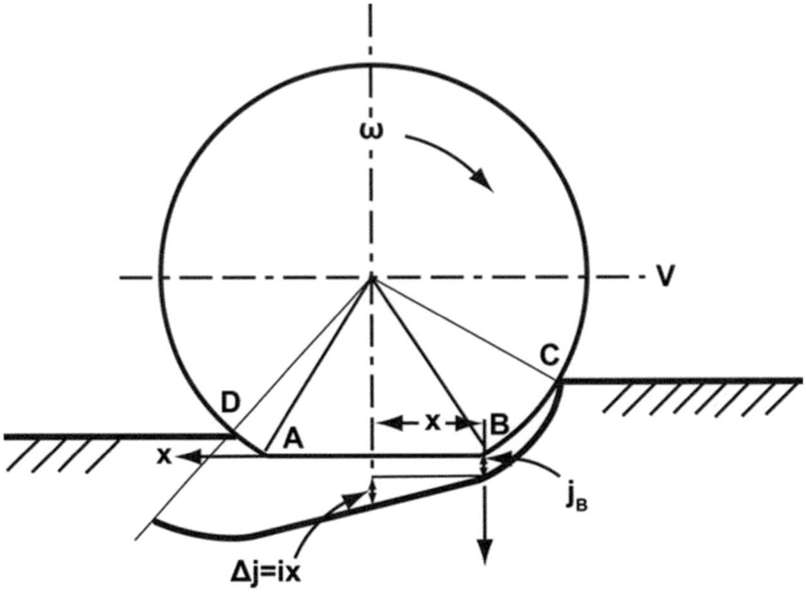

Fig. 3.9 Development of shear displacement under a tire operating in the elastic mode [2]

on the tire. It should be noted that in order to develop a more complete tire-terrain interaction model, the effect of the shear stress on the vertical load should be taken into consideration by

$$W = rb\left[\int_0^{\theta_0} p(\theta)\cos\theta d\theta + \int_0^{\theta_0} \tau(\theta)\sin\theta d\theta\right] \tag{3.54}$$

and on the drawbar pull by

$$F_d = rb\left[\int_0^{\theta_0} \tau(\theta)\cos\theta d\theta - \int_0^{\theta_0} p(\theta)\sin\theta d\theta\right] \tag{3.55}$$

and on the tire torque by

$$M_z = r^2 b\int_0^{\theta_0} \tau(\theta)d\theta \tag{3.56}$$

Similar to the free rolling performance, if the critical pressure, p_{gcr}, is less than the ground pressure, p_{gr}, then the tire operates in the elastic mode as shown in Fig. 3.9.

During an elastic mode, the AB portion of the tire circumference is flattened, and the shear displacement is developed along BC. The flat portion AB is referred to as the rigid tire earlier and the slip velocity is considered constant. The increase in shear displacement along section AB, Δj, is proportional to the slip of the tire s and the distance x between the point in question and the point B, and is expressed by

$$\Delta j = sx \tag{3.57}$$

The total cumulative shear displacement j_x at a distance x from point B is expressed by

$$j_x = j_B + sx \tag{3.58}$$

where j_B is the shear displacement at point B. After determining the displacement along the tire-terrain interface, the corresponding shear stress distribution is defined using the appropriate shear stress-shear displacement relationship. The traction force can then be determined by integrating the horizontal component of the shear stress over the entire contact area.

One of the most important parameters in tractive performance is the longitudinal stiffness. The longitudinal tire stiffness of the truck tire is defined as the tire's ability to recover maximum traction after experiencing 100% slip conditions through the application of a rapid angular acceleration to the tire's center. The longitudinal slip is calculated using Eq. 3.59, where s is the longitudinal slip, v and ω are the longitudinal and rotational velocities at the center of the tire, respectively, and r is the effective rolling radius of the tire:

$$s = \left(1 - \frac{v}{r\omega}\right) * 100 \tag{3.59}$$

The longitudinal tire stiffness can be computed using different computational models. During the simulation procedure, the tire is first inflated to the desired inflation pressure; then the desired vertical load is applied to the center of the tire. Rapid angular acceleration ($\dot{\omega}$) is then applied to the tire's center until the desired steady-state speed is reached. Due to the quick angular acceleration, the tire experiences 100% slip conditions at the beginning of the simulation. The longitudinal and angular velocities at the center of the tire, in addition to the longitudinal force at the contact patch, are computed. The simulation results are used to analyze the longitudinal force against the longitudinal slip ratio. From the relationship between the longitudinal force, f_x, and slip, s, the tire longitudinal stiffness, C_s $\left(\frac{\partial F_x}{\partial s}|_{s=0}\right)$, and the peak coefficient of adhesion, μ_p, are determined.

Figure 3.10a shows the variation of the longitudinal force as a function of longitudinal slip percentage at different vertical loads for a truck tire running over sandy loam at 756 kPa inflation pressure. It can be noticed that the longitudinal force reaches a maximum value for a longitudinal slip between 15 and 40% depending on the operating conditions. The longitudinal force increases as the applied vertical load increases, and the longitudinal force reaches its peak earlier for a higher vertical load. On the other side, the peak longitudinal force is slightly affected by the inflation pressure; as the inflation pressure increases, the peak longitudinal force decreases. This is due to the fact that the higher the inflation pressure, the lower the contact area is.

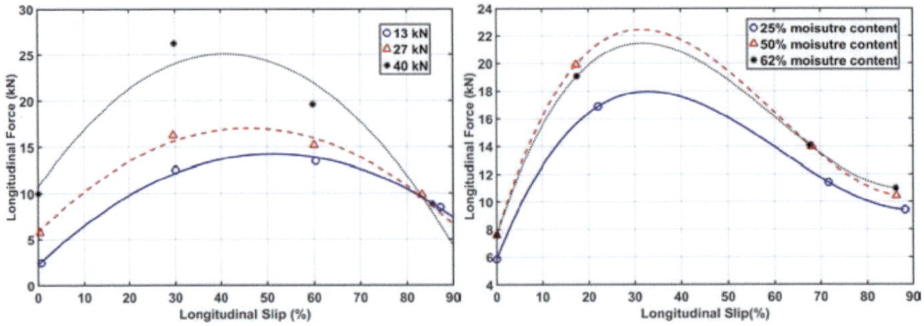

(a) Effect of vertical load at 756 kPa and 50% moisture content

(b) Effect of moisture at 586 kPa and 40 kN load

Fig. 3.10 Longitudinal force as a function of longitudinal slip for several sandy loam with different moisture contents

Furthermore, the effect of the moisture content in sandy loam on the longitudinal force is observed in Fig. 3.10b. Figure 3.10b shows the variation of the longitudinal force as a function of longitudinal slip percentage at different moisture contents for a truck tire running over sandy loam with different moisture contents at 40 kN vertical load and 586 kPa inflation pressure. The simulations are done at different moisture contents by generating different soils. This figure shows that the 50 and 62% moisture content soil has a similar longitudinal force at 0% slip, and the 25% moisture content has less initial longitudinal force. As the longitudinal slip increases, the longitudinal force increases for all moisture content. However, the 62% moisture content records a slightly lower force than that of the 50% moist sandy loam.

3.4 Cornering Performance

The same strategy used to derive the rolling resistance equations can be used to derive the forces acting on a steered tire. Figure 3.11 shows the static stress distribution under a non-rolling rigid tire.

Applying Bekker's approach, the static stress, $p(\theta)$, at any static contact angle, θ, along the tire-terrain contact line is calculated by

$$p(\theta) = R^n \left(\frac{k_c}{b} + k_\phi \right) (\cos\theta - \cos\theta_f) \tag{3.60}$$

where k_c and k_ϕ are the cohesive and frictional moduli, respectively, and n is the sinkage exponent. However, for the purpose of dynamic analysis, the rotation motion should be taken into consideration as shown in Fig. 3.12.

The entry angle, θ_f, and the exit angle, θ_r, of the tire-terrain contact will be different due to the forward motion caused by the longitudinal velocity, v. The entry angle can be calculated from the total sinkage h of the tire, while the exist angle can be calculated with the recovered "exist" sinkage, h_e, by [10]

$$\theta_f = \operatorname{acos}\left(1 - \frac{h}{R_0}\right) \tag{3.61}$$

$$\theta_r = \operatorname{acos}\left(1 - \frac{h_e}{R_0}\right) \tag{3.62}$$

In order to have a more accurate model, the two-dimensional model is extended to a three-dimensional in-plane model with a slip angle, α, as shown in Fig. 3.13.

In order to compute the normal stress, $\sigma(\theta)$, a combination of Bekker's formula and the model is needed and can be estimated for $\theta_m \leq \theta \leq \theta_f$ by

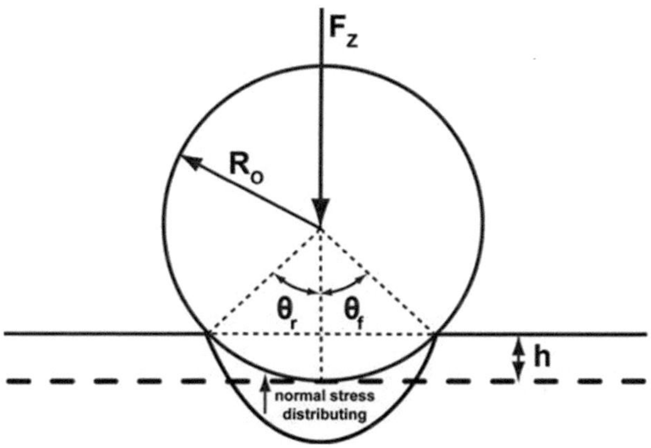

Fig. 3.11 Static stress distribution under a non-rolling rigid tire [2]

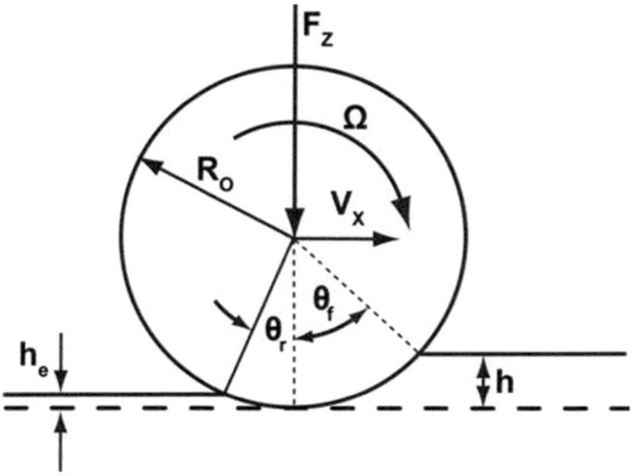

Fig. 3.12 Tire-terrain entry and exit angles when rolling [2]

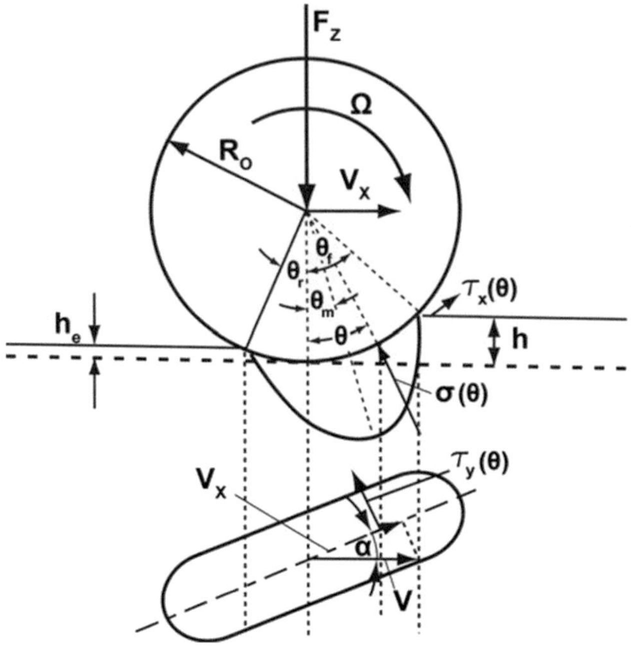

Fig. 3.13 Forward rolling and steered tire-terrain contact [2]

$$\sigma(\theta) = R_0^n \left(\frac{k_c}{b} + k_\phi\right) \left(\cos\theta - \cos\theta_f\right)^n \tag{3.63}$$

and for $\theta_r \leq \theta \leq \theta_m$ by

$$\sigma(\theta) = R_0^n \left(\frac{k_c}{b} + k_\phi\right) \left(\cos\left(\theta_f - \frac{\theta - \theta_r}{\theta_m - \theta_r}\left(\theta_f - \theta_m\right)\right) - \cos\theta_f\right)^n \tag{3.64}$$

The angle θ_m is the tire angle at which the maximum normal stress occurs. The angle θ_m is found experimentally [11] by

$$\theta_m = (a_0 + a_1\kappa)\,\theta_f \tag{3.65}$$

where a_0 and a_1 are parameters that depend on the tire-terrain interaction, and κ is the longitudinal slip of the tire. Janosi and Hanamoto's [12] expression to determine the longitudinal shear stress, $\tau_x(\theta)$, is given by

$$\tau_x(\theta) = (c + \sigma(\theta)\tan\phi)\left(1 - e^{-j_x(\theta)/k_x}\right) \tag{3.66}$$

where c is the terrain cohesion and ϕ is the angle of internal friction of the terrain, and k_x is the shear deformation modulus in the x direction. In a similar manner, we can determine the lateral shear stress, $\tau_y(\theta)$, by

$$\tau_y(\theta) = (c + \sigma(\theta)\tan\phi)\left(1 - e^{-j_y(\theta)/k_y}\right) \tag{3.67}$$

where k_y is the shear deformation modulus in the y-direction, and the longitudinal shear displacement along the contact area, j_x, can be estimated [13] by

$$j_x(\theta) = R_0\left(\theta_f - \theta - (1-\kappa)(\sin\theta_f - \sin\theta)\right) \tag{3.68}$$

Similarly, the lateral shear displacement along the contact area, j_y, can be estimated as a function of the slip angle, α, by

$$j_y(\theta) = R_0\,(1-\kappa)(\theta_f - \theta)\tan\alpha \tag{3.69}$$

On the other hand, the longitudinal shear deformation modulus, k_x, is defined by

$$k_x = k_{x0} + k_{x1}\alpha \tag{3.70}$$

where k_{x0} and k_{x1} are parameters related to the shear modulus, and the lateral shear modulus, k_y, can also be defined as a function of the slip angle α by

$$k_y = k_{y0} + k_{y1}\alpha \tag{3.71}$$

The forces acting on a rigid tire due to terrain can now be calculated by integrating the appropriate stresses along the contact line. The longitudinal force, F_x, is defined by

$$F_x = bR_0\int_{\theta_r}^{\theta_f}\left[\tau_x(\theta)\cos\theta - \sigma(\theta)\sin\theta\right]d\theta \tag{3.72}$$

The lateral force, F_y, is defined by

$$F_y = bR_0\int_{\theta_r}^{\theta_f}\tau_y(\theta)d\theta \tag{3.73}$$

The vertical load, F_z, is defined by

$$F_z = bR_0\int_{\theta_r}^{\theta_f}\left[\tau_x(\theta)\sin\theta + \sigma(\theta)\sin\theta\right]d\theta \tag{3.74}$$

The rolling resistance moment, M_y, is defined by

$$M_y = bR_0^2\int_{\theta_r}^{\theta_f}\tau_x(\theta)d\theta \tag{3.75}$$

And the self-aligning moment, M_z, is defined by

$$M_z = bR_0^2 \int_{\theta_r}^{\theta_f} \tau_y(\theta) \sin\theta d\theta \tag{3.76}$$

? Example 3.3

Assuming a rigid ring tire with 315 mm width is running over dry sand (refer to Table 2.1) and given that the total sinkage h of the tire is 250 mm, the existing sinkage of the tire h_e is 20 mm, the rolling radius R_0 is 450 mm, and a slip angle 12°, calculate the longitudinal and lateral shear stress as a function of θ knowing that $\theta_m = 3°$, $\theta_f = 10°$. Assume that the tire is free rolling and the longitudinal slip is zero percent, and that the shear deformation modulus in x and y directions is 26 MPa (hint: assume that the sinkage exponent is 1 instead of 1.1).

Solution. First, start by finding the entry angle (θ_f) and exit angle (θ_r):

$$\theta_f = 10° = 0.174533 \text{ rad}$$
$$\theta_r = a\cos(1 - \frac{h_e}{R_0}) = 0.299 \text{ rad}$$

Based on the relationship between the angles, choose the correct formula for the normal stress case: $\theta_m \leq \theta \geq \theta_f$. The next step is to find the normal stress $\sigma(\theta)$ using Eq. 3.63:

$$\sigma(\theta) = R_0^n \left(\frac{k_c}{b} + k_\phi\right) (\cos\theta - \cos\theta_f)^n$$
$$\sigma(\theta) = 0.45^1 \left(\frac{0.99}{0.315} + 1528.43\right) (\cos\theta - \cos 0.174533)^1$$
$$\sigma(\theta) = 636.314 \cos\theta - 0.98) \text{ kPa}$$

Next, find the longitudinal shear displacement $j_x(\theta)$:

$$j_x(\theta) = R_0 \left(\theta_f - \theta - (1-\kappa)(\sin\theta_f - \cos\theta)\right)$$
$$j_x(\theta) = 0.45\,(0.174533 - \theta - (1-0)(0.17365 - \sin\theta))$$
$$j_x(\theta) = 0.45(8.83 \times 10^{-4} - \theta + \sin\theta)$$

Next, find the lateral shear displacement $j_y(\theta)$:

$$j_y(\theta) = R_0(1-\kappa)(\theta_f-\theta)\tan\alpha$$
$$j_y(\theta) = 0.45(1-0)(0.174533-\theta)\tan 0.21$$
$$j_y(\theta) = 0.0166942-(0.09565)\theta$$

Next, find the longitudinal shear stress $\tau_x(\theta)$:

$$\tau_x(\theta) = (c+\sigma(\theta)\tan\phi)\left(1-e^{-\frac{j_x(\theta)}{k_x}}\right)$$

$$\tau_x(\theta) = (1.04+(636.314(\cos\theta-0.98))\tan 0.489)\left(1-e^{-\frac{0.45(8.83\times10^{-4}-\theta+\sin\theta)}{26\times10^6}}\right)$$

Finally, find the lateral shear stress $\tau_y(\theta)$:

$$\tau_y(\theta) = (c+\sigma(\theta)\tan\phi)\left((1-e^{-\frac{j_y(\theta)}{k_x}}\right)$$
$$\tau_y(\theta) = (1.04+338\cos\theta-331)\left(1-e^{-\frac{0.017-(0.096)\theta}{26106}}\right)$$

The cornering characteristics of a tire can be determined computationally by performing the cornering test. The cornering stiffness, C_α, is defined as the ability of the tire to resist deformation in the shape while a vehicle is undergoing a cornering operation; furthermore, cornering stiffness can be determined from the lateral force applied on the contact area. Equation 3.77 defines the cornering stiffness relationship to the slip angle and the lateral force. For a small slip angle, the cornering stiffness is considered linear, while for a slip angle greater than 2° the relationship can be highly nonlinear. The cornering stiffness concept is used to approximate the interaction between side and circumferential tire forces, where F_l is the lateral force of the tire:

$$C_\alpha = \frac{F_l}{\alpha} \tag{3.77}$$

Figure 3.14 shows a schematic of a tire with a slip angle. The contact forces F_x and F_y are on the global x-y axis, while the lateral force F_l is on the local y' axis. The lateral force can be written as a function of F_x and F_y as follows:

$$F_l = F_x\cos\alpha + F_y\sin\alpha \tag{3.78}$$

During the cornering characteristics simulation virtual test, the tire is first inflated to the desired inflation pressure, then the vertical load is applied to the center of the tire. A constant longitudinal speed of 10 km/h is then applied to the center of the tire, and the model is kept running for 2 s. The contact forces are computed on the contact patch between the tire-ground and tire-water. The tire is pre-steered to several slip angles between 0° and 12°; this will allow one to examine the influence of the slip angle on tire operational performance.

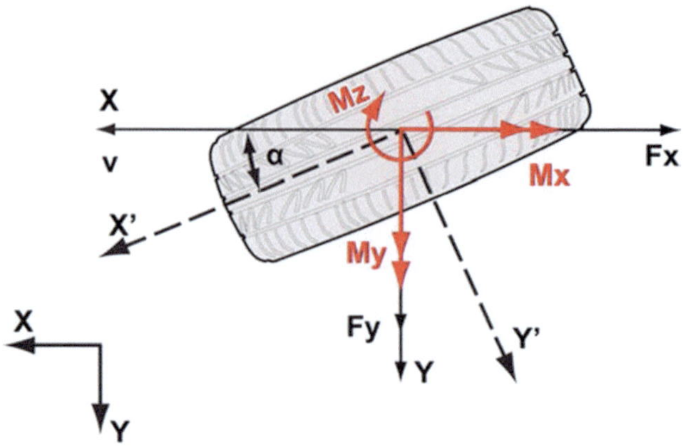

Fig. 3.14 Schematic of a tire with slip angle

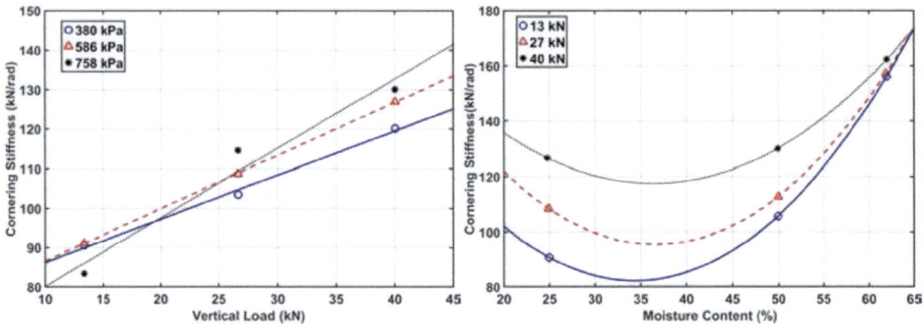

(a) Cornering stiffness as a function of vertical load (b) Cornering stiffness as a function of moisture

Fig. 3.15 Cornering stiffness as a function of vertical load and moisture content for tire running over sandy loam

The cornering stiffness of the tire is determined at several operating conditions. Figure 3.15a shows the variation of the cornering stiffness as a function of vertical load at different inflation pressures of tire running over sandy loam with 25% moisture. It is observed that the cornering stiffness increases as the vertical load increases at a given inflation pressure and moisture content. Additionally, the concerning stiffness of the tire also increases as the inflation pressure increases at a given vertical load and moisture content. At a high vertical load, the contact area between the tire and the soil increases which causes a higher lateral force and thus higher stiffness.

Furthermore, Fig. 3.15b shows the variation of the cornering stiffness as a function of moisture content at a tire inflation pressure of 586 kPa and different vertical loads. The cornering stiffness increases as the moisture content increases at a given vertical load and inflation pressure. The increase in moisture content increases the sinkage

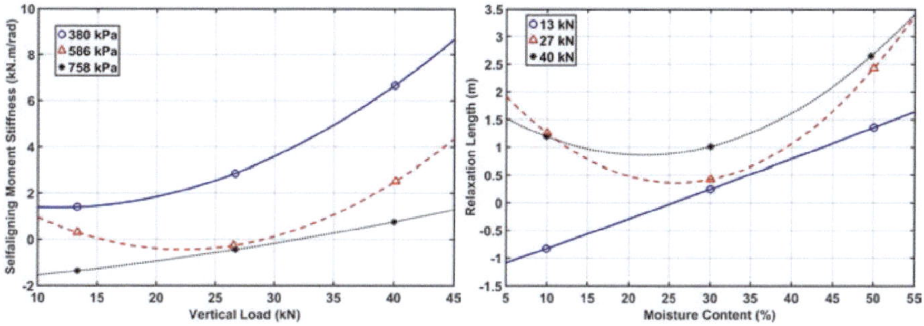

(a) Selfaligning moment stiffness as a function of vertical load

(b) Selfaligning moment stiffness as a function of moisture

Fig. 3.16 Self-aligning moment stiffness as a function of vertical load and moisture content for tire running over sandy loam with 25% moisture content

of the tire in the soil and thus increases the contact area which increases the lateral force causing a higher cornering stiffness.

The self-aligning moment stiffness as a function of vertical load at different inflation pressures for the tire running over sandy loam with 25% moisture content is presented in Fig. 3.16a.

The self-aligning moment stiffness increases as the vertical load increases at the given inflation pressure and moisture content. Furthermore, the self-aligning moment stiffness decreases as the inflation pressure increases at the given vertical load and moisture content. The increases in inflation pressure reduce the contact area between the tire and soil which reduces the required moment to align the tire.

The variation of the self-aligning moment stiffness as a function of moisture content for an inflation pressure of 586 kPa and different vertical loads is presented in Fig. 3.16b. The self-aligning moment stiffness has a nonlinear relationship with the moisture content. The self-aligning moment stiffness reduces as the moisture content increases until the moisture content reaches a value between 40 and 45%, then the self-aligning moment stiffness starts increasing as the moisture content increases. This nonlinear relationship requires further investigation at low moisture content.

Figure 3.17a shows the variation of the relaxation length as a function of vertical load at different inflation pressure for a tire running over sandy loam with 25% moisture content. The relaxation length increases as the vertical load increases at a given inflation pressure and moisture content. Moreover, the relaxation length decreases as the inflation pressure increases at a given vertical load and moisture content.

Figure 3.17b shows the variation of the relaxation length as a function of moisture content for an inflation pressure of 586 kPa and different vertical loads. The relaxation length is observed to have a nonlinear relationship as a function of the moisture content. The relaxation length reduces as the moisture content increases until the moisture content reaches a value between 40 and 45%, then the relaxation length increases as the moisture content increases. This observation requires further

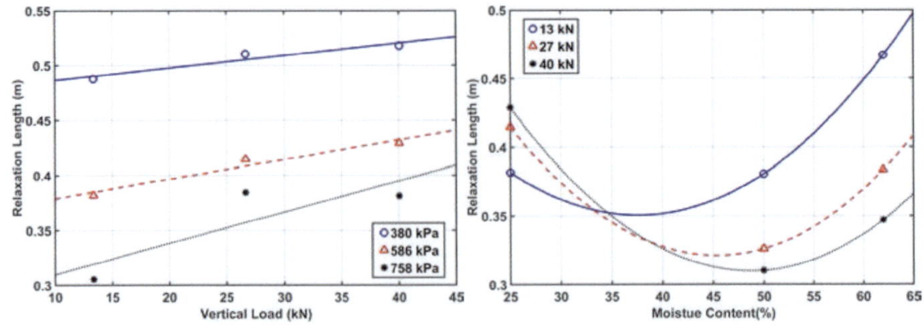

(a) Relaxation length as a function of vertical load (b) Relaxation length as a function of moisture

Fig. 3.17 Relaxation length as a function of vertical load and moisture content for a tire running over sandy loam

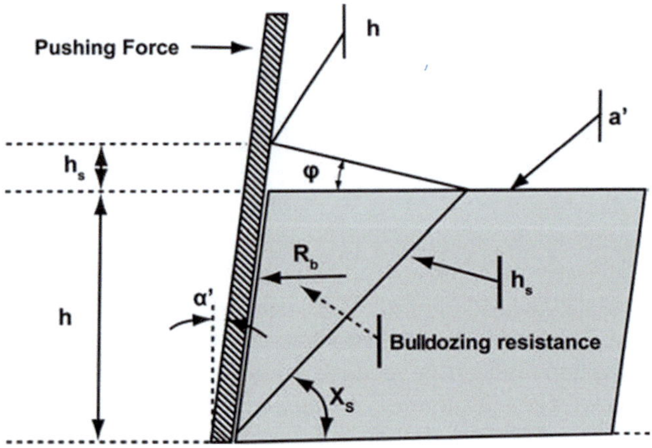

Fig. 3.18 The Hegedus model for estimating the bulldozing resistance of a vertical plate [2]

investigation, especially at low moisture content. It should be noted that in some cases tire bulldozing has an important effect on the tire-terrain interaction. The bulldozing effect becomes more vital when a tire is moving over the terrain causing an additional bulldozing resistance force.

Figure 3.18 shows the Hegedus model [14] for estimating the bulldozing resistance of a vertical plate. It should be noted that there are several methods to estimate the bulldozing resistance force due to soil mechanics and geotechnical principles. The Hegedus model assumes that a vertical plate pushes the terrain by pushing the blade. A linear destructive phase is then developed and swollen terrain is generated.

The destructive angle, X_c, is located between the terrain horizontal level and is estimated by

$$X_c = 45^\circ - \frac{\phi}{2} \tag{3.79}$$

where ϕ is the angle between the horizontal terrain level and the swollen soil, h_0. The swollen soil height, h_0, is estimated by

$$h_0 = \frac{h(\cot X_c - \tan\alpha')}{\tan\alpha' + \cot\phi} \tag{3.80}$$

where α' is the angle of the blade; assuming that the plate represents the tire in this case, the angle α' represents the inclination angle of the tire with the terrain. The bulldozing resistance, R_b, is estimated by

$$R_b = \frac{\cot X_c - \tan X_c + \phi}{1 - \tan\alpha' \tan X_c + \phi}\left[hc + \frac{1}{2}\rho h^2\left(\cot X_c - \tan\alpha' + \frac{(\cot X_c - \tan\alpha')^2}{\tan\alpha' + \cot\phi}\right)\right] \tag{3.81}$$

where ρ is the terrain density. The bulldozing resistance force of the tire depends on the sinkage of the tire into the terrain and the slip angle, α. The bulldozing force, F_b, of a tire with sinkage, h, moving in a lateral direction can be computed by

$$F_b = \int_{\theta_r}^{\theta_f} R_b(h)d\theta = \int_{\theta_r}^{\theta_f} R_b\left(R - h(\theta)\cos\theta\right)d\theta \tag{3.82}$$

The total lateral force, F_{ytot}, can be calculated by adding the bulldozing force to the lateral shear force

$$F_{ytot} = F_b \sin\alpha + F_y \tag{3.83}$$

? Example 3.4

Using Example 3.3 for the same condition, calculate the total lateral force of the rigid tire wheel rolling over dry sand, given that the bulldozing resistance is 0.4.

Solution. The total bulldozing force F_b is calculated as

$$F_b = \int_{\theta_r}^{\theta_f} R_b(h)d\theta$$

$$F_b = \int_{0.3}^{0.17} 0.4(0.25)d\theta$$

$$F_b = (0.1)(0.174533 - 0.29925)$$

$$F_b = -0.01247\ \text{N}$$

Next, the lateral shear force, F_y, is calculated by taking the lateral shear stress to integrate to find the lateral shear force, assuming the exponential portion of the lateral shear stress function tends toward 0. Thus

$$\tau_y(\theta) = (1.04 + (338.33(\cos\theta - 0.98)$$

The lateral force becomes

$$F_y = bR_0 \int_{\theta_r}^{\theta_f} \tau_y(\theta) d\theta$$
$$F_y = 0.315(0.45) \int_{0.3}^{0.17} 1.04 + (338.33(\cos\theta - 0.98) d\theta$$
$$F_y = 0.061287 \text{ kN}$$

The total lateral force thus becomes

$$F_{ytot} = F_b(\sin\alpha) + F_y$$
$$F_{ytot} = -0.01247(\sin 0.20944) + 0.061287$$
$$F_{ytot} = 0.05869 \text{ kN}$$

3.5 Multi-pass Effect and Characterization

The effect of multi-pass is vital when a tire passes over a specific patch of terrain, which a previous tire has passed over. This creates different terrain properties due to compaction, which causes the tire model to store both the elastic and plastic deformations of each tire.

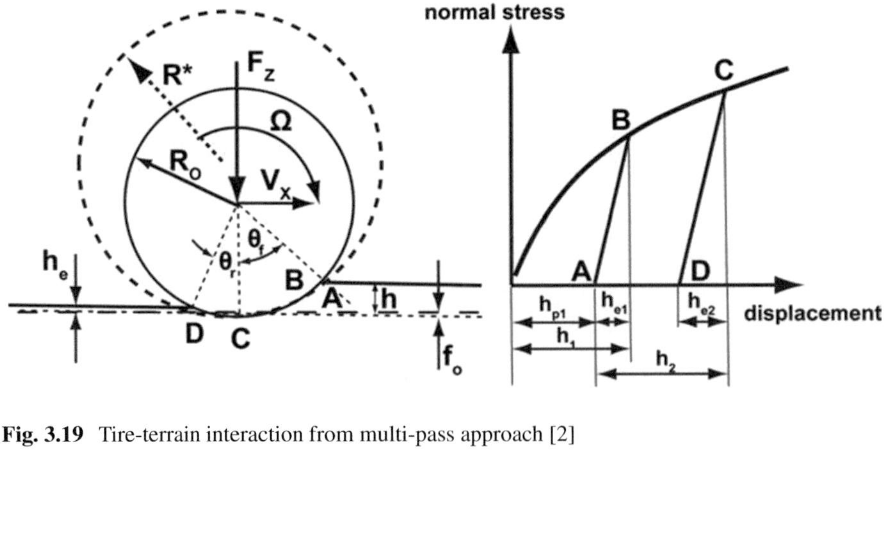

Fig. 3.19 Tire-terrain interaction from multi-pass approach [2]

(a) Truck tires model setup over sand with gravel soil

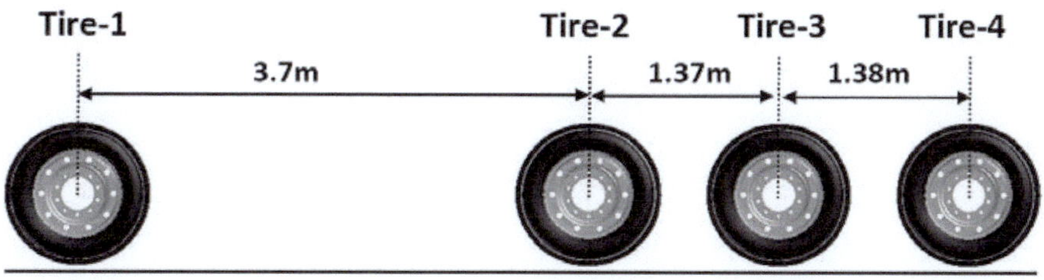

(b) Schematic of the multi-tire locations

Fig. 3.20 Truck tire model setup running over gravelly soil with normal displacement contour [1]

Figure 3.19 shows the normal pressure characteristics due to the multi-pass approach. When a tire passes over a patch of deformed terrain due to a previous tire, the normal pressure calculation should account for the plastic deformation history. For example, if two tires are rolling successively over the same patch of terrain, the first tire will have a total deformation, h_1, and an elastic part, h_{e1}, while the second tire that will pass over the same patch will first have an elastic deformation from A to B which is equal to the elastic deformation of the first tire, h_{e1}. And then it will continue to follow the normal pressure characteristics to point C, which will cause another plastic deformation, h_{p2}. The second plastic deformation, h_{p2}, will be equal to the total deformation h_2 subtracted from the elastic deformation h_{e2} as shown in Fig. 3.19.

Figure 3.20a shows the simulation setup of the tire-gravelly soil interaction model. The model consists of a hard surface which the tires are set onto, and a soil box which contains gravelly soil particles. This model is considered to be a bicycle model that has four tires set at a distance similar to that of the four axles of the tested truck. Each of these tires is subjected to a vertical load equivalent to half of the axle load. The first and fourth tires are free rolling non-driven tires, and the second and third tires are driven and are representing the first-second drive axles. The distance between the four tires is similar to the distance between truck axles as shown in Fig. 3.20b.

In order to evaluate and validate the results obtained from the simulation model, physical testing was carried out at Volvo Groups Truck Technology facility in

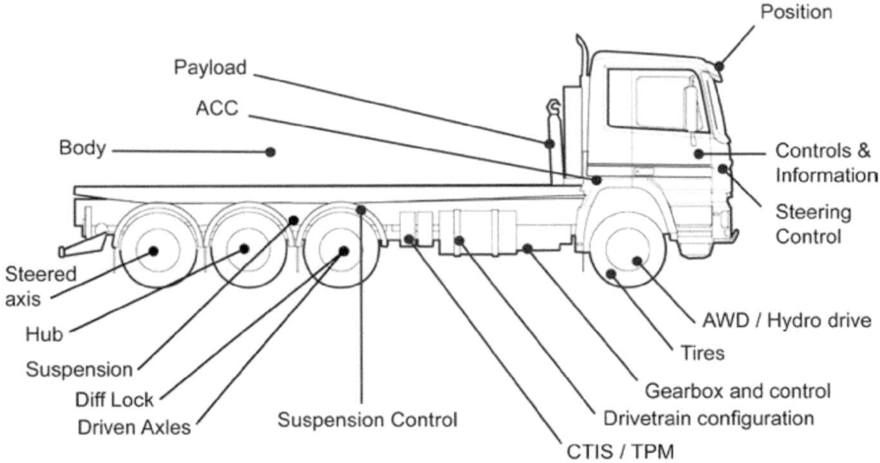

(a) Truck used during testing with part labels

(b) Testing zones available in the site

Fig. 3.21 Testing area and truck in Fjärås, Göteborg [1]

Göteborg, Sweden. The physical testing was performed and repeated at several operating conditions. The traction performance and soil compaction were analyzed.

The truck used in this test is a Volvo F-series medium 8 $\times$ 4 tag rigid as shown in Fig. 3.21a. The lead is mechanical with front suspension and the rear suspension is air. The truck is generally used for very badly maintained roads or off-road less than 5%. The truck can carry up to 44 tonnes of gross combined weight, out of which the front axle carries 8 tonnes and the rear axles carry 27 tonnes. The driving grades are 98% of drive distance with a maximum grade of 16%.

Furthermore, Fig. 3.21b shows the testing area which consists of two zones A1 and A2. Zone A1 has a soft terrain with 110–150 mm depth, while zone A2 is also soft terrain but with 150–250 mm depth. The soft terrain is identified as gravelly soil, and the material properties along with the geomechanical properties were provided by [15]. The inclination of both zones ranges 2–4%, and for the purpose of this research it is considered minimal.

The traction tests were performed at two different settings; in setting one, the traction control system was off and the truck was loaded, while in setting two the traction control system was off and the truck was unloaded. The measurements were performed for reaction and acceleration from terrain and through-axle in addition to the terrain compaction and change of characteristics of the terrain after multiple compactions. It should be noted, however, that due to limitations of sensors and equipment, only the driven axles (axles two and three) were equipped with transducers. Both driven axles were equipped with transducers to measure all 6 motions of the tire; namely the longitudinal, lateral and vertical forces; overturning moment, driving torque and self-aligning moment; in addition to the truck forward speed and the angular velocity of each drive tire.

The effect of soil compaction is well demonstrated in the rolling resistance coefficient of the two drive axles during a given inflation pressure, load, and speed. In order to further investigate the effect of soil compaction on tire rolling resistance, Fig. 3.22a, b are presented.

Figure 3.22a shows the variation of the rolling resistance coefficient as a function of the tire position for an unloaded and loaded truck. The effect of soil compaction on the tire rolling resistance is noticed when comparing the first and second drive axles. The second drive axle has a lower rolling resistance coefficient in comparison to that of the first drive axle for both loaded and unloaded truck cases. The difference in the rolling resistance coefficient between the first and second driving axles varies between 8 and 17% for a loaded and unloaded truck, respectively. It is observed that for an unloaded truck, the effect of soil compaction is higher than that of a loaded truck; this is due to the fact that at a low vertical load, the tire sinks less in the soil and thus does not displace the soil particles as much as in the case of a fully loaded truck.

Figure 3.22b shows the contact area between the tire and the gravelly soil as a function of the tire position for the unloaded and loaded truck. It is observed that

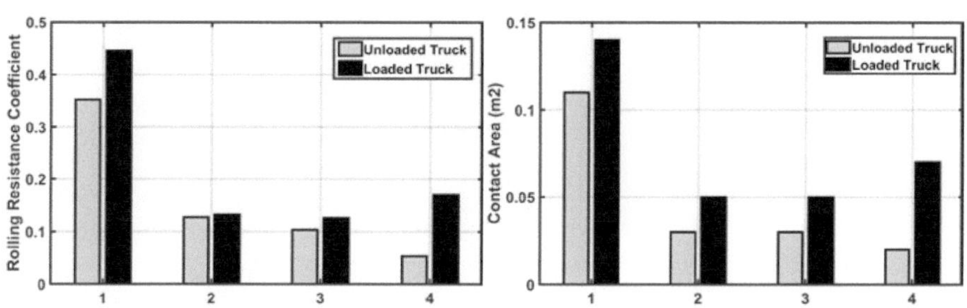

(a) Rolling resistance coefficient (b) Contact Area

Fig. 3.22 Interaction parameters as a function of tire position for unloaded and loaded truck running over gravelly soil [16] tire-1 has the highest contact area leading to the highest rolling resistance coefficient. This is due to the fact that the soil was fresh before the truck ran over it; also, tire-1 has more load than that of the drive tires. Furthermore, the first and second drive tires have similar contact areas leading to a similar rolling resistance coefficient, which further explains the results from Fig. 3.22a. The fourth tire, which is tire-4, has a lower contact area than that of the drive tires due to the hardness of the soil (compaction), thus leading to a lower rolling resistance coefficient.

3.6 Moisture Effect and Characterization

The terrain-dependent in-plane and out-of-plane rigid ring model parameters of the tire running over sand with different moisture contents are determined and investigated in this section.

Figure 3.23 shows the moist sand displacement during a rolling resistance run at 586 kPa inflation pressure and 10% moisture content. The maximum recorded displacement is around 400 mm at 40 kN vertical load, while the lowest recorded displacement is around 180 mm at 13 kN vertical load. It should be noted that almost all soil boxes have a zero displacement at the bottom of the box to avoid soil penetration.

Since there is no dry sandy loam, the properties of the sandy loam at 25% are used to apply the layering technique to increase the moisture content to 62%. The predicted rolling resistance coefficients of the tire running over sandy loam modeled using the first (interpolation) and second (moisturizing) techniques are compared at different inflation pressures and vertical loads.

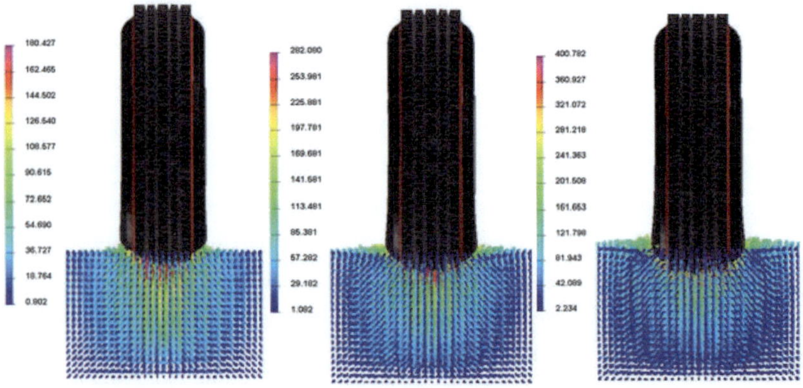

(a) 13 kN vertical load (b) 27 kN vertical load (c) 40 kN vertical load

Fig. 3.23 Tire moist soil interaction at different vertical loads and 586 kPa inflation pressure and 10% moisture content [1]

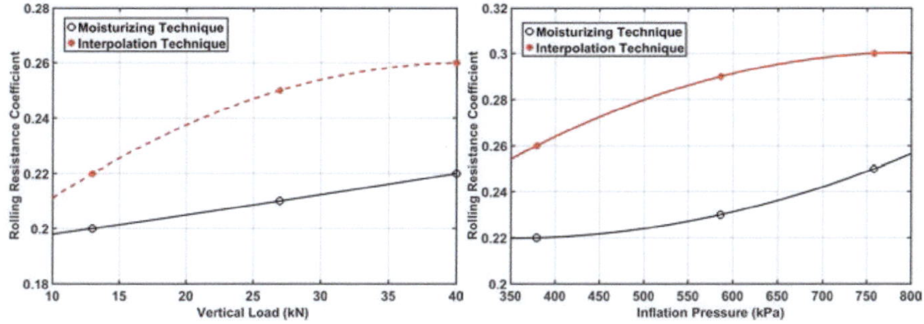

(a) Rolling resistance coefficient as a function of vertical load

(b) Rolling resistance coefficient as a function of inflation pressure

Fig. 3.24 Rolling resistance coefficient as a function of vertical load and inflation pressure for both techniques [1]

Figure 3.24a shows the variation of the rolling resistance coefficient as a function of vertical load at 379 kPa inflation pressure for a tire running over sandy loam with 62% moisture content using both moisturizing techniques. The rolling resistance coefficient increases as the vertical load increases for both techniques. However, the moisturizing technique tends to give a lower rolling resistance coefficient than that of the interpolation technique at given operating conditions. For instance, for the tire running over 62% sandy loam at 13 kN vertical load and 379 kPa inflation pressure, the difference between the interpolation and moisturizing technique is around 10%.

Figure 3.24b shows the variation of the rolling resistance coefficient as a function of inflation pressure at 40 kN vertical load for a tire running over sandy loam with 62% moisture content using both moisturizing techniques. The rolling resistance coefficient increases as the inflation pressure increases for both techniques. However, similar to the results shown in Fig. 3.24a, the moisturizing technique tends to give a lower rolling resistance coefficient than that of the interpolation technique at given operating conditions. For example, for the tire running over sandy loam with 62% moisture content at 40 kN vertical load and 379 kPa inflation pressure, the difference between the interpolation and moisturizing technique is around 14%.

In general, the difference between the moisturizing and interpolation technique ranges from 10 to 20% depending on the operating conditions. Figure 3.25a shows the variation of the rolling resistance coefficient as a function of the moisture content at 758 kPa inflation pressure and different vertical loads at a given speed of 10 km/h. The rolling resistance coefficient increases as the moisture content increases; this is due to the higher tire sinkage at higher moisture content. On the other side, the rolling resistance coefficient increases as the load increases at the given inflation pressure and moisture content.

It is noted that the variation of the angular velocity at constant linear longitudinal velocity results in longitudinal slip which indicates that the rolling resistance is not calculated at a freely rolling tire.

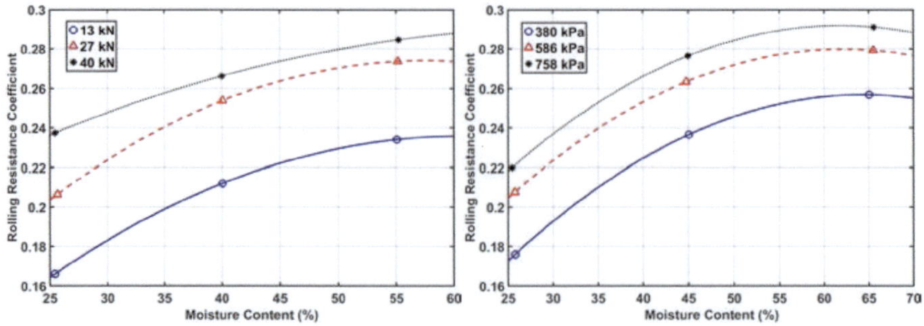

(a) Effect of vertical load at 758 kPa (b) Effect of inflation pressure at 40 kN

Fig. 3.25 Rolling resistance coefficient as a function of moisture content at different inflation pressures and vertical loads

To further understand the relationship between the rolling resistance coefficient and the tire operating conditions, Fig. 3.25b is presented. Figure 3.25b shows the variation of the rolling resistance coefficient as a function of the moisture content at a given vertical load of 40 kN and different inflation pressures at a given longitudinal speed of 10 km/h. It is shown from this figure that as the inflation pressure increases, the rolling resistance coefficient increases for constant soil moisture and vertical load.

This result is due to the fact that the increase of the inflation pressure increases the tire sinkage leading to more resistance forces. The same conclusion mentioned in the previous paragraph is also noticed here. The rolling resistance coefficient increases as the moisture content increases at a given vertical load and inflation pressure. This relationship is due to the fact that the contact area between the tire and sandy loam increases as the moisture content increases.

3.7 Fuel Economy and Transport Efficiency

The fuel economy of off-road vehicles depends on multiple factors including the fuel consumption characteristics of the engine, transmission characteristics, internal resistance of the running gear, operating speed, and external resistance and drawbar pull.

The output power, P, can be calculated as follows:

$$P = \frac{\left(\sum R + F_d\right) v}{(1 - i)\eta_t} \tag{3.84}$$

where R are the resistive forces in N, F_d is the drawbar pull in N, v is the speed in km/h, i is the slip, and η_t is the transmission efficiency. The fuel consumed per hour of operation, u_h, is determined by

$$u_h = P u_s \tag{3.85}$$

where u_s is the specific fuel consumption of the engine in kg/kW.h. It should be noted that the rolling resistance and slip of an off-road vehicle over a given terrain affect the power requirements and depend on the design of the running gear and vehicle configuration.

The relative cost of transporting a unit payload on a given route is considered to be the best way to compare off-road transporter vehicles. The relative cost of transporting a load involves multiple parameters including the performance, fuel consumption characteristics, load factor, and customer's preference. Transport productivity is defined as the product of payload, and the average speed through a specific region is used as a criterion for evaluating the performance of off-road transportation.

The average speed of a vehicle can be measured experimentally for an existing vehicle, while for a vehicle under development, it depends on the terrain condition and varies from one patch to another. Some of the factors that affect the average speed are (but are not restricted to) the tractive performance, obstacle negotiation, mobility, and vehicle vibration.

The transport efficiency, η_{tr}, is defined as the ratio of the transport productivity by the corresponding power input to the system:

$$\eta_{tr} = \frac{W_p v}{P} \tag{3.86}$$

where W_p is the payload and P is the power input to the system.

Problems

1. A pneumatic truck tire 315/80R22.5 is to be installed on an off-road truck. The tire has a radius of 546 mm, a section height of 266 mm, and a width of 315 mm. The tire is to carry a load of 22 kN, and the vehicle is to operate on a soil with pressure-sinkage parameters $n = 1$ and $k_\phi = 600$ kN/m^3. The vehicle analyst is proposing two inflation pressures 300 and 500 kPa. The average ground pressure at the given load and 300 kPa inflation pressure is 100 kPa, while that of 500 kPa inflation pressure is 250 kPa. Compare the sinkage and compaction resistance of the truck tire at the proposed inflation pressures.
2. In order to increase the traction of a wheeled vehicle, a device with 20 lugs has been attached to a narrow rim. The outside diameter of the device measured from the lug tip is 1.27 m. The lugs are 30 cm wide and penetrate vertically 30 cm into the ground. Estimate the tractive effort per lug developed in the vertical position in sandy loam with a cohesion of 2 kPa and an angle of shear resistance of 25°. Note that $\gamma_s = 17$ kN/m^3, assume that the lug is smooth, and neglect the friction and adhesion between the lug and sandy loam.
3. In order to increase the traction of an off-road wheeled vehicle over wet soil, a steel cage with 16 smooth lugs is attached. The outside diameter of the steel

cage across the tips of the lugs is 1.2 m, and lugs are 25 cm wide. The lugs also penetrate the soil at the vertical position with 15 cm depth. Calculate the tractive effort per lug in the vertical position and the corresponding torque, developed in a wet soil with $c = 13.79$ kPa, $\phi = 7°$, and $\gamma_s = 20$ kN/m^3. Also calculate the corresponding driving torque required.

4. An inclined bulldozer blade with an angle of 60° to the horizontal is used to remove a layer of soil 10 cm deep. The blade has a width of 5 m, and its surface has an angle of friction with the soil of 24° and no adhesion. The soil has an angle of internal shearing resistance of 25°, a cohesion of 3.79 kPa, and a weight density of 16,286 N/m^3. The values of the parameters characterizing blade-soil interaction $K_{p\gamma}$ and K_{pc} are 10 and 2, respectively. Estimate the resultant force on the blade required to make the initial cut without surcharge. Also identify the magnitudes and directions of the horizontal and vertical components of the resultant force exerted on the blade by the soil.

References

1. El-Sayegh Z (2020) Modeling and analysis of truck tire-terrain interaction. PhD thesis
2. Wong JY (2008) Theory of ground vehicles. Wiley
3. Osman MS (1964) The mechanics of soil cutting blades. Trans ASAE 9(4):318–328
4. Terzaghi K (1965) Theoretical soil mechanics. Wiley
5. McKyes E (1985) Soil cutting and tillage. Elsevier
6. Krick G (1969) Radial and shear stress distribution beneath rigid wheels and pneumatic tyres on yielding soils with regard to tyre deformation. J Terramech 6(3):73–98
7. El-Sayegh Z, El-Gindy M, Johansson I, Öijer F (2018) Improved tire-soil interaction model using fea-sph simulation. J Terramech 78:53–62
8. Bekker MG (1969) Introduction to terrain-vehicle systems. part i: The terrain. part ii: The vehicle. Technical report, DTIC Document
9. Bekker MG (1975) Motion resistance of pneumatic tyres
10. van Oosten JJ, Koppenaal J (2015) A tire soft-terrain model for vehicle dynamics analysis. In: International tyre colloquium, 4th, 2015, Guildford, United Kingdom
11. Yoshida K, Watanabe T, Mizuno N, Ishigami G (2003) Terramechanics-based analysis and traction control of a lunar/planetary rover. In: Field and service robotics. Springer, pp 225–234
12. Janosi Z (1961) The analytical determination of drawbar pull as a function of slip for tracked vehicles in defarmable soils. In: Proceedings of 1st International Conference of ISTVS. Turin, 1961
13. Wong J-Y, Reece AR (1967) Prediction of rigid wheel performance based on the analysis of soil-wheel stresses part i. performance of driven rigid wheels. J Terramech 4(1):81–98
14. Hegedus E (1960) A simplified method for the determination of bulldozing resistance. US Army Ordnance Tank-Automotive Command
15. Prakash S (2018) Nonlinear finite element analysis of shrinking reinforced concrete slabs-on-ground
16. El-Sayegh Z, El-Gindy M, Johansson I, Öijer F (2020) Development and validation of off-road tire-gravelly soil interaction using advanced computational techniques. J Terramech 91:45–51

Chapter 4
Road Vehicle Stability and Control

The handling characteristics of a vehicle can be evaluated based on the vehicle response to steering input. The control of the direction of motion of a vehicle and its ability to stabilize its direction of motion against external disturbances are the basic issues in vehicle handling.

The handling characteristics of a vehicle can be evaluated based on the vehicle response to steering input. The control of the direction of motion of a vehicle and its ability to stabilize its direction of motion against external disturbances are the basic issues in vehicle handling.

A system (spring-mass system with or without linear damper) is considered to be stable if it returns to a position of equilibrium within a finite time of the disturbances. In the case of a simplified linear vehicle model (2-DOF), the equations of motion can be reduced to a set of linear differential equations with constant coefficients. Root Routh's stability criteria method can be used to examine the system stability, as will be described in the section below.

4.1 One Degree of Freedom Spring-Mass-Damper System

Figure 4.1 shows the one degree of freedom mass-spring-damper system.

The equation of motion of the mass-spring-damper system can be written as follows:

$$m\ddot{x} + c\dot{x} + kx = F(t) \tag{4.1}$$

In the case of free vibration, the $F(t)$ will be equal to zero and we can assume the solution of the equation is as follows:

$$x = x_0 e^{\lambda t} \tag{4.2}$$

M. El-Gindy and Z. El-Sayegh, *Road and Off-Road Vehicle Dynamics*,
https://doi.org/10.1007/978-3-031-36216-3_4

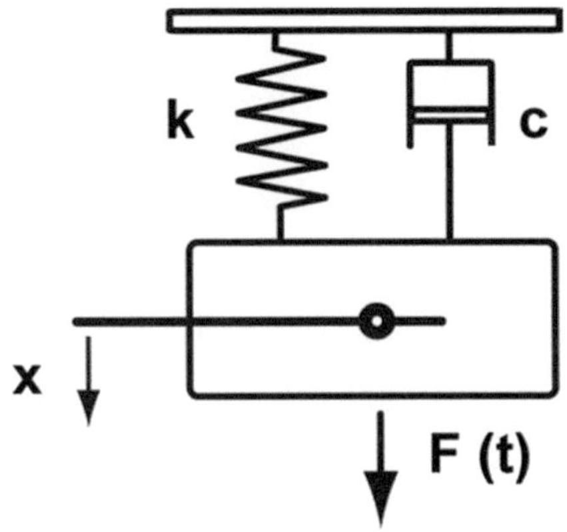

Fig. 4.1 One degree of freedom mass-spring-damper system

where x_0 is the initial position of the mass, t is the time, and λ is the root of the characteristics equation of the system described in Eq. 4.1. Substituting Eq. 4.2 in 4.1 and setting $F(t) = 0$, then the characteristic equation can be written as follows:

$$m\lambda^2 + c\lambda + k = 0 \tag{4.3}$$

which can be rewritten as follows

$$\lambda^2 + A\lambda + B = 0 \tag{4.4}$$

where $A = c/m$ and $B = k/m$. The roots of the characteristics Eq. 4.4 can be determined as

$$\lambda_{1,2} = -\frac{A}{2} \pm \sqrt{\left(\frac{A}{2}\right)^2 - B} \tag{4.5}$$

$$= -\frac{c}{2m} \pm \sqrt{\left(\frac{c}{2m}\right)^2 - \frac{k}{m}} \tag{4.6}$$

$$= R(\lambda) + J(\lambda) = a + bi \tag{4.7}$$

where $i = \sqrt{-1}$, a is the real part, and b is the imaginary part of λ. By substituting in Eq. 4.2, the following are the possible four solutions depending on the sign of the real part, a, and if the imaginary part b has a positive value or zero.

Figure 4.2 shows the possible four cases of the system behavior based on the obtained values of the system roots, P_1 and P_2.

- Case (a). If the solution of Eq. 4.7 led to zero value of the imaginary part, b, and negative value for the real part, a, the system will be exponentially stable as shown in Fig. 4.2a.

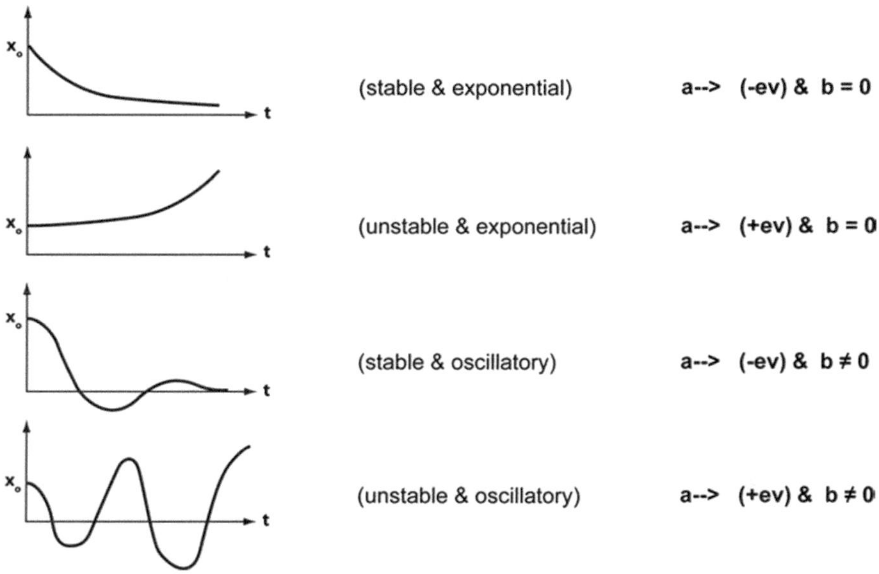

Fig. 4.2 Possible solutions for a simple one degree mass-spring-damper system

- Case (b). If the solution of Eq. 4.7 led to zero value of the imaginary part, b, and positive value for the real part, a, the system will be exponentially unstable as shown in Fig. 4.2b.
- Case (c). If the solution of Eq. 4.7 led to non-zero value of the imaginary part, b, and negative value for the real part, a, the system will be oscillatory stable as shown in Fig. 4.2c.
- Case (d). If the solution of Eq. 4.7 led to non-zero value of the imaginary part, b, and positive value for the real part, a, the system will be oscillatory unstable as shown in Fig. 4.2d.

4.2 Simplified Two-Axle Vehicle (Bicycle Model)

Figure 4.3 shows a two-axle (Bicycle model) of a two-axle vehicle. The front tire is assumed to represent a combination of the left and right wheels, while the steering input, δ, is the average of the left and right steering inputs. Also, the slip angles α_f and α_r are the average slip angles of the left and right slip angles of the front and rear tires, respectively.

The vehicle weight is W and the mass is $W/g = m$. The longitudinal speed, u, is assumed constant. The two degrees of freedom of the vehicle model are the yaw rate, r, and the sideslip velocity, v. The cornering forces at the front and rear tires

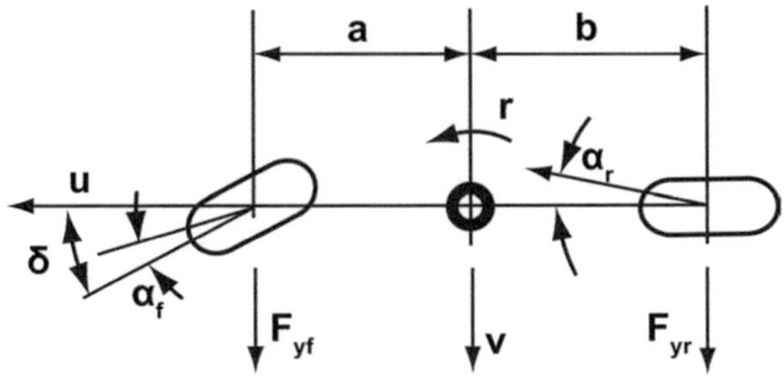

Fig. 4.3 Two degrees of freedom bicycle model

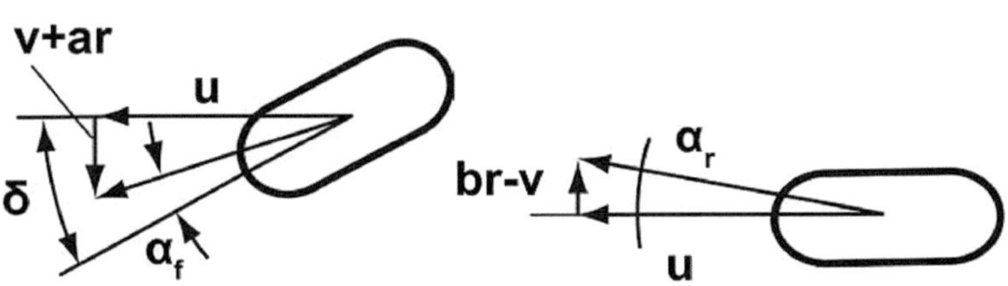

(a) Front tire slip angle (b) Rear tire slip angle

Fig. 4.4 Tire front and rear side slip angles

during cornering are F_{yf} and F_{yr}, respectively. Assuming that the steering angle, δ, is small, the font cornering force F_{yf} is assumed to be equal to $F_{yf}\cos\delta$.

The slip angle of the front tire shown in Fig. 4.4a of the bicycle model is determined as follows:

$$\alpha_f = \delta - \tan^{-1}\left(\frac{v + ar}{u}\right) \tag{4.8}$$

$$\approx \delta - \left(\frac{v + ar}{u}\right) \tag{4.9}$$

The slip angle of the rear tire shown in Fig. 4.4b of the bicycle model is determined as follows:

$$\alpha_r \approx -\left(\frac{v - br}{u}\right) \tag{4.10}$$

The linear cornering forces at the front and the rear tires can be determined as follows:

$$F_{yf} = C_{F_{\alpha f}}\alpha_f \tag{4.11}$$

$$F_{yr} = C_{F_{\alpha r}}\alpha_r \tag{4.12}$$

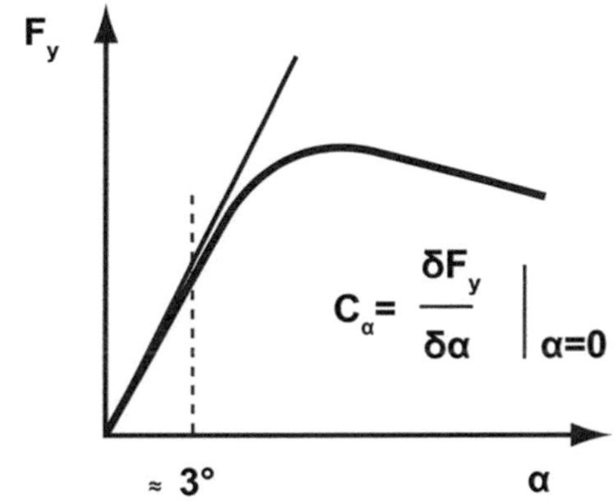

Fig. 4.5 Cornering stiffness determination

where $C_{F_{\alpha f}}$ and $C_{F_{\alpha r}}$ are the cornering stiffness of the front and rear tires. The cornering stiffness is the slope of the tangent of the cornering force versus slip angle, at zero slip angle, as shown in Fig. 4.5.

The equation of motion in the lateral direction can be written as follows:

$$m\left(\ddot{v} + ur\right) = F_{yf} + F_{yr} \tag{4.13}$$

The equation of motion in the yaw direction can be written as follows:

$$I\dot{r} = aF_{yf} - bF_{yr} \tag{4.14}$$

Substituting F_{yf} and F_{yr} from Eqs. 4.11 and 4.12 in Eqs. 4.13 and 4.14, then Eq. 4.13 can be rewritten as follows:

$$m\left(\ddot{v} + ur\right) = -\left(C_{\alpha f} + C_{\alpha r}\right)\frac{v}{u} - \left(aC_{\alpha f} - bC_{\alpha r}\right)\frac{r}{u} + C_{\alpha f}\delta \tag{4.15}$$

And Eq. 4.14 can be written as

$$I\dot{r} = -\left(aC_{\alpha f} - bC_{\alpha r}\right)\frac{v}{u} - \left(a^2C_{\alpha f} - b^2C_{\alpha r}\right)\frac{r}{u} + aC_{\alpha f}\delta \tag{4.16}$$

Assuming that $\dot{v} = pv$ and that $\dot{r} = pr$, where p is a derivative operator $p = \frac{d}{dt}$. Substituting in Eqs. 4.15 and 4.16, the equations of motion can be rewritten as follows:

$$\left[mp + \frac{C_{\alpha f} + C_{\alpha r}}{u}\right]v + \left[mu + \frac{aC_{\alpha f} - bC_{\alpha r}}{u}\right]r = C_{\alpha f}\delta \tag{4.17}$$

And

$$\left[Ip + \frac{a^2C_{\alpha f} + b^2C_{\alpha r}}{u}\right]r + \left[\frac{aC_{\alpha f} - bC_{\alpha r}}{u}\right]v = aC_{\alpha f}\delta \tag{4.18}$$

Rewriting these equations in matrix form as follows:

$$\begin{bmatrix} A_1 & B_1 \\ A_2 & B_2 \end{bmatrix} \begin{bmatrix} \frac{v}{\delta} \\ \frac{r}{\delta} \end{bmatrix} = \begin{bmatrix} C_{\alpha f} \\ aC_{\alpha f} \end{bmatrix} \tag{4.19}$$

where

$$A_1 = mp + \frac{C_{\alpha f} + C_{\alpha r}}{u} \tag{4.20}$$

$$A_2 = \frac{aC_{\alpha f} - bC_{\alpha r}}{u} \tag{4.21}$$

$$B_1 = mu + \frac{aC_{\alpha f} - bC_{\alpha r}}{u} \tag{4.22}$$

$$B_2 = Ip + \frac{a^2 C_{\alpha f} + b^2 C_{\alpha r}}{u} \tag{4.23}$$

The characteristic equation of the equations of motion is the determinate of the coefficient matrix of Eq. 4.19 and can be written as

$$A_1 B_2 - A_2 B_1 = 0 \tag{4.24}$$

Substituting A_1, A_2, B_1, and B_2, the characteristic equation can be written as

$$p^2 + \left[\frac{a^2 C_{\alpha f} + b^2 C_{\alpha r}}{uI} + \frac{C_{\alpha f} + C_{\alpha r}}{mu} \right] p + \left[\frac{C_{\alpha f} C_{\alpha r} (a+b)^2}{mIu^2} - \frac{aC_{\alpha f} - bC_{\alpha r}}{I} \right] = 0 \tag{4.25}$$

or in the form of second-order algebraic equation as follows:

$$p^2 + Ap + B = 0 \tag{4.26}$$

The roots of the above characteristic equation can be determined as

$$p_{1,2} = -\frac{A}{2} \pm \sqrt{\left(\frac{A}{2}\right)^2 - B} \tag{4.27}$$

The state variables v and r of the bicycle model can be assumed as follows:

$$v = v_0 e^{pt} \tag{4.28}$$

$$r = r_0 e^{pt} \tag{4.29}$$

where r_0 and v_0 are the initial yaw rate and sideslip velocity.

4.2.1 Vehicle Stability Analyses

By examining Eqs. 4.28 and 4.29, it is obvious that the roots p_1 and p_2 must be negative in order to ensure the stability of the vehicle and mitigation of the sideslip velocity and yaw rate as the time t increases. One of the roots may change its sign from negative to positive if the term B in Eq. 4.27 became negative. The term A is always positive. In order to determine the stability boundary of this system B may change its sign from positive to negative for a particular vehicle depending on the speed u. The speed at which the sign change occurs will be called critical speed u_{cr} and can be determined when $B = 0$, as follows:

$$\frac{C_{\alpha f} C_{\alpha r} (a+b)^2}{m I u^2} - \frac{a C_{\alpha f} - b C_{\alpha r}}{I} = 0 \tag{4.30}$$

The critical speed can then be written as

$$u_{cr} = \sqrt{\frac{C_{\alpha f} C_{\alpha r} (a+b)^2}{m \left(a C_{\alpha f} - b C_{\alpha r}\right)}} \tag{4.31}$$

By examining Eq. 4.31, it can be noted that if $bC_{\alpha r} < aC_{\alpha f}$, the critical speed will have a real number and indicates that at this speed the vehicle is oversteering. If the $bC_{\alpha r} = aC_{\alpha f}$, the critical speed will be infinity, which means that the vehicle is insensitive to the variation of the speed and becomes neutral steer. If $bC_{\alpha r} > aC_{\alpha f}$, the critical speed will not exist because the value under the square root will be negative and becomes imaginary. In this case, the vehicle will change its mode to understeering and the critical speed will become a characteristic speed u_{ch} as follows:

$$u_{ch} = \sqrt{\frac{C_{\alpha f} C_{\alpha r} (a+b)^2}{m \left(-a C_{\alpha f} + b C_{\alpha r}\right)}} \tag{4.32}$$

Equation 4.31 can be rewritten as using the terms defined in Fig. 4.6

$$u_{cr} = \sqrt{\frac{l^2}{\left(\frac{F_{zr}}{C_{\alpha r}} - \frac{F_{zf}}{C_{\alpha f}}\right)}} \tag{4.33}$$

From Fig. 4.6, it can noted that $l = a + b$ and that

$$F_{zf} = \frac{ma}{l} \tag{4.34}$$

$$F_{zr} = \frac{mb}{l} \tag{4.35}$$

Equation 4.31 can then be rewritten in the following format:

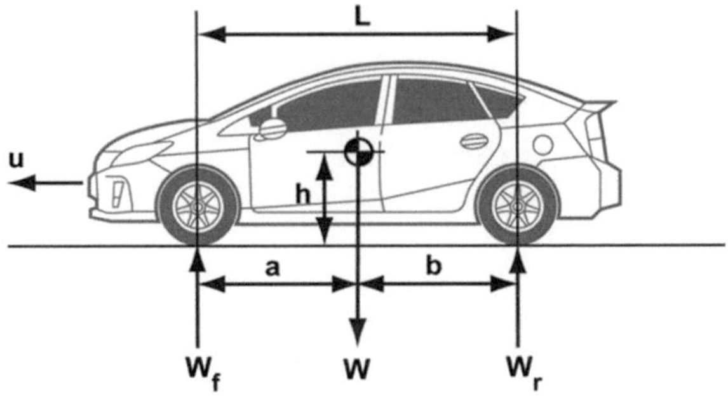

Fig. 4.6 Weight distribution

Table 4.1 Understeer/Neutral Steer/Oversteer conditions

Understeer	Neutral steer	Oversteer
$\alpha_f > \alpha_r$	$\alpha_f = \alpha_r$	$\alpha_f < \alpha_r$
$aC_{\alpha f} < bC_{\alpha r}$	$aC_{\alpha f} = bC_{\alpha r}$	$aC_{\alpha f} > bC_{\alpha r}$
$\frac{F_{zf}}{C_{\alpha f}} > \frac{F_{zr}}{C_{\alpha r}}$	$\frac{F_{zf}}{C_{\alpha f}} = \frac{F_{zr}}{C_{\alpha r}}$	$\frac{F_{zf}}{C_{\alpha f}} < \frac{F_{zr}}{C_{\alpha r}}$
$K_{us} > 0$	$K_{us} = 0$	$K_{us} < 0$

$$u_{cr} = \sqrt{\frac{l^2}{-K_{us}}} \tag{4.36}$$

where

$$K_{us} = \frac{F_{zf}}{C_{\alpha f}} + \frac{F_{zr}}{C_{\alpha r}} \tag{4.37}$$

K_{us} is called "Understeer Coefficient" and it is a measure of the understeer/neutral steer/understeer conditions. From the above equations, the understeer/neutral steer/oversteer conditions as a function of the basic vehicle design parameters can be summarized in Table 4.1.

4.2.2 *Steady-State Steering Response*

In order to achieve steady-state condition in turns at constant speed, u, and steering input, δ, the following conditions must be satisfied:

$$\dot{r} = \dot{v} = 0 \tag{4.38}$$

which means that p in Eqs. 4.17 and 4.18 must be replaced by zero. In this case, these equations can be rewritten in matrix format as follows:

$$\begin{bmatrix} A_{s1} & B_{s1} \\ A_{s2} & B_{s2} \end{bmatrix} \begin{bmatrix} \frac{v}{\delta} \\ \frac{r}{\delta} \end{bmatrix} = \begin{bmatrix} F_1 \\ T_1 \end{bmatrix} \tag{4.39}$$

where

$$A_{s1} = \frac{C_{\alpha f} + C_{\alpha r}}{u} \tag{4.40}$$

$$A_{s2} = \frac{aC_{\alpha f} - bC_{\alpha r}}{u} \tag{4.41}$$

$$B_{s1} = mu + \frac{aC_{\alpha f} - bC_{\alpha r}}{u} \tag{4.42}$$

$$B_{s2} = \frac{a^2C_{\alpha f} + b^2C_{\alpha r}}{u} \tag{4.43}$$

By applying Cramer's rule, the sideslip velocity response to steering input can be determined as follows:

$$\left(\frac{v}{\delta}\right)_{ss} = \frac{\begin{vmatrix} F_1 & B_{s1} \\ T_1 & B_{s2} \end{vmatrix}}{\mid A \mid} \tag{4.44}$$

Similarly, the yaw rate response to steering input can be written as follows:

$$\left(\frac{r}{\delta}\right)_{ss} = \frac{\begin{vmatrix} A_{s1} & F_1 \\ A_{s2} & T_1 \end{vmatrix}}{\mid A \mid} \tag{4.45}$$

where the determinant A is given as follows:

$$\mid A \mid = A_{s1}B_{s2} - A_{s2}B_{s1} \tag{4.46}$$

By solving Eq. 4.44, the steady-state side slip velocity response to steering input can be determined as follows:

$$\left(\frac{v}{\delta}\right)_{ss} = \frac{u\left(blC_{\alpha f}C_{\alpha r} - amu^2C_{\alpha f}\right)}{C_{\alpha f}C_{\alpha r}l^2 - mu^2\left(aC_{\alpha f} - bC_{\alpha r}\right)} \tag{4.47}$$

Similarly, the yaw rate response to steering input, Eq. 4.45, can be rewritten as follows:

$$\left(\frac{r}{\delta}\right)_{ss} = \frac{ulC_{\alpha f}C_{\alpha r}}{C_{\alpha f}C_{\alpha r}l^2 - mu^2\left(aC_{\alpha f} - bC_{\alpha r}\right)} \tag{4.48}$$

The equation of interest to be analyzed is the yaw rate response to steering input, shown above. From this equation, several vehicle responses to steering input can be derived. For instance, the steady-state acceleration response can be obtained by multiplying the right-side and left-side terms as follows:

$$\left(\frac{A_y}{\delta}\right)_{ss} = \left(\frac{r}{\delta}\right)_{ss} u \tag{4.49}$$

while the vehicle curvature response during steady-state negotiation of a circle of radius R can be obtained as follows:

$$\left(\frac{1/R}{\delta}\right)_{ss} = \left(\frac{r}{\delta}\right)_{ss} \frac{1}{u} \tag{4.50}$$

If the speed $u = 0$ in the above equation, the Ackerman condition can be satisfied as follows:

$$\left(\frac{1/R}{\delta}\right)_{ss.u=0} = \frac{1}{l} \text{ or } \delta = \frac{l}{R} \tag{4.51}$$

The above equation can be represented in Fig. 4.7, where "O" is the turning radius at zero longitudinal speed u. At speeds higher than zero, the vehicle curvature response to steering input of the linear bicycle vehicle model in cases of understeer, neutral steer, and oversteer can be presented in Fig. 4.8.

In Fig. 4.8, it can be noticed that as the speed increases, the curvature response to steering input increases until the vehicle speed reaches the critical speed, u_{cr}; at this speed, the turning radius will approach zero and the curvature response will approach infinity. At this point, the vehicle lost its stability and started to spin about the vertical axis passing through the center of gravity. In the case of the neutral steer vehicle, the curvature response will not be sensitive to the increase of the speed and the curvature response will remain constant, i.e., the vehicle will be able to move in a circle of radius R at constant steering input and speed. In cases of understeering, the increase in speed will result in an increase of the turning radius as constant steering input until the vehicle speed reaches its characteristic speed, at which the steering input $\delta = 2l/R$. At this point, the vehicle will lose its stability or controllability.

? Example 4.1

A car has a weight of 22 kN and a wheelbase of 3 m. 60% of the vehicle weight is on the front axle, and the remaining 40% is on the rear axle, under stationary conditions:

1. If the cornering stiffness of the left and right front tires is 40 kN/rad and that of the left and right rear tires is 40 kN /rad, determine the steady-state handling characteristics of the vehicle.
2. If the left and right front tires are replaced by radial-ply tires with a cornering stiffness of 50 kN/rad per tire, and the left and right rear tires remain unchanged,

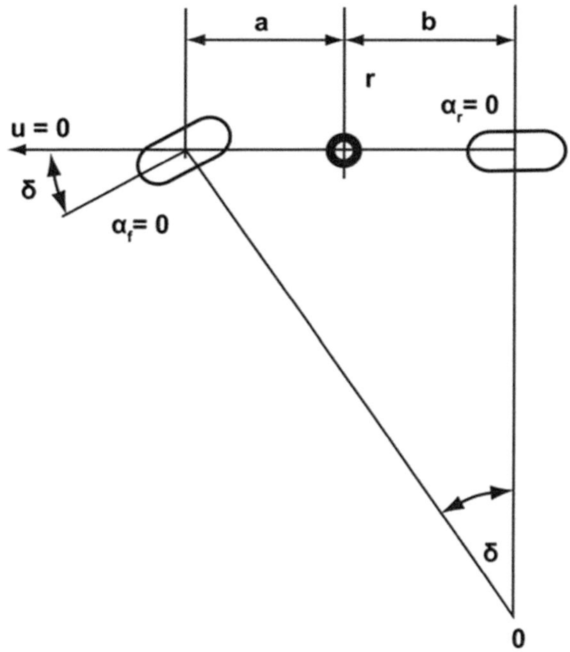

Fig. 4.7 Ackerman steering at zero speed

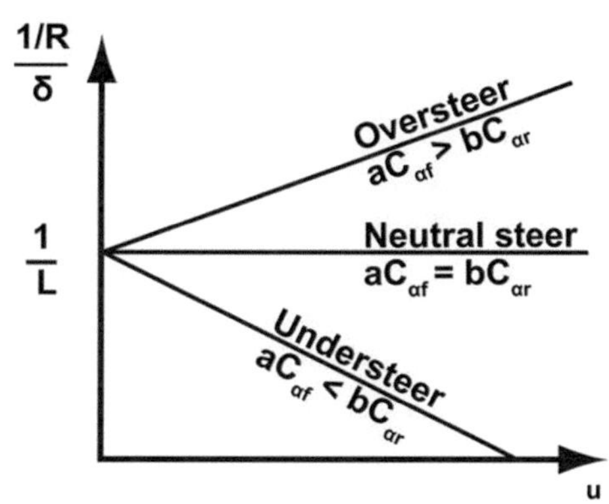

Fig. 4.8 Curvature response to steering input

determine the steady-state handling characteristics of the vehicle under these conditions.

Solution. To Find Center of Gravity: Wheelbase (l) = 3.0 m. Center of Gravity is 20% toward front due to 60% front axle weight and 40% rear axle weight.

$$3/2 = 1.5$$
$$1.5 \times (1 - 0.2) = 1.2\,\text{m}$$

The center of gravity (COG) is 1.2 m from the front axle.
Front axle cornering stiffness ($C_{\alpha f}$) = 40 kN/rad.
Rear axle cornering stiffness ($C_{\alpha r}$) = 40 kN/rad.
Distance from front axle to COG (a) = 1.2 m.
Distance from front axle to COG (b) = 3.0 m − 1.2 m = 1.8 m.

Next, determine the understeer/ oversteer cases; the understeer coefficient is

$$\begin{aligned} K_{us} &= \frac{F_{zf}}{C_{\alpha f}} - \frac{F_{zr}}{C_{\alpha r}} \\ &= \frac{221.2/3}{(240} - \frac{221.8/3}{240} \\ &= -0.055\,\text{rad} \end{aligned}$$

Due to the understeer coefficient being negative, the vehicle is in oversteer. Steady-State Handling Characteristics if front tires were replaced with radial-ply using the Understeer Coefficient Formula:

$$\begin{aligned} K_{us} &= \frac{F_{zf}}{C_{\alpha f}} - \frac{F_{zr}}{C_{\alpha r}} \\ &= \frac{221.2/3}{(250} - \frac{221.8/3}{240} \\ &= -0.077\,\text{rad} \end{aligned}$$

Due to the understeer coefficient being negative, the vehicle is still in oversteer.

4.3 Handling Diagram

The handling or directional behaviors of a road vehicle during steady-state turns can be evaluated using what is called the "Handling Diagram" as shown in Fig. 4.9.

The handling diagram is a plot of the steady-state lateral acceleration $a_y/g(u^2/gR)$ versus one of the following quantities: $(\alpha_r - \alpha_f)$ or $(lr/u - \delta)$ or $(l/R - \delta)$, where

- l is the vehicle wheelbase (m),
- r is the vehicle yaw rate (rad/s),
- R is the turning Radius (m),
- a_y is the steady-state lateral acceleration (m/s^2),
- g is the gravitation (m/s^2),
- δ is the average front tires steering angle (rad),
- α_f is the average front tires slip angle (rad), and
- α_r is the average rear tires slip angle (rad).

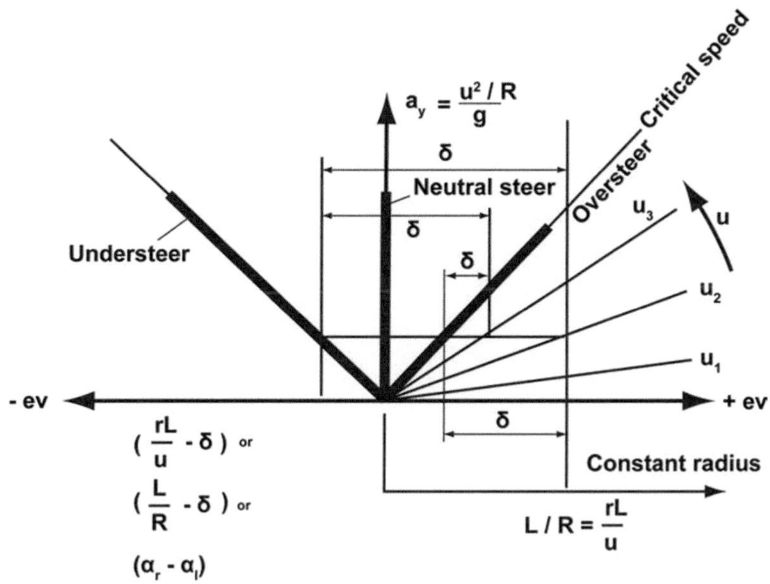

Fig. 4.9 Linear vehicle model handling and speed diagram

It can be easily proven that the three quantities $(\alpha_r - \alpha_f)$ or $(lr/u - \delta)$ or $(l/R - \delta)$ are equal.

Figure 4.9 shows the linear handling and speed diagram. The left side of the X-axis is negative and the right side is positive. The vertical axis is the steady-state lateral acceleration, a_y/g (u^2/gR). The thick lines are called handling diagrams. If the handling diagram lies in the right side, the slope of the handling diagram will be negative, the vehicle is understeering. If the handling diagram is vertical, the vehicle is neutral steering. If the handling diagram is in the positive side, the vehicle is oversteering. From the handling diagram, the understeer coefficient K_{us} is defined as –1/Slope of the handling diagram and can be written as follows:

$$K_{us} = -\frac{l/R - \delta}{a_y/g} \tag{4.52}$$

It can be noticed that the sign of the K_{us} will determine whether the vehicle is understeering (K_{us} is positive), neutral steering ($K_{us} = 0$), or oversteering (K_{us} is negative). The right side of Fig. 4.9 shows superimposed plots called speed lines. The slopes of these speed lines are determined from the calculation of the ratio of $(u^2/gR)/(l/R)$; therefore, the slope is a function of the speed u and equal to u^2/gl, which is dimensionless. If the slope of the speed line is equal to the slope of the oversteered handling line, the speed u becomes equal to the critical speed u_{cr}. To

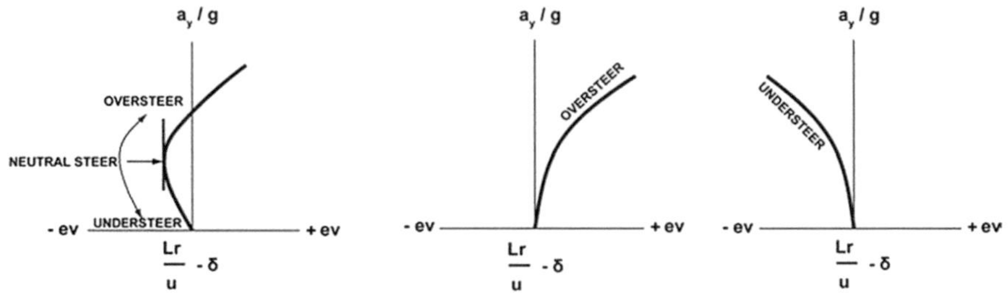

Fig. 4.10 Nonlinear handling

prove that, equate the slope of the speed line with the slope of the oversteered linear handling diagram, which is $-1/K_{us}$, and the speed line slope u^2/gl, the results will be

$$u = u_{cr} = \sqrt{gl/k_{us}} \tag{4.53}$$

Figure 4.9 is giving important information regarding the required steering input δ as a function of the speed to negotiate a given circle of radius R. As speed increases (slope of the speed line increases), the driver of the oversteered vehicle will need to keep reducing the steering input to maintain the turning radius R. If the speed reached the critical speed u_{cr}, oversteered vehicle will be able to negotiate the turn at δ equal to zero. While in the case of the neutral steered vehicle, the change of the vehicle speed or the slopes of the speed lines will not change the required steering input to negotiate the turn. If the vehicle is understeering and the speed increases, the steering input will keep increasing until the vehicle speed reaches the characteristic speed u_{ch}; at this point, the steering input will be equal to $\delta = 2l/R$.

The above analysis is valid for linear vehicle models such as the bicycle linear model. In the case of analyzing a nonlinear vehicle model which is closer to the reality, the handling diagram will be also nonlinear and can be used to determine the understeer steer/neutral steer/oversteer vehicle characteristics as a function of the level of the steady-state lateral acceleration. The steady-state lateral acceleration can be achieved at constant steer input and constant speed. A sample of a nonlinear handling diagram is shown in Fig. 4.10.

The nonlinear handling diagram may take any of the shapes shown in Fig. 4.10. The most common one is the case (a), in which the vehicle starts to be understeered at a low level of lateral acceleration (low speed or large turning radius). As the lateral acceleration increases, the vehicle becomes neutral steer and then starts the transition to oversteering as the lateral acceleration increases. As the vehicle enters the oversteer region, the driver must be very careful because the vehicle becomes very sensitive to steering inputs and can negotiate turns at high speeds with small steering input. In the other two cases of (b) and (c), the vehicle does not show a transition from one state to another, i.e., from understeer to oversteer or from oversteer to understeer but remains understeer as in case (b) or remains oversteer as in case (c) at all levels of the

Table 4.2 Yaw rate gain and lateral acceleration gain as a function of speed

Speed of vehicle (km/h)	Yaw rate gain (rad/s/rad)	Lateral acceleration gain (g/rad)
10	0.04312	0.01221
20	0.08461	0.04791
30	0.12304	0.10452
40	0.15732	0.17819
50	0.18680	0.26447
60	0.21123	0.35886
70	0.23070	0.45727
80	0.24556	0.55627
90	0.25632	0.65323
100	0.26355	0.74627

lateral accelerations. Vehicles with a long wheelbase may show behavior similar to case (b) and vehicles with a short wheelbase may show behavior similar to case (c).

? Example 4.2

A vehicle has a weight of 20 kN and a wheelbase of 3.2 m. The ratio of the distance between the center of gravity of the vehicle and the front axle to the wheelbase is 0.45. The cornering stiffness of each of the front and rear tires is 38 kN/rad. The average steering gear ratio (ξ_s) is 20. Determine the steady-state yaw rate and the lateral acceleration responses to the steering input of the vehicle with respect to the steering wheel angle at forward speed ranging from 10 to 100 km/h.

Solution. The cornering stiffness of front and rear axle tires is 38 kN/rad and 38 kN/rad. The first step is to define the understeer coefficient, the yaw rate gain formula, and the lateral acceleration gain formula:

$$K_{us} = \frac{F_{zf}}{C_{\alpha f}} - \frac{F_{zr}}{C_{\alpha r}}$$

$$G_{yaw} = \frac{r}{\delta \xi_s} = \frac{u}{l + \frac{k_{us} u^2}{g} \xi_s} \tag{4.54}$$

$$G_{acc} = \frac{a_y/g}{\delta \xi_s} = \frac{u^2}{(gl + k_{us} v^2)\xi_s} \tag{4.55}$$

The comparison table is listed below:

Table 4.2 and Fig. 4.11 show the variation of the yaw rate gain and the lateral acceleration gain as a function of vehicle speed. The yaw rate gain and the lateral

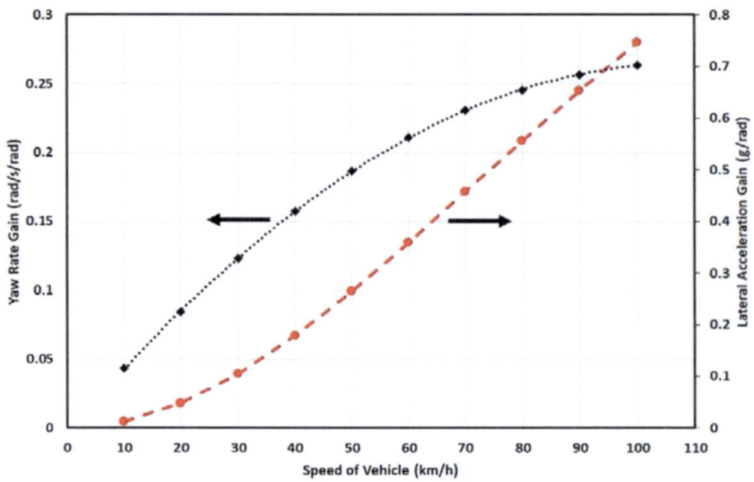

Fig. 4.11 Yaw rate gain and lateral acceleration gain as a function of speed

acceleration gain with respect to steering wheel angle are presented in Eqs. 4.54 and 4.55, respectively.

4.4 Simplified Three-Axle Articulated Vehicle (Bicycle Model)

In this section, the steady-state handling characteristics of a two-unit articulated vehicle would be analyzed. Practically, this two-unit vehicle can be a tractor/semi-trailer, bus/pony-trailer, and a car towing a trailer or a caravan, a truck/pony-trailer. All these vehicle configurations consist of a lead unit (tow unit) and a trailing unit (towed unit), which are joined together through a pintle hook, king pin, or a fifth wheel. The dynamic response of such a combination is dependent on their design parameter, the type and location of the articulation mechanism, and the operating conditions (payload, speed, and road conditions).

To understand the basic dynamics of an articulated vehicle, a simplified model of a three-axle tractor/semi-trailer will be presented and analyzed under steady-state conditions (constant speed and steering input). This vehicle consists of a two-axle tractor and one-axle trailer. For the sake of simplification, the articulation point is assumed to be located at the center of the tractor rear axle. Practically, and for stability reasons, the location of the articulation point (fifth wheel) of a semi-trailer should be located ahead of the center of the tractor rear axle.

In this analysis, the simplified linear tractor/semi-trailer model consists of 3-DOF (degrees of freedom), namely tractor yaw rate, sideslip (lateral) velocity, and articulation angle between the tractor and the semi-trailer.

The handling characteristics of the vehicle will be analyzed under the following five conditions:

1. Both tractor and semi-trailer are understeer.
2. The tractor is understeer, while the semi-trailer is oversteer.
3. The tractor is oversteer, while the semi-trailer is understeer.
4. Both the tractor and semi-trailer are oversteer, and the ratio of the understeer coefficients of the tractor to that of the semi-trailer is less than the ratio of the semi-trailer wheelbase to the tractor wheelbase.
5. Both the tractor and semi-trailer are oversteer, and the ratio of the understeer coefficients of the tractor to that of the semi-trailer is greater than the ratio of the semi-trailer wheelbase to the tractor wheelbase.

4.5 Handling Characteristics of Articulated Three-Axle Tractor-Semi-Trailer

Figure 4.12 shows the simplified three degrees of freedom linear model of the two-axle tractor and one-axle semi-trailer.

The equations of motion can be composed as follows for the tractor:

$$m(\dot{v} + ur) = F_{y1} + F_{y2} - Y \tag{4.56}$$

$$I\dot{r} = aF_{y1} - bF_{y2} + Yd \tag{4.57}$$

The equations of motion for the semi-trailer are

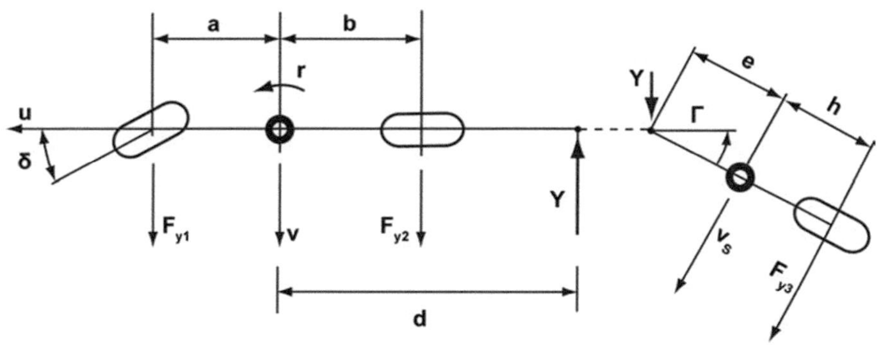

Fig. 4.12 Three-axle tractor-semi-trailer bicycle model

$$m_2(\dot{v_s} + ur_s) = F_{y3} + Y \tag{4.58}$$

$$I_s\dot{r_s} = Ye - hF_{y3} \tag{4.59}$$

The analysis of the vehicles is performed at constant forward speed and steady-state steering maneuvers, i.e., the following assumptions will be applied $r = r_s$, $\dot{v} = \dot{v_s} = 0$, and $\dot{r} = \dot{r_s} = 0$.

Since the vehicle has 3-DOF, the equations of motion can be reduced to three equations by substituting

$$Y = F_{y3}\frac{h}{e} \tag{4.60}$$

Then the equations become

$$m_1ur = F_{y1} + F_{y2} - F_{y3}\frac{h}{e} \tag{4.61}$$

$$0 = aF_{y1} - bF_{y2} + F_{y3}\frac{h}{e} \tag{4.62}$$

$$m_2ur = F_{y3}\frac{e-h}{e} \tag{4.63}$$

where

$$F_{y1} = C_{\alpha 1}\alpha_1;\ F_{y3} = C_{\alpha 2}\alpha_2;\ F_{y3} = C_{\alpha 3}\alpha_3 \tag{4.64}$$

and

$$\alpha_1 = \delta - \left(\frac{v + ar}{u}\right) \tag{4.65}$$

$$\alpha_2 = -\left(\frac{v - br}{u}\right) \tag{4.66}$$

$$\alpha_3 = -\left(\frac{v_s - hr}{u}\right) \tag{4.67}$$

v_s can be determined from the hitch kinematics as follows for a small angle as shown in Fig. 4.13:

$$v_s + er = v - rd + u\Gamma \tag{4.68}$$

or

$$v_s = v - (e + d)r + u\Gamma \tag{4.69}$$

The trailer's sideslip angle becomes

$$\alpha_3 = -\left(\frac{v - (e + h + d)r}{u}\right) - \Gamma \tag{4.70}$$

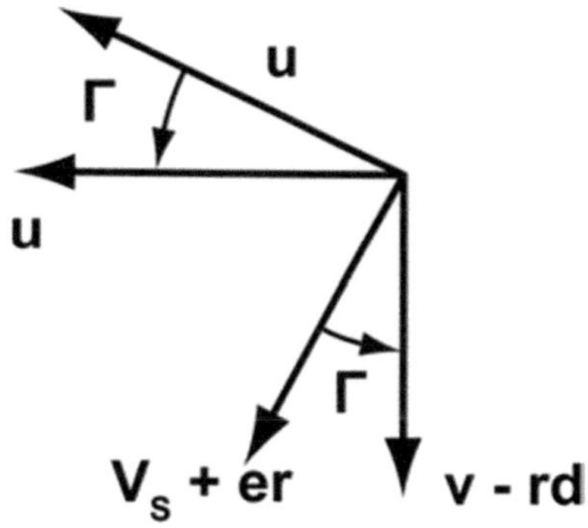

Fig. 4.13 Small angle Γ definition

The steady-state behavior can be examined using the articulation angle response to steering input $(\Gamma/\delta)_{ss}$, which can be derived from the following relationships for the tractor:

$$\alpha_1 - \alpha_2 = \delta - (a+b)\frac{r}{u} \tag{4.71}$$

$$= \delta - \frac{l_t}{R} \tag{4.72}$$

$$\delta = (\alpha_1 - \alpha_2) + \frac{l_t}{R} \tag{4.73}$$

Similarly, the relationships for the semi-trailer are

$$\alpha_2 - \alpha_3 = \Gamma - \frac{e+h+d-b}{R} \tag{4.74}$$

$$= \Gamma - \frac{l_s}{R} \tag{4.75}$$

$$\Gamma = (\alpha_2 - \alpha_3) + \frac{l_s}{R} \tag{4.76}$$

The articulation angle response to steering input can be obtained from the ratio of the above equations as follows:

$$\left(\frac{\Gamma}{\delta}\right)_{ss} = \frac{\frac{l_s}{R} + K_{us}\frac{u^2}{gR}}{\frac{l_t}{R} + K_{ut}\frac{u^2}{gR}} \tag{4.77}$$

$$= \frac{l_s\left(1 + \frac{K_{us}}{l_s}\frac{u^2}{g}\right)}{l_t\left(1 + \frac{K_{ut}}{l_t}\frac{u^2}{g}\right)} \tag{4.78}$$

where K_{us} and K_{ut} are the semi-trailer and tractor understeering coefficients, respectively. It should be noted that

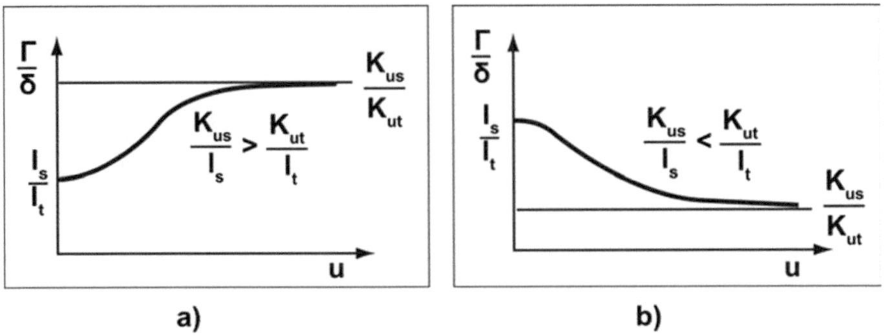

Fig. 4.14 Steady-state responses in case of both tractor and semi-trailer are understeering

$$K_{us} = \frac{\alpha_2 - \alpha_3}{u^2/gR} \tag{4.79}$$

$$K_{ut} = \frac{\alpha_1 - \alpha_2}{u^2/gR} \tag{4.80}$$

where R is the turning radius and g is the gravitation (9.81 m/s^2). It should be noted that K_{us} and K_{ut} are positive for understeering, zero for neutral, and positive for oversteering (Fig. 4.14).

There are several possible situations that may occur during steady-state motion which are listed below:

1- Both tractor and trailer are understeered, meaning $K_{us} > 0$ and $K_{ut} > 0$.

In this case, both the tractor and semi-trailer reached steady-state articulation angle response as the speed, u, increases. But the articulation angle is highly dependent on the ratio of the understeering coefficient to the wheelbase. The end value of the $(\Gamma/\delta)_{ss}$ remains positive but it is equal to the ratio of the semi-trailer to tractor understeer confidence K_{us}/K_{ut}. In conclusion, both cases (a) and (b) show stable cases in spite of the variation of the end values of the steady-state articulation angle at high speed. It is always recommended to have a match between tractors and semi-trailers having understeer coefficients to ensure the stability of the combination.

2- Tractor is understeered ($k_{ut} > 0$) and the trailer is oversteered ($k_{us} < 0$). In this case, the semi-trailer articulation angle will become steady, but negative as the speed of the tractor increases above a critical speed.

This situation is highly undesirable during turning at high speeds. The back end of the trailer may hit the guardrails if turning to the left or cross the island if turning to the right as shown in Fig. 4.15.

3- Both the tractor and trailer are oversteered meaning $K_{us} < 0$ and $K_{ut} < 0$. Figure 4.16 shows the worst case scenario when both the tractor and semi-trailer are oversteering.

Jackknifing at constant high speed may occur in either the left or right direction; closing of the semi-trailer to the tractor causes significant damage and most likely

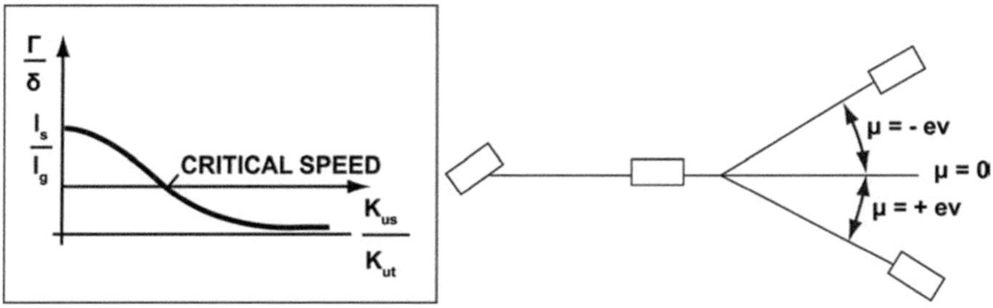

Fig. 4.15 Steady-state response in case of understeering tractor and oversteering semi-trailer

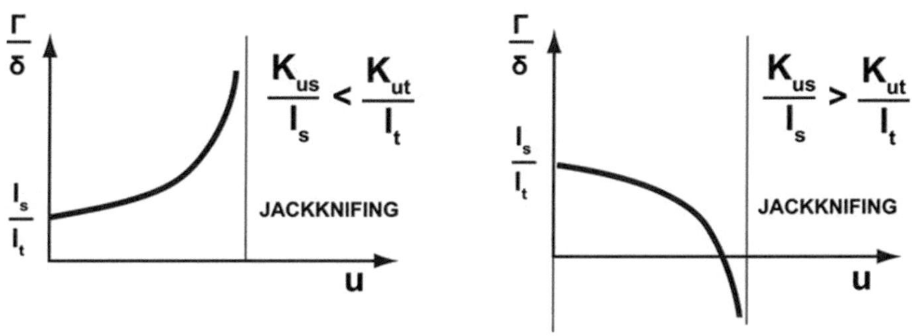

Fig. 4.16 Steady-state response in case of oversteering of both tractor and semi-trailer

rollover too. This kind of jackknifing is different than the jackknifing caused by lock-up of the tractor rear axle(s) due to heavy braking or spinning of the tractor and then closing to the semi-trailer. The steady-state jackknifing is a very alarming situation in case the match of the tractor and semi-trailer is wrong.

4.6 Handling Characteristics of Three-Axle Truck

Figure 4.17 shows a simplified model of a straight truck has tandem drive axles. The handling characteristics of the analysis of the handling diagram are more involved in comparison to a single rear axle vehicle. The complexity is the fact of multiple or infinite number of handling diagrams depending on the level of lateral acceleration during steady-state maneuvers.

For example, if we perform the tests at constant speed u_1 and then repeat the tests at different speeds u_2, this will result in two different handling diagrams as shown in Fig. 4.18.

This phenomenon is explained using the relationship between the steering angle and the lateral acceleration as follows:

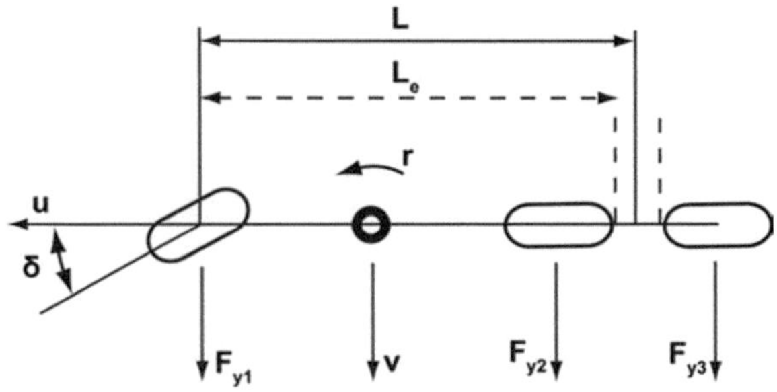

Fig. 4.17 Bicycle model of a three-axle straight truck

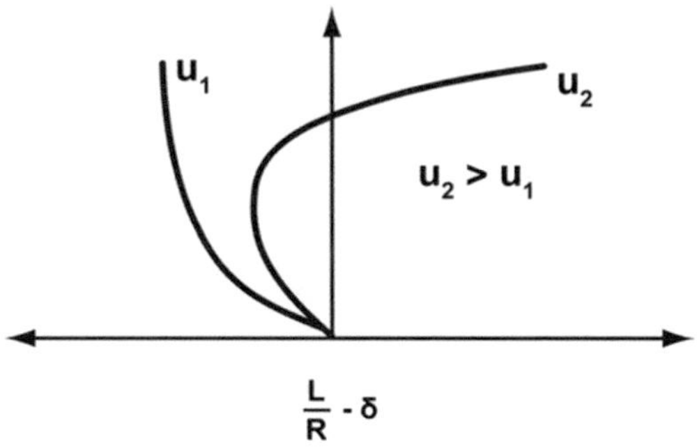

Fig. 4.18 Handling diagram for three-axle straight truck at different speeds

$$\delta = \frac{l_e}{R} + K_{us}\frac{u^2}{gR} \tag{4.81}$$

where

$$l_e = l + \Delta \tag{4.82}$$

$$\Delta = f(C_{\alpha 1}, \text{vehicle parameters}, u) \tag{4.83}$$

where l_e is the equivalent wheelbase that should be used to calculate and plot the handling diagram and Δ represents the modification of the geometric wheelbase l (distance between the front axle to the mid point of the tandem axle).

4.7 Nonlinear Handling Diagram for a Tandem-Axle Tractor

For maneuvers, which results in high lateral acceleration and side-to-side load transfer, the lateral force generated by the tires is a nonlinear function of slip angle:

$$\frac{l}{R} - \delta = \psi\left(a_y, \frac{1}{u^2}\right) \tag{4.84}$$

In the case of a tandem axle tractor, the handling diagram is sensitive to forward speed. The following can be concluded:

1. For a two-axle vehicle, a unique handling diagram (valid for a range of forward speeds) can be constructed and it is possible to examine the directional stability of the vehicle over an entire range of operating speeds.
2. For a vehicle with a tandem axle (three-axle), there is a family of handling curves and the vehicle directional stability can be examined from the handling diagram constructed at different constant speeds.

4.8 Evaluation of Handling Characteristics and Performance Measures of Single and Articulated Vehicles

This section presents the various evaluation methods to evaluate the handling characteristics and performance measures for single and articulated vehicles.

4.8.1 Evaluation of Handling Characteristics of Road Vehicles

To evaluate the handling characteristics of a road vehicle, various types of field or virtual tests can be conducted:

- Constant Radius Test,
- Constant Speed Test, and
- Constant Steer Angle Test.

During the tests, the steer angle, forward speed, and yaw velocity (rate) of the vehicle are usually measured. Yaw velocity is measured by a rate-gyro or determined by the lateral acceleration divided by vehicle forward speed. Lateral acceleration is measured by an accelerometer or determined by yaw velocity multiplied by vehicle forward speed. Based on the relationship between the steer angle and yaw velocity, the handling characteristics of the vehicle can be evaluated.

1- Constant Radius Test

In this test, the vehicle is driven along a curve with a constant radius at various speeds. The steer angle δ with the corresponding lateral acceleration is measured. The handling behavior of the vehicle can then be determined from the slope of the steer angle-lateral acceleration curve as shown below using Fig. 4.19.

$$\frac{d\delta}{d(a_y/g)} = K_{us} \tag{4.85}$$

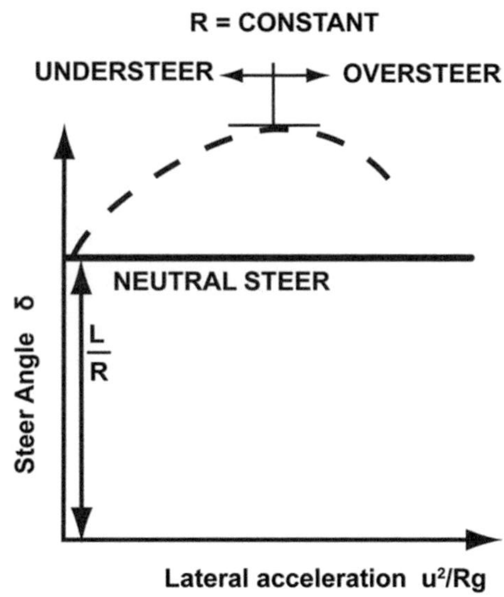

Fig. 4.19 Assessment of handling characteristics by constant radius test [1]

This indicates that the slope of the curve represents the value of the understeer coefficient. The following handling characteristics can be obtained:

- If the steer angle required to maintain the vehicle on a constant radius turn is the same for all forward speeds, the vehicle is neutral steer.
- The vehicle is considered to be understeer when the slope of the steer angle-lateral acceleration curve is positive, which indicates the value of the understeer coefficient K_{us} being greater than zero.
- The vehicle is considered to be oversteer when the slope of the curve is negative, which indicates the value of the understeer coefficient K_{us} being less than zero.

When the understeer coefficient $K_{us} < 0$, which means the slip angle of the front tire α_f being less than that of the rear tire α_r, the steer angle δ required to negotiate a given curve decreases with an increase of vehicle forward speed. A vehicle with this handling property is said to be "oversteer" (Fig. 4.20).

2- Constant Speed Test

In this test, the vehicle is driven at a constant forward speed at various turning radii. For a constant speed turn, the slope of the curve is given by

$$\frac{d\delta}{d(a_y lg)} = \frac{gl}{u^2} + K_{us} \tag{4.86}$$

Figure 4.21 shows the constant speed test results. It can be noted that when the slope of the curve is zero, it indicates that the oversteer vehicle is operating at the critical speed as shown by

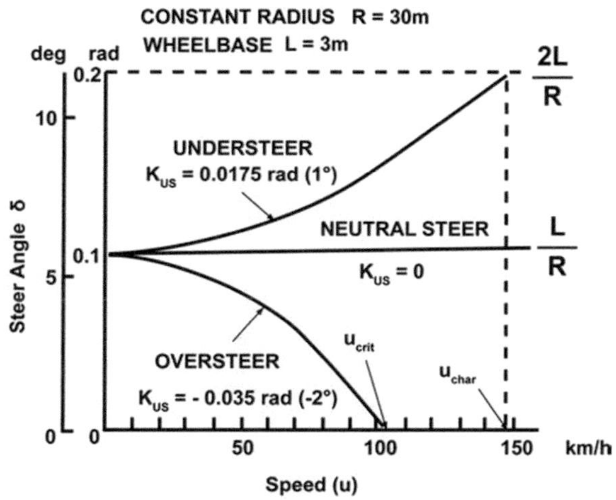

Fig. 4.20 Constant radius test results [1]

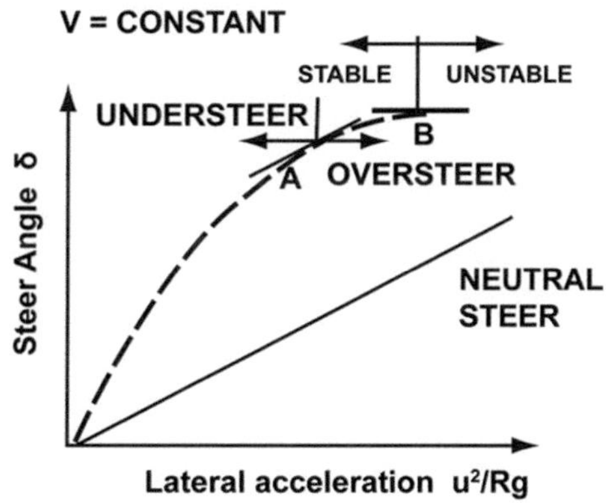

Fig. 4.21 Assessment of handling characteristics by constant speed test [1]

$$\frac{gl}{u^2} + K_{us} = 0 \text{ and } u^2 = \frac{gl}{-K_{us}} = u_{crit}^2 \tag{4.87}$$

This indicates that the slope of the curve represents the value of the understeer coefficient. The following handling characteristics can be obtained:

- For neutral steer, the value of the understeer coefficient K_{us} will be zero and the slope of the steer angle-lateral acceleration line will be a constant of gl/u^2.
- For understeer when the slope of the steer angle-lateral acceleration curve is greater than that, which indicates that the value of the understeer coefficient K_{us} is positive.
- For oversteer when the slope of the curve is less than that, which indicates that the value of the understeer coefficient K_{us} is negative.

3- Constant Steer Angle Test

In this test, the vehicle is driven with a fixed steering wheel angle at various forward speeds. The lateral accelerations at various speeds are measured. From the

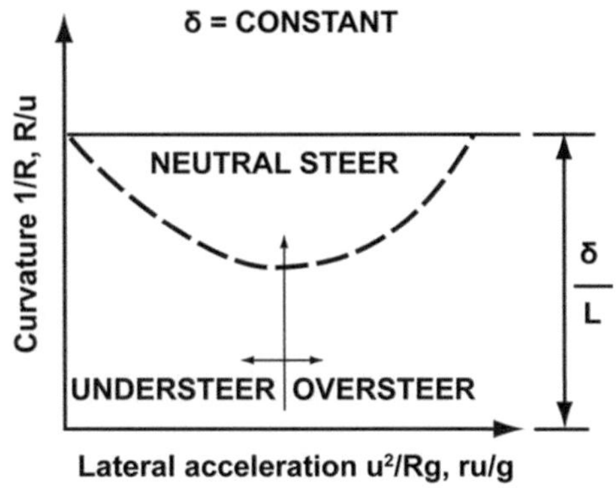

Fig. 4.22 Assessment of handling characteristics by fixed steer angle test [1]

test results, the curvature $1/R$, which can be calculated from the lateral acceleration and forward speed by $1/R = a_y/u^2$, is plotted against lateral acceleration.

$$\frac{d(1/R)}{d(a_y/g)} = -\frac{K_{us}}{l} \tag{4.88}$$

From Fig. 4.22, it can be noted that the following handling characteristics can be achieved:

- For neutral steer, the value of the understeer coefficient K_{us} will be zero, and the slope of the curvature-lateral acceleration curve is zero.
- For understeer when the slope of the curvature-lateral acceleration curve is negative, which indicates that the value of the understeer coefficient K_{us} is positive.
- For oversteer when the slope of the curvature-lateral acceleration curve is positive, which indicates that the value of the understeer coefficient K_{us} is negative.

Table 4.3 is summarizing and comparing these three methods for evaluating the handling characteristics of a vehicle.

It should be mentioned that the weight distribution of the vehicle and the cornering stiffness of the tires are the prime factors of controlling the steady-state handling characteristics of a vehicle.

- Front-engine, front-wheel-drive car tend to exhibit understeer behavior.
- Rear-engine, rear-wheel-drive car tend to have oversteer characteristics

4.8.2 Evaluation of Performance Measures of Heavy Trucks

In this section, an example of how the performance measures of a heavy truck, regardless of its number of axles, are evaluated is shown. As it was explained if

Table 4.3 Comparison between various handling test methods

Constant radius test	Constant speed test	Constant steer angle test
The simplest and requires little instruction	It is more representative of the actual road behavior of a vehicle t than the constant radius test	It is easy to execute
The steer angle of the front tire and forward speed are the only essential parameters to be measured during the test	As the driver usually maintains a more or less constant speed in a turn	Both the constant speed and constant steer angle tests would require the measurement of the lateral or yaw velocity

the truck or the lead unit has more than 2 axles, it will be a family of handling diagrams depending on testing speed. Therefore, the examination of the handling diagram will be performed only at 100 km/h. Regardless of the number of axles, the low-speed and high-speed performance evaluation will be performed at 8.5 km/h and 100 km/h, respectively. The physical testing of heavy trucks can be very costly; therefore, virtual testing using reliable design parameters and high end multibody simulation package can be acceptable. The suggested performance measures can be summarized as follows:

1. Handling performance.
2. Static Rollover Threshold (SRT).
3. Dynamic rollover stability, including
    a. Load Transfer Ratio (LTR).
    b. Rearward Amplification (RWA).
4. Low-Speed Friction Demand (LFD).
5. Offtracking, including
    a. Low-Speed Offtracking (LOF).
    b. Steady-state High-Speed Offtracking (HOF).
    c. Transient Offtracking (TOF).
6. Lateral Low-Speed Friction Utilization (LFU).
7. Lateral High-Speed Friction Utilization (HFU).

Table 4.4 shows the suggested target values for the performance measures benchmark.

Definitions, and method for evaluation, of these measures are described in detail in the following sections.

Table 4.4 Performance measure benchmarks

Performance measures	Performance measure benchmark
Static rollover threshold	≥ 0.4 g
Load transfer ratio	≤ 0.60
Rearward amplification	≤ 1.6
Friction demand	≤ 0.10
Lateral friction utilization (low speed)	≤ 0.80
High-speed offtracking	≤ 0.46 m
Low-speed offtracking	≤ 6.00 m
Transient offtracking	≤ 0.8 m
Handling performance (point 1)	≥ 0.20 g's
Handling performance (point 2)	≥ -4.52 deg/g
Handling performance (point 3)	$\geq 0.50 \leq 2.00$ deg/g

4.8.2.1 Handling Measures

Description: In this measure, handling performance is evaluated by using the Three-Point steady-state conditions.

Method of Evaluation:

First Point: The lateral acceleration at which the transition from understeer to oversteer during a ramp-steer maneuver at a vehicle speed of 90 km/h should be equal or higher than 0.2 g. The ramp-steer rate of 2.0 deg/s will be applied at the steering wheel, yielding an approximate rate of increase in lateral acceleration between 0.01 and 0.02 g's per second.

Second Point: This concerns the understeer coefficients at 0.3 g's, a coefficient that must be higher than the critical understeer coefficient to ensure stability.

Third Point: This concerns the understeer coefficient at 0.1 g's. The understeer coefficient in this case should be in the range of 0.5 to 2 deg/g to ensure reasonable controllability of the vehicle within its common operating range.

The calculations needed to evaluate the understeer coefficient used in the construction of the handling diagram are based on a constant vehicle speed of 100 km/h, using $[(Lr/u - \delta_{sw}/N_g), A_y]$, where δ_{sw} is the steering wheel angle, N_g is the steering box gear ratio, and r is the steady-state tractor yaw rate. The reason for using the nominal front axle steering angle (δ_{sw}/N_g) instead of the actual front axle steer angle, δ, is to account for the understeer attributable to the steering system compliance. Accordingly, the understeer coefficient of interest, K_u, expressed as "degrees per g", is defined by

$$K_u = \frac{d\left(\delta_{sw}/N_g - lr/u\right)}{dA_y} \tag{4.89}$$

The pass/fail criterion is addressed by comparing K_u with the critical understeer coefficient, K_{ucr}, which can be expressed as $-gl/u^2$, where u is the vehicle speed ($u = 100\,\text{km/h}$), l is the tractor geometric wheelbase, and g is the acceleration due to gravity (9.81 m/s^2). If the value of K_u is greater than the target value K_{ucr}, the vehicle will pass the criterion.

4.8.2.2 Static Rollover Threshold (SRT)

Description: The common definition of static rollover threshold has been "the level of lateral acceleration, in units of g's, beyond which overturn occurs in a steady turn". A precise interpretation of this definition does not permit the assessment of rollover threshold when the yaw divergence of a vehicle occurs at a level of lateral acceleration that is less than the rollover threshold value. It is well known that some vehicles exhibit yaw divergence at lateral acceleration levels lower than their rollover threshold. Apparently, they cannot achieve a steady-state turn at these high lateral accelerations. In cases such as this, a static roll model or a physical test must be used for evaluation purposes in place of a time-series model of vehicle response. The static rollover threshold, therefore, is defined as the maximum lateral acceleration level in g's beyond which static rollover of a vehicle occurs.

Method of Evaluation: The recommended method for the evaluation of the static rollover threshold is to apply ramp steering wheel input rate of 2.0 deg/s at 100 km/h and calculate the lateral acceleration at which all the inner wheels lift off the ground except the front axle wheel.

4.8.2.3 Dynamic Rollover Stability Measures

The dynamic rollover stability, in the form of the load transfer ratio and rearward amplification ratio, is evaluated with a time-series simulation model such as the UMTRI Yaw/Roll Model. The path-follow method used in the Vehicle Weights and Dimensions Study is replaced by a path-follow lane-change of the front axle of a tractor. This change allows for a more complete assessment of vehicle performance, particularly about the design of the tractor.

Description: The dynamic rollover stability measures can be described as follows:

1. Load Transfer Ratio (LTR): The LTR is defined as the ratio of the absolute value of the difference between the sum of the right wheel loads and the sum of the left wheel loads, to the sum of all the wheel loads. For vehicles with trailer units decoupled in roll, such as the A-Train, load transfer ratio calculations apply only to the independent units. The front steering axle is excluded from the calculations because of its relatively high roll compliance.

2. Rearward Amplification (RWA): The rearward amplification ratio is a frequency-dependent measure, defined as the ratio of the peak (positive or negative) lateral acceleration at the center of gravity of the rearmost trailer to the amplitude of controlled lateral acceleration of 0.15 g at the center of front axle of the lead unit (tractor).

Method of Evaluation: These two measures are obtained during a rapid high-speed path-change maneuver conducted at 100 km/h, yielding a lateral acceleration amplitude at the center of the front axle of a tractor of 0.15 g's within time period constraints of 3.0 s. The recommended maximum values for the rearward amplification and load transfer ratio are 2.0 and 0.6, respectively [2].

4.8.2.4 Low-Speed Friction Demand Measure (LFD)

Description: The low-speed friction demand measure describes the tire friction levels required at the drive axles of a tractor or straight truck during a tight-radius, low-speed, path-following turn. On slippery surfaces, the friction demand at the tractor's rear axles may exceed that which is available from the road surface if the trailer has a widely spaced axle group. Excessive friction demand is a contributing factor to tractor jackknife and results in excessive tire wear at low-speed maneuvers [3].

Method of Evaluation: This measure characterizes the tire/road friction level demanded of the drive axle group of a tractor or straight truck during a 90-degree turn at a vehicle speed of 8.25 km/h (5.1 mph). During the maneuver, the center of the front steer axle tracks an arc with a 12.8-m radius (approximately a 14-m outside-wheel-path radius). The target maximum allowable value for LFD is 0.1.

The friction demand is calculated as follows:

$$LFD = \left| \frac{Fd_y}{\cos\Gamma Fd_z} \right| \tag{4.90}$$

where

- Fd_y is the sum of the tractor tridem drive axles cornering forces;
- Fd_z is the sum of the tractor tridem drive axles vertical loads;
- Γ is the articulation angle between the tractor and the first semi-trailer.

4.8.2.5 Offtracking Measures

Description: The offtracking measures describe the steady-state low-speed, high-speed, and transient high-speed offtracking.

Method of Evaluation:

1. Low-speed Offtracking (LOF): Low-speed offtracking is evaluated at a vehicle speed of 8.25 km/h using a tight turn maneuver with a radius of 12.8 m (measured to the center of front steer axle). Offtracking is defined as the maximum lateral displacement of the centerline of the last axle of the vehicle from the path taken by the steer axle. A value of 6.0 m is considered the maximum allowable amount of low-speed offtracking.
2. Steady-State High-Speed Offtracking (HOF): The high-speed offtracking measure is obtained when the vehicle is operated in a shallow turn with a radius of 393 m, at a speed of 100 km/h, and is thus attaining a lateral acceleration level of 0.2 g. A threshold value of 0.46 m has been identified to depict the condition in which a minimal clearance of 0.15 m remains between the trailer tires and the outside of a 3.66-m wide conventional traffic lane, when a 2.44-m-wide tractor follows a path down the centerline of the lane.
3. Transient High-Speed Offtracking (TOF): The transient high-speed offtracking measure is computed following the execution of the high-speed path-following lane-change maneuvers used to determine the dynamic roll stability measures. A threshold value of 0.8 m has been selected as being representative of the maximum allowable amount of TOF.

4.8.2.6 Lateral Friction Utilization (LFU)

Description: The lateral friction utilization measures describe the steady-state low-speed and high-speed friction utilized by the axle groups of the vehicle combinations.

Method of Evaluation:

1. Low-speed Friction Utilization (LFU): Low-speed friction utilization is evaluated at a vehicle speed of 8.25 km/h using tight turn maneuvers with a radius of 12.8 m (measured to the center of front steer axle). The measure is calculated at each axle group and should not exceed 80% for all axles.
2. High-Speed Friction Utilization (HFU): This measure is obtained during a rapid high-speed path-change maneuver conducted at 100 km/h, yielding a lateral acceleration amplitude at the center of the front axle of a tractor of 0.15 g's within time period constraints of 3.0 s. The measure is calculated at each axle group and should not exceed 80% for all axles.

The low- and high-speed friction utilization is calculated as follows:

$$FU = \left| \frac{F_y}{\mu_p F_z} \right| \times 100 \tag{4.91}$$

where

- $F_y$is the cornering force at each axle,
- $F_z$is the vertical load at each axle, and
- μ_p is the peak tire/road coefficient of adhesion.

Problems

1. A passenger car weighs 20.02 kN and has a wheelbase of 279.4 cm. The center of gravity is 127 cm behind the front axle. If a pair of radial-ply tires, each of which has a cornering stiffness of 45.88 kN/rad, are installed in the front, and a pair of bias-ply tires, each of which has a cornering stiffness of 33.13 kN/rad, are installed in the rear, determine whether the vehicle is understeer or oversteer. What would happen to the steady-state handling characteristics (understeer/oversteer) of the vehicle, if the front and rear tires were interchanged? Also, calculate the characteristic speed or the critical speed of the vehicle as appropriate.
2. A sports car weighs 9.919 kN and has a wheelbase of 2.26 m. The center of gravity is 1.22 m behind the front axle. The cornering stiffness of each front tire is 58.62 kN/rad and that of each rear tire is 71.36 kN/rad. The steering gear ratio is 20:1. Determine and plot the steady $\neq$ state yaw rate (velocity) and lateral acceleration gains to steering wheel angle of the vehicle in the speed range of 10 to 160 km/h.

References

1. Jo YW (2008) Theory of ground vehicles. John Wiley & Sons
2. John W, Paul M (2007) Safety analysis of a double & triple b-train carrying loaded containers. In: Report prepared for saskatchewan highways and transportation
3. Paul S, Fancher Jr, Arvind M (1989) Safety implications of trucks designed to weigh over 80,000 pounds. SAE Trans 145–155

Chapter 5
Vehicle Rollover Dynamics

Although a technically correct definition of rollover would be the state at which the overall center of gravity (CG) of the vehicle has moved laterally past the vehicle's "balance point", researchers typically define the rollover point or rollover threshold as the state where the load from one side of the vehicle has transferred to the other side. This more conservative definition of rollover is standard in the literature for the main reason of safety.

5.1 Simple Rigid Model

The cause of vehicular rollover is apparent in even the simplest analytical models. For instance, in the process of analyzing more complex models, Ervin [1] examines the static roll performance of the completely rigid vehicle shown in Fig. 5.1. The vehicle is modeled as lying on a flat surface and is made up of a single axle rigidly attached to the vehicle body (i.e., no suspension) and with rigid tires. The model has a single DOF (the roll angle ϕ) and is general enough to describe any type of rigid multi-axle vehicle, e.g., passenger cars, utility vehicles, or heavy articulated or non-articulated vehicles.

As the vehicle negotiates a turn (in this case a right-hand turn), a lateral force is developed, Wa_y on the vehicle, which may be modeled as being applied to the CG of the vehicle. Note that a_y is measured in g's unit and therefore the units of Wa_y are still those of force. In addition, lateral friction at the tires is not shown in the diagram. Summing the moments on the vehicle about the tire/road interface centered between the tires and invoking equilibrium reveals (for small angles) as shown in the equation below:

$$Wa_y h + (F_1 - F_2)\, T + W h \phi = 0. \tag{5.1}$$

M. El-Gindy and Z. El-Sayegh, *Road and Off-Road Vehicle Dynamics*,
https://doi.org/10.1007/978-3-031-36216-3_5

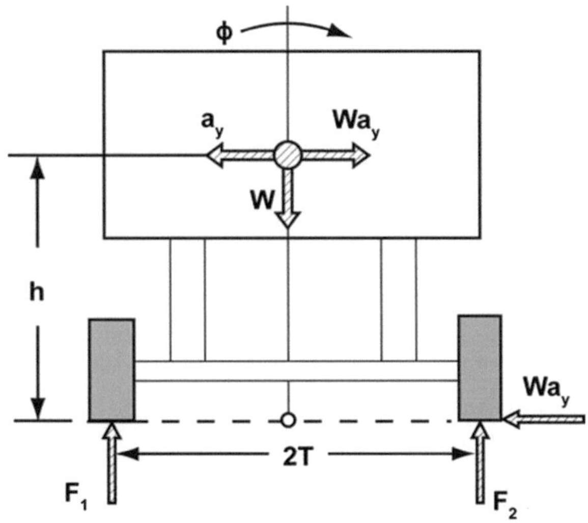

Fig. 5.1 Rigid vehicle model

Ervin refers to the terms from left to right side as the primary overturning moment, the restoring moment, and the lateral displacement moment. This equation may be rearranged into a more convenient form as shown below

$$W A_y h = (F_2 - F_1)\, T - W h \phi \tag{5.2}$$

Note how the lateral displacement moment term caused by roll of the vehicle subtracts from the restoring moment term available at the tires. Ervin refers to the summation of the two terms as the net stabilizing moment and points out that rollover occurs when the left side of Eq. 5.2 is greater than the right side, i.e., the primary overturning moment is greater than the net restoring moment. However, to examine the so-called roll threshold of this vehicle, we note that since this vehicle is rigid, the only way for it to exhibit a non-zero roll angle (non-zero ϕ) is for all the weight to have shifted to one side of the vehicle. Therefore, during a right-hand turn and at the instant the vehicle in Fig. 5.1 starts to roll, $F_1 = 0$, $F_2 = W$, and $\phi = 0$. Since the roll threshold is defined as the value of lateral acceleration at this point, solving Eq. 5.2 for the lateral acceleration yields

$$A_y = \frac{T}{h} \tag{5.3}$$

It should be noted that the moment required to complete the rollover $W(T - h\phi)$ is reducing in a linear manner as the roll angle is increasing; therefore, less moment is required to complete the rollover. Therefore, the static roll threshold of a rigid vehicle is simply the ratio of half-track width to CG height that is commonly referred to as the

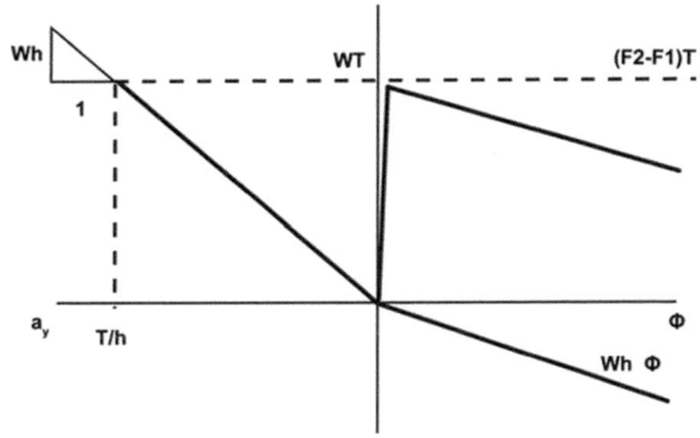

Fig. 5.2 Rollover diagram for a rigid model

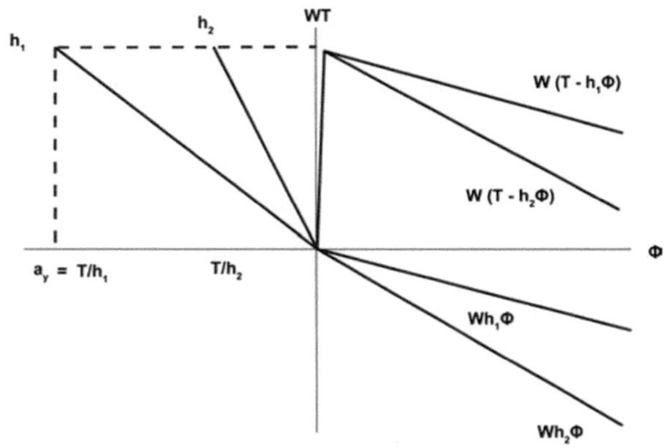

Fig. 5.3 Effect of increasing the center of gravity height on SRT

Track Width Ratio (TWR). Although the rigid vehicle model is extremely simplified, the TWR provides a useful starting point for more complex rollover models.

Figure 5.3 shows the effect of increasing the center of gravity height on the static rollover threshold (SRT) of the rigid model. It can be seen that if the center gravity height is increased from h_1 to h_2, the SRT will be reduced from T/h_1 to T/h_2.

? Example 5.1

A vehicle weighs 20 kN. The center of gravity height is 100 cm. The suspensions and tires are assumed to be rigid. The average track width is 150 cm. Determine the static rollover threshold (SRT) and plot the rollover moment diagram up to a roll angle of 10 degrees. For the following cases:

1. the vehicle center of gravity height is increased from 100 cm to 150 cm;
2. the track width of the vehicle is increased from 150 cm to 180 cm;
3. the total weight is increased from 20 kN to 40 kN.

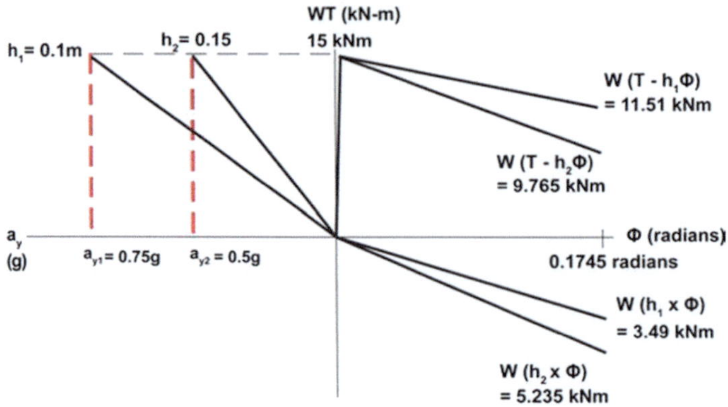

Fig. 5.4 Effect of increasing the center of gravity height on SRT

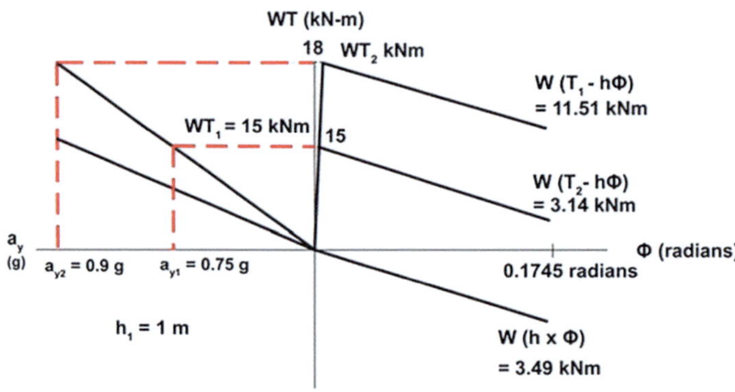

Fig. 5.5 Effect of increasing the vehicle track width on SRT

Solution. For case 1 the center of Gravity Height Changes, determine Static Rollover Threshold given:

$$h_1 = 1\,\text{m};\ h_2 = 1.5\,\text{m};\ T = 1.5/2 = 0.75\,\text{m}$$
$$W = 20000\,\text{N};\ \phi = -0.1745\,\text{rad}$$

SRT $= (a_y)$ when $F_1 = 0$, $F_2 = W$, and $\phi = 0$ from Eq. 5.1

$$a_{y1} = \frac{0.5T}{h_1} = \frac{0.75}{1} = 0.75\,\text{g}$$
$$a_{y2} = \frac{0.5T}{h_2} = \frac{0.75}{1.5} = 0.5\,\text{g}$$

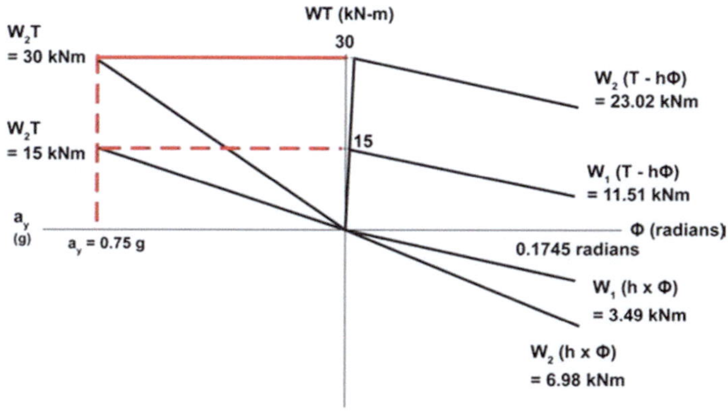

Fig. 5.6 Effect of increasing the vehicle track width on SRT

Then, the new SRT is seen to be 0.5 g. For the graph, the important points are calculated as

$$
\begin{aligned}
WT &= 20000 \times 1.5 \times 0.5 = 15000 = 15\,\text{kNm} \\
W(T - h_1\phi) &= 20000(0.75 - 1 \times 0.1745) = 13429.2\,\text{Nm} = 11/51\,\text{kNm} \\
W(T - h_2\phi) &= 20000(0.75 - 1.5 \times 0.1745) = 19.765\,\text{kNm} \\
W(h_1\phi) &= 20000(1 \times 0.1745) = 3.49\,\text{kNm} \\
W(h_2\phi) &= 20000(0.15 \times 0.1745) = 5.235\,\text{kNm}
\end{aligned}
$$

For case 2, the track width changes, in order to determine the static rollover threshold given:

$$
\begin{aligned}
h = 1\,\text{m};\ T_1 = 0.755\,\text{m};\ T_2 = 0.9\,\text{m} \\
W = 20000\,\text{N};\ \phi = -0.1745\,\text{rad}
\end{aligned}
$$

SRT = (a_y) when $F_1 = 0$, $F_2 = W$, and $\phi = 0$ from Eq. 5.1

$$
\begin{aligned}
a_{y1} &= \frac{T_1}{}\text{h} = \frac{0.75}{1} = 0.75\,\text{g} \\
a_{y2} &= \frac{T_2}{}\text{h} = \frac{0.9}{1} = 0.9\,\text{g}
\end{aligned}
$$

The new SRT is seen to be $0.9\,\text{m/s}^2$; for the graph, the important points are calculated:

$$\begin{aligned} WT_1 &= 20000 \times 1.5 \times 0.5 = 15000 = 15\,\text{kNm} \\ WT_2 &= 20000 \times 1.8 \times 0.5 = 18000 = 18\,\text{kNm} \\ W(T_1 - h\phi) &= 20000(0.75 - 1 \times 0.1745) = 11.51\,\text{kNm} \\ W(T_2 - h\phi) &= 20000(.9 - 1 \times 0.1745) = 3.14\,\text{kNm} \\ W(h\phi) &= 20000(1 \times 0.1745) = 3.49\,\text{kNm} \end{aligned}$$

For case 3, the vehicle weight changes, to determine the static rollover threshold given:

$$\begin{aligned} h = 1\,\text{m};\ T &= 0.75\,\text{m};\ W_1 = 20000\,\text{N} \\ W_2 &= 40000\,\text{N};\ \phi = -0.1745\,\text{rad} \end{aligned}$$

SRT $= (a_y)$ when $F_1 = 0$, $F_2 = W$, and $\phi = 0$ from Eq. 5.1

$$a_{y1} = a_{y2} = \frac{0.5T}{h_1} = \frac{0.75}{1} = 0.75\,\text{g}$$

The new SRT is seen to be 0.75 g; for the graph, the important points are calculated:

$$\begin{aligned} W_1T &= 20000 \times 1.5 \times 0.5 = 15000 = 15\,\text{kNm} \\ W_2T &= 20000 \times 1.5 \times 0.5 = 15000 = 30\,\text{kNm} \\ W_1(T - h\phi) &= 20000(0.75 - 1 \times 0.1745) = 11.51\,\text{kNm} \\ W_2(T - h\phi) &= 40000(0.75 - 1 \times 0.1745) = 23.02\,\text{kNm} \\ W_1(h\phi) &= 20000(1 \times 0.1745) = 3.49\,\text{kNm} \\ W_2(h\phi) &= 40000(1 \times 0.1745) = 6.98\,\text{kNm} \end{aligned}$$

5.2 Compliant Suspension Model

Incorporating a spring suspension between the axle and the body as well as compliant tires, the vehicle is modeled as rolling in a single roll plane, as shown in Fig. 5.2. The major difference between this and the previous rigid model is that the vehicle is able to roll (non-zero ϕ measured from vertical) without all the weight being shifted to one side. Note that ϕ_1 is the roll angle of the vehicle body, or sprungmass. Equation 5.1 still applies, and we see that with a non-zero roll angle, the lateral displacement moment ($Wh\phi_T$) subtracts and therefore reduces the net restoring moment. This in effect reduces the roll threshold of the vehicle to

$$A_y = \frac{T}{h} - \phi_T \tag{5.4}$$

Fig. 5.7 Vehicle model with compliant suspension and tires

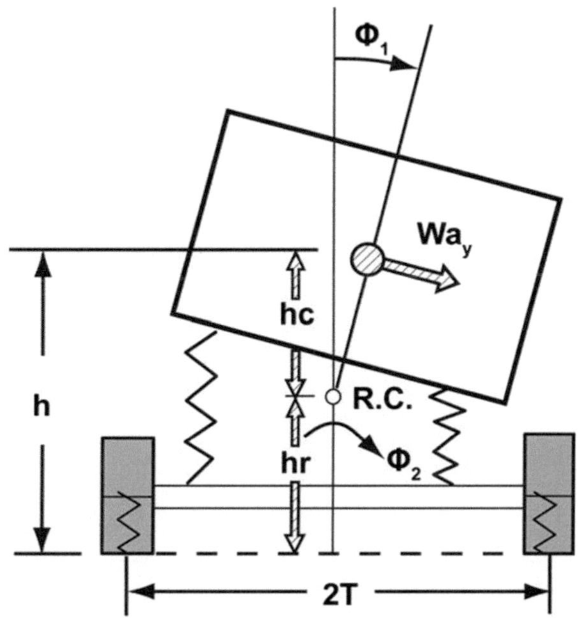

where $\phi_T = \phi_1 + \phi_2$ is the lift-off angle, i.e., the maximum angle the vehicle may roll through until one side of the vehicle "lifts off" the ground. Ervin suggests that this analysis may be extended to vehicles with more than one axle, i.e., multi-axle vehicles as long as the suspension and tire stiffness at each axle is "uniformly proportional" to the load carried by that axle.

Therefore, though multi-axle, this is a lumped suspension model and the vehicle will still roll in a single roll plane with a maximum lateral acceleration given by Eq. 5.4. Figure 5.3 is a graphical representation of this multi-axle lumped suspension model with the value of roll moment plotted on the vertical axis, stabilizing or restoring moments plotted to the right, and lateral acceleration plotted to the left (Fig. 5.7).

The net restoring moment may be considered a combination of the suspension moment (from weight shifting in a turn causing a roll angle ϕ) and lateral displacement moment caused by lateral movement of the vehicle CG. As ϕ increases, the slope of the net restoring moment curve stays positive until it reaches the roll threshold value ϕT and defines the maximum lateral acceleration the vehicle may withstand before rolling over, i.e., the roll threshold. This is the point where the restoring moment (due to the suspension) saturates, i.e., is at its maximum. Past this point, the net restoring moment takes the negative slope of the lateral displacement moment which is still increasing with increasing roll angle. Notice that the roll threshold $(A_y)_{max}$ is less than the roll threshold found with the rigid model, T/h. The suspension roll center is an imaginary but convenient point that the suspension may be thought of as rolling about. Another definition is a point on a vertical plane passing

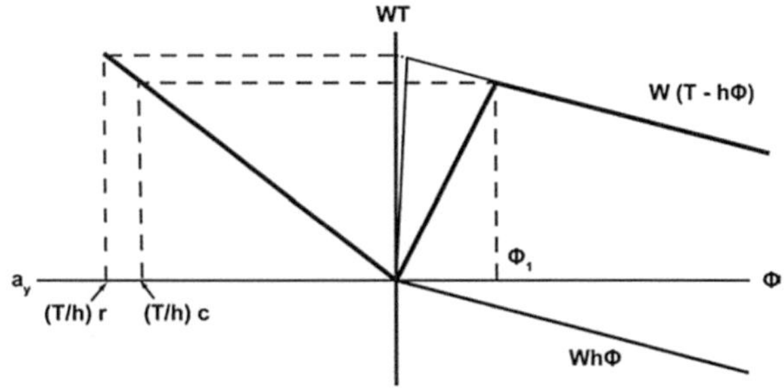

Fig. 5.8 Effect of suspension and tires compliances

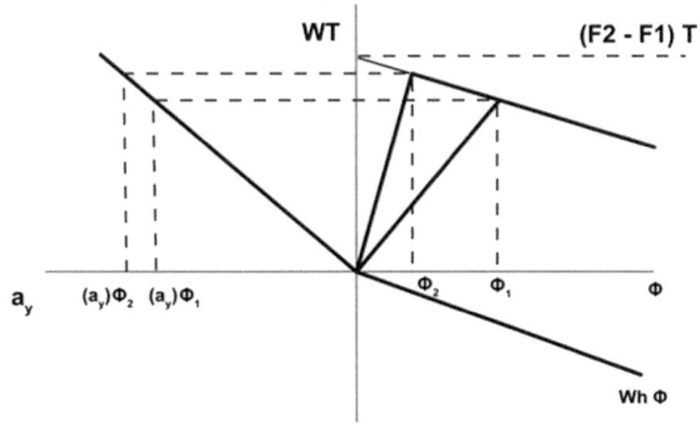

Fig. 5.9 Effect of suspension roll center height

through wheel centers where a force may be applied without causing suspension roll (Figs. 5.8 and 5.9).

Likewise, the tire roll center is another imaginary point at which the axles or unsprung mass may be thought of as rolling about. Therefore, for vehicles with large suspension roll heights (h_r), such as utility vehicles or heavy trucks, the lateral displacement moment ($Wh_s\phi$) is larger per degree of roll angle ϕ. Consequently, these vehicles have reduced roll threshold levels relative to those vehicles with smaller values of suspension roll heights.

By examining these two following cases:

1. If $h = h_r$ (i.e., $h_c = 0$), the sprung mass will roll because of tire compliance and the suspension will have no effect on the roll angle of the sprung mass (the suspension becomes rigid). In this case the angle at which the rollover occurs is (ϕ_2) which can be calculated as follows: $Wa_{y\phi_2}h = (W - 0)T - Wh\phi_2$; or $\phi_2 = \frac{T - a_{y\phi_2}h}{h}$
2. If $h = h_c$ (i.e., $h_r = 0$), the sprung mass will roll because of the suspension and the tire compliance. In this case, there will be no effect on the roll angle of the sprung mass (the suspension becomes rigid). The angle at which the rollover occurs is (ϕ_2) which can be calculated as follows: $Wa_{y\phi_1}h = (W - 0)T - Wh\phi_1$; or $\phi_1 = \frac{T - a_{y\phi_1}h}{h}$.

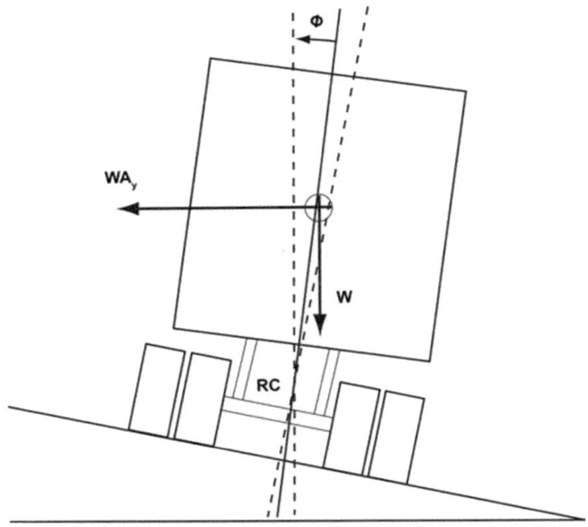

Fig. 5.10 Effect of superelevation on lateral displacement moment (adapted from [3])

5.3 Effect of Superelevated Roadway

Typically, curved transitions on roadways, such as on-ramps and exit-ramps, are designed with super elevations to reduce the effect of lateral acceleration on the roll response of the vehicle traversing the curve. As may be seen in Fig. 5.10, the super elevation has the effect of reducing or possibly even changing the sign of the lateral displacement moment (by reduction of the moment arm) thereby increasing the effective roll threshold of the vehicle [2].

5.4 Rollover of Single Vehicle

Typically, passenger cars have relatively low values of suspension roll center heights and therefore high levels of roll threshold. As a matter of fact, the value of lateral acceleration that a passenger vehicle may support is governed not by the roll threshold, but by the coefficient of friction at the tire/road interface as described by Allen et al. [4]. Therefore, a passenger car will most likely slide laterally before it will roll over (assuming it is not a case of tripped rollover which will be examined later). Targeting the roll stability of passenger cars and utility vehicles and using rollover risk data obtained from an U.S. government report, Allen et al. ran a variety of statistical regressions attempting to correlate various vehicle parameters with rollover propensity. As expected, they found a strong correlation between the rollover rate and TWR. In addition, a strong correlation between the wheelbase ratio (wheelbase divided by CG height) and TWR was found, which indirectly implies a relation between wheelbase ratio and rollover frequency. Finally, the authors conducted field tests with three vehicles: a light-utility vehicle, a subcompact, and "intermediate sized" sedan. These vehicles were believed to have characteristics representative of the vehicles selected for the government report. More specifically, their levels of rollover propensity, defined by the stability metrics obtained in the first part of their research (TWR, etc.), spanned the range of the vehicles selected for the government report. The vehicles were exposed to a variety of common steady-state and transient maneuvers, including limit-performance maneuvers designed to determine vehicle characteristics in extreme situations. A computer simulation was performed, and the results were compared and validated with the field test data. Results of the field tests and computer simulations indicate that the TWR does indeed have a severe effect on a vehicle's roll stability. Specifically, the light-utility vehicle (which has the lowest TWR) was found to roll with two of the limit-performance maneuvers, while the passenger cars were found to be "quite stable". In addition, the longitudinal and lateral load transfer ratio (LTR) was believed to have a severe effect on vehicle directional stability.

In a follow-up paper by Allen et al. [5], the range of test vehicles was increased to twelve for field testing and simulation (the number was later increased to twenty-nine, but these additional vehicles were only for statistical analyses). The types of vehicles included in the test matrix also increased, including utility vehicles, pickups, vans, and front and rear-wheel drive passenger cars. This time, the effect of the LTR, roll stiffness distribution, and throttle settings on vehicle stability during limit maneuvers was examined in more detail. In addition, side-pull tests were conducted on the twelve vehicles calculating the side force "required for rollover". This measure was used in conjunction with the TWR to calculate efficiency factors which illustrate a particular vehicle's resistance to rollover. Results of the analysis of roll stiffness distribution indicated that percentage roll stiffness is typically shifted toward the front axle of the vehicles examined, thereby increasing understeer capabilities. This was particularly true for rear-wheel drive vehicles, which require more inherent understeer characteristics in the suspension to make up for their lack of drive traction at the front wheels. Conversely, due to the drive traction available at the front wheels and more weight (typically) on the front axles, the vehicles investigated had less percentage roll stiffness toward the front than the rear-wheel drive vehicles. This is because front-wheel-drive vehicles require less suspension understeer characteristics.

In the same fashion, Garrott and Heydinger [3] attempted to correlate single-vehicle accident data from the state of Michigan with a wide range of parameters, both vehicular and environmental. The vehicle information included Tilt-Table Ratios (TTR) and parameters such as wheelbase length and whether or not the vehicle was front-wheel drive or rear-wheel drive. In addition, various dynamic measures, such as understeer at various levels of lateral acceleration and yaw rate steady-state gain, were included and calculated solely by computer simulation, using a wide array of steering maneuvers. The dynamic measures were what they had hoped to correlate with the vehicle accident data involving rollovers. However, results showed that six of the most important variables were environmental and static vehicle parameters rather than any of the dynamic metrics they had hoped to correlate. It was concluded that none of the simulated vehicle response metrics were good predictors of rollover propensity for a particular vehicle.

Likewise, Hinch et al. [1] attempted to correlate accident data with both vehicular and environmental metrics. However, they extended the range of both the accident data (taken from five states instead of just one) and the static vehicle metrics, including the TTR, TWR (they refer to the TWR as the Static Stability Factor or SSF), and the Side-Pull Ratio (SPR). Some of the environmental variables were whether the accident happened in a rural or non-rural area, whether the driver was under the influence of alcohol or drugs, and whether the accident occurred on a curved roadway, among others. In addition, they attempted to correlate two other dynamic metrics designed for tripped rollover situations. These are the Rollover Prevention Metric (RPM) and the Critical Sliding Velocity (CSV). The RPM is defined as the difference between the initial translational energy of the vehicle and the rotational energy after tripping divided by the initial translational energy (then multiplied by 100 to get a percentage). Mathematically, this is defined as

$$\%RPM = \frac{E_i - E_r}{E_i} \times 100 \tag{5.5}$$

where E_i is the initial kinetic energy and E_r is the rotational energy after the vehicle has been tripped. An interesting result is obtained when Eq. 5.5 is simplified. It turns out that the RPM is independent of velocity and is only a function of the vehicle's mass, moment of inertia, and height to the CG.

The CSV is defined as the "minimum lateral velocity required to initiate rollover when the vehicle is in a tripping orientation". The equation for the CSV is given in the paper and is only a function of track width, moment of inertia, CG height, and the coefficient of restitution. Initially, Hinch et al. wished to correlate the CSV and RPM to the incidents of tripped rollover from the accident data of the five states. Unfortunately, the accident data was not specific enough to separate tripped rollover from the non-tripped rollover cases, and the two metrics were therefore used to examine their effect (if any) on all the rollover accidents. Other stability metrics included wheelbase length, percentage of vehicle weight on the rear axle (typically affecting oversteer and yaw stability), and a braking stability metric which measures the deceleration of a vehicle required to lift the rear axle off the road. Using the TWR as the reference, a strong correlation was obtained between the TWR and not only the two other static stability measures (TTR and SPR), but the RPM as well. In the second phase of the research, only the Michigan data files were used for statistical correlation since it had the largest sample size and reporting methods seemed to be more consistent and detailed than other states. Hinch et al. were interested in predicting rollover occurrences between like vehicles using the various environmental and stability metrics. The best model, correlating with an R^2 value of 0.80, was found with the TTR, vehicle class/drive configuration, accidents per registered vehicle, whether the accident occurred in a rural area, whether the roadway was curved, whether the accident occurred on or off the road, the age of the driver, whether the driver was under the influence of drugs or alcohol, and the presence or absence of anti-lock brakes.

5.5 Simplified Rollover Model for Two-Axle Vehicle

Figure 5.11 shows a simplified linear two-axle vehicle with three masses. Sprung mass m_S and front and rear unsprung masses m_{UF} and m_{UR}, respectively. The front and rear track widths are T_F and T_R. The roll axis is connecting the front and rear suspension roll centers Rf and Rr. The height of the sprung mass center of gravity from the roll axis is d (called the lever arm) and assumes that the vehicle is in steady-state turn and the lateral acceleration A. The heights of the unsprung masses centers of gravity above ground are H_{UF} and H_{UR}.

The roll angle of the sprung mass about the roll axis ϕ can be determined by equating the input moment at the center of gravity of the sprung mass (point A) with the resisting moment from the suspension and tires' roll stiffness k_{sf} and k_{sr} as follows:

The input moments at point A:

$$M_s = m_s A d \cos\phi + m_s g d \sin\phi = m_s A d + m_s g d \tag{5.6}$$

The total resisting moment from the front and rear suspensions and tires' spring effects in addition to sway bars if installed:

$$M_\phi = k_s \phi = \left(k_{sf} + k_{sr}\right)\phi \tag{5.7}$$

By equating Eqs. 5.6 and 5.7, the angle ϕ can be determined:

$$\phi = \frac{m_s A d}{k_s - m_s g d} \tag{5.8}$$

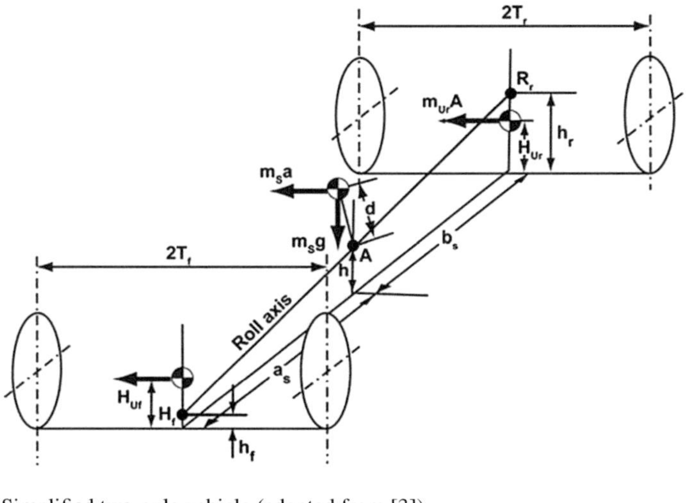

Fig. 5.11 Simplified two-axle vehicle (adapted from [3])

The load transfer from one side to another during turning at a given speed and subjected to lateral acceleration, A, has the following three components:

1. due to suspension roll stiffness,
2. due to centrifugal force at the lateral acceleration of the sprung mass, and
3. due to centrifugal force at the lateral acceleration of the unsprung masses of the front and rear axles.

The equations for these three components can be summarized next:

1. Load transfer (δ) due to Front Suspension's Roll Stiffness (FSS) and Rear Suspension's Roll Stiffness (RSS) is shown below:

$$\Delta FSS = \frac{k_f \phi}{T_f} = \frac{\frac{k_f m_s A d}{T_f}}{k - m_s g d} \tag{5.9}$$

$$\Delta RSS = \frac{k_r \phi}{T_r} = \frac{\frac{k_r m_s A d}{T_r}}{k - m_s g d} \tag{5.10}$$

   where T_f and T_r are the half-track width of the front and rear axles, respectively. The distribution of the spung mass, m_s, which is located at the sprung mass center of gravity can be calculated as follows:

$$m_{sf} = \frac{m_s b_s}{l} \tag{5.11}$$

$$m_{st} = \frac{m_s a_s}{l} \tag{5.12}$$

   where a_s and b_s are the distance between the sprung mass center of gravity and the front and rear axles, respectively. l is the vehicle wheelbase. The distributed masses m_{sf} and m_{sr} are located at the front and rear suspensions' roll centers, respectively.
2. Load transfer due to the distributed Front Sprung mass (FSP) and Rear Sprung mass (RSP) centrifugal forces as seen below:

$$\Delta FSP = \frac{m_{sf} A h_f}{T_f} \tag{5.13}$$

$$\Delta RSP = \frac{m_{sr} A h_r}{T_r} \tag{5.14}$$

   where h_f and h_r are the heights of the front and rear suspensions' roll centers, respectively.
3. Load transfer due to the Front Unsprung mass (FUS) and Rear Unsprung mass (RUS) centrifugal forces as seen below:

$$\Delta FUS = \frac{m_{Uf} A h_{Uf}}{T_f} \tag{5.15}$$

$$\Delta RUS = \frac{m_{Ur} A h_{Ur}}{T_r} \tag{5.16}$$

where m_{Uf} and m_{Ur} are the front and rear unsprung masses and h_{Uf} and h_{Ur} are the front and rear unsprung masses heights.

The total load transfer from one side to another at the front and rear axles during turning can be calculated as follows:

Front: $LTR_F = \Delta FSS + \Delta FSP + \Delta FUS$ Rear: $LTR_R = \Delta RSS + \Delta RSP + \Delta RUS$

? Example 5.2

A light truck weighs 40,000 N and has a wheelbase of 4.0 m. The sprung mass center of gravity is 1.5 m behind the front axle. The height of the sprung mass center of gravity is 1.5 m. The front and rear suspensions roll stiffnesses are 6810 N.m/deg and 10,000 N.m/deg, respectively. The lever arm is 0.50 m. The track widths of the front and rear axles are 90 cm and 95 cm, respectively. The front and rear unsprung masses are 250 kg and 400 kg, respectively. The heights of the unsprung masses' centers of gravity are 0.30 m and 0.40 m, respectively. The heights of the front and rear suspension roll centers are 0.10 m and 0.20 m, respectively. If the vehicle was negotiating a turn of radius of 100 m at a constant speed of 100 km/h, determine the following:

1. The sprung mass roll angle.
2. The front and rear load transfer ratios.
3. The sprung mass roll angle if the front and rear axles are switched.
4. Similar to part 3 in this example if the front and rear axles are switched, calculate the LTR at the "new" front and rear axles and comment on the results. Assume the masses and track widths are switched as well as all center of gravity heights and roll center heights.
5. If the front roll stiffness increased to 400 kN.m and the rear roll stiffness reduced to 200 N.m, what will be the effect on the following?

   a. Roll angle.
   b. Vehicle handling characteristics.

Solution. Given:

$W = 40000\,\text{kg}$, wheelbase $= 4\,\text{m}$, $k_f = 6810\,\text{Nm/deg}$, $k_r = 10000\,\text{Nm/deg}$, $T_f = 0.9\,\text{m}$,
$T_r = 0.95\,\text{m}$, $R = 100\,\text{m}$, $V = 100\,\text{km/h}$, $d = 0.5\,\text{m}$, $h_f = 0.1\,\text{m}$, $h_r = 0.2\,\text{m}$, $m_{uf} = 250\,\text{kg}$,
$m_{ur} = 500\,\text{kg}$, $h_{uf} = 0.3\,\text{m}$, $h_{ur} = 0.4\,\text{m}$

1. Sprung Roll Mass; the sprung mass roll angle can be determined using Eq. 5.8:

$$\phi = \frac{m_s A d}{k_s - m_s g d}$$

The lateral acceleration A is $A = \frac{U^2}{R} = 27.8^2 100 = 7.7 m/s^2$. The sprung mass m_s is calculated as

$$m_s = \frac{40,000}{9.81} - (250 + 400) = 3427.5\,\text{kg}$$

Substituting in Eq. 5.8, the roll angle can be calculated as follows:

$$\phi = \frac{3427.5 \times 7.7 \times 0.5}{[6810 + 10,000] - 3427.5 \times 9.81 \times 0.5} = 4.1^\circ$$

2. The front and rear load transfer ratios; the front and rear load transfer can be calculated from the equation below

Front: $LTR_F = \Delta FSS + \Delta FSP + \Delta FUS$ Rear: $LTR_R = \Delta RSS + \Delta RSP + \Delta RUS$

The load transfer due to front and rear suspensions roll stiffness, 5.6:

$$\Delta FSS = \frac{k_f \phi}{T_f} = \frac{68104.1}{9} = 3102.3\,\text{N}$$
$$\Delta RSS = \frac{k_r \phi}{T_r} = \frac{100004.1}{9.5} = 4315.8\,\text{N}$$

Unsprung masses for front and rear:

$$m_{sf} = \frac{m_s b_s}{l} = \frac{3427.5 \times (4 - 1.5)}{4} = 2142.2\,\text{kg}$$
$$m_{sr} = \frac{m_s a_s}{l} = \frac{3427.5 \times 1.5}{4} = 1285.3\,\text{kg}$$

Load transfer due to the distributed sprung mass centrifugal forces as seen in Eq. 5.8:

$$\Delta FSP = \frac{m_{sf}Ah_f}{T_f} = \frac{2142.2 \times 7.7 \times 0.1}{0.9} = 1832.8\,\text{N}$$

$$\Delta RSP = \frac{m_{sr}Ah_r}{T_r} = \frac{1285.3 \times 7.7 \times 0.2}{0.95} = 2083.4\,\text{N}$$

Load transfer due to the front and rear unsprung masses centrifugal forces as seen below:

$$\Delta FUS = \frac{m_{Uf}Ah_{Uf}}{T_f} = \frac{250 \times 7.7 \times 0.3}{0.9} = 641.7\,\text{N}$$

$$\Delta RUS = \frac{m_{Ur}Ah_{Ur}}{T_r} = \frac{400 \times 7.7 \times 0.4}{0.95} = 1296.8\,\text{N}$$

Front: $LTR_F = \Delta FSS + \Delta FSP + \Delta FUS = 31023.3 + 1832.8 + 641.7 = 33497.8\,\text{N}$
Rear: $LTR_R = \Delta RSS + \Delta RSP + \Delta RUS = 3157.9 + 2083.4 + 1296.84 = 44663.1\,\text{N}$

3. The sprung mass roll angle if the front and rear are switched; the sprung mass roll angle can be determined using

$$\phi = \frac{m_s Ad}{k_s - m_s gd}$$

The lateral acceleration A: $A = \frac{U^2}{R} = \frac{27.8^2}{100} = 7.7\text{m/s}^2$. The sprung mass m_s:

$$m_s = \frac{40,000}{9.81} - (250 + 400) = 3427.5\,\text{kg}$$

Substituting in Eq. 5.8, the roll angle can be calculated as follows:

$$\phi = \frac{3427.5 \times 7.7 \times 0.5}{[6810 + 10,000] - 3427.5 \times 9.81 \times 0.5} = 4.1\,\text{deg}$$

Even by changing the front and rear axles, the sprung mass roll angle stays constant.

4. The front and rear load transfer ratio if the front and rear axles of the vehicle are switched. The front and rear load transfer can be calculated as
Front: $LTR_F = \Delta FSS + \Delta FSP + \Delta FUS$ Rear: $LTR_R = \Delta RSS + \Delta RSP + \Delta RUS$
The load transfer due to front and rear suspensions roll stiffness is calculated as

$$\Delta FSS = \frac{k_f\phi}{T_f} = \frac{10000 \times 4.1}{0.95} = 43157.8\,\text{N}$$

$$\Delta RSS = \frac{k_r\phi}{T_r} = \frac{6810 \times 4.1}{0.9} = 31023.3\,\text{N}$$

Unsprung masses for front and rear:

$$m_{sf} = \frac{m_s b_s}{l} = \frac{3427.5 \times 1.5}{4} = 1285.3\,\text{kg}$$

$$m_{sr} = \frac{m_s a_s}{l} = \frac{3427.5 \times (4 - 1.5}{4} = 2142.2\,\text{kg}$$

Load transfer due to the distributed sprung mass centrifugal forces as seen below:

$$\Delta FSP = \frac{m_{sf} A h_f}{T_f} = \frac{1285.3 \times 7.7 \times 0.2}{0.95} = 2083.4\,\text{N}$$

$$\Delta RSP = \frac{m_{sr} A h_r}{T_r} = \frac{2142.2 \times 7.7 \times 0.1}{0.9} = 1832.8\,\text{N}$$

Load transfer due to the front and rear unsprung masses centrifugal forces:

$$\Delta FUS = \frac{m_{Uf} A h_{Uf}}{T_f} = \frac{400 \times 7.7 \times 0.4}{0.95} = 1296.8\,\text{N}$$

$$\Delta RUS = \frac{m_{Ur} A h_{Ur}}{T_r} = \frac{250 \times 7.7 \times 0.3}{0.9} = 641.7\,\text{N}$$

Front: $LTR_F F = \Delta FSS + \Delta FSP + \Delta FUS = 43157.9 + 2083.4 + 1296.84 = 44663.1\,\text{N}$
Rear: $LTR_R = \Delta RSS + \Delta RSP + \Delta RUS = 31023.3 + 1832.8 + 641.7 = 33497.8\,\text{N}$
It was seen that switching the front and rear axles resulted in the exact same values for the load transfer ratios as before only on the switched axles.

5. If the front roll stiffness increased to 400 kN.m and the rear roll stiffness reduced to 200 N.m, what will be the effect on the following?

   a. Roll angle; the sprung mass roll angle can be determined using $\phi = \frac{m_s A d}{k_s - m_s g d}$.

   The lateral acceleration A: $A = \frac{U^2}{R} = \frac{27.8^2}{100} = 7.7\text{m/s}^2$. The sprung mass m_s:

$$m_s = \frac{40,000}{9.81} - (250 + 400) = 3427.5\,\text{kg}$$

   Substituting in Eq. 5.6, the roll angle can be calculated as follows:

$$\phi = \frac{3427.5 \times 7.7 \times 0.5}{[400,000 + 200] - 3427.5 \times 9.81 \times 0.5} = 0.034^\circ$$

   The new roll angle is seen to be much smaller than before at 0.034° compared to 4.1°.

b. Handling characteristics; by having a very low roll angle, the vehicle will be unstable and susceptible to rollover. This overall is not good for handling and vehicle operation due to the very high stiffness of the front axle when compared with the back (400,000 Nm/deg versus 200 Nm/deg).

5.6 Rollover of Articulated Vehicles

The handling dynamics of an articulated vehicle, such as a tractor semi-trailer, differ from that of a non-articulated vehicle significantly as reported in literature [6]. Among other reasons, the ability for the trailer to articulate or pivot relative to the tractor contributes an additional mass that the driver must be concerned with. In addition, the response of the trailer to inputs from the tractor, such as steering maneuvers, is typically amplified and lags behind the response of the tractor making it difficult to control. Many times, this causes stability problems and a trailer may start to roll excessively or begin to jackknife (yaw instability) before the driver is aware of the problem and therefore may not have time to take corrective action.

As described earlier, Ervin [2] derived the steady-state equations of motion for a multi-axle vehicle rolling in a single roll plane with the maximum lateral acceleration given by Eq. 5.4. It was also noted that the analysis was only valid for vehicles with roll stiffness at each axle that is "uniformly proportional" to the load carried by that axle. However, Ervin observes that in practice, roll stiffness may vary significantly at each axle. This implies Eq. 5.2 is not valid for these vehicles. Therefore, he examined the effect on the vehicle of individual axles with non-uniform suspension stiffness relative to the load carried. Referring to the individual roll stiffness as R_{axle1}, R_{axle2}, and R_{axle3} with $R_{axle1} < R_{axle2} < R_{axle3}$ (typically this is the case), the results are shown graphically in Figure 5 for a three-axle tractor semi-trailer. As the truck undergoes a lateral acceleration, A_y, the first wheel to lift off is the inside wheel of the trailer axle (axle 3) at a roll angle ϕ_3 of the trailer. This is the point where the rear axle produces its maximum restoring moment W_3T_3 where W_3 and T_3 are the weight and track width of axle 3, respectively.

The saturation of the axle 3 restoring moment effectively reduces the slope of the net restoring moment. As A_y increases further, the inside wheel of axle 2 begins to lift off at a roll angle ϕ_2 due to its intermediate level of roll stiffness R_{axle2}. This again has the effect of reducing the net restoring moment even further and this time allows the slope to become negative. The roll angle ϕ_2 is the point of instability of the entire vehicle since the net restoring moment is unable to react to the increasing lateral acceleration and impending rollover ensues. The roll angle ϕ_2 therefore defines the roll threshold of the vehicle $(A_y)_{max}$. Finally, the suspension at the tractor axle, axle 1, saturates at a roll angle ϕ_1, reducing the slope of the net restoring moment even further. It is also clear from the analysis that treating the individual axles with separate roll stiffness reveals a roll threshold $(A_y)_{max}$ less than that when the vehicle is treated with the lumped suspension model as was done previously. Furthermore, both of the roll thresholds are less than that of the rigid model (Fig. 5.12).

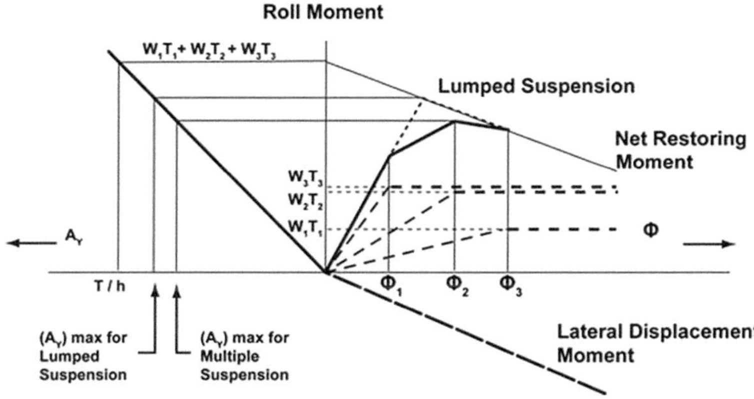

Fig. 5.12 Roll moment diagram of multiple suspension multi-axle vehicle (adapted from [3])

The longitudinal and lateral distribution of cargo also has a profound effect on the roll stability of a tractor semi-trailer. Ervin shows that a forward shift of goods actually decreases the roll threshold of the vehicle. However, he points out that this may increase the truck's yaw stability as previous research suggests [7]. In addition, if the cargo is laterally offset relative to the centerline of the vehicle, a static moment is produced which will increase or decrease the truck's roll threshold depending on which side the load is offset on and in which direction the truck is turning. For example, a truck turning to the right with a cargo offset to the left (thereby causing a static moment to the left) will increase the lateral displacement moment ($Wh\phi$), decrease the net restoring moment, and therefore decrease the roll threshold of the vehicle. Conversely, if the truck makes a left-hand turn with the cargo still to the left, the static moment acts to increase the net restoring moment and therefore provides an increase in the roll threshold of the vehicle.

5.6.1 Static Roll Threshold

Trying to correlate rollover accident data with various vehicle parameters or stability metrics for heavy trucks is a more difficult task than with automobiles, light trucks, and sport utility vehicles. This is for no other reason than the inconsistent reporting of the details in single-vehicle heavy truck rollover accidents. Specifics about the type of truck, load carried, suspension properties, etc., are not always recorded well and therefore make it difficult to correlate heavy truck parameters with accident data. However, one may expect that the TWR of an articulated heavy vehicle has the same first-order effect on the roll stability as it does with non-articulated vehicles. One may also assume a strong correlation between the TWR and the static roll threshold (SRT) as found with non-articulated vehicles. The SRT may be measured in a variety of ways including static tests such as the Tilt-Table or Side-Pull, a quasi-steady-state turning test, or finally mathematically using simulation software such as the Static

Roll Model ([8, 9]) or the Yaw/Roll Model [10]. Indeed, the SRT is considered the single best measure of a particular vehicle's roll stability.

The Static Roll Model, developed at the University of Michigan's Transportation Research Institute (UMTRI), is typically preferred for the calculation of SRTs over other models simply because of its ease of use. Although the results are not as accurate as those that may be obtained with more complex software, such as the Yaw/Roll Model, it requires fewer input parameters and still gives a first-order approximation of the SRT. Typically, accurate input parameters are difficult and/or expensive to collect. Therefore, El-Gindy and Hosamel-deen [11] conducted a sensitivity analysis of the Static Roll Model to separate those vehicle parameters that have a significant effect on the results from those that do not. The less significant parameters may therefore be approximated without affecting the results and eliminating many of the costs. The sensitivity analysis consisted of analyzing a "baseline" vehicle, a five-axle tractor semi-trailer, and adjusting the various parameters individually by a "reasonable percentage" while keeping the others constant at their baseline value. The parameters examined are as follows: **Tractor Frame:** Torsional stiffness and coulomb friction.

Trailer: Combined structural and fifth-wheel stiffness.

Fifth Wheel: Lash.

Tires: Vertical stiffness, lateral stiffness, and overturning stiffness.

Suspension:

Tractor: Spring rates (front and rear), and lash (front and rear).

Trailer: Spring rates, lash, auxiliary roll stiffness, and roll center heights.

Dimensions: CG heights of sprung masses and track widths.

El-Gindy and Hosamel-deen found it useful to separate the results on the sensitivity analysis into three different groups (A, B, and C) of parameters. Group A consists of parameters that had virtually no effect on the results and may be eliminated from the actual model of the system. Group B contains parameters that are required in the model, but the accuracy of the parameters is not very important. Finally, Group C consists of those parameters, which are not only vital to the model, but the accuracy of measured values for these parameters is very important. The results are as follows:

Group A (may be ignored in the model)

- Tractor frame torsional stiffness.
- Tractor frame coulomb friction.
- Overturning stiffness of the tires.
- Lash in the fifth-wheel plates.

Group B (important to model/accuracy of measurement not too important)

- Tractor front spring rates.
- Trailer combined structural and fifth-wheel stiffness.
- Lateral stiffness of the tires.

Group C (important to model/accuracy of measurement very important)

- Tire vertical stiffness.
- Tractor rear spring rates.

- Trailer spring rates.
- Lash in tractor front and rear suspension and trailer suspension.
- Auxiliary roll stiffness.
- Roll center height.
- Sprung mass center of gravity height.
- Wheel track widths.

One of the most frequently cited relationships between heavy truck rollover and SRT was developed by Ervin [12] and is shown in Fig. 5.6. The accident data was obtained from the Bureau of Motor Carrier Safety and describes the rollover occurrence of heavy trucks in single-vehicle accidents (SVAs). Using software developed at UMTRI that was a precursor to the Static Roll Model, Ervin was able to estimate a SRT for each heavy truck using the reported gross vehicle weight (GVW) and using typical values of tire, suspension, and geometric parameters. It is clear from the graph that there exists a strong relationship between a heavy truck's roll threshold and rollover occurrence in an SVA. One of the most notable observations is that a fully loaded truck with a roll threshold between 0.40 and 0.45 g is five to seven times as likely to roll over than an empty vehicle with a roll threshold of 0.65 g.

Preston-Thomas and El-Gindy [13] examined the feasibility of requiring a minimum SRT value of heavy trucks for operation in Canada and possibly all of North America. Data collected from tilt-table tests by Preston-Thomas [13] and Woodrooffe [14] is summarized for a variety of vehicles, including dump trucks, cement mixers, tractor semi-trailers, and a B-train tank truck. The results support the previous suggestions for a minimum SRT ranging from 0.38 to 0.42 g, depending on the type of heavy truck. However, the authors point out the difficulty that may arise in enforcing a minimum SRT due to the wide range of heavy trucks and loads and the "power of special-interest groups". Preston-Thomas and El-Gindy suggest that the policy of requiring a minimum SRT may be successful if started by enforcing the policy on a single type of heavy trucks, such as petroleum-hauling trucks. It is believed petroleum-hauling trucks are a practical place to start since they are all relatively similar in their tank design and typically have lower roll thresholds than that of heavy trucks, and rollover of these trucks is particularly dangerous due to their hazardous cargo. In addition, it is believed that the manufacturers and operators are willing to spend money to improve the safety of these vehicles.

5.7 Factors Affecting Roll Stability

Ervin [12] examined, via simulation, the sensitivity of a heavy truck's roll threshold from variations in size and weight parameters. The goal was to examine the effect on a truck's roll threshold due to legislative changes in requirements of the size and weight variables. Specifically, the effect of changes in axle loading, gross vehicle weight, width, payload CG height, and lateral offset of the payload was examined quantitatively. A summary of the key results is as follows:

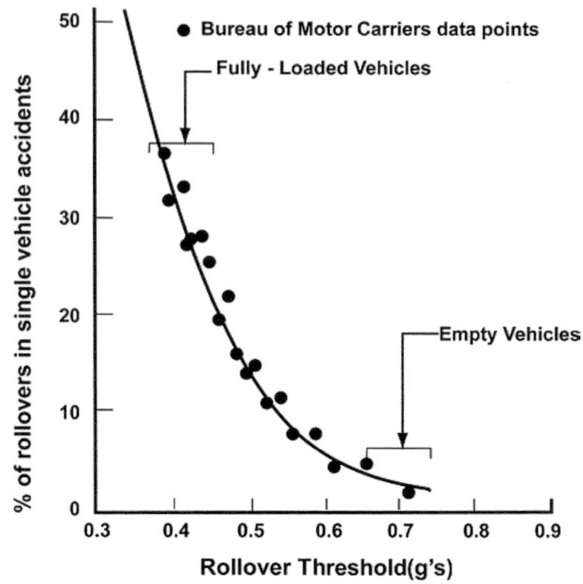

Fig. 5.13 Rollover occurrence versus estimated static roll threshold [6]

Axle Loading—Increases in axle load limit decrease roll threshold and the percentage decrease in roll threshold is approximately in proportion to the percentage increase in axle load limit. In addition, the roll threshold is reduced as the tractor's load is shifted toward the front axle (for a fixed value of gross vehicle weight) thereby increasing the load on the tractor's steering axle. However, as the loading increases, the effect of the load distribution on the roll threshold is not as great as the payload weight or CG height (Fig. 5.13).

Gross Vehicle Weight (GVW)—In all simulations, it was clear that increases in GVW decrease the roll threshold.

Vehicle Width—The width of the vehicle may be observed in three different ways: the width of the trailer bed, suspension width, and track width of the tires. All three were examined for their effect on the roll threshold and it was concluded that in general, increases in the widths have a very beneficial effect on the roll threshold of the vehicle.

Payload CG Height—Increases in payload CG height greatly reduce the roll threshold of the vehicle. In addition, the payload CG height does not seem to create differences of roll threshold between the front and rear trailer units of a double combination. Lateral Offset of Payload CG—It was clear that increases in payload CG lateral offset significantly reduce the roll threshold of the vehicle. Note that these results were also concluded [15].

In a study by El-Gindy [16], various performance measures, which are used to evaluate the dynamic stability of commercial vehicles, are reviewed. Specifically, the paper reviews performance measures used originally in the Canadian Vehicle Weights and Dimension Study of 1984 (sponsored by the Roads and Transportation Association of Canada, or RTAC). The study was an attempt at reforming the various regulatory principles governing commercial vehicle transport and safety. In addition, improvements to these original RTAC measures were subsequently suggested and some new performance measures are reviewed as well. El-Gindy notes that some heavy vehicles can experience yaw instability during a maneuver at lateral accelerations lower than their static rollover threshold. Therefore, the typical definition of the static roll threshold, which states that the SRT is the lateral acceleration at which rollover occurs in a steady turn, cannot apply to these cases of yaw divergence. He therefore proposed a new definition of the SRT as "the maximum lateral acceleration level in g's beyond which static rollover of a vehicle occurs". In addition, he suggests the use of a "validated static roll model" or experimental testing (e.g., tilt-table test) to calculate an SRT and avoid the use of dynamic models such as the Yaw/Roll Model since it does not accurately assess the SRT of vehicles that demonstrate yaw divergence at levels below their rollover threshold.

El-Gindy also reviews two dynamic roll stability measures, the Load Transfer Ratio (LTR) and Rearward Amplification (RWA), and a damping measure that was not included in the original set RTAC measures known as the Yaw Damping Ratio (YDR). Though the YDR is not a direct measure of roll stability, it is included because it may have a substantial indirect effect on the roll response of an articulated vehicle.

Load Transfer Ratio (LTR)—The LTR is defined as "the ratio of the absolute value of the difference between the sum of the left wheel loads and the sum of the right wheel loads, to the sum of all the wheel loads" [17], or

$$LTR = \frac{\left|\sum_{i=1}^{n} (Fr_i - Fl_i)\right|}{\sum_{i=1}^{n} (Fr_i + Fl_i)} \tag{5.17}$$

where Fr_i and Fl_i are the right and left vertical loads, respectively, of the ith axle and where n is the number of axles. The LTR is a useful measure of how close the vehicle is to rollover. It has values ranging from 0 to 1 with an $LTR = 0$ signifying the most stable state with all loads balanced and $LTR = 1$ at rollover with all the weight shifted to one side of the vehicle. Typically, the tractor front axle is neglected in the LTR calculation due to its high roll compliance and therefore negligible effect on the roll response of the vehicle [18]. In addition, the LTR is calculated for individual units that are coupled in roll. For example, a tractor and its semi-trailer are coupled in roll and therefore considered a single unit. Likewise, a B or C-train combination is considered a single unit since the tractor, semi-trailer, and full-trailer are all coupled in roll. However, a combination such as an A-train is decoupled in roll at the full-trailer and the LTR must therefore be calculated for both the tractor and semi-trailer combination as a unit and then for the full-trailer separately. For safety, El-Gindy recommends that the LTR does not exceed 0.6.

Rearward Amplification (RWA)—RWA is a measure of the severity of the rear-most trailer's "reaction" to inputs from the tractor. It is a frequency-dependent measure and is defined as the ratio of the peak lateral acceleration (positive or negative) of the CG of the rearmost trailer to the amplitude of controlled lateral acceleration of 0.15 g measured at the center of the front axle of the tractor [19]. The RWA measure may be applied to single or multiple trailer configurations, and it is recommended that it does not exceed 2.2.

Yaw Damping Ratio (YDR)—The YDR is a measure of the rearmost trailer's ability to dampen the lateral acceleration oscillations. Therefore, a trailer with a relatively small YDR may have exceptionally large oscillations resulting in peak lateral accelerations that may exceed the roll threshold of the vehicle. Though this measure was not one of the original RTAC measures, El-Gindy and others believe it is important since excessive oscillations may lead to an accident. The YDR is evaluated from an 80° pulse input at the steering wheel over a time period of 0.1 s and at a vehicle speed of 100 km/h. The response plot will be a sinusoidal decay (assuming the system is underdamped) and the YDR is simply found using the logarithmic decrement technique. Therefore,

$$YDR = \frac{\delta}{\sqrt{4\pi^2 + \delta^2}} \tag{5.18}$$

where

$$\delta = ln\left(\frac{x_1}{x_2}\right) \tag{5.19}$$

and x_1, x_2 are two successive amplitudes on the response plot. El-Gindy recommends a target value of 0.15 at a vehicle speed of 100 km/h.

5.8 Anti-Roll Suspensions

As described in a previous section, rollover of automobiles from severe handling maneuvers is unlikely to occur due to the limited friction available at the tire/road interface. The vehicles are therefore more likely to slide laterally than rollovers. Research into anti-roll suspensions for automobiles is extensive, but typically the reasoning behind it is less for safety and more for ride comfort or handling performance. Therefore, only anti-roll suspensions for heavy vehicles will be addressed here. Ironically, little research has been conducted into anti-roll suspensions for heavy trucks though they have the greatest rollover risk compared with non-heavy vehicles.

The most basic anti-roll suspension typically just includes a passive anti-roll bar (also referred to as a stabilizer bar). Points A and B are attached to the axle while the horizontal section of the anti-roll bar is attached to the body of the vehicle. As the body of the vehicle rolls through some angle ϕ, and since points A and B cannot move, the anti-roll bar twists thereby creating a reaction torque M_r to help stabilize the vehicle. An active anti-roll suspension includes some kind of torsional actuator

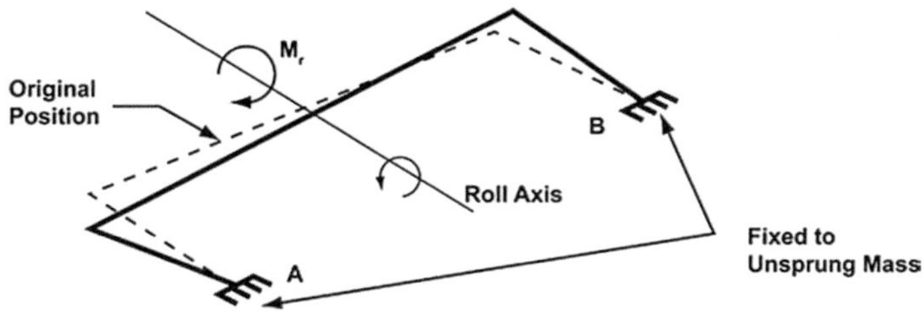

Fig. 5.14 Typical anti-roll (stabilizer) bar [6]

that can dynamically alter the overall stiffness of the system depending on control inputs from an on-board computer (Fig. 5.14).

Li et al. [20] investigate the performance of an active anti-roll suspension for heavy trucks using the Yaw/Roll Model. Initially, a simple linear model of an articulated vehicle, based on work from Segel [19], is used to develop baseline controller gains using feedback during random and step-steer inputs. The controller is then implemented into the more complex Yaw/Roll model for simulation. The results are then compared to that of two other vehicles with the same properties: one with passive suspension and one with a passive suspension and a passive anti-roll bar. The response of the trailer to a step-steer input is shown in Fig. 5.8 for all three types of suspension. It is evident that the active suspension is able to roll the trailer into the turn thereby increasing its roll threshold. It was concluded that an actuator with a bandwidth between 1 and 5 Hz is required and the hydraulic power supply must be able to produce 10 kW for active roll control of the trailer (Fig. 5.15).

Lin et al. [22] continue their investigation of active anti-roll suspensions with optimal roll control. Specifically, they examine two control laws: Lateral Acceleration Feedback (LAFB) using proportional and derivative gains and a controller optimized by the standard Linear Quadratic Regulator (LQR) technique. As in their previous work, a simple linear model [23] of a single-unit articulated vehicle was used. The gains for both the LAFB and LQR controller were tuned using stochastic input data (power spectrums) that was believed to cover the range of typical steering angle input by a driver. Results indicate that the roll threshold of the vehicle was increased by as much as 66% and lateral load transfer was decreased by 34% (for a random steering input) by use of the optimized active anti-roll bar. Figure 5.9 shows the vehicle response to a step-steer input for the passive suspension and the LAFB and LQR active suspensions. It is evident from the plot that both active suspensions perform similarly though the LAFB controller responds slightly faster than LQR, but with a less damped response. In addition, the LQR controller is able to roll the vehicle somewhat further into the turn (reducing the roll threshold). However, the LAFB controller is preferred by Lin et al. over the LQR controller for active roll control of an articulated vehicle due to its simple transducer requirements (Fig. 5.16).

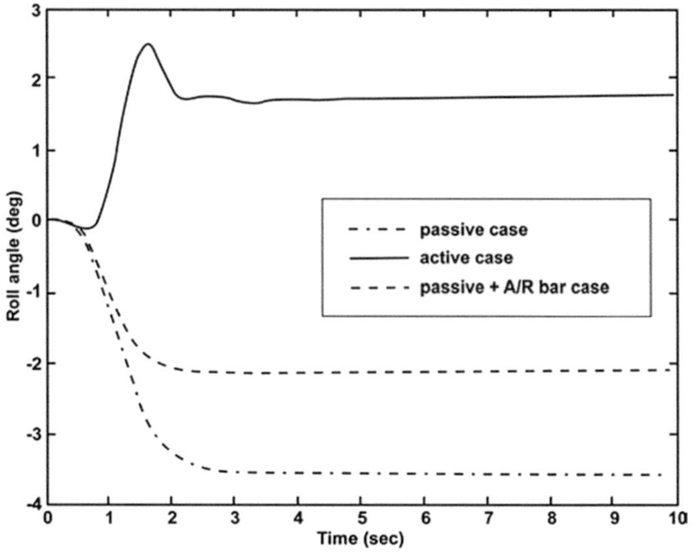

Fig. 5.15 Roll angle responses of trailer with three kinds of suspension for a step input of steering angle [21]

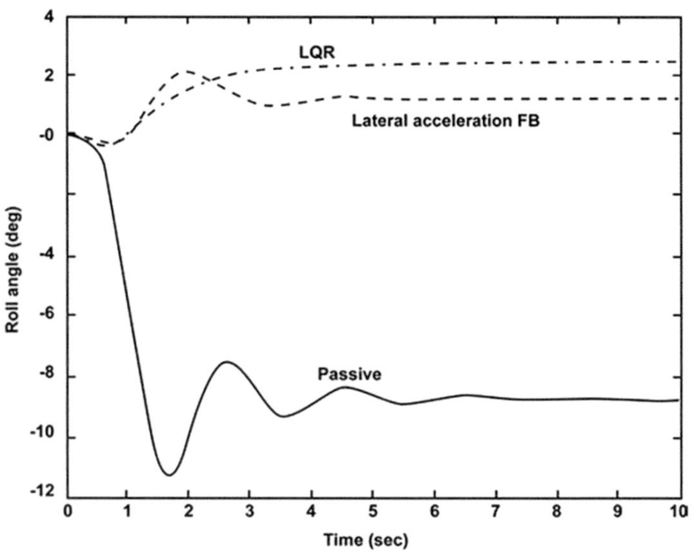

Fig. 5.16 Roll angle responses of trailer with optimized suspensions for a step input of steering angle [24]

5.9 Liquid Versus Rigid Cargo

Up to this point, the roll stability of vehicles with only a fixed CG position, i.e., rigid cargo, has been examined. However, it should be intuitive that a moving CG position, either laterally, longitudinally, vertically, or a combination of all three, can have a significant effect on the roll threshold of a vehicle. Therefore, a review of some of the research into this effect is warranted.

Ranganathan et al. [24, 25] completed an extensive investigation into the effect of liquid cargo on the roll stability of the vehicle. Three different tank designs, circular, modified oval, and modified square shown in Fig. 5.17, were examined and the corresponding equations of describing CG position were solved for various inclination angles of the tank. These quasi-steady-state equations for the movement of liquid within the tank were then integrated into a validated steady-state roll model of the vehicle originally designed for rigid cargo. The solution of the algebraic equations involved incrementally updating specific matrices to account for the movement of the liquid at specific intervals. Finally, the effect of compartmentation within the tank on roll stability, which is sometimes used to minimize the lateral and longitudinal slosh [26], was investigated.

Results indicate that circular tanks are superior to the modified oval and square tanks at minimizing load transfer due to shift of cargo CG position from changes in lateral acceleration. This is evident from Fig. 5.17 where lateral CG translation versus lateral acceleration for a fixed tank tilt angle of 5° is plotted for various fill levels of an inviscid fluid. Likewise, the vertical translation of cargo CG was significantly higher for the modified oval and square tanks than for the circular tank.

Figure 5.18 is a plot of roll threshold versus percent fill level of a circular tank for liquid and equivalent rigid load. It is evident that the liquid cargo has a significant effect on the roll threshold of the vehicle due to the severe lateral translation of the CG over that for the equivalent rigid load. It was estimated that liquid cargo at a 50% fill condition could reduce the roll threshold by 0.10 g compared with that of an equivalent rigid load. Similarly, results for the modified oval and square tanks

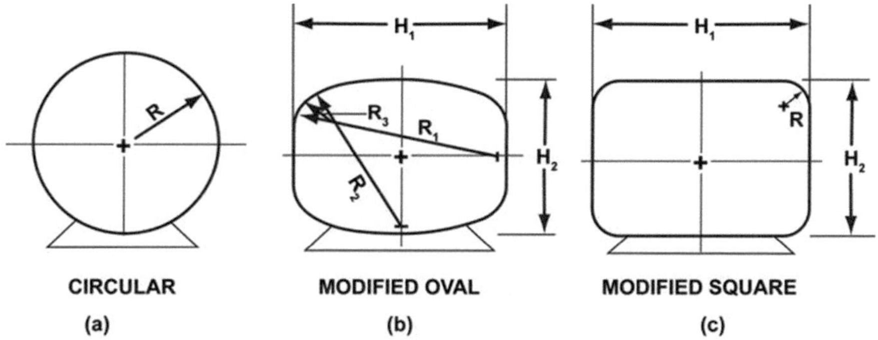

Fig. 5.17 Tank cross-sections in general purpose transportation [27]

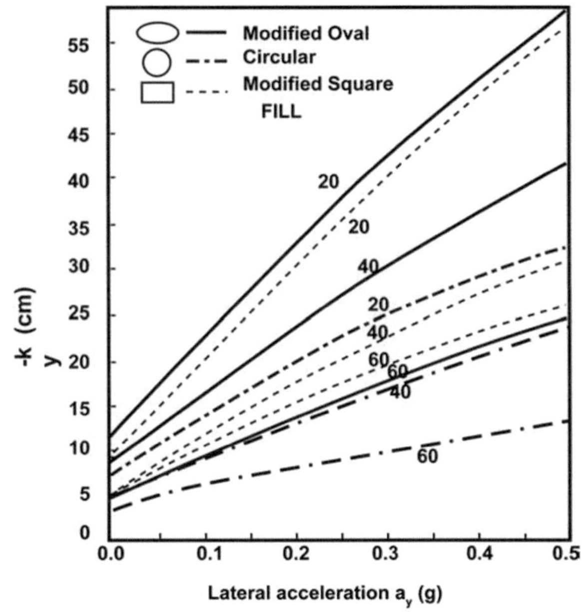

Fig. 5.18 Comparison of rollover acceleration limits for a partially filled cleanbore circular tank vehicle and equivalent rigid cargo vehicle [27]

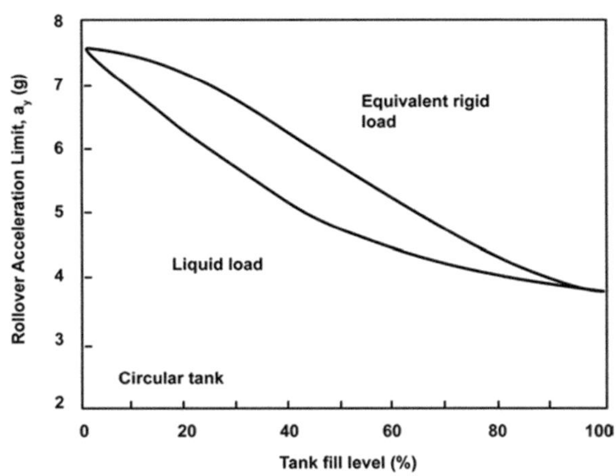

Fig. 5.19 Comparison of rollover acceleration limits for a partially filled cleanbore circular tank vehicle and equivalent rigid cargo vehicle [27]

indicate a 0.15 and 0.20 g reduction in roll threshold over that for an equivalent rigid load, respectively.

Figures 5.18 and 5.19 are plots of roll threshold versus the fill percentage of the cargo for liquid and an equivalent rigid cargo for circular and modified oval tanks, respectively. Note that the total payload weight is fixed and loads on the composite axles are kept constant for these plots. The density of the cargo was altered to keep the weight constant though the fill percentage has changed.

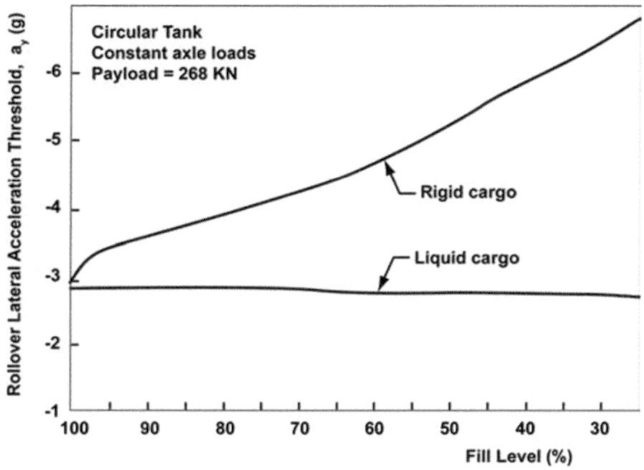

Fig. 5.20 Comparison of rollover lateral acceleration limits of partially filled circular tank vehicle and equivalent rigid cargo vehicle when composite axle loads are held constant [27]

The circular tank data in Fig. 5.20 shows that the roll threshold of the liquid-filled tank vehicle is virtually unaffected by the change in percent fill. Therefore, even though the CG height of the liquid cargo is being reduced with the reduction in percent fill, the excessive lateral offset of the CG has the effect of canceling any gain in roll stability from this. As expected, the roll threshold of the rigid cargo vehicle increases with a reduction in percent fill due to a decrease in CG height and the less significant effect of lateral CG translation. However, Fig. 5.16 shows that for the modified square tank, the lateral translation of the CG outweighs the roll stability gains caused from a reduction in CG height. Similar results were obtained for the modified oval tank.

Compartmentation of the tank, like the four-compartment tank shown in Fig. 5.21, was shown to have a significant effect on the roll threshold of the vehicle. Using a modified oval compartmented tank for simulation purposes, it was concluded that for a 50% fill condition, the highest rollover threshold is obtained when compartment II is partially filled. However, the exact amount of fill is not mentioned in the paper. For fill levels of about 50%, it is suggested that compartments II or III may be partially filled to generate the highest roll threshold value.

Although this research involved simulation with an articulated vehicle, many of the results may possibly be extended to non-articulated tank vehicles, though further research would have to verify that proclamation. Furthermore, future research into the effect of dynamic load transfers from the movement of liquid cargo would be of interest. It is plausible that the dynamic load shifts may even have a more severe effect on the roll stability of the vehicle and the driver may have a difficult time compensating for the momentum and oscillations of the liquid cargo as it moves from side to side in the tank (Fig. 5.22).

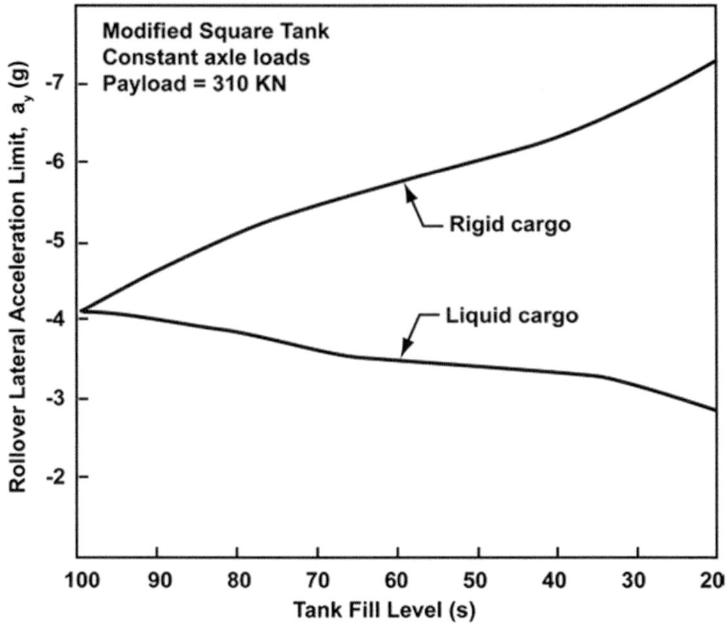

Fig. 5.21 Comparison of rollover lateral acceleration limits of partially filled modified square tank vehicle and equivalent rigid cargo vehicle when composite axle loads are held constant [27]

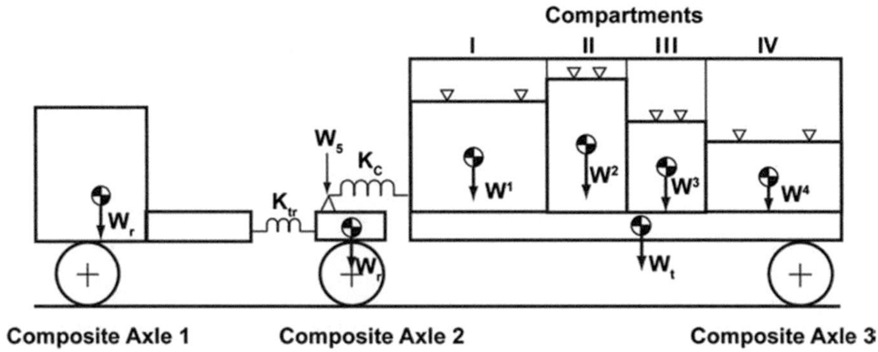

Fig. 5.22 Representation of the tractor semi-trailer axles and sprung weights with tank compartmentation [27]

5.10 Warning Systems/Predicting Rollover

Obviously, without any control devices to help stabilize a vehicle in an unstable situation, it is up to the driver to take the corrective action. To do this, it is necessary that the driver receives some kind of warning of an impending rollover and that it is in enough time for the driver to react. Therefore, the warning device must be able to dynamically predict a rollover situation quickly and correctly. The correctness of the prediction may be examined in a couple of different ways. If the warning system fails to signal the driver in a rollover situation, i.e., a false negative, catastrophic results may occur.

On the other hand, if the warning system alerts the driver to an imminent rollover though the vehicle is in no real danger of rollover, i.e., a false positive, then the driver may feel that the warning system is too liberal at giving a warning and may opt to ignore one in the future where the danger of rollover is real. Therefore, it is clear that a rollover-warning system must be able to predict all situations of imminent rollover and give few, if any, false warnings to ensure that the warnings are taken seriously at all times. In addition, a common observation among researchers is that for a warning device to be feasible for heavy articulated vehicles, it would need to be as insensitive as possible to the many different parameters inherent with commercial trucking, such as payload CG heights and suspension properties, among others.

Rakheja and Piché [27] attempted to establish a set of rollover stability criteria for an articulated vehicle for use as input parameters to a rollover-warning device. In addition, they investigated the phenomenon of jackknife instability and aimed to develop a similar set of stability criteria for use in a jackknife-warning device. However, although jackknifing is an unstable condition of an articulated vehicle that sometimes leads to rollover, this review is limited to pure cases of rollover and will therefore not address instability due to jackknifing further.

The authors note that previous research has pinned the onset of rollover instability on a variety of dynamic factors including high levels of lateral acceleration, rapid increases in roll angle, roll frequency of the trailer approaching its resonant roll frequency, and excessive spring deflection in the suspension. However, they also point out that many of these parameters are "extremely sensitive to vehicle design and operational factors" and require significant instrumentation to measure them. The procedure for developing a warning device is therefore two-fold:

1. Identification of vital motion cues or dynamic response parameters related to the onset of vehicle instabilities.
2. System development including online acquisition of motion cues and generation of warning to the driver.

In addition, the dynamic response parameters from step 1 must be "relatively insensitive to variations in design and operating factors" and "directly measurable". To identify the key response variables and use common values of vehicle parameters such as weights, dimensions, tires, and suspension for a five-axle tractor semi-trailer, simulations were run using various computer models. Furthermore, they note that rollover may occur during simple steady-state (static) cornering or from high-speed dynamic maneuvers such as a lane-change or obstacle avoidance. Therefore, both types of maneuvers (static and dynamic) must be examined to establish the key response variables.

5.10.1 Steady-State Cornering

Since the static roll threshold (SRT) of a vehicle is a good measure of a vehicle's roll threshold during steady-state cornering, the Static Roll Model was used. The SRTs of 72 different vehicle configurations were analyzed to decide which parameters most affected the SRT. Using the TWR of a rigid vehicle (T/h) and the calculated SRTs for the 72 configurations, Rakheja and Piché developed a list of "compliance factors". These compliance factors are a convenient measure of the reduction in the TWR for a rigid vehicle due to the inclusion of compliance in the vehicle model. They are calculated simply by the ratio of the SRT to the TWR as shown in the equation below

$$C = \frac{SRT}{TWR} = \frac{SRT}{T/h} \tag{5.20}$$

Values for the compliance factors ranged from 0.64 to 0.74 and 0.68 to 0.75 for the 2.44 and 2.59 m wide vehicles, respectively. In addition, they concluded that the SRT calculated with the Static Roll Model is solely dependent on CG height for a given track width and was relatively insensitive to various vehicle parameters such as suspension properties, tractor frame torsional stiffness, articulation mechanisms, and trailer structure. Although most of these results agree with previous conclusions of the sensitivity analysis performed on the Static Roll Model by El-Gindy and Hosamel-deen, their opinion on the sensitivity to suspension properties differs strongly. Specifically, they found the suspension properties of the trailer unit to have a first-order effect on the roll stability of the vehicle. Nevertheless, Rakheja and Piché suggest two plans that can be used to detect the onset of roll instability during steady-state cornering; one when the CG height is known and one when it is not.

5.10.1.1 Known Center of Gravity Height

Rakheja and Piché suggest that the impending rollover is quite easy to detect when the CG height is known. From their results, it was concluded that the compliance factors were relatively unaffected by suspension type and CG height. Therefore, the mean and standard deviation of the compliance factors were calculated with values of 0.72 and 0.03, respectively. A threshold compliance factor, C_t, was subsequently developed and calculated from

$$C_t = \frac{C_m - 3\sigma}{\alpha} \tag{5.21}$$

where C_m is the mean compliance factor, σ is the standard deviation, and α is a safety factor. As an example, a threshold compliance factor of 0.57 is realized using a safety factor of 1.1. Finally, using the known values of track width and CG height, a "safe value" of lateral acceleration may be calculated from Eq. 5.20 with $C = C_t$.

5.10.1.2 Unknown Center of Gravity Height

Rakheja and Piché point out that from previous studies [28], the CG height of a fully laden trailer lies somewhere around 2 meters. Therefore, using the compliance factor developed for a trailer with a CG height of 2.03 m as in their study, a "safe limit" of lateral acceleration may be calculated from

$$A_{safe} = \frac{C_{2.03} \times TWR}{\alpha} = \frac{C_{2.03} \times T/h}{\alpha} \tag{5.22}$$

where α is the safety factor and $C_{2.03}$ is the compliance factor for a vehicle with a CG height of 2.03 m (possibly the average of the 2.03 m CG height compliance factors for all types of suspensions computed). Consequently, monitoring of the lateral acceleration of the trailer is considered a good indicator of impending rollover during steady-state cornering.

5.10.2 High-Speed Directional Maneuvering

As noted previously, rollover of a heavy vehicle in a dynamic maneuver may occur at lateral acceleration levels significantly below that of the SRT. Therefore, a separate set of dynamic response parameters must be found for the warning device to accurately predict roll instability. The Yaw/Roll Model was used to examine the response of the various vehicle configurations described before at 100 km/h in both a severe lane-change maneuver and an obstacle avoidance maneuver. As with steady-state cornering (and the definition of the SRT), wheel lift-off defines the onset of rollover for dynamic maneuvers as well. Therefore, Rakheja and Piché again investigate the wheel loads as a possible indicator.

However, wheel loads are difficult to measure directly, and as an alternative, suspension loads are examined. Using the 2.03 m CG height, simulations are performed for different types of suspensions at the semi-trailer axles. Results indicate that the suspension load measurement is a good indicator of impending rollover for mechanical spring suspensions, but not for air spring suspensions. Therefore, since the response measure needs to be common for all types of suspensions, it is concluded that the measurement of suspension loads "does not provide a reliable indication of the tire lift-off and thus the onset of vehicle rollover". Another indirect measure of tire loads is the axle roll angle which may be measured relatively easily and in real time. Simulations are performed for various CG heights and trailer widths in both lane-change and obstacle avoidance maneuvers.

It is concluded that the axle roll angle may be used as an indicator of imminent rollover. In addition, and probably most importantly, additional simulations suggest that it is possible to stabilize the vehicle with a corrective steering maneuver once the axle roll angle indicates a state of rollover. Specific results indicate safe values (using a safety factor of 1.1) for axle roll angles for 2.44 and 2.59 m wide trailers as approximately 1.4 and 1.3°, respectively.

In an investigation into the frequency a heavy vehicle approaches its roll threshold in actual service, George [29] performed extensive research with five different tractor semi-trailers on loan from various transport companies in Australia. All-important weights and dimensions of the vehicles were measured. In addition, a tilt-table was used to measure separate SRTs of the tractor and the trailer and to also estimate the CG height of each trailer. The five tractor semi-trailers were then rigged with data collection equipment to measure in real time the lateral acceleration and yaw rate of the tractor (measured in the cabin) and lateral acceleration at the rear of the trailer (at the chassis level). Additionally, all five vehicles were equipped with strain link transducers to measure lateral load transfer. All vehicles were then put back into their normal operation, though the drivers were instructed to stay off the highways as much as possible to increase the amount of data generated from the use of back roads.

The software TruckDas, an "event triggered" package developed by the Australian Road Research Board Ltd. (ARRB), was used to collect and process all the data. The event trigger to start logging data was the lateral acceleration of the tractor and the level was set to 40% of the vehicle's measured SRT. Once triggered (an "event"), TruckDas recorded the time and magnitude of lateral acceleration separately for the tractor and the trailer from a zero or steady-state level to their peak lateral acceleration. In addition, TruckDas recorded the mean steer path information, estimated road slope, vehicle speed, and the time and distance measured from when the vehicle left the terminal.

Results from the 268 events indicated that the vehicles were operated "most of the time" at a level of 52% of the individual truck's roll threshold. Figure 5.23 is a histogram of the normalized roll-limit (ratio of lateral acceleration measured to the roll threshold of the vehicle) versus the number of occurrences. George notes, however, that the results are skewed due to the fact that data for lateral acceleration levels under 40% of the SRT are not recorded (because of the event trigger being set at 40% of the SRT). However, it is not clear to this author why there appears data below 40%-normalized roll-limit in Fig. 5.23. Nevertheless, he proceeds to calculate the "preferred" level of lateral acceleration from knowledge of the individual lateral acceleration levels and corresponding vehicle speeds with the relationship below

$$A_c = \frac{v^2}{R} \tag{5.23}$$

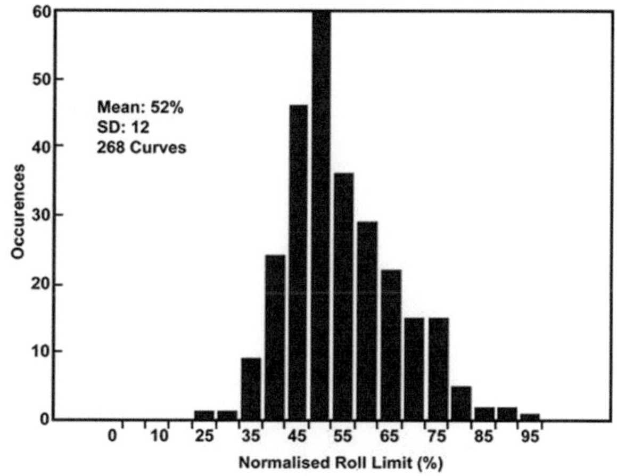

Fig. 5.23 Test vehicles distribution to the roll-limit [30]

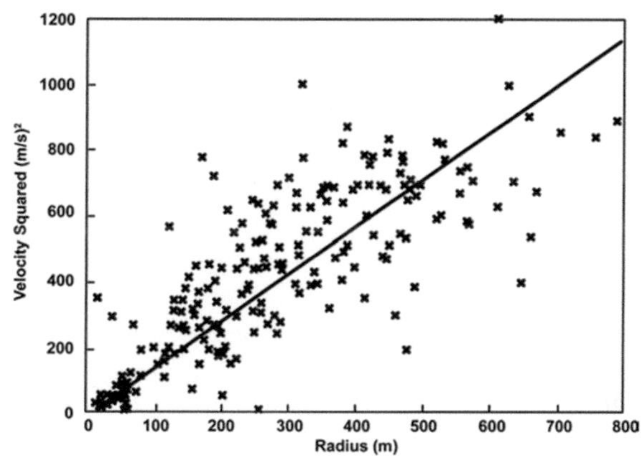

Fig. 5.24 Driver's preferred choice of vehicle speed and steer path [30]

where A_c is the centripetal acceleration, v is the speed, and R is the radius of curvature of the turn. The radius of curvature was calculated for each level of lateral acceleration and vehicle speed, then plotted versus the square of the speed as shown in Fig. 5.24. Analyzing the data as a whole, it reveals the vehicle speed and steer maneuver producing a lateral acceleration that the drivers feel most comfortable with. A linear regression was applied to the data and resulted in a "preferred" lateral acceleration level of 1.42 m/s^2, or 0.145 g at the tractor. Furthermore, in order to determine the amount of time after a warning of impending rollover that a driver had to take corrective action, the maximum lateral acceleration attained and the time for each of the events to occur were plotted and shown in Fig. 5.25.

Again, it is not clear why data appears below the 40% roll threshold trigger level. As may be seen in the plot, the data above 75% of the roll-limit has been separated into three distinct categories: those peak accelerations occurring within 2 s from a zero or steady-state value (A'), between 2 and 6 s (B'), and those peaks occurring

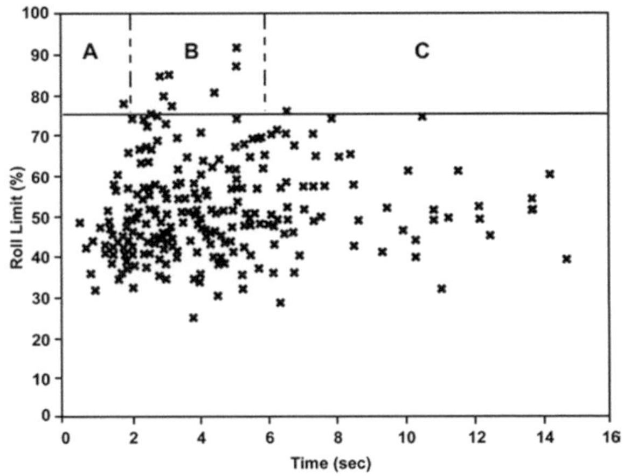

Fig. 5.25 Critical recorded times to reach maximum lateral acceleration from zero to steady-state acceleration [30]

in greater than 6 s (C'). It is clear from Figure 20 that most events over the 75% roll-limit occurred within 2 to 6 s (section B') of a zero or steady-state level of lateral acceleration. In addition, eleven events out of the 268 recorded exceeded the 75% roll-limit, where five events exceeded 80%. George notes that although this may seem fairly infrequent, this means that on average a heavy truck comes to within 20% of its roll threshold once every 715 km the vehicle travels. Furthermore, if these numbers are an average representation of the Australian trucking fleet as a whole (using data from the Australian Bureau of Statistics, 1988), each heavy truck will approach its roll-limit approximately 100 times a year.

Analysis was further carried out to examine the possibility of using the 75% of the roll-limit as the trigger to warn the driver, i.e., when the lateral acceleration reaches 75% of the roll-limit during a particular maneuver, a signal is sent to the driver to warn of impending rollover. Therefore, for each of the eleven events recorded that exceeded the 75% roll-limit, the individual slopes of the acceleration/time curves like that in Fig. 5.26 were calculated.

The curve was then extended until it reached the roll threshold for that particular vehicle. The time between the occurrence of peak lateral acceleration and the estimated time that the roll threshold was met was recorded. In other words, this is the time from when the lateral acceleration peaked during the experiment to when it would cross the roll threshold for that particular vehicle, assuming the lateral acceleration keeps increasing at the same rate. However, it is not clear exactly how George calculated the slope from Fig. 5.26, though it appears he used the peak acceleration value and the one preceding it. Finally, a relationship between the time to warn the driver and the level of lateral acceleration was obtained as shown in Fig. 5.27. It is clear from the plot that according to this research, the more severe the maneuver that creates high levels of lateral acceleration, the less time there is to warn the driver.

A hypothesis that may be drawn from this is that drivers do not intentionally approach the roll threshold of their vehicles since the occurrences happen so quickly

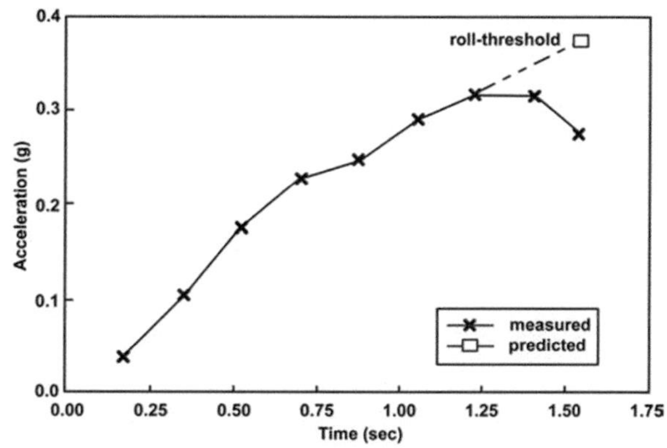

Fig. 5.26 Method of predicting the time to the roll threshold [30]

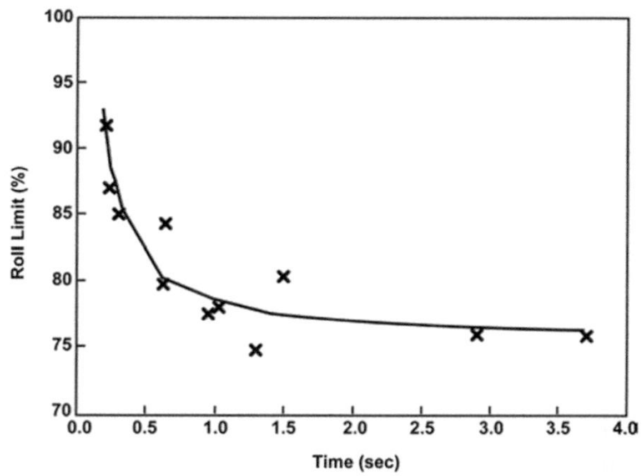

Fig. 5.27 Relationship of the predicted time from the percentage of the roll-limit to the roll threshold [30]

and may be caused by unexpected maneuvers such as accident avoidance. For example, Fig. 5.27 shows that the event that reached 92% of the roll-limit had approximately a 0.25 s window before the vehicle reached 100% (rollover). This short amount of time is due to the high slope (rate of change of lateral acceleration) found from Fig. 5.26 for this particular event.

Consequently, the driver approached the 92% roll-limit point very quickly and therefore probably unintentionally. Conversely, if we examine one of the events that peaked at about 75% of the roll-limit, the slope from Fig. 5.26 would be much less and therefore the time window to complete rollover has a value of approximately 3 s. In this case, the driver slowly approached the 75% roll-limit of the vehicle, then made a corrective action to reduce the lateral acceleration, probably all the while feeling in complete control. It should be stated, however, that George makes some general conclusions from Fig. 5.27 that this author believes are somewhat questionable.

Specifically, George states that "providing a warning signal when a vehicle is at 75% of its roll-limit would allow a time period greater than 3 s for the driver to act". This cannot be deduced merely from Fig. 5.27. This plot simply shows that for the vehicles that reached 75% of the roll-limit, but increased no further, there is approximately 3 s to warn the driver. However, for the vehicle that reached 92% of its roll-limit, nothing can be concluded from Fig. 5.22 about the time between when the vehicle reached the 75% roll-limit point and the 92% roll-limit point. The slope of the curve like that in Fig. 5.21 would need to be determined for the event that reached the 92% roll-limit to estimate the amount of warning time available. Therefore, to state that there are 3 s available to warn the driver when the vehicles reach 75% of their roll-limit does not seem justifiable based solely on the data presented in Fig. 5.26.

Perhaps a plot of the time between when the vehicle reaches the 75% roll-limit point and the peak lateral acceleration for the individual vehicles versus the percent roll-limit would be useful in determining the time available to warn the driver. Though this author believes some of the conclusions made are not entirely accurate, George has clearly shown the importance of the rate that the lateral acceleration changes on predicting impending rollover.

In a study by Liu et al. [31], an alternate version of the Load Transfer Ratio, called the Roll Safety Factor (RSF), is presented as a possible stability measure for predicting rollover. However, the RSF only differs from the LTR version in Ervin and Guy [32] in that no absolute value is taken, i.e.,

$$RSF = \frac{\sum_{i=1}^{n}(Fr_i - Fl_i)}{\sum_{i=1}^{n}(Fr_i + Fl_i)} \tag{5.24}$$

where ith is the axle number and n is the number of axles. As stated previously, the LTR, or RSF in this case, typically neglects the contribution from the front tractor axle. Liu et al. specify the RSF for this case as RSFs where, defining the front tractor axle as axle 1,

$$RSF = \frac{\sum_{i=2}^{n}(Fr_i - Fl_i)}{\sum_{i=2}^{n}(Fr_i + Fl_i)} \tag{5.25}$$

Note that the RSF (and RSFs) may take on values between –1 and 1, giving not only the magnitude of relative instability, but also the direction the vehicle is rolling.

Using a single-axle analytical model, they were able to identify which dynamic variables were directly related to the RSF. Specifically, the RSF was related to the axle roll angle (ϕ_u), the sprung mass roll angle (ϕ_s), the lateral acceleration of the sprung mass (A_y), and a parameter frequently referred to as the steering factor which is the product of the front wheel steering angle (δ_f) and the square of the vehicle forward velocity (v). Note the steering factor derived from the steady-state cornering equation for a two-axle vehicle is

$$\delta_f = \frac{L}{R} + \frac{K_u A_y}{g} \tag{5.26}$$

Table 5.1 Dependency of rollover indicators on vehicle design parameters

Rollover indicator	Vehicle weight	Track width	Tire properties	CG height	Suspension properties	Handling properties	Reliability measurability
RSF							1 / Poor
ϕ_u	x	x	x				2 / Good
A_y	x	x	x	x	x		3 / Good
ϕ_s	x	x	x	x	x		3 / Good
$\delta_f v^2$	x	x	x	x	x	x	4 / Good

where L is the wheelbase of the vehicle, R is the radius of curvature of the turn, K_u is the understeer coefficient, A_y is the lateral acceleration at the CG of the vehicle, and g is the acceleration due to gravity. From the equation for centripetal acceleration, $A_y = v^2/R$, we may substitute for R in Eq. 5.26 and rearrange

$$\delta_f v^2 = A_y \left(L + \frac{K_u v^2}{g} \right) \tag{5.27}$$

giving us the steering factor on the left-hand side. Relating these variables to the RSF through the simple vehicle model enabled them to investigate how dependent the variables are on the various vehicle parameters and their level of measurability.

Liu et al. [33] also assesses reliability by how dependent or independent that indicator is on the various vehicle design parameters, where independence is desired. Table 5.1 lists the results of their analysis in order of their reliability from best to worst. It is clear that the RSF is independent of all listed design parameters and therefore it is the most reliable of the indicators. However, load transfer (and therefore the RSF) is typically difficult to measure resulting in a poor' measurability rating.

On the other hand, the axle roll angle is typically easy to measure but depends on vehicle parameters such as weight, track width, and tire properties. Likewise, the lateral acceleration and sprung mass roll angle are relatively easy to measure but depend on CG height and suspension properties in addition to those that the axle roll angle depends upon. Finally, the steering factor is also relatively easy to measure but depends upon all the listed design parameters, making it the least reliable rollover indicator. Using the Yaw/Roll Model and a "typical" five-axle tractor semi-trailer, they conducted a more thorough sensitivity analysis between the RSFs and a more complete set of rollover indicators. The set of rollover indicators included separate tractor and trailer lateral accelerations, tractor and trailer roll angles, axle roll angles, and the steering factor. It was also of interest to examine the phase, or lag/lead time, of the various indicators in relation to the RSFs, since knowledge of the phase indicates the predictive power of the indicator on an impending rollover situation.

Three different maneuvers were simulated for the vehicle including a steady-state cornering maneuver and two transient ones: trapezoidal steering and sinusoidal steering. In general, results indicated an inverse relationship with predictive power, i.e., large lead times, and strength of correlation with the RSFs for all types of simulated maneuvers. For example, one of the stronger correlations existed between the trailer lateral acceleration and the RSFs, but since the two were nearly in phase with one another, no lead time was present. On the other hand, the steering factor had the largest lead time (of the order of 0.7 s) but had the worst correlation with the RSFs. It was concluded that none of the proposed indicators by themselves were good candidates for use with the RSFs in an early warning device, but perhaps a combination of the indicators may be used successfully.

5.11 Active Rollover Prevention Control Strategies

Accidents involving heavy trucks (tractor semi-trailers, etc.) have also contributed significantly to the number of injuries and fatalities on North American highways over the past few decades. In addition, monetary expenses, consisting of damage to roadways and environmental cleanup from hazardous spills, are at an all-time high. A 1988 report from the National Highway Traffic Safety Administration (NHTSA) showed rollover occurred in 52% of the heavy vehicle accidents where the driver was killed. Also, an earlier report by Ervin et al. [7] concluded that rollover of heavy vehicles was responsible for 95% of the bulk spillage of hazardous materials. To make matters worse, the number of heavy trucks on the roads in North America increases every year and economic demands continuously put pressure on regulators to increase the size and weight limits on heavy trucks. Of course, the reason companies desire increases in size and weight limits is obvious, but these decisions seem to be made without much regard to the reduction in rollover stability this may cause. Therefore, the frequency of rollover accidents in the years to come will almost certainly grow unless measures are taken to compensate for the reduction in the roll stability due to the changes in size and weight regulations.

The handling dynamics of an articulated vehicle, such as a tractor semi-trailer, differ from that of a non-articulated vehicle significantly. Among other reasons, the ability for the trailer to articulate or pivot relative to the tractor contributes an additional mass that the driver must be concerned with. In addition, the response of the trailer to inputs from the tractor, such as steering maneuvers, is typically amplified and lags behind the response of the tractor making it difficult to control. Many times this causes stability problems and a trailer may start to roll excessively or begin to jackknife (yaw instability) before the driver is aware of the problem and therefore may not have time to take corrective action.

El-Gindy [11] conducted a sensitivity analysis of the Static Roll Model to separate those vehicle parameters that have a significant effect on the static rollover threshold (SRT) of a five-axle tractor-semi-trailer. Preston-Thomas [14] examined the feasibility of requiring a minimum SRT value of heavy trucks for operation in Canada and possibly all of North America. Data collected from tilt-table tests by Preston-Thomas [30] is summarized for a variety of vehicles, including dump trucks, cement mixers, tractor semi-trailers, and a B-train tank truck. The results support the previous suggestions for a minimum SRT ranging from 0.38 to 0.42 g, depending on the type of heavy truck.

In an investigation into the frequency a heavy vehicle approaches its rollover threshold in actual service, George [29] performed extensive research with five different tractor semi-trailers on loan from various transport companies in Australia. All-important weights and dimensions of the vehicles were measured. In addition, a tilt-table was used to measure separate SRT of the tractor and the trailer and to also estimate the CG height of each trailer. The five tractor semi-trailers were then rigged with data collection equipment to measure in real time the lateral acceleration and yaw rate of the tractor (measured in the cabin) and lateral acceleration at the rear of the trailer (at the chassis level). Additionally, all five vehicles were equipped with strain link transducers to measure lateral load transfer. All vehicles were then put back into their normal operation, though the drivers were instructed to stay off the highways as much as possible to increase the amount of data generated from the use of back roads. The software TruckDas, an "event triggered" package developed by the Australian Road Research Board Ltd. (ARRB), was used to collect and process all the data. The event trigger to start logging data was the lateral acceleration of the tractor and the level was set to 40% of the vehicle's measured SRT. Once triggered (an "event"), TruckDas recorded the time and magnitude of lateral acceleration separately for the tractor and the trailer from a zero or steady-state level to their peak lateral acceleration. In addition, TruckDas recorded the mean steer path information and estimated road slope, vehicle speed, and the time and distance measured from when the vehicle left the terminal.

Preston-Thomas and Woodrooffe [30] conducted a feasibility study on the production of an on-board rollover-warning device for heavy trucks. Using previous data from a feasibility study performed by Sparks and Berthelot [34], it was estimated that a warning device would need to cost at most $633 for the device to be cost effective if based on the economic factors alone. Note that this price was based on the economic benefits gained from eliminating all of the "preventable" and 25% of the "potentially preventable" heavy truck rollover accidents as defined by Sparks and Berthelot. Preston-Thomas and Woodrooffe conclude that a rollover-warning device using the LTR as the sole indicator of the incipient rollover was only economically feasible if incorporated with an existing on-board weigh scale system. The cost of on-board weigh scale systems at the time of the report ranged from $4000 to $6000 and they believe that the incremental cost of a warning device incorporated with one would be small.

An existing commercial warning device called the Stabe-Alert : Stability Monitoring and Alarm System, developed by an apparently now defunct company called Roadway Safety Systems, Inc., is also presented in the report. This Stabe-Alert system uses speed sensors and a strain-gauged disk load sensor [35] for measurement of load transfer. A brochure for the system is also included in the report. Finally, suggestions for future research including full-scale testing of a device on a suitable tractor semi-trailer are described.

The application of control systems to vehicle dynamics has begun to make it feasible, in theory at least, to achieve such benefits as improving the heavy vehicle's directional and roll stability. The main goal of the controller design is to achieve the predefined desired handling performance of a vehicle in a robust fashion. Thus, the importance of the selection of a suitable control strategy is obvious. In the case of examining the articulated vehicles, the definition of the desired trajectory to be followed by the vehicle units is not as straightforward as in the case of a single vehicle where a virtual model following principle can be used [36].

Lin et al. [22] investigate the performance of an active anti-roll suspension for heavy trucks using the Yaw/Roll Model. Initially, a simple linear model of an articulated vehicle, based on work from Segel [19], is used to develop baseline controller gains using feedback during random and step-steer inputs. The controller is then implemented into the more complex Yaw/Roll model for simulation. The results are then compared to that of two other vehicles with the same properties: one with passive suspension and one with a passive suspension and a passive anti-roll bar. It is evident that the active suspension is able to roll the trailer into the turn thereby increasing its roll threshold. It was concluded that an actuator with a bandwidth between 1 and 5 Hz is required and the hydraulic power supply must be able to produce 10 kW for active roll control of the trailer.

Lin et al. [22] continue their investigation of active anti-roll suspensions with optimal roll control. Specifically, they examine two control laws: lateral acceleration feedback (LAFB) using proportional and derivative gains and a controller optimized by the standard linear quadratic regulator (LQR) technique. As done in their previous work, a simple linear model [19] of a single-unit articulated vehicle was used. The gains for both the LAFB and LQR controller were tuned using stochastic input data (power spectrums) that was believed to cover the range of typical steering angle input by a driver. Results indicate that the roll threshold of the vehicle was increased by as much as 66% and lateral load transfer was decreased by 34% (for a random steering input) by use of the optimized active anti-roll bar.

In research conducted by Palkovics [21], a method called Active Unilateral Braking Control (AUBC) at the tractor's rear axles was used to stabilize yaw divergence of a five-axle tractor semi-trailer during simulations. Use of the AUBC system, which produced a yaw torque at the rear tandem axle, was able to not only minimize yaw divergence of the tractor (and reduce the risk of jackknifing), but to also help reduce roll divergence of the trailer. A 3-DOF linear model of the tractor semi-trailer [37] was initially used for the optimization of the AUBC system with a standard LQR technique. The controller was then implemented onto a 34-DOF nonlinear model using inputs such as force measured at the kingpin and articulation angle. It is believed that an advantage of the AUBC system is that the inputs (kingpin force and articulation angle) are readily available from current anti-lock brake systems (ABS) and the output of the controller (brake force) may be adjusted by valves controlling mean brake pressure. In other words, the system requirements are relatively simple and are available as long as the truck has been implemented with ABS. Figures 5.28 and 5.29 are vertical wheel load plots for a passive and an LQR optimized AUBC controlled vehicle, respectively, for a path-follow evasive maneuver conducted at 110 km/h.

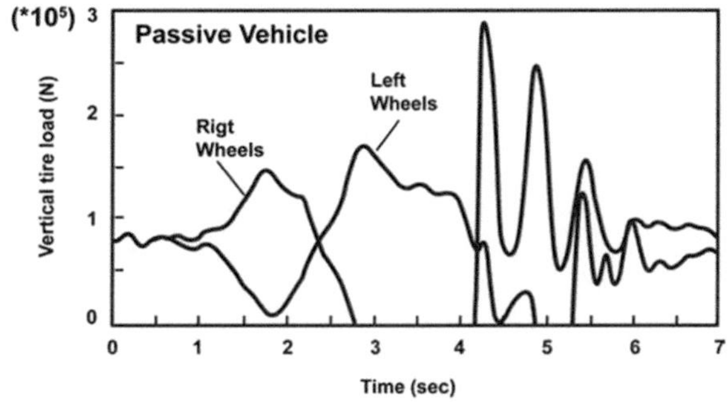

Fig. 5.28 Vertical tire loads for a passive vehicle during an evasive maneuver at 110 km/h

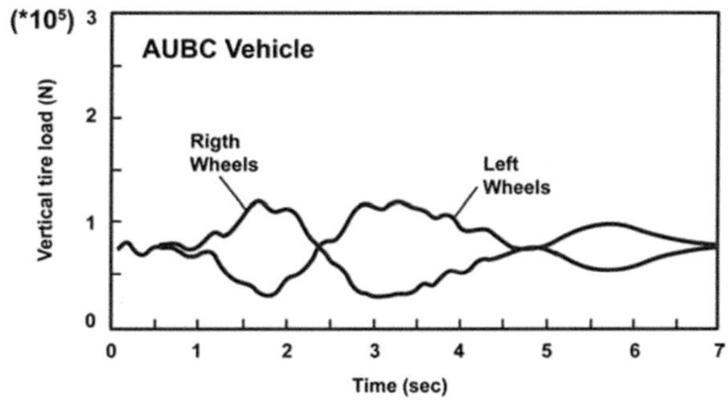

Fig. 5.29 Vertical tire loads for an AUBC controlled vehicle during an evasive maneuver at 110 km/h

In 1998, El-Gindy et al. [38] have examined various control strategies of a truck/full-trailer (See Fig. 5.28) using the LQR approach. The optimal state-feedback controller was determined by minimizing a given performance index. The results obtained through an LQR controller are promising; however, practical realization requires careful consideration. The results of the simulation showed that the Rearward Amplification Ratio of a Truck/full-trailer could be improved without significant change of the driver control or deviation from the desired path. These results also showed that active yaw control at either the truck or trailer would not be desirable due to the strong effect on the vehicle path and driver control without much reduction of the rearward amplification.

Predicting rollovers is a difficult task and a dependable method of doing so has yet to be developed. Evaluating the instantaneous roll stability of a heavy vehicle (to predict rollover) typically involves using a combination of stability metrics, such as the static roll threshold and lateral load transfer, but with limited success. Therefore, the major difficulty seems to be in deciding which stability metrics need to be used. Active yaw control systems seem to be the practical solution to prevent or reduce the rollover rates of heavy trucks. The traditional concepts used previously to design a yaw controller or active unilateral braking control systems also seem to be inefficient; however, in this research, the controller is designed to limit a performance measure, such as lateral load transfer or lateral acceleration within safe limits.

Problems

1. A vehicle weighs 30 kN. The center of gravity height is 120 cm. The suspensions and tires are assumed rigid. The average track width is 1800 cm. Determine the static rollover threshold (SRT) and plot the rollover moment diagram up to roll angle of 45°.
2. Determine the static rollover threshold and plot the roll moment diagram for the vehicle described in problem 1 for the following cases:

    a. The vehicle center of gravity height is increased to 150cm and 200 cm.
    b. The track width is increased to 2100 cm and reduced to 1500 cm.
    c. The total weight is increased to 40 kN and reduced to 10 kN.

3. A light truck weighs 40 kN and has a wheelbase of 400 cm. The sprung mass center of gravity is 150 cm behind the front axle. The height of the sprung mass center of gravity is 150 cm. The front and rear suspensions roll stiffness are 200 kN.m/rad and 400 kN.m/rad, respectively. The lever arm is 50 cm. The track widths of the front and rear axles are 900 cm and 950 cm, respectively. The front and rear unsprung masses are 250 kg and 500 kg, respectively. The heights of the unsprung masses center of gravities are 30 cm and 40 cm, respectively. If the vehicle was negotiating a turn of radius of 100 m at a constant speed of 100 km/h. Determine the following:

    a. The sprung mass roll angle.
    b. The front and rear load transfer ratios.
    c. The sprung mass roll angle if the front and rear are switched.
    d. Use case 3 to calculate the LTR at the front and rear axles and comment on the results.
    e. If the front roll stiffness increased to 400 kN m and the rear roll stiffness reduced to 200 N m, what will be the effect on the following?
        i. Roll angle.
        ii. Vehicle handling characteristics.

References

1. Ervin RD (1986) The dependence of truck roll stability on size and weight variables. Int J Veh Design 7(5–6):192–208
2. Douglas WB, Bohdan TK (1991) Effects of horizontal-curve transition design on truck roll stability. J Transp Eng 117(1):91–102
3. Riley Garrott W, Heydinger GJ (1992) An investigation, via simulation, of vehicle characteristics that contribute to steering maneuver induced rollover. Technical report, SAE Technical Paper
4. Allen RW, Szostak HT, Theodore JR, David HK (1990) Field testing and computer simulation analysis of ground vehicle dynamic stability. SAE Trans 102–125
5. Allen RW, Szostak HT, Rosenthal TJ, Klyde DH, Owens KJ (1991) Characteristics influencing ground vehicle lateral/directional dynamic stability. SAE Trans 336–361

6. Goldman RW, El-Gindy M, Kulakowski BT (2001) Rollover dynamics of road vehicles: Literature survey. Int J Heavy Veh Syst 8(2):103–141
7. Ervin RD (1979) The yaw stability of tractor-semitrailers during cornering. Final report, Technical report
8. MacAdam CC (1982) A computer-based study of the yaw/roll stability of heavy trucks characterized by high centers of gravity. SAE Trans 4052–4073
9. Hossam A (1985) (Hossam Ahmed) El-Gindy, Roads, Transportation Association of Canada, and Jo Yung Wong. In: Users Guide to the UMTRI Models, Computer Simulation of Heavy Vehicle Dynamic Behaviour
10. Chalasani M (1983) Road tanker design: its influence on the risk and economic aspects of transporting gasoline in michigan
11. El-Gindy M, Hosamel-deen YH (1989) Sensitivity parametric analysis of umtri static roll model. Int J Veh Design 10(2):187–209
12. Robert DE (1983) The influence of size and weight variables on the roll stability of heavy duty trucks. SAE Trans 629–654
13. Liu PJ, Subhash R, Ahmed AKW (1998) Dynamic rollover threshold of articulated freight vehicles. Int J Heavy Veh Syst 5(3–4):300–322
14. Preston-Thomas J (1991) Measured rollover thresholds of three-axle and four-axle cement mixers and dump trucks. In: NRCC, Institute of Mechanical Engineering, Ground Transportation Technology
15. Billing JR, Lam CP, Couture J (1989) Development of regulatory principles for multi-axle semitrailers. In: Proceedings of the Second International Symposium on Heavy Vehicle Weights and Dimensions
16. El-Gindy M (1995) An overview of performance measures for heavy commercial vehicles in north america. Int J Veh Design 16(4–5):441–463
17. Woodrooffe JH, El-Gindy M (1992) Application of handling and roll stability performance measures for determining a suitable tractor wheelbase. In: Proceedings: International technical conference on the enhanced safety of vehicles, vol 1992. National Highway Traffic Safety Administration, pp 30–35
18. Paul SF (1986) A factbook of the mechanical properties of the components for single-unit and articulated heavy trucks. phase i. final report. Technical report
19. Segel L (1956) Theoretical prediction and experimental substantiation of the response of the automobile to steering control. Proceed Inst Mech Eng: Auto Div 10(1):310–330
20. Zheng-zhong LI, Dong-ya HOU, Pu ZRAO (2001) Graphs and data processing system in modern route surveying [j]. In: Journal of Liaoning Technical University (Natural Science Edition), vol 2
21. Palkovics L, El-Gindy M (1995) Design of an active unilateral brake control system for five-axle tractor-semitrailer based on sensitivity analysis. Veh Syst Dyn 24(10):725–758
22. Lin RC, Cebon D, Cole DJ (1996) Optimal roll control of a single-unit lorry. Proceed Inst Mech Eng Part D: J Auto Eng 210(1):45–55
23. Thomas DG (1997) Vehicle dynamics. Warren dale
24. Ranganathan R, Rakheja S, Sankar S (1989) Kineto-static roll plane analysis of articulated tank vehicles with arbitrary tank geometry. Int J Veh Design 10(1):89–111
25. Ranganathan R, Rakheja S, Sankar S (1989) Steady turning stability of partially filled tank vehicles with arbitrary tank geometry
26. Bauer HF (1981) Dynamic behaviour of an elastic separating wall in vehicle containers: Part 1. Int J Veh Design 2(1):44–77
27. Subhash R, Alain P (1990) Development of directional stability criteria for an early warning safety device. SAE Trans 877–889
28. Miller DWG, Barter NF (1973) Roll-over of articulated vehicles. In: Presented at the 1973 Conference-Vehicle Safety Legislation: Its Engineering and Social Implications, vol C203/73
29. George RM (1992) Behaviour of articulated vehicles on curves. In: Proceedings: International Technical Conference on the Enhanced Safety of Vehicles, vol 1992. National Highway Traffic Safety Administration, pp 331–337

30. Preston-Thomas J, Woodrooffe JHF (1990) Feasibility study of a rollover warning device for heavy trucks
31. Liu PJ, Rakheja S, Ahmed AKW (1997) Detection of dynamic roll instability of heavy vehicles for open-loop rollover control. SAE Trans 632–639
32. Ervin RD, Guy Y (1986) Influence of weights and dimensions on the stability and control of heavy trucks in canada-part 1
33. Wang L, Liu X, Liu Y-l, Zeng F-F, Ting W, Yang C-L, Shen H-Y, Li X-P (2010) Correlation of pepsin-measured laryngopharyngeal reflux disease with symptoms and signs. Otolaryngol Head Neck Surg 143(6):765–771
34. Sparks GA, Berthelot C (1989) The cost/benefit analysis of a rollover warning device for large trucks. SPARKS and Associates Ltd
35. Barnett JD, West R (1983) A new load sensor for truck self weighing systems. Technical report, SAE Technical Paper
36. Nagai M, Nishizawa Y, Sawa A (1988) Control of 4-wheel-steering car (experimental study on virtual model following control). Soc Instrum Control Eng 1:489–490
37. Gibbons GR (1960) Survey of the modern nonresident motorist statutes, a U Fla L Rev 13:257
38. El-Gindy M, Lewis S, Mrad N (1998) Active control of a truck full-trailer's rearward amplification. Int J Heavy Veh Syst 5(3–4):277–299

Chapter 6
Road Vehicle Tractive Performance

In this chapter, road vehicle performance characteristics are discussed. The two main characteristics are the tractive and braking efforts. Tractive effort is developed by the tires in order to overcome the resisting forces acting on a vehicle which determines the performance potential of a vehicle. The major external forces acting on a single or two-axle vehicle are:

- Aerodynamic forces.
- Grade resistance.
- Inertia force due to acceleration.
- Rolling resistance.
- Draw bar resistance (if a vehicle is towing a trailer).

6.1 Maximum Tractive Effort

Figure 6.1 shows the external forces acting on a two-axle vehicle running over a flat surface. The aerodynamic resistance, R_a acts in the longitudinal direction, in addition to the rolling resistance of the front and rear tires, R_{rf} and R_{rr}, respectively. The drawbar load, R_d, the grade resistance, R_g, and the tractive effort of the front and rear tires, F_r and F_r also act in the longitudinal direction.

The equation of motion for a single vehicle accelerating on a flat surface in the longitudinal axis is defined as

$$\frac{W}{g}a = F_f + F_r - R_{rf} - R_{rr} - R_a \tag{6.1}$$

where

- a Linear acceleration of the vehicle (m/s^2).
- g Acceleration due to gravity (m/s^2).
- F_f Tractive effort at front tire (front-wheel drive) (kN).

M. El-Gindy and Z. El-Sayegh, *Road and Off-Road Vehicle Dynamics*,
https://doi.org/10.1007/978-3-031-36216-3_6

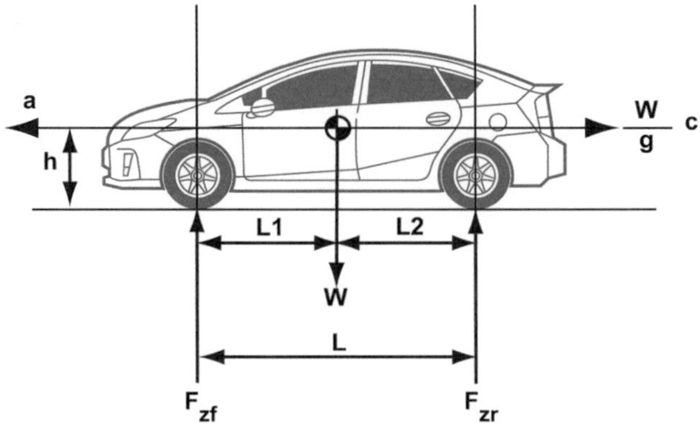

Fig. 6.1 Forces acting on a two-axle vehicle [1]

- F_r Tractive effort at rear tire (rear-wheel drive) (kN).
- R_r Rolling resistance (kN).
- R_{rf} Rolling resistance at front tires (kN).
- R_{rr} Rolling resistance at rear tires (kN).
- W Vehicle weight (kN).

Equation 6.1 can be rewritten as

$$F_f - F_r - \left(R_a + R_{rf} + R_{rr} + \frac{aW}{g} \right) = 0 \tag{6.2}$$

where $R = R_a + R_{rf} + R_{rr}$, and $R_r = R_{rf} + R_{rr}$. It can also be written as

$$F = R + \frac{aW}{g}. \tag{6.3}$$

The normal load on the front axle, F_{zf}, can be determined as follows:

$$F_{zf} = \frac{Wl_2 - haW/g + R_a}{L}. \tag{6.4}$$

Similarly, the normal load on the rear axle, F_{zr}, can be determined by summing moments about B

$$F_{zr} = \frac{Wl_1 + haW/g + R_a}{L}. \tag{6.5}$$

Assuming $h_a = h_d = h$, Eqs. 6.4 and 6.5 can be rewritten as

$$F_{zf} = \frac{l_2}{L}W - \frac{h}{L}\left(\frac{aW}{g} + R_a\right) \tag{6.6}$$

$$F_{zr} = \frac{l_1}{L}W + \frac{h}{L}\left(\frac{aW}{g} + R_a\right). \tag{6.7}$$

The normal force on the front and rear axles can be rewritten as

$$F_{zf} = \frac{l_2}{L}W - \frac{h}{L}(F - R_r) \tag{6.8}$$

$$F_{zr} = \frac{l_1}{L}W + \frac{h}{L}(F - R_r). \tag{6.9}$$

The maximum tractive effort that the tire-ground contact can support is determined in terms of the coefficient of road adhesion, μ, and vehicle parameters. For a rear-wheel drive vehicle

$$F_{rmax} = \mu F_{zr} = \mu\left[\frac{l_1}{L}W + \frac{h}{L}(F_{max} - R_r)\right] \tag{6.10}$$

and

$$F_{rmax} = \frac{\mu W(l_1 - f_r h)/L}{1 - \mu h/L}. \tag{6.11}$$

For a front-wheel-drive

$$F_{fmax} = \mu F_{zf} = \mu\left[\frac{l_2}{L}W - \frac{h}{L}(F_{max} - R_r)\right] \tag{6.12}$$

and

$$F_{fmax} = \frac{\mu W(l_2 + f_r h)/L}{1 + \mu h/L}. \tag{6.13}$$

It should be noted that F_{rmax} is greater than F_{fmax} because of the increase of F_{zr} and reduction of F_{zf} due to the load transferred from the front axle to rear axle. Also, note that the maximum tractive force of all wheel drive will be higher than that of only front-wheel or rear-wheel drive. In the case of all wheel drive, the maximum tractive force will be

$$F_{max} = F_{fmax} + F_{rmax}. \tag{6.14}$$

6.2 Aerodynamic Forces and Moments

With the growing emphasis on:

- Fuel Economy.
- Reduction of the undesirable exhaust emissions.

It is important to optimize the vehicle power required. To achieve this, it is vital to reduce:

- Aerodynamics.
- Rolling resistance.
- Inertia resistance (potentially, vehicle weight).

It should be noted that, for vehicles cruising at a speed higher than 88 km/h (55 mph), the power required to overcome the air resistance is greater than that required to overcome rolling and mechanical (power train) resistance. The major sources of the aerodynamic resistance can be distributed as follows:

- 90% air flow over the exterior of the vehicle body.
- 10% air flow through the engine radiator and ventilation openings.

The external aerodynamic resistance comprises the following components:

1. Pressure drag: which arises from the component of normal pressure on the vehicle body acting against the motion of the vehicle.
2. Skin friction: which is the shear stress in the boundary layer adjacent to the external surface of the vehicle body.

Of these two components, the pressure drag forms 90% of the total external aerodynamic resistance for passenger cars. The skin friction is proportionally increasing and the length of vehicle body increases. The symbols used in this section are listed below:

- A_r, Frontal or projected area in (m^2);
- C_D, Coefficient of aerodynamic resistance, which is a function of the vehicle design and operating factors;
- C_L, Aerodynamic lift coefficient, which is a function of the attack angle, ground/chassis clearance, and spoiler design;
- C_{LF}, Front aerodynamic lift coefficient, which is a function of the attack angle, ground/chassis clearance, and front spoiler design;
- C_{LR}, Rear aerodynamic lift coefficient, which is a function of the attack angle, ground/chassis clearance, and rear spoiler design;
- C_M, Aerodynamic pitching coefficient, which is a function of the vehicle wheelbase, frontal area, attack angle, ground/chassis clearance, and spoiler design;
- m_v, Vehicle mass in (kg);
- R_a, Aerodynamic resistance force (N);
- R_L, Aerodynamic lift force (N);
- V_r, Vehicle speed relative to the wind. Coefficient of aerodynamic lift coefficient, which is a function of the attack angle, ground/chassis clearance, and spoiler design;
- ρ, Air mass density which is equal to 1.23 kg/m^3 in performance calculation.

The aerodynamic resistance, R_a, is defined as

$$R_a = \frac{\rho}{2} C_D A_f V_r^2. \tag{6.15}$$

Table 6.1 Summary of radial-ply and bias-ply tire characteristics

Vehicle type	C_D	$A_r(\text{m}^2)$
Honda Civic 1.2	0.37–039	1.72
Honda Accord 1.8 EX	0.4–0.42	1.88
Ford Escort 1.3 GL	0.39–0.41	1.83
BMW 728i	0.42–0.44	2.13
Porsche 924	0.31–0.33	1.80
Mercedes 500 SEL	0.36–0.37	2.16
Saab 900 Turbo 16	0.34–0.36	2.05
VOLVO 740 GLE	0.40–0.42	2.16

Frontal Area in, m for passenger cars with a mass range of 800–2000 kg can be approximated by

$$A_f = 1.6 + 0.00056(m_v - 765). \tag{6.16}$$

The aerodynamic lift, R_L, is defined by

$$R_L = \frac{\rho}{2} C_L A_f V_r^2 \tag{6.17}$$

and the aerodynamic pitching moment, M_a, is defined by

$$M_a = \frac{\rho}{2} C_M A_f L_c V_r^2. \tag{6.18}$$

Table 6.1 shows example values of aerodynamic resistance coefficient, C_D, and frontal area, A_f for some passenger cars.

6.3 Power Plant and Transmission Characteristics

It is well known that internal combustion engines have less favorable performance characteristics (i.e., quite far from the desired ideal torque-speed and power-speed characteristics), and can be used only with a suitable transmission. Despite this shortcoming, it has found the widest application in vehicles because of its:

- Relatively high power to weight ratio.
- Good fuel economy.
- Low cost.
- Easiness to start.

The performance of a vehicle is limited by the smaller of the following two factors:

1. The first factor is the maximum tractive effort that the tire-ground contact can support. This factor limits the performance in low gears (high reduction ratios, and high engine torque), where the limit of the maximum tractive effort is determined by the nature of the tire-road adhesion.
2. The second factor is the tractive effort that the engine torque with a given transmission can provide. Usually, this factor limits the performance in high gears (low reduction ratios, and low engine torque), where the maximum tractive effort is determined by the engine and transmission characteristics.

The ideal performance characteristics of a power plant and the actual characteristics of two types of power plant systems will be discussed. These two types are manual transmission (manual gear shifts); and automatic transmission (automatic gear shifts).

Also, the advantages and disadvantages of each system will be discussed. Some details about the components of each system will be analyzed and described. Before studying the automatic transmission characteristics, the concept of the Hydraulic Coupling and the Hydraulic Torque Converter will be discussed. The match between a given engine and transmission is an extremely important issue to be considered to determine the final performance characteristics of a power plant.

6.3.1 *Manual Gear Transmission*

The principal requirements for the transmission are:

- to achieve the desired maximum vehicle speed with an appropriate engine;
- to be able to start, fully loaded, in both forward and reverse directions on a steep gradient, and to be able to maintain a high speed on a gentle slope, in high gear for passenger cars;
- to properly match the characteristics of the engine to achieve the desired operating fuel economy and acceleration characteristics.

The gear ratio of the highest gear is calculated by

$$\xi_n = \frac{n_e r(1-s)}{V_{max}\xi_{ax}} \tag{6.19}$$

where s is assumed 3% (or .03) to reflect the effect of the longitudinal elastic deformation of the radius of the tire, r. The maximum tractive effort for a rear-wheel drive vehicle on maximum grade θ_{max} (see Fig. 6.2) is calculated as

$$W \sin\theta_{max} = \frac{\mu W(l_1 - f_r h)/L}{1-\mu h/L} - f_r W \tag{6.20}$$

the lower gear ratio, ξ_1, is calculated as

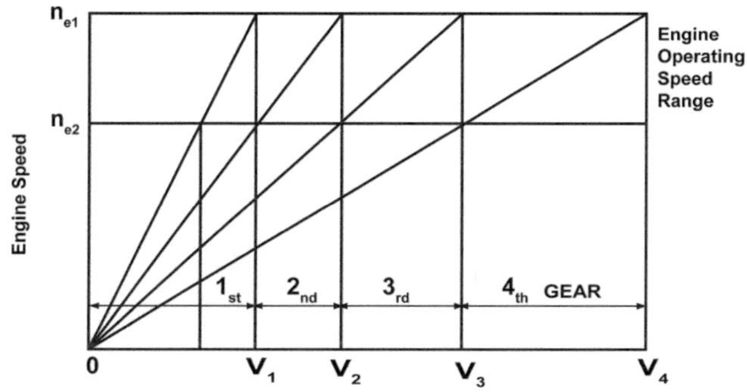

Fig. 6.2 Selection of gear ratio based on geometric progression rule for truck transmission

$$\xi_1 = \frac{W(\sin\theta_{max} + f_r)r}{M_{emax}\xi_{ax}\eta_t}. \tag{6.21}$$

The method for selecting the gear ratios for the intermediate gears between the highest and the lowest is, to a great extent, dependent upon the type of vehicle (heavy commercial vehicles or passenger cars). For heavy commercial vehicles, the gear ratios are usually arranged in a geometric progression. The basis for this is to have the engine operating within the same speed range in each gear, as shown in Fig. 6.2. This would ensure that in each gear, the operating fuel economy is similar.

The four-speed gearbox

$$\frac{\xi_2}{\xi_1} = \frac{\xi_3}{\xi_2} = \frac{\xi_4}{\xi_3} = \frac{n_{e2}}{n_{e1}} = K_g. \tag{6.22}$$

It should be noted that Eq. 6.22 represents the shift based on geometric progression rules, which is applied only in the case of large transmissions. For passenger cars, this rule may not be accurate and the shift must be approximated. Furthermore, the tractive effort of a vehicle is calculated by

$$F = \frac{M_e\xi_0\eta_t}{r} \tag{6.23}$$

and the vehicle's speed

$$V = \frac{n_e r}{\xi_0}(1 - i). \tag{6.24}$$

Figure 6.3 shows the variations of the mechanical efficiency with the input speed for a three-speed automatic gearbox. The transmission is connected to an engine operating at the wide open throttle and developing a maximum torque of 407 N.m [2]. In vehicle performance predictions, as a first approximation, the following average values for the mechanical efficiency of the major subsystems in the transmission may

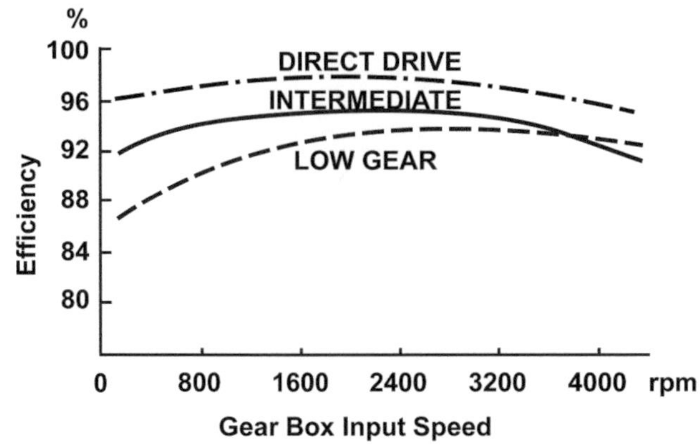

Fig. 6.3 Mechanical efficiency of a three-speed automatic gearbox at wide open throttle [1]

be used: for the gearbox-direct drive 98%, for the gearbox-indirect drive 95%, and for the drive axle 95%.

6.3.1.1 Prediction of Vehicle Performance with Manual Transmission

As previously described, the maximum tractive effort is determined either by the tire-road adhesion limit; or by power plant characteristics; or both.

Figure 6.4 [1] will be used as an example to illustrate the procedure for predicting acceleration and gradability characteristics.

The NET tractive (trust) effort versus speed is calculated by subtracting the total resistance (in this case, the $R_a + R_r$) from the gross tractive. The Net tractive effort curves are shown in Fig. 6.4 as solid lines. In this case, the maximum speed can be determined from the intersection between the tractive effort curve at high gear and the speed axis (x-axis) (100 mph or 162 km/h). In this case, the value of the tractive effort at any speed lower than the maximum indicates the surplus effort (or the available trust which can be used to either accelerate or overcome a grade resistance at a given gear shift). For example, if the vehicle speed is 80 km/h, it can maintain this speed on an uphill of 7.5% gradability.

An additional useful vertical axis representing the gradability can be added to the chart to determine the maximum grad and speed at a given gearshift. In order to complete this chart, the tire-road adhesion limits (the tractive effort a tire-road can support) in case of variable road conditions (dry, wet, or icy road having different coefficients of adhesion) can be plotted on the same chart. Additional curves representing various roads and conditions (dry and wet asphalt) are shown in Fig. 6.4. In principle, the potential performance of a vehicle is determined and limited by the smallest values of the developed tire-road adhesion or engine-transmission output drive torque and characteristics:

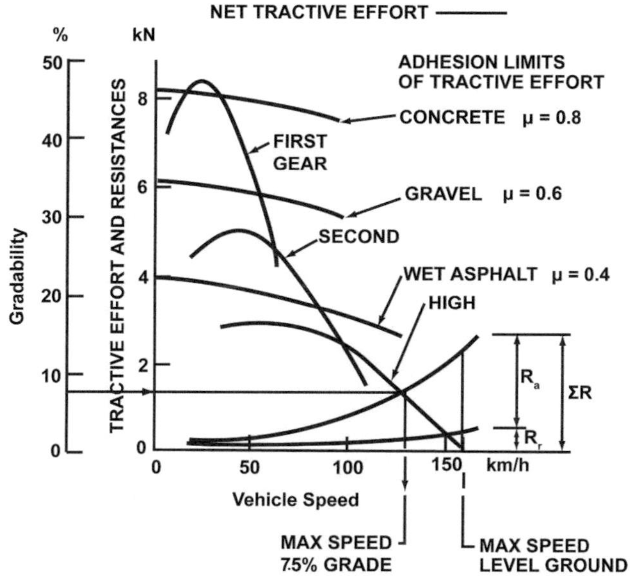

Fig. 6.4 Performance characteristics of a passenger car with a three-speed manual transmission [1]

- The maximum effort as determined by the engine torque and transmission characteristics with the second gear engaged is 5.5 kN.
- Whereas the maximum effort on wet asphalt that the tire-road can support with the second gear is only 4 kN (at speed = 0.0).

This shows that with second gear and at a speed less than approximately 95 km/h, the tractive effort of the vehicle on wet asphalt is limited by the tire-road adhesion, not by the engine torque. In a case like this, it is recommended to shift to the third gear to produce less engine torque, or use partial throttle opening.

6.3.1.2 Gradability

The tractive effort has to overcome grade resistance, rolling resistance, and aerodynamic resistance. Thus, the tractive force is defined by

$$F = W \sin\theta + R_r + R_a. \tag{6.25}$$

The maximum grade (θ) a vehicle can negotiate at a constant speed, therefore, is determined by the net tractive effort available at that speed

$$G = \frac{1}{W}(F - R_r - R_a) = \frac{F_{net}}{W}. \tag{6.26}$$

Acceleration-Time and Distance-Time The acceleration, speed, and distance of a vehicle can be calculated based on Newton's second law. As the net trust (net tractive effort) and the resisting forces are determined, the maximum acceleration and the time required to reach a certain speed can be determined. In this calculation the effect of the inertia of the rotating components on vehicle acceleration must be taken into account when the acceleration is computed. This is due to the engagement of several rotating components to the drive wheels, such as the transmission, the propeller shaft (for rear-wheel drive vehicles), and the wheels themselves. The time-speed and time-distance relationships are of prime interest and can be derived using the equations of motion of the vehicle. As described previously, the net trust (tractive effort) is a function of the speed, therefore, the time required to accelerate a vehicle from one speed to another, or the distance that the vehicle travels, can be determined only using a numerical method to integrate the time and speed equations.

It should be noted that the time required to accelerate the vehicle should include the time delay due to the gear shifts during acceleration, typically the time delay can be estimated based on the type of transmission (either manual or automatic). For example, the time delay for a manual transmission shift is in the range of 1–2 s, and for an automatic transmission, it is in the range of 0.5–1 s. Gradability, G is defined as: "The maximum grade a vehicle can negotiate at a given speed." The maximum grade, G, a vehicle can negotiate is equal to

$$G = \frac{1}{W}(F - R_r - R_a) = \frac{F_{net}}{W}. \tag{6.27}$$

Also, $G = \tan\theta$, where θ is the slope angle. G can be presented as a percentage. The grade resistance for a small slope angle can be assumed equal to WG.

The vehicle acceleration can be determined from the equation of motion

$$F = \sum R = F_{net} = \gamma_m ma \tag{6.28}$$

where

$$\gamma_m = 1 + \frac{\sum I_w}{mr^2} + \frac{\sum I_1\xi_1^2}{mr^2} + \cdots + \frac{\sum I_n\xi_n^2}{mr^2}. \tag{6.29}$$

i is the mass moment of inertia of the rotating transmission gears at ith shift. $(kg.m^2)$, I_w is the wheel mass moment of inertia $(kg.m^2)$, F is the net trust (tractive effort) (kN), m is the vehicle mass (kg), γ_m is the mass factor $(kg.m^2)$, r is the wheel radius, and V is the vehicle speed $(km/h$ or $m/s)$.

For a passenger car

$$\gamma_m = 1.04 + 0.0025\xi_0^2. \tag{6.30}$$

The equation of motion can be rewritten as follows:

$$\gamma_m m \frac{dV}{dt} = F \sum R = F_{net} \tag{6.31}$$

and

$$dt = \frac{\gamma_m m dV}{F_{net}} \tag{6.32}$$

or in other words, $F_{net} = f(V)$. The time-speed relationship thus becomes

$$t = \gamma_m m \int_{V_1}^{V_2} \frac{dV}{f(V)}. \tag{6.33}$$

The distance S that the vehicle travels during an acceleration period from speed V_1, to V_2 can be calculated by integrating the following equation:

$$S = \int_{V_1}^{V_2} \frac{VdV}{F_{net}/\gamma_m m} = \gamma_m m \int_{V_1}^{V_2} \frac{VdV}{f(V)}. \tag{6.34}$$

? Example 6.1

A vehicle weighs 21.24 kN, including the four road wheels. Each of the wheels has a rolling radius of 33 cm and a radius of gyration of 25.4 cm and weighs 224.6 N. The engine develops a torque of 325 N.m at 3500 rpm. The equivalent mass of the moment of inertia of the parts rotating at engine speed is 0.733 kg.m^2. The transmission efficiency is 85% and the total reduction ratio of the driveline in the third gear is 4.28 to 1. The vehicle has a frontal area of 1.86 m^2 and the aerodynamic drag coefficient is 0.38. The coefficient of rolling resistance is 0.02. Determine the acceleration of the vehicle on a level road under these conditions.

Solution. The mass factor, γ_m for the vehicle in the third gear can be calculated by

$$\gamma_m = 1 + \frac{\sum I_w + \sum I\xi^2}{mr^2} \tag{6.35}$$

$$= 1.084. \tag{6.36}$$

The thrust of vehicle F is determined by

$$F = \frac{M_e \xi_0 \eta_t}{r} = 3585\,\text{N}. \tag{6.37}$$

The vehicle speed, V can be calculated by

$$V = \frac{n_e r}{\xi_0}(1 - i). \tag{6.38}$$

Assume that $i = 3\%$, the vehicle speed becomes $V = 98.7$ km/h. The total resistance of the vehicle is the sum of the aerodynamic resistance, R_a and the rolling resistance R_r:

$$\sum R = R_a + R_r = 755\,\text{N}. \tag{6.39}$$

The acceleration a of the vehicle can be determined as

$$a = \frac{F - \sum R}{\gamma_m m} = 1.2\,\text{m/s}^2. \tag{6.40}$$

6.3.2 *Automatic Gear Transmission*

6.3.2.1 Hydraulic Coupling (HC)

The Hydraulic Torque Converter (HTC) is a fluid coupling which provides a flexible link between the engine and the automatic transmission. The hydraulic torque converter has different characteristics than the Hydraulic Coupling (HC). The Hydraulic Coupling simply contains two components:

- PUMP (or Impeller): which is attached to the output shaft of the engine and acts as a flywheel at the same time. The pump converts the mechanical energy into flow energy.
- TURBINE: which is attached to the input shaft of the transmission and converts the flow energy into mechanical energy.

The Hydraulic Coupling input torque (engine torque) is equal to the turbine torque (HC output torque). No torque amplification can be obtained from the HC. Understanding the performance of the HC is necessary to understand the performance of a Hydraulic Torque Converter (HTC).

Simplified HC Mathematical Model Capacity Factor: An indication of the ability of the Hydraulic Torque Converter to absorb or transmit torque.

Clutch Point: This is the point at which the lock-up clutch is activated and engages the pump with the turbine. At this point, the whole torque components are rotating together at the engine speed and the engine torque is transmitted directly to the transmission (Fig. 6.5).

$$Q_a = 2\pi R_a a V_{ca}\ (\text{m}^3/\text{s}) \tag{6.41}$$

$$Q_b = 2\pi R_b a V_{cb}\ (\text{m}^3/\text{s}) \tag{6.42}$$

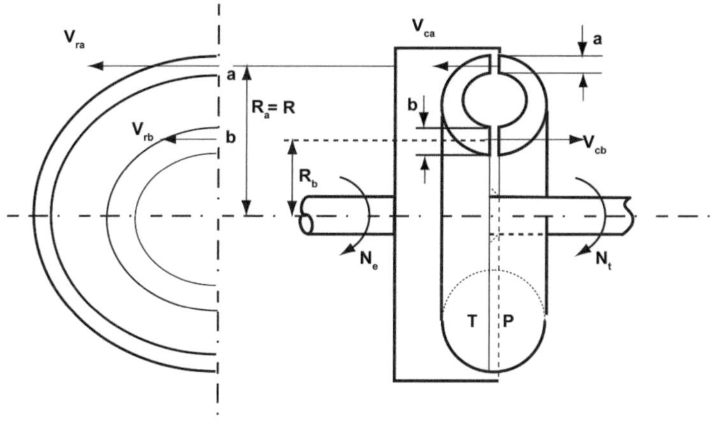

Fig. 6.5 Schematic of a hydraulic coupling system

where V_{ca} and V_{cb} are the circulation velocities and ρ is the fluid density. Using the continuity equation ($Q_a = Q_b$), it is desired that ($V_{ca} = V_{cb} = V_c$), then the relationship becomes

$$aR_a = bR_b \tag{6.43}$$

torque calculations

$$F = Q\rho V_r. \tag{6.44}$$

The force at points a and b, F_a and F_b are

$$F_a = 2\pi\rho R_a a V_c V_{ra} \tag{6.45}$$

$$F_b = 2\pi\rho R_b a V_c V_{rb} \tag{6.46}$$

where V_{ra} and V_{rb} are the rotational velocities, the torque acting on the pump is expressed as

$$T_p = F_a R_a - F_b R_b \tag{6.47}$$

while the torque acting on the turbine is expressed as

$$T_t = F_b R_b - F_a R_a. \tag{6.48}$$

It can be noticed that the torque of the pump and turbine are equal but opposite. It should be noted that there are no losses in torque when transmitted. Therefore, the torque acting on both the turbine and pump are equal and there is no torque magnification.

Substituting in either T_p or T_t equations, we get

$$T = 2\pi\rho V_c \left[R_a^2 a V_{ra} - R_b^2 b V_{rb}\right]. \tag{6.49}$$

Since $R_b = \frac{a}{b} R_a$ and $V_{ra} = \omega_p R_a$, $V_{rb} = \omega_t R_b$, then

$$T = 2\pi\rho V_c R_a^3 a \left[1 - \frac{R_b^2}{R_a^2}\frac{\omega_t}{\omega_p}\right]\omega_p. \tag{6.50}$$

Let

$$\alpha = \frac{R_a}{R_b} \tag{6.51}$$

$$V_c = \frac{V_c}{V_{ra}}\omega_p R_a \tag{6.52}$$

$$a = \frac{a}{R_a} R_a. \tag{6.53}$$

Then the torque equation becomes

$$T = 2\pi\rho \frac{a}{R_a} R^5 \frac{V_c}{V_{ra}} \omega_p^2 \left[1 - \alpha^2 \frac{\omega_t}{\omega_p}\right] \tag{6.54}$$

since, $\eta = \frac{T_t N_t}{T_p N_p} = \frac{N_t}{N_p} = \frac{\omega_t}{\omega_p}$ in case the turbine torque and the pump torque are equal but opposite, then the relationship between V_c and V_{ra} are given by

$$\frac{V_c}{V_{ra}} = K\sqrt{S(2 - S)} \tag{6.55}$$

where K is a constant and S is defined a the slip ratio which is equal to

$$S = \frac{\omega_p - \omega_t}{\omega_p} = 1 - \frac{N_t}{N_p} = 1 - \eta. \tag{6.56}$$

Substituting the ratio V_c/V_{ra} equation, we get

$$\frac{V_c}{V_{ra}} = K\sqrt{1 - \eta^2}. \tag{6.57}$$

The torque equation becomes

$$T = \rho\left[2\pi \frac{a}{R_a} K\left(1 - \alpha^2\eta\right)\sqrt{1 - \eta^2}\frac{4\pi^2}{2^5.60^2}\right] D^5 N_p^2. \tag{6.58}$$

Rearranging the equation

$$T = T_p = \lambda\rho D^5 N_p^2. \tag{6.59}$$

The pump pressure, P_p, is thus defined as

$$P_p = \lambda\rho D^5 N_p^3. \tag{6.60}$$

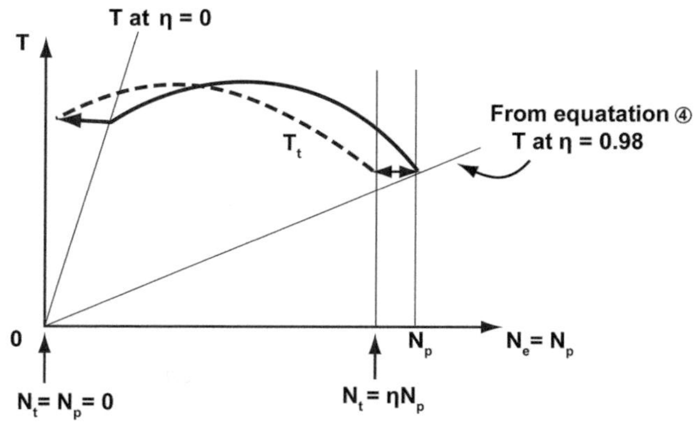

Fig. 6.6 Engine (pump) torque and turbine torque (output torque) versus engine speed (pump speed)

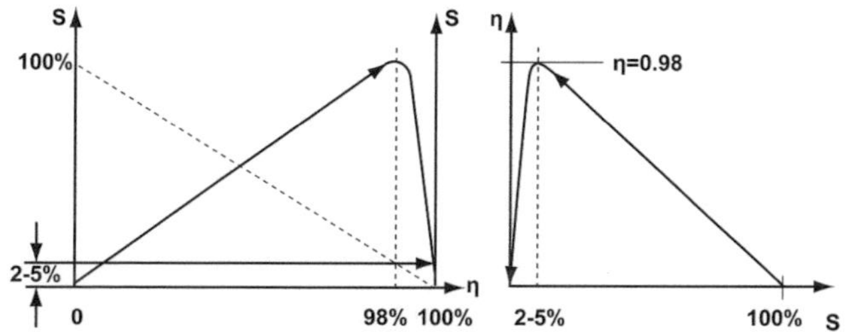

Fig. 6.7 **a** Slip versus efficiency; **b** efficiency versus slip

Figure 6.6 shows the pump and turbine torque versus the engine speed (pump speed) of a HC. The function of the pump is to convert the mechanical energy into flow energy, while the function of the turbine is to convert the flow energy into mechanical energy.

Figure 6.7 shows the slip versus efficiency and the efficiency versus slip. It is shown that the efficiency of a HC drops rapidly to zero as the slip $\left(1 - \frac{N_t}{N_p}\right)$ reduces to 2–3%. At this point, the pump and turbine must be locked and rotate as one unit.

6.3.2.2 Hydraulic Torque Converter

The Hydraulic Torque Converter (HTC) is one of the main components of the current automatic transmission. The HTC consists of a pump, a turbine, and a stator (reactor). The stator provides a reaction on the fluid circulation in the converter and is mounted on a free wheel (one way clutch) and acts as a torque amplifier, which enables the turbine to generate output torque higher than the input torque (pump torque). The stator will start to rotate only if the angular speed of the pump and the turbine are close to being equal. In this case, the stator will rotate at the same speed as the turbine and the HTC will be converted to a hydraulic coupling (HC) without any magnification of the input torque (i.e., the input torque is equal to the output torque).

Also, the HTC is equipped with a lock-up clutch, which is automatically activated and engages the turbine with the pump, so that they rotate together as one unit. This activation occurs when the HTC efficiency starts to drop (at turbine/pump relative slip of 2–3%). This process improves the transmission performance and hence the fuel economy. The advantages of incorporating an HTC into the automatic transmission can be summarized as follows:

- When properly matched, it will not stall the engine.
- It provides flexible coupling between the engine and the drive wheels.
- Together with a suitable selected multi-speed gearbox, it provides torque-speed characteristics that approach the ideal.

The disadvantage of the automatic transmission over the manual transmission can be summarized as follows:

- Fuel consumption with an automatic transmission is higher than that with a manual transmission by 1–11 % for both city and highway driving.
- High maintenance and initial costs due to its complex design and control units.
- Driver has no control over the gear shift.

In general, the principal requirements for transmission are:

1. To achieve the desired maximum vehicle speed with an appropriate engine.
2. To be able to start moving a fully loaded vehicle, in both forward and reverse directions, on a steep gradient, typically 33% (1 in 3), and to be able to maintain a speed of 88–96 km/h on a gentle slope, such as 3%, in high gear (low reduction ratio or direct drive) for passenger cars.
3. To properly match the characteristics of the engine to achieve the desired operating fuel economy and acceleration characteristics.

The symbols used in this section are listed below:

- B_0 is the pressure at the engine air intake.
- B_v is the vapor pressure (represents the air humidity).
- C_{sr} is the speed ratio which is equal to the output speed/input speed.
- C_{tr} is the torque ratio which is the output torque/input torque.
- i is the tire slip ratio.
- K_g is the gearbox factor.
- K_{tc} is the capacity factor (size factor) of a hydraulic coupling.
- K_e is an engine capacity factor (size factor).
- n_{e1} is the engine speed corresponding to 10% higher than the engine speed at maximum power (rad/s).
- n_{e2} is the engine speed corresponding to maximum torque (rad/s).
- n_{tc} is the output speed of the hydraulic torque converter (rad/s).
- M_e engine torque $(N.m)$.

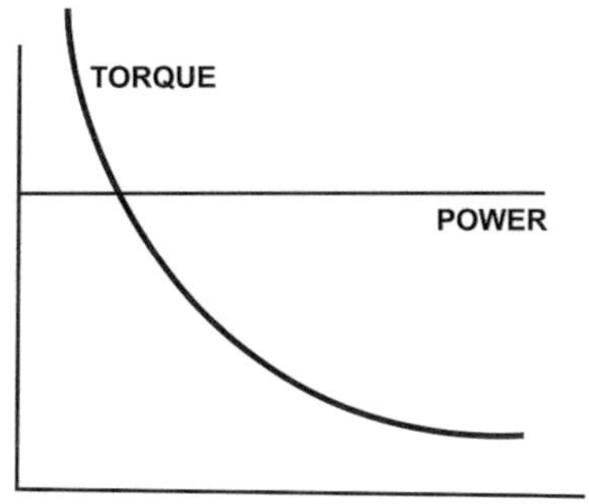

Fig. 6.8 Ideal performance characteristics for vehicular power plants [1]

- M_{emax} is the maximum engine torque-speed corresponding to maximum vehicle speed $(N.m)$.
- P is the engine power at a given ambient temperature, T.
- P_e is the engine power at B_e (101.32 kPa) and T_e (15.5 °C or 60 °F).
- V_{max} is the desired maximum vehicle speed (m/s).
- r is the tire rolling radius.
- T_e is the engine intake air temperature (should be 15.5 °C).
- T is the ambient temperature.
- ξ_n is the gear reduction ratio for 0 shifts transmission. If $\xi_n = 1$, the gearbox in direct drive.
- ξ_{ax} is the gear reduction ratio in the drive axle.
- ξ_o is the overall reduction ratio (transmission gear ratio x drive axle gear ratio).
- η_t is the overall transmission efficiency.
- η_c is the efficiency of a hydraulic torque converter = output speed × output torque/input speed × input torque equals $C_{sr} \times C_{tr}$.

Figure 6.8 shows the ideal performance characteristics for vehicular power plants. It can be seen that the output torque varies with speed, hyperbolically, which will provide the vehicle a high effort at low speeds, where demands for acceleration, drawbar pull or grade climbing capability are high.

The internal combustion engine has less favorable performance characteristics and can be used only with suitable transmission. Despite this shortcoming, it has found the widest application in automotive vehicles to date because of its relatively high power to weight ratio, good fuel economy, low cost, and easiness to start (Fig. 6.9).

The characteristics of a gasoline engine and a diesel engine are presented in Fig. 6.10a, b, respectively. The internal combustion engine starts operating smoothly at a certain speed (the idle speed). A good combustion quality and maximum engine torque are reached at an intermediate engine speed. As speed increases further, the mean effective pressure decreases because of growing losses in the air-induction manifolds, and the engine torque also declines. Power output, however, increases with an increase in speed up to the point of maximum power. Beyond this point, the

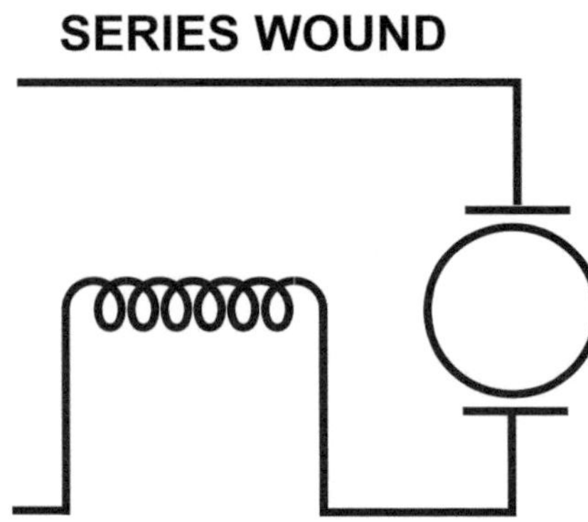

Fig. 6.9 Torque-speed characteristics diagram of a series-wound electric motor [1]

engine torque decreases more rapidly with an increase in speed. This results in a decline in power output. In vehicular applications, the maximum permissible speed of the engine is usually set just above the speed of the maximum power output. Vehicles designed for traction, such as agricultural and industrial tractors, usually operate at much lower engine speeds since the maximum torque, and not power, determines the limits to their tractive performance. To limit the maximum operating speed, engines for heavy-duty vehicles are often equipped with a governor.

The engine power, P, under a given atmospheric condition for a gasoline engine is given by

$$P = \frac{P_0(B_a - B_v)}{B_0}\sqrt{\frac{T_0}{T}} \tag{6.61}$$

and for a diesel engine, it is given by

$$P = \frac{P_0(B_a - B_v)}{B_0}\frac{T_0}{T}. \tag{6.62}$$

The engine performance is considerably affected by the atmospheric conditions. The effects of engine inlet temperature and ambient pressure on engine performance are shown in Fig. 6.11 [3]. It can be seen that if the engine air inlet temperature is higher and the ambient pressure is lower than the reference conditions, the power output of the engine will be lower.

The four-speed automatic transmission (ZF 4 HP22) with HC is shown in Fig. 6.12. Where the torque converter drives the clutches for the selection of the individual gears.

Where

- i is the transmission ratio;
- 1 is the input;
- 2 is the lock-up clutch;
- 3 is the torque converter;
- 4 to 10 are the clutches;

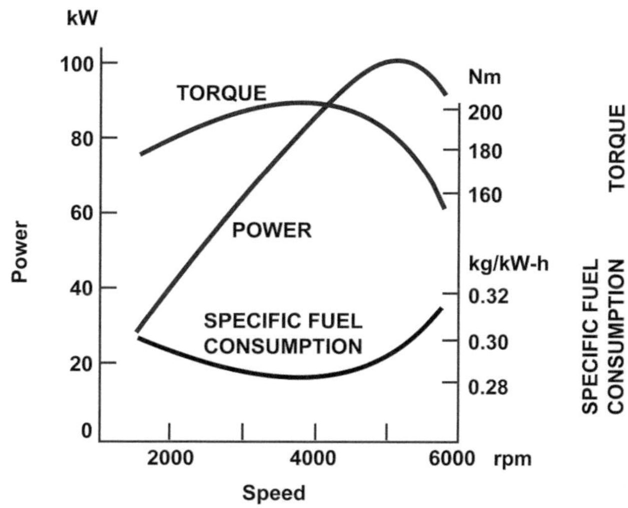

(a) gasoline engine

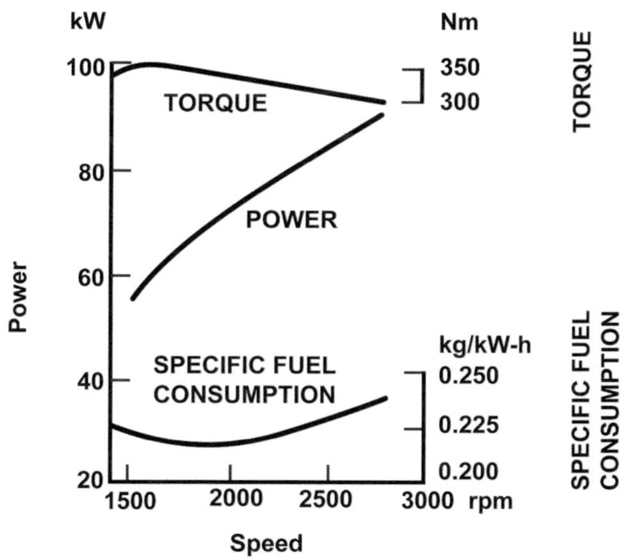

(b) diesel engine

Fig. 6.10 Performance characteristics of various engine types [1]

- 11 to 14 are the one way clutches;
- 15 to 17 are the planetary gears;
- 18 is the output.

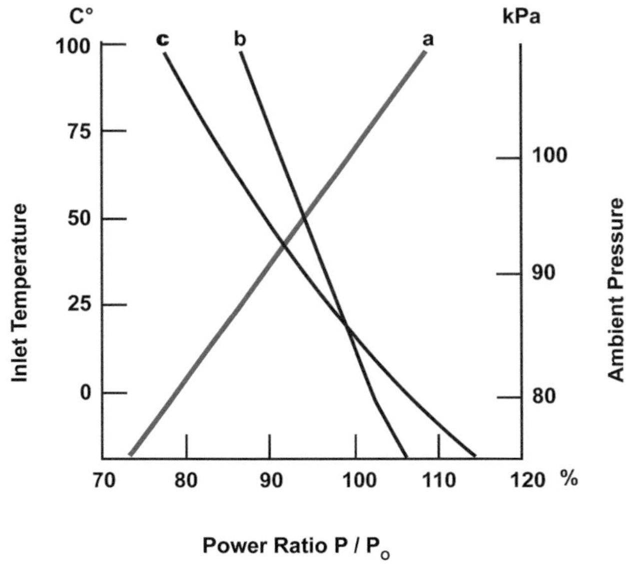

Fig. 6.11 Effect of atmospheric conditions on engine power. Curve a—power ratio versus ambient pressure. Curve b—power ratio versus intake temperature for gasoline engines. Curve c—power ratio versus intake temperature for diesel engines

Automatic transmission can be integrated with a retarder for buses, trucks, and special purpose vehicles as shown in Fig. 6.13 (ZF Ecomat 5 HP 500).

Where 1 is the hydrodynamic torque converter with lock-up clutch, 2 is the hydrodynamic retarder, 3 is the five-speed planetary-gear unit, 4 is the oil pump, and 5 is the transmission control.

Figure 6.14 shows the transmission control system. Label 1 is the selector lever with position switch, 2 is the program switch, 3 is the kickdown switch, 4 is the throttle-valve angle sensor, 5 is the duration-of-injection signal, 6 is the engine-speed signal, 7 is the transmission, 8 is the output-speed sensor, 9 is the pressure regulator, 10 is the solenoid valves, 11 is the electronic control unit, 12 is the malfunction indicator, and 13 is the engine-torque reduction by retarding of ignition.

The performance characteristics of a torque convener are usually described in terms of the following four parameters:

- Speed ratio, C_{sr}, is equal to the output speed/input speed.
- Torque ratio, C_{tr}, is equal to output torque/input torque.
- Efficiency, η_c, is equal to output speed $\times$ output torque/input speed $\times$ input torque $= C_{sr}C_{tr}$.
- Capacity factor (size factor), K_{tc}, is equal to speed/$\sqrt{torque}$.

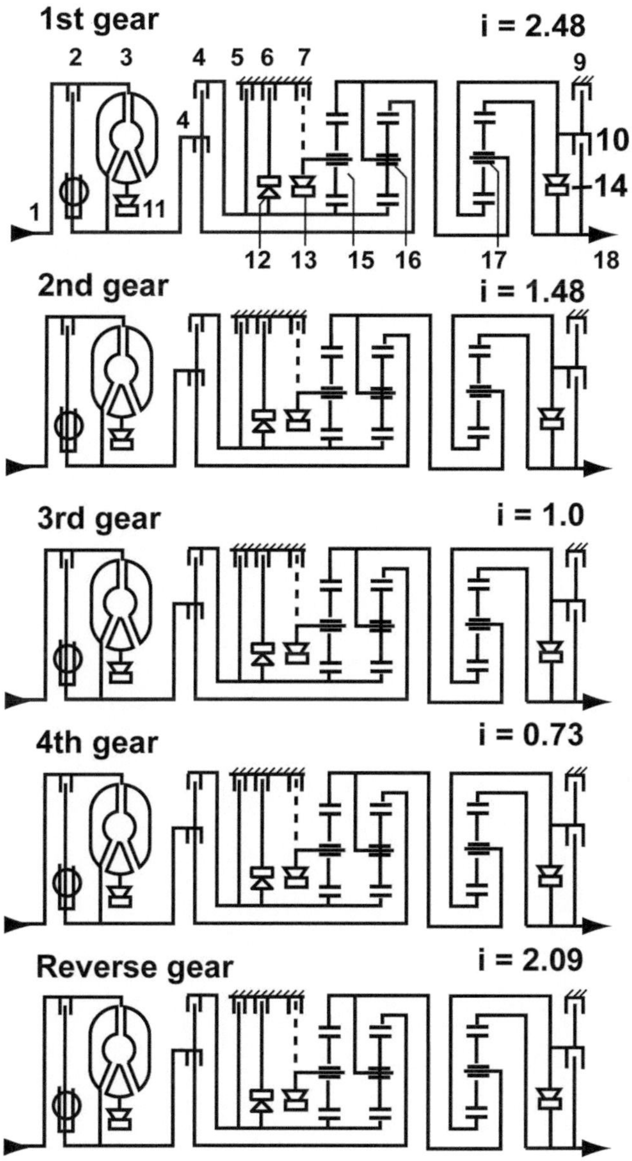

Fig. 6.12 Four-speed automatic transmission [4]

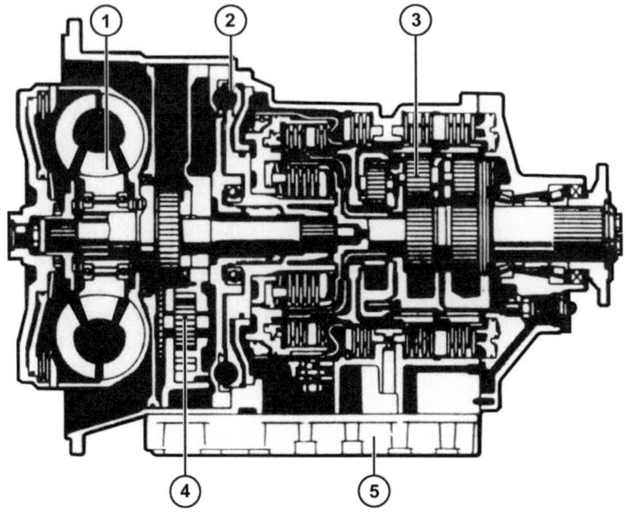

Fig. 6.13 Cross-section of a typical automatic transmission with retarder (2) [4]

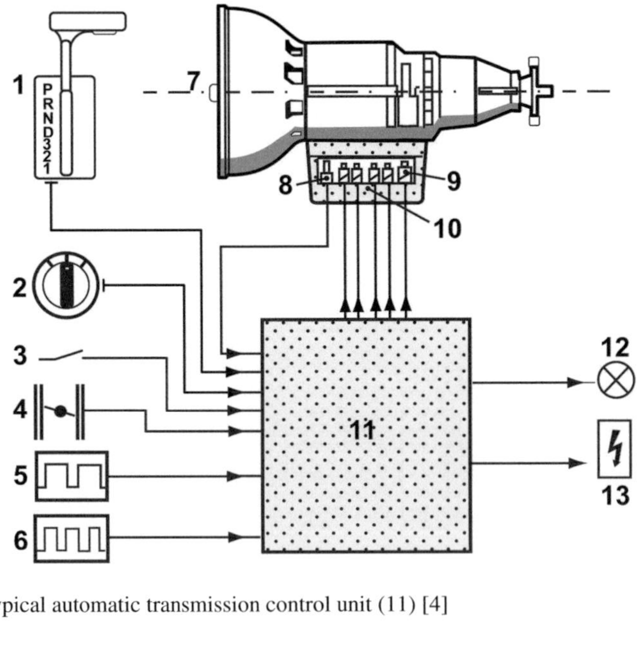

Fig. 6.14 Typical automatic transmission control unit (11) [4]

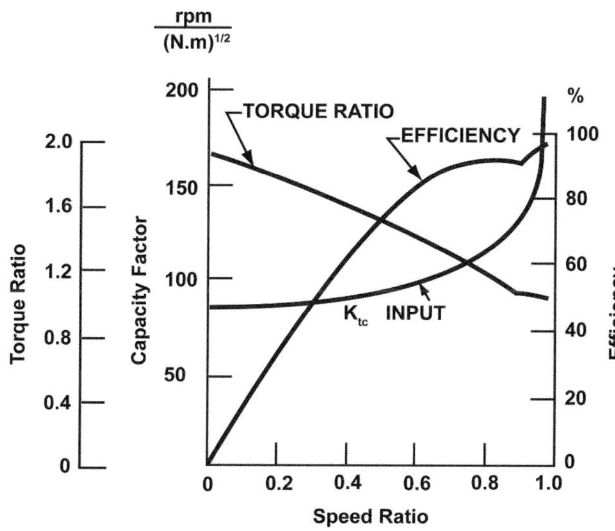

Fig. 6.15 Performance characteristics of a torque convener [2]

The engine capacity factor, K_e is defined as

$$K_e = \frac{n_e}{\sqrt{M_e}}. \tag{6.63}$$

The output torque, M_{tc} is defined by

$$M_{tc} = M_e C_{tr} \tag{6.64}$$

while the output-speed converter, n_{tc} is defined by

$$n_{tc} = n_e C_{sr}. \tag{6.65}$$

The tractive effort of a vehicle is defined as

$$F = \frac{M_{tc}\xi_0\eta_t}{r} = \frac{M_e C_{tr}\xi_0\eta_t}{r} \tag{6.66}$$

the speed of the vehicle can also be expressed as

$$V = \frac{n_{tc}r}{\xi_0}(1-i) = \frac{n_e C_{sr} r}{\xi_0}(1-i). \tag{6.67}$$

Representative performance characteristics of the torque converter are shown in Fig. 6.15, in which the torque ratio, efficiency, and input capacity factor, which is the ratio of the input speed to the square root of the input torque, are plotted against

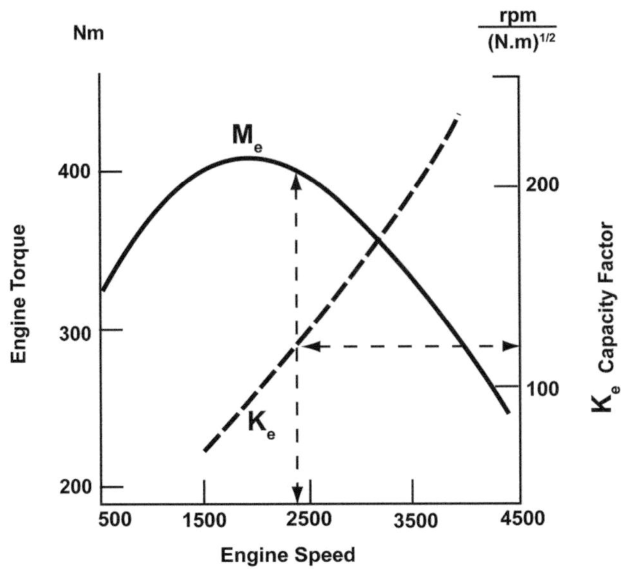

Fig. 6.16 Capacity factor of an internal combustion engine [2]

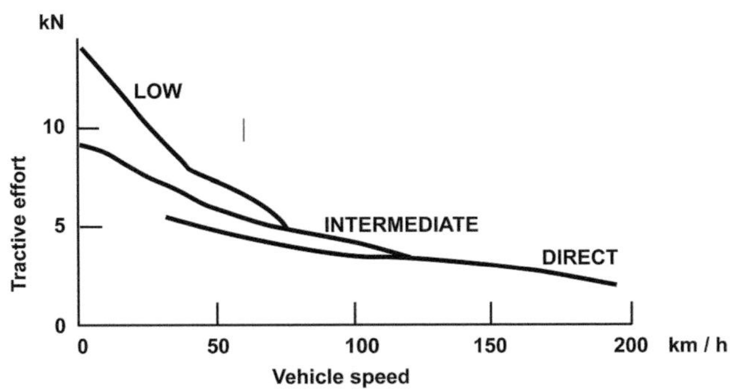

Fig. 6.17 Tractive effort-speed characteristics of a passenger car with a three-speed automatic transmission [2]

the speed ratio [2]. The torque ratio of the converter reaches a maximum at stall condition where the speed ratio is zero.

The variation of the capacity factor with speed for a particular engine is shown in Fig. 6.16. To achieve the proper matching, the engine and the converter should have a similar range of capacity factors.

Figure 6.17 shows the variation of the tractive effort with speed for a passenger car equipped with a torque converter and a three-speed gearbox [2].

It should be mentioned that the efficiency of a torque converter is low over a considerable range of speed ratio, as shown in Fig. 6.14. To improve the overall efficiency of the automatic transmission and hence fuel economy, a "lock-up" clutch is incorporated in the torque converter. It is programmed to engage in a predetermined vehicle speed range. When the "lock-up" clutch is engaged, the engine power is directly transmitted to the output shaft of the torque converter.

? Example 6.2

An engine with torque-speed characteristics as shown in Fig. 6.16 is coupled with a torque convener with characteristics as shown in Fig. 6.15. Determine the output speed and output torque of the torque convener when the engine is operating at 2450 rpm with an engine output torque of 393 N.m (290 lb.ft).

Solution The engine capacity factor, K_e, is calculated first

$$K_e = \frac{n_e}{\sqrt{M_e}} = 123\,\text{rpm}/(\text{N.m})^{1/2}. \tag{6.68}$$

It is noted that both the capacity factor and the torque converter are equal, thus

$$K_{tc} = 123\,\text{rpm}/(\text{N.m})^{1/2}. \tag{6.69}$$

From Fig. 6.15, the speed ratio at 123 is $C_{sr} = 0.9$ and the torque ratio is $C_{tr} = 1.02$. The output speed of the torque converter, n_{tc}, is

$$n_{tc} = 0.9 \times 2450 = 2205\,\text{rpm}. \tag{6.70}$$

The output torque of the torque converter, M_{tc}, is

$$M_{tc} = 393 \times 1.02 = 400\,\text{N.m}. \tag{6.71}$$

The efficiency of the torque converter under this operating condition can be calculated as

$$\eta_c = 0.9 \times 1.02 = 91.8\%. \tag{6.72}$$

6.4 Fuel Economy

Factors affecting fuel economy of a vehicle:

1. Engine operation and characteristics.
2. Transmission type and characteristics.
3. Weight of the vehicle.
4. Aerodynamic resistance.
5. Rolling resistance of the tires.
6. Driving cycle (conditions).
7. Driver behavior.

Effect of engine operation and characteristics

It is noted that vehicle operations at low engine speed and high torque are always more economical than higher speed and lower torque settings with the same power input.

Effect of Transmission:

1. Most cars may gain approximately a 20% saving of fuel if a vehicle is operated using an overdrive gear of 30% less reduction ratio than the top gear, which can be used to run the vehicle at the same speed.
2. Using the overdrive gear rather than the top drive gear to run the vehicle at the same speed will result in lower engine speed and better fuel economy, but the reserved power to overcome additional resistance will be less.
3. The improvement of fuel economy under steady-state cruising conditions is an exploitation of the fact that for the same power input, the internal combustion engine is always more economical to operate at low speed and high torque than a higher speed and lower torque setting.
4. Vehicles with automatic transmission consume 1% to 11% more fuel than those with manual transmission, for both city and highway driving.

Effect of Vehicle Weight: Reduction of vehicle weight will result in improving fuel economy. This is because the power required to accelerate the vehicle is proportional to its weight. It is estimated that reduction of vehicle weight by 1 kg will result in a reduction of the fuel consumption by about 7.24×10^{-5} L/km.

Effect of Tire Rolling Resistance: It is found that a 10% reduction in tire rolling resistance will result in an approximate saving of 2% in fuel consumption.

Effect of Aerodynamic Resistance: Reducing the aerodynamic resistance will result in the reduction of fuel consumption. For example, reducing the aerodynamic resistance coefficient from 0.5 to 0.3 will reduce fuel consumption by approximately 23% at a steady speed of 96 km/h.

Driving Cycle (conditions): Fuel consumption for driving in the city with slow speeds and frequent "stop and go" is substantially higher than that for driving on the highway with steadier and higher speeds.

Standard Evaluation of Fuel Economy: Fuel economy is characterized by fuel consumed for a given distance traveled under the same driving cycle. To provide a common basis for comparing the fuel economy of different vehicles, the EPA (Environmental Protection Agency) of the USA has devised both city and highway driving cycles as follows:

1. EPA City driving cycle consists of 10 "stop and go" driving segments within 766 s and with a maximum speed of 60 mph (96 km/h).
2. EPA highway driving cycle consists of four segments to simulate the driving conditions on:

   - Local roads,
   - Collector lane,
   - Principal arterial, and
   - Minor arterial.

Within 765 s and maximum speed of 60 mph (96 km/h).

Both the data from the EPA City and highway driving cycles are combined and presented based on a common indicator called CAFE (Corporate Average Fuel Economy) expressed in miles per gallon (mpg). This indicator is established according to the following formula:

$$Mpg_{composite} = \frac{1}{(0.55/citympg) + (0.45/highwaympg)}. \tag{6.73}$$

In the USA, all automotive manufacturers are required to conduct fuel economy tests based on EPA cycles.

6.5 Electric and Hybrid Vehicles

This chapter will discuss different types of hybrid electric vehicles. Three different architectures will be discussed: Internal Combustion (IC), Battery Electric, and Fuel Cell. It will also detail three different layouts for hybrid vehicles: Series, Parallel, and Series-Parallel.

Each section will briefly describe the layout of each architecture, the relevant efficiencies for the flow of power, and the advantages and disadvantages of the type of vehicle.

6.5.1 Conventional IC Engine Vehicles

Internal Combustion (IC) vehicles are powered by an internal combustion engine that burns a combustible fuel to create power. This engine is connected to a mechanical transmission that transmits power to the wheels.

There are four efficiency values that are of interest:

1. η_{W-T} (Well-Tank Efficiency) is the efficiency of producing fuel and transporting it to the fuel tank.η_{W-T} for fossil fuels is about 84%.
2. η_{eng} (Engine Efficiency) is between 0–40% depending on driving conditions and the engine type such as diesel or gasoline.
3. η_{T-W} (Tank to Wheel Efficiency) is the efficiency of the drivetrain. η_{T-W} is 20% for diesel engine-powered drivetrain and is 17% for gasoline engine-powered drivetrain.
4. η_{W-W} (Well to Wheel Efficiency) is the overall efficiency of all the steps between producing the fuel and the energy being used at the wheels.

$$\eta_{W-W} = \eta_{W-T} \times \eta_{T-W}. \tag{6.74}$$

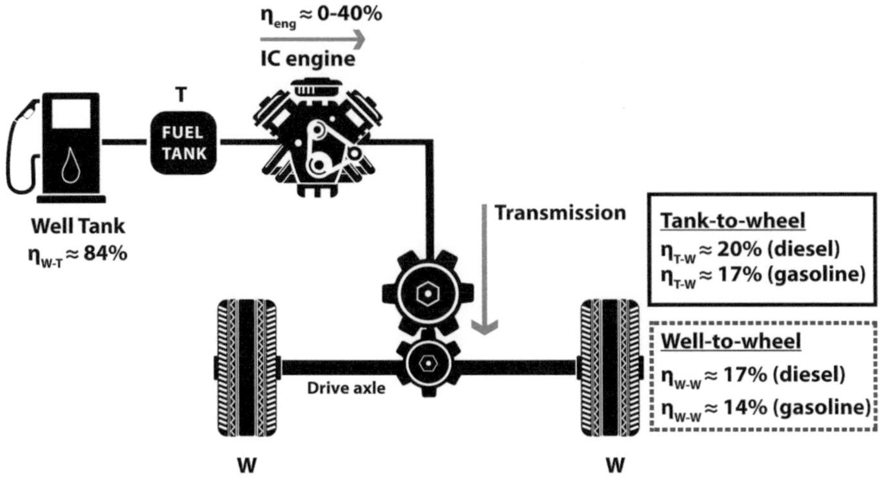

Fig. 6.18 Conventional vehicle architecture and energy flow [5]

η_{W-W} is about 17% for diesel vehicles and 14% for gasoline vehicles. The advantages of an internal combustion vehicle are the long-range resulting from the high energy density of fuels compared to batteries (lithium-ion batteries hold only 15.3% of the energy of an equivalently sized amount of gasoline) [6] and quick refueling. Factors that reduce the capacity of a battery, such as age or temperature change, do not affect a fuel tank. A fuel tank will never shrink with age, and gasoline stored in the fuel tank will not disappear unless it is punctured. The disadvantage of an internal combustion engine is that it does not provide its maximum torque output instantaneously and must run continuously even when the vehicle is idling, in addition, internal combustion engines produce emissions that harm the environment. There is also no system in place that allows the vehicle to reclaim the energy lost during braking, so kinetic energy is lost every time the vehicle slows or stops (Fig. 6.18).

6.5.2 *Battery Electric Vehicles*

Battery Electric Vehicles (BEV) store their energy using a battery and are refueled from an external source of electricity such as the power grid. Electricity from the battery is used to power a motor connected to the drive axle by a mechanical transmission (Fig. 6.19).

The efficiency values for BEVs are:

1. η_{grid} (Grid efficiency) is the efficiency for the process of generating electricity and distributing it to the charger. η_{grid} is estimated to be 40%. This value is the

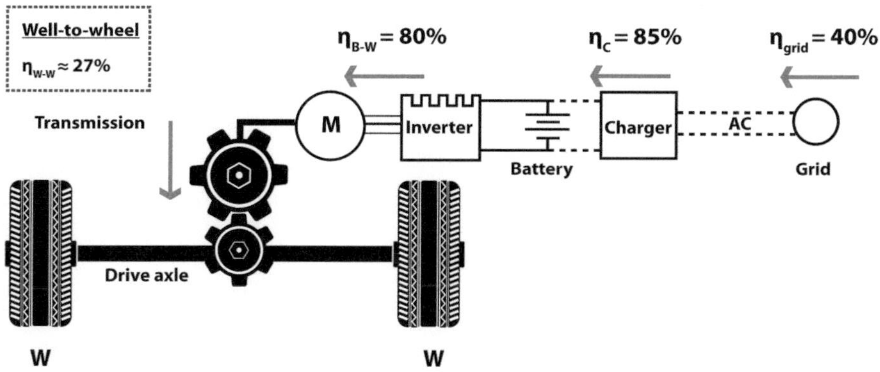

Fig. 6.19 Battery electric vehicle architecture and energy flow [5]

result of different methods of power generation such as fossil fuel, renewable, and nuclear.

2. η_C (Charger efficiency) for electric vehicles is around 85%.
3. η_{B-W} (Battery to Wheel efficiency) is around 80%.
4. η_{W-W} (Well to Wheel efficiency) is around 27%.

BEVs have a very high efficiency compared to other vehicle architectures and some significant performance benefits. Electric motors installed in these vehicles are highly efficient and able to produce their maximum amount of torque from a standstill allowing for impressive acceleration performance. The motors also do not draw power when not in use compared to IC vehicles allowing for high energy savings in stop-start traffic. BEVs also take advantage of regenerative braking to recover energy during braking that is normally lost on an IC vehicle. Incredibly, an EV is able to reach over 80% Battery to Wheel efficiency when the energy regained from regenerative braking is considered [7]. These advantages come at the cost of using batteries that have decreasing charge capacity over their lifespan, reduction of charge due to changes in temperature, and long charge times.

The adoption of BEVs is supported by their ability to use the existing electrical grid to recharge. This is very important from the perspective of a consumer because they will be able to recharge their vehicle from any electrical outlet whether it is in their own home or their place of work.

6.5.3 Hybrid Electric Vehicles

Hybrid vehicles are vehicles that use two or more different sources of power to drive the vehicle. In particular, Hybrid Electric Vehicles (HEV) generally combine an IC engine with the systems of an EV. HEVs are an exciting prospect because they allow the design of vehicles that are more efficient, produce fewer emissions, and have higher performance. Their systems allow the engineer to capture the energy that was once wasted and reuse it later. If the mechanical characteristics of HEV design are understood well, anything from a highly efficient family car to a high-performance race car may be designed. HEVs have the following advantages over conventional vehicles:

- No idling by turning off the IC engine when it is not needed—means fuel is no longer consumed when the vehicle is not moving.
- Regenerative braking reduces the energy wastage from braking by using the electric motor on a vehicle as a generator during braking. This system allows the vehicle to convert kinetic energy normally dissipated as heat by the brakes back to electrical energy stored in the vehicle's batteries that can then be reused to propel the vehicle. This type of braking also reduces wear and tear on mechanical brakes.
- The capability to operate the IC engine and electric motor in their most efficient conditions during stop-start, low torque, and low-speed traffic conditions.
- High amounts of low-speed torque are available from electric motors.

Compared to a BEV, an HEV is also able to extend the operating life of its batteries by only partially depleting and recharging its batteries compared to a BEV that relies on its batteries as its sole source of power.

There are also **Plug-In Hybrid Vehicles (PHEVs)** where the vehicle's battery may be recharged from the electrical grid like a BEV, and there is a higher emphasis on using the batteries for short-distance driving. A PHEV can operate in a Charge Depleting mode where it operates as a BEV for short distances, and also in a Charge-Sustaining mode where it behaves as an HEV with the batteries being charged by the IC engine.

For the purposes of this chapter, there are three basic types of HEV architectures that will be discussed. These architectures combine the IC engine with the EV systems in different ways to power the vehicle. These architectures are Series, Parallel, and Series-Parallel, and the commonality between all of them is that the IC engine shares a connection to the EV systems.

It should be noted, however, that the design of HEVs does not have to be limited to the three basic architectures. High-performance prototype race cars have also begun to use hybrid technology to achieve success in endurance racing. These prototype cars are fitted with electric motors acting as Motor Generator Units (MGU) on their front axles and are powered by their IC engines at their rear axles. As with other HEVs, they generate electricity during braking, but experimental methods of energy recovery are also tested on them, such as using exhaust gases to generate power.

The usage of hybrid systems is not limited to road vehicles either, and massive vehicles such as attack submarines use diesel-electric hybrid systems to travel silently underwater without the use of a noisy IC engine.

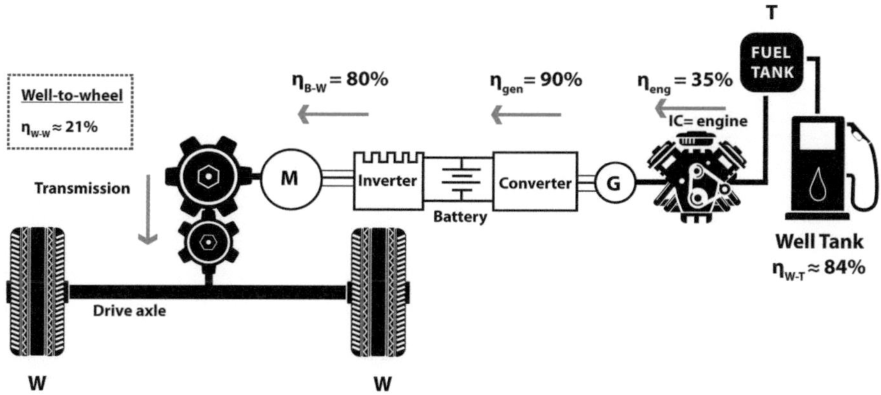

Fig. 6.20 Series hybrid electric vehicle architecture and energy flow [5]

6.5.3.1 Series HEVs

The simplest way of describing the Series HEV architecture is that it is a BEV with an IC engine installed to act as a generator. The role of the IC engine is to generate electrical power that can be used to charge the battery or drive the electric motor. As such, the IC engine may be smaller and designed to operate as efficiently as possible for charging the batteries. Only the electric motor is connected to the vehicle's mechanical transmission to power the drive axle.

This architecture should not be mixed with IC-Electric drive architectures such as Diesel-Electric transmission because those architectures only use electricity as a method of transmitting power from the engine to electric traction motors rather than as a secondary power. A Series HEV is also able to operate as a BEV when the IC engine is not in operation (Fig. 6.20).

1. η_{W-T} (Well-Tank Efficiency) is about 84%. This value is used at the start of the process because the sole source of energy input to the system is the fuel tank.
2. η_{eng} (Engine Efficiency) is about 35%. The Series HEV architecture is able to operate its IC engine at a higher efficiency than a conventional vehicle because it is only used for power generation. This allows the engine to be operated in the most efficient power ranges, and only when required through the use of start-stop systems.
3. η_{B-W} (Battery to Wheel Efficiency) is around 80%.
4. η_{W-W} (Well to Wheel Efficiency) is around 21%. While this value is not as high as a BEV's Well to Wheel Efficiency, it is still higher than that of an IC vehicle.

The primary advantage of the Series HEV architecture is that the engine is decoupled from the wheels and it has high flexibility to operate at its best efficiency spot, which improves its efficiency and fuel economy.

The Series HEV architecture is highly advantageous for low-speed vehicles with frequent starting and stopping. These systems can yield up to 35% better fuel

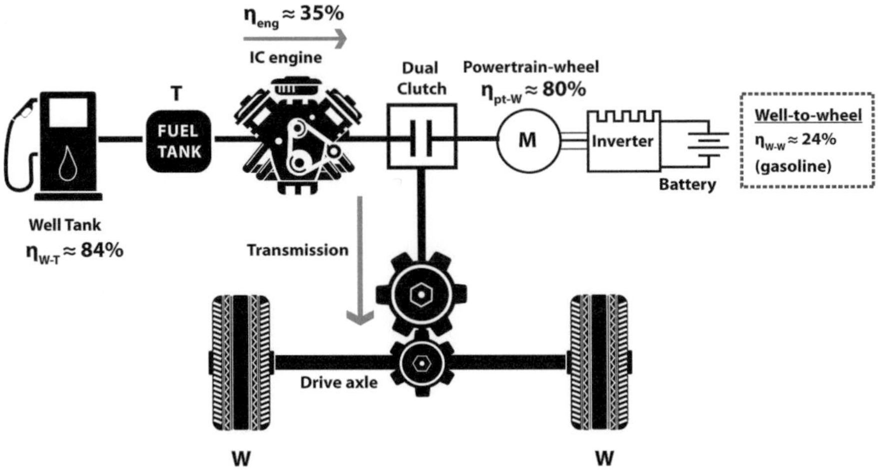

Fig. 6.21 Parallel hybrid electric vehicle architecture and energy flow

economy compared to diesel buses [5]. Since the internal combustion engine is only used for power generation, these buses are able to use a smaller diesel engine designed for pickup trucks instead of a larger, less efficient diesel engine.

From the perspective of a start-stop scenario going from one stop to another, a series HEV is able to use the high torque available from electric motors to rapidly accelerate from a stop and efficiently drive for long distances on electric power. Whenever the need arises for the batteries to be charged, the IC engine may be used at its most efficient settings to generate electricity before being turned off again. When the vehicle has reached its next stop, it is able to use regenerative braking to charge its batteries and recoup energy that would have been lost on a conventional vehicle.

6.5.3.2 Parallel HEVs

A Parallel HEV has its IC engine and an electric motor connected in parallel to the vehicle's transmission by a dual clutch system, so either the IC engine can drive the vehicle or the electric motor or both. The engine may also be connected to the electric motor to charge the battery. This architecture may be likened to an IC vehicle with electric motors assisting [8] (Fig. 6.21).

The efficiency values for a Parallel HEV are similar to those of other HEVs.

1. η_{W-T} (Well-Tank Efficiency) is about 84%.
2. η_{eng} (Engine Efficiency) is around 35%.
3. η_{pt-W} (Powertrain to Wheel Efficiency) is around 80%.
4. η_{W-W} (Well to Wheel Efficiency) is around 24%.

The Parallel HEV architecture is a simple method of allowing the vehicle to be propelled by either its IC engine or the electric motor. This architecture is also able to achieve higher Well to Wheel efficiency than a Series HEV by coupling the IC engine directly to the electric motor, avoiding the inefficiency caused by powering a generator with an IC engine [9].

The intent of this architecture is to allow the electric motor to propel the vehicle at low speeds and the IC engine to propel the vehicle at high speeds [9]. The batteries may be recharged through the use of regenerative braking, or by connecting the electric motor to the transmission to act as a generator while the IC engine is in use. In this way, the IC engine and electric motor are only used when they are the most efficient option.

In a start-stop scenario where the vehicle travels from one red light to another, the IC engine of a Parallel HEV may remain off while the vehicle waits for a green light. When it is time to set off, the vehicle may accelerate using either its IC engine, electric motor, or a combination of both, depending on the driving conditions. The vehicle may then continue to the next red light using either its IC engine, electric motor, or a combination, depending on power requirements. If the vehicle is climbing a hill, for example, the electric motor may be used to assist the IC engine so that both are running at the most optimal conditions. When the vehicle reaches the next red light, the electric motor will be used in regenerative braking mode to recharge the battery and prevent as much energy as possible from being dissipated as heat through the brakes.

6.5.3.3 Series-Parallel HEVs

A Series-Parallel HEV connects the IC engine and the electric motor together using a Continuously Variable Transmission (CVT) involving a planetary gearbox. This architecture allows the IC engine to charge the battery like a Series HEV, and also power the drive axle through the transmission like a Parallel HEV. The electric motor is also able to power the vehicle directly. The Toyota Prius is an example of a Series-Parallel HEV (Fig. 6.22).

1. η_{W-T} (Well-Tank Efficiency) is about 84%.
2. η_{eng} (Engine Efficiency) is around 35%.
3. η_{pt-W} (Powertrain to Wheel Efficiency) is around 80%.
4. η_{W-W} (Well to Wheel Efficiency) is around 24%.

The Series-Parallel architecture is the most complicated and allows the vehicle to distribute power between the IC engine, electric motor, and drive shaft simultaneously. When high power output is required, such as for rapid acceleration, the IC engine may be used with the electric motor to propel the vehicle. The vehicle may also be run as a series electric vehicle when it is more efficient to do so, and the IC engine will be operated at the most efficient settings to charge the battery.

Consider the example of a vehicle starting at a traffic light and stopping at the next one again. When the vehicle is idling at the red light, the engine can either be

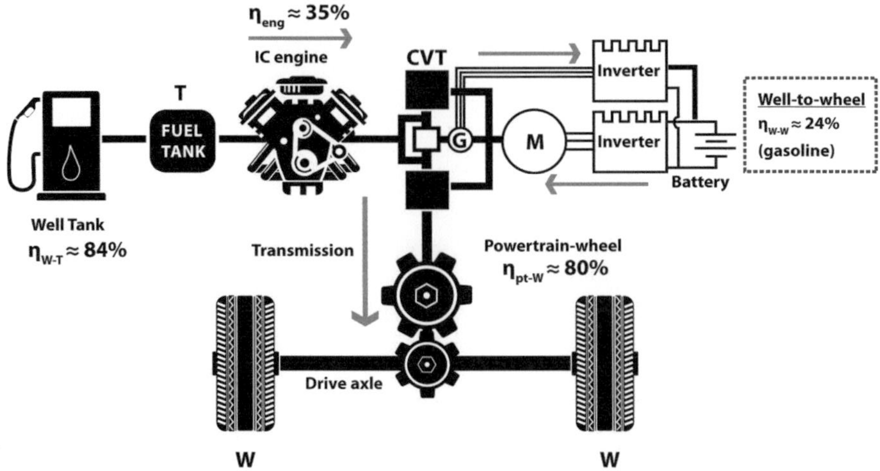

Fig. 6.22 Series-parallel hybrid electric vehicle architecture and energy flow [5]

completely off or used to charge the battery. When the light turns green, the vehicle can rapidly accelerate thanks to the electric motor being able to generate its maximum amount of torque from a standstill compared to an IC engine that only produces its maximum torque at medium to high RPMs. The engine can then be shut off or run at its most efficient setting to charge the batteries in series mode as the vehicle changes to electric drive to maintain the vehicle's speed till the next redlight, resulting in fuel savings. When the vehicle slows down for the red light, the electric motor will act as a generator and recuperate energy back into the battery.

6.5.4 Fuel Cell Electric Vehicles

A fuel cell electric vehicle depends on the process of Hydrogen (H_2) molecules reacting with Oxygen (O_2) to generate electricity. This process generates electricity to power the vehicle with only heat and water emitted as byproducts. The battery may be recharged during braking or by the fuel cell (Fig. 6.23).

1. η_{W-T} (Well-Tank Efficiency) is about 60%.
2. η_{fc} (Fuel Cell Efficiency) is around 58%.
3. η_{pt-W} (Powertrain to Wheel Efficiency) is around 78%.
4. η_{W-W} (Well to Wheel Efficiency) is around 27%.

The Well to Wheel Efficiency is calculated with the following equation:

$$\eta_{W-W} = \eta_{W-T} \times \eta_{fc} x \eta_{pt-W}. \tag{6.75}$$

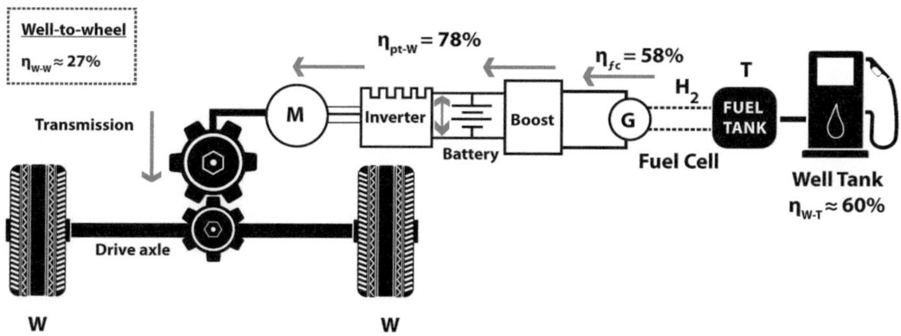

Fig. 6.23 Fuel cell electric vehicle architecture and energy flow [5]

The most important advantage of hydrogen cars is that they have a higher range than BEVs due to the higher energy density of hydrogen with zero greenhouse gas emissions as a byproduct of their working cycle. In the same fashion as a series HEV, they only consume hydrogen fuel when power generation is required.

FCEVs may also be refueled rapidly, although they require special infrastructure to handle the pressurized hydrogen fuel. This has limited the widespread adoption of FCEVs by creating a negative feedback loop where the cost to implement the infrastructure cannot be financially justified to service a small number of FCEVs, and it is hard for the consumer market to adopt FCEVs due to a lack of infrastructure.

FCEV technology has found a niche application in railway vehicles and is able to provide reliable long-distance transport economically. Trains operate in a limited area along set routes, so it is easier to construct the required infrastructure along set points compared to passenger road vehicles that must have the versatility to travel anywhere on the road network.

6.6 Fuel Economy for Electric and Hybrid Electric Vehicles

The overall on-board powertrain and well-to-wheel efficiencies for the various vehicles are summarized below (Table 6.2)

The vehicle architectures with the best well-to-wheel efficiency are the BEV and FCEV at 27%. However, it is important to note that the BEV has the most efficient powertrain efficiency at 80%. It is by combining the BEV's powertrain efficiency with an IC vehicle that HEVs are able to have higher powertrain efficiency of 25% for Gasoline Series HEV and 28% for Gasoline Parallel HEV compared to just 17% for gasoline vehicles and 20% for diesel vehicles. The well-to-wheel efficiency is also improved with Gasoline Series HEVs at 21% and Gasoline Parallel HEVs at 24% compared to 14% for a conventional gasoline car.

Table 6.2 Drivetrain efficiency companion for different fuel vehicle [8]

Fuel	Powertrain efficiency (%)	Well-to-wheel efficiency (%)
Gasoline SI	17	14
Diesel CI	20	17
BEV	80	27
Gasoline series HEV	25	21
Gasoline parallel HEV	28	24
Hydrogen FCEV	45	27

6.7 Batteries for Electric and Hybrid Vehicles

In this section, different battery types and battery packs will be discussed.

6.7.1 Battery Types and Battery Packs

Knowledge of batteries is key to the design of an efficient HEV because they are directly linked to the range, performance, and efficiency of the electrical drive. Common types of batteries include Lead Acid (PbA), Nickel Metal Hydride (NiMH), and Lithium-Ion (Li-ion), which will be discussed in this section. They may also be connected in different ways, such as series, parallel, and series-parallel circuits, to meet certain power requirements. The batteries that will be discussed generate electrical power by using a chemical reaction known as an **Oxidation-Reduction (redox)** reaction. This reaction is a combination of two types of chemical reactions:

Oxidation: An oxidation reaction is when a molecule loses an electron. A common example would be when a piece of metal rusts.

Reduction: A reduction reaction is when a molecule gains an electron. An example of this is electroplating or electrolysis.

This redox reaction takes place in what is known as a Voltaic Cell, and a battery is made up of one or more of these **voltaic cells**. The purpose of a voltaic cell is to combine these two reactions in a controlled manner so that the transfer of electrons can be harnessed to generate an electrical current that may be used to power electrical devices.

There are three basic components in a battery:

1. Anode: The negative terminal where the oxidation reaction occurs during discharge,
2. Cathode: The positive terminal where the reduction reaction occurs during discharge, and

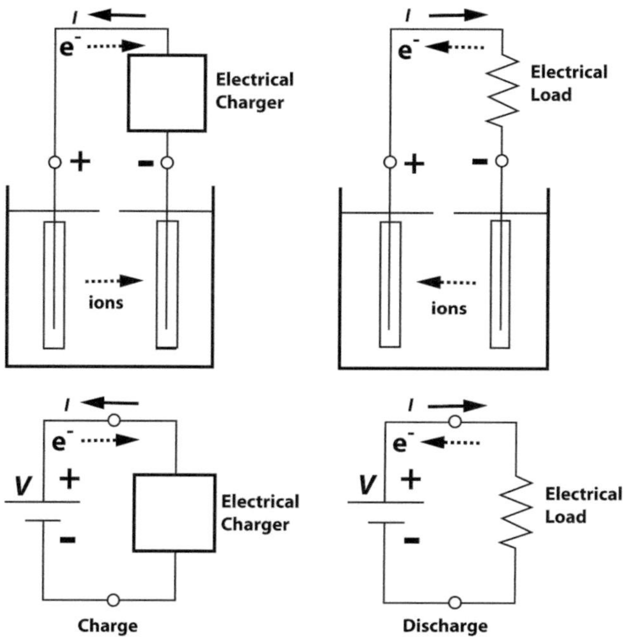

Fig. 6.24 Basic battery cell layout

3. Electrolyte: A solution containing ions submerges the anode and cathode. It allows ionic charge to flow between the anode and cathode to allow the redox reaction to continue.

There are two different types of batteries:

Primary Batteries: Batteries that release power by undergoing an irreversible chemical reaction.

Secondary Batteries: Batteries that release power by undergoing a reversible chemical reaction. All rechargeable batteries are secondary batteries, and these are the type used in HEVs.

In the diagram below, the redox reaction is visualized showing the flow of electrons between the anode and cathode when the battery is charging and discharging. When discharging, the anode oxidizes, which releases an electron to the cathode, which accepts the electron in a reduction reaction. This movement of electrons creates the current. When charging, the reaction is reversed (Fig. 6.24).

Another way of thinking about how a rechargeable battery works is by saying the anode rusts during the oxidation reaction to produce power. When the battery is recharged, electrolysis occurs where the rust is turned back into the anode's alloy in a reduction reaction. Of course, the condition of an anode can't be completely restored with every cycle, which is what leads to a reduction of capacity when a battery is used.

Electrical current can be described quantitatively with these units:

Amps (A): This unit measures the amount of electrical current electrons passing through. In fluid dynamics, this would be the same as the flow rate.

Volt (V): This unit measures the difference of potential that would drive one ampere of current against one ohm resistance. This can be likened to the dynamic pressure of a flow in fluid dynamics. An interesting note is that this unit was named after Alessandro Volta, the inventor of the battery.

When used in any type of electric vehicle, many batteries are connected together to meet the required voltage and current required. The voltage of a battery is measured in Volts (V) and measures the electric potential of the battery. The current of a battery is measured in Amperes (A).

6.7.1.1 Series

When one or more batteries are connected in a series, their voltage is added together.

$$V_T = V_1 + V_2 + \dots. \tag{6.76}$$

6.7.1.2 Parallel

When one or more identical batteries are connected in parallel, the output voltage will be the same, and the current will be the sum of the current output of the batteries.

$$V_T = V_1 = V_2 = \dots \tag{6.77}$$

$$A_T = A_1 + A_2 + \dots. \tag{6.78}$$

6.7.1.3 Series-Parallel

Batteries in electric vehicles are most commonly connected in a Series-Parallel circuit where series circuits of batteries are connected in parallel. In this case, the total voltage would be equal to the total voltage of one series of batteries, and the current would be equal to the sum of the current output of each series of batteries (Fig. 6.25).

6.7.1.4 Battery Chemistries and Comparison

The development of batteries has been continuous since their invention by Alessandro Volta to improve their capabilities. From the development of rechargeable batteries to Lithium-Ion batteries, the objective has always been to find different ways of storing electrical energy in denser, more durable, and longer-lasting batteries.

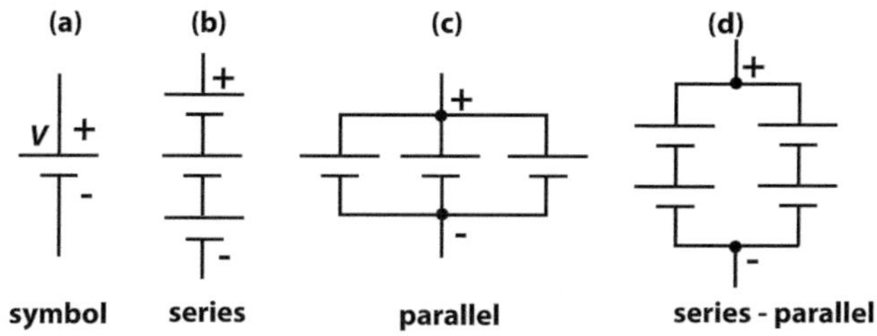

Fig. 6.25 **a** Battery symbol **b** series **c** parallel **d** series-parallel

The purpose of a battery is to store energy, and may be evaluated in a quantitative manner by understanding the following concepts:

Watts (W): Watts is the measurement of electrical power and is calculated by multiplying the voltage of a current by its amperage. It may be calculated by multiplying a power source's voltage with its current.

Joules (J): This is the unit measuring one unit of energy and is a measure of how much energy it can store. Energy is the product of multiplying power by time.

$$\text{Energy} = \text{Power} \times \text{Time}. \tag{6.79}$$

This can be expressed in units as: 1 J [Joule] = 1 Ws [watt-second].

Watt-Hour (Wh): This unit is the one commonly used for the amount of energy stored in a battery and is a derivative of the above relationship between Joules and watt-seconds. As batteries are expected to store huge amounts of energy, the unit of Ws is not large enough to quantify the energy stored in a battery, so watt-hour is used instead. It is a product of multiplying 1 Ws by 3600 s in an hour.

$$1\,\text{Wh} = 3600\,\text{Ws} = 3600\,\text{J}.$$

Kilowatt-Hour (kWh) is also used as a unit for quantifying the energy in a battery.

$$1\,\text{kWh} = 1000\,\text{Wh}.$$

The watt-hour and kilowatt-hour may also be used to characterize the specific energy or how power dense a battery is in watt-hours per kilogram (Wh/kg).

Ampere-Hour (Ah): This is the unit used for quantifying the capacity of a battery and is a product of current (I) and time. Related to this unit is the Coulomb (C), which is the measure of charge. The resulting equations are:

$$\text{Charge} = \text{Current} \times \text{Time or } Q = I \times t. \tag{6.80}$$

This is expressed in units as: 1 As (Amp-second) = 1 C

1 Ah = 3600 As = 3600 C The capacity may also be related back to energy by multiplying it with battery voltage:

$$1\,\text{J} = 1\,\text{As} \times 1\text{V}. \tag{6.81}$$

Battery Types For electric vehicles, there are three types of batteries of interest: lead acid, nickel metal hydride, and lithium-ion.

The first rechargeable battery was invented by Gaston Planté in 1859. It was known as the "pile secondaire" in French or "secondary battery" in English, and used lead alloys for its anode and cathode [10] dipped in an acidic solution. This type of battery is known as a lead acid (PbA) battery and was used in early electric vehicles such as cars and submarines. However, these types of batteries have very low energy density and are very heavy. Later, Nickel Metal Hydride (NiMH) batteries were designed which have more power density.

The most common type of battery in use on electric vehicles at present is Lithium-Ion (Li-ion) batteries. Li-ion batteries generate about 3–4 V per cell using a cathode made of lithium metal oxide (LiMO2) and an anode made of carbon such as graphite carbon.

These batteries have no memory effect (the capacity decreases from partial charging), but they still lose capacity over time and use.

It is important to understand the parameters of batteries and the characteristics of different chemistries of batteries because different vehicles have different requirements. BEVs will require batteries suitable for being used as a sole power source, while HEVs require batteries optimized for a high number of discharges (Table 6.3).

When describing how full a battery's capacity is, there are two terms to understand:

State of Charge (SOC): This is the battery capacity remaining and is usually expressed as a percentage. This is equivalent to the information given by a battery bar on a phone or electronic device.

Depth of Discharge (DOD): This is the value of the portion of the battery that has already been discharged and can also be expressed as a percentage.

The SOC and DOD are both opposite sides of the same coin, and if you consider a partially filled cup of juice, the SOC is the portion of the glass that contains juice, while the DOD is the portion of the glass that is empty.

6.7.2 *Battery Models*

Consider the battery pack for the Tesla Model S. This vehicle has an 84 kWh battery pack that generates 22.8 V of electricity. It is able to accomplish this by connecting six 3.8 V cells in series that add up to 22.8 V. There are 74 series of cells connected in parallel in each series to form a battery module and sixteen modules that make up a battery pack (Figs. 6.26 and 6.27).

Table 6.3 Timeline of hybrid electric vehicles development with battery type and pack parameters [10]

Vehicle	Vehicle weight (kg)	Battery weight (kg)	Chemistry	Rated energy (kWh)	Specific energy (Wh/kg)	Cell/pack nominal voltage (V)	Rated power (kW)	Specific power (W/kg)	P/E
1996 GM EV1	1400	500	PbA	17	34	2/312	100	200	6
1999 GM EV1	1290	480	NiMH	29	60	1.2/343	100	208	3
1997 Toyota Prius	1240	53	NiMH	1.8	34	1.2/274	20	344	11
2008 Tesla Road-ster	1300	450	Li-ion	53	118		185	411	3
2011 Nissan Leaf	1520	294	Li-ion	24	82	3.75/360	80	272	3
2011 Chevy Volt	1720	196	Li-ion	17	87	3.75/360	110	560	6
2012 Tesla Model S	2100	650	Li-ion	85	157		270	500	3
2017 Chevy Volt	1624	440	Li-ion	60	136	3.75/360	150	341	3

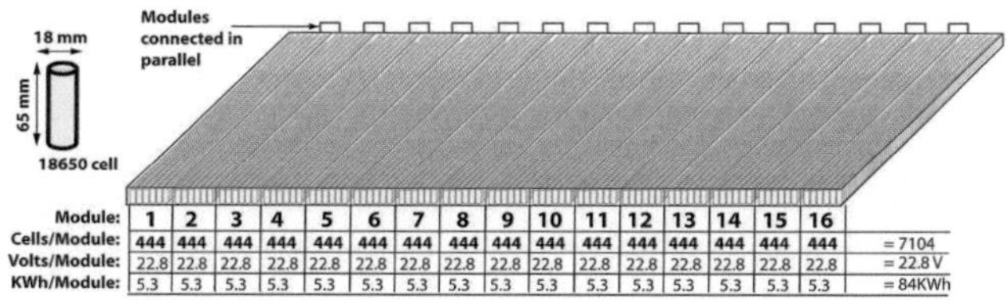

Fig. 6.26 Schematic of Tesla model S battery pack [11]

6.8 AC Machines

6.8.1 Introduction to AC Machines

Alternating Current (AC) Machines are machines that use electromagnetic force to create mechanical work. In particular, knowledge of AC traction motors is important

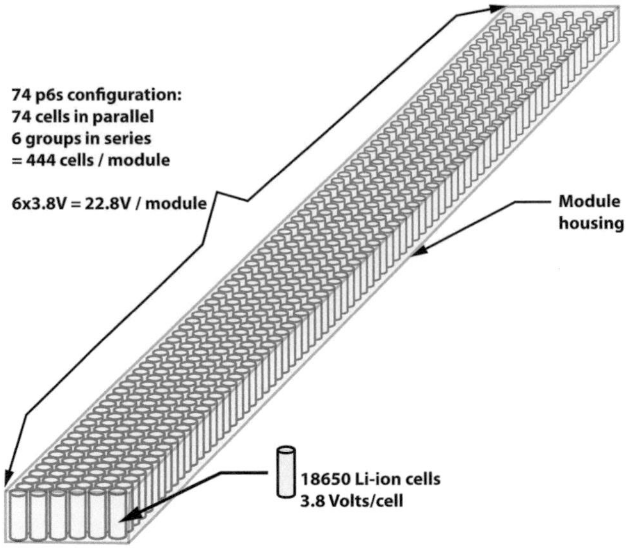

Fig. 6.27 Schematic of Tesla model S battery module [11]

because they are one component of HEVs that cannot be overlooked. These machines are highly efficient in their operation and contribute greatly to the efficiency of an HEV architecture.

The greatest advantage of AC traction is that, unlike IC Engines that require high rotations per minute to reach their maximum torque output, an AC machine is able to produce this instantly from a standstill. This characteristic proves exciting whether you are designing an economical car meant for city driving with frequent starting and stopping or a high-performance race car capable of reaching 300 km/h within 8 s.

AC machines rely on Electromagnetic Induction to move. This is the physical phenomenon where a magnetic field can be created by passing an electrical current through a series of conductive windings that make a coil. In short, if a wire is wrapped into a coil and then has a current passed through it, the coil turns into a magnet (Fig. 6.28).

The above illustration shows a conductive wire that has been wound into a coil with current being passed through it. The result is the creation of a magnetic field passing through the center of the coil with a north and south pole. This magnetic field can be illustrated by drawing the Magnetic Flux which is the characterization of how a magnetic field flows through space.

A magnetic field can also be passed through a coil to induce a current through it, so the opposite can happen where magnetic flux passes through the center of a coil to induce a current through the windings. This behavior is characterized by Lenz's Law that states the magnetic field generated by the induced current will oppose the polarity of the magnet field that is being passed through the center of the coil.

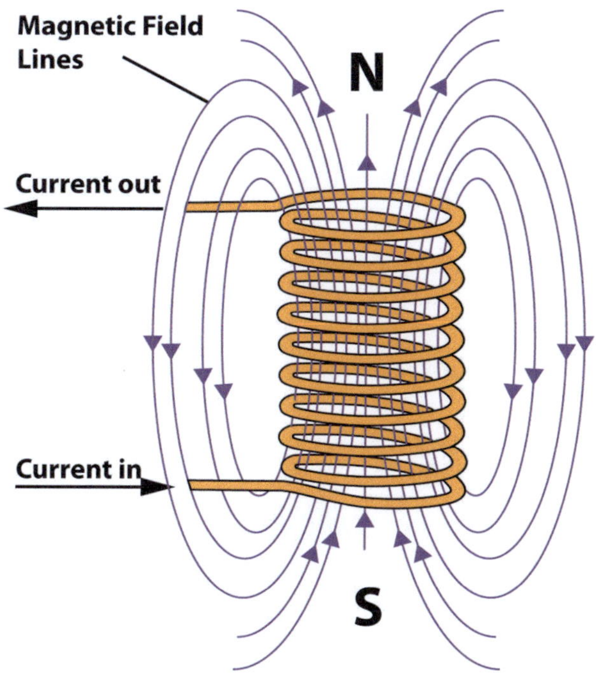

Fig. 6.28 Induction coil and magnetic flux diagram

The opposing magnetic fields result in a physical force that repels the source of the two fields from each other in the same way that two magnets repel each other when pushed together with two of the same poles facing each other.

6.8.2 The Operation of AC Machines

The famed AC electricity pioneer, Nikola Tesla, took advantage of electromagnetic induction to invent the world's first three-phase AC motor, where a rotating magnetic field was used to rotate a shaft. These motors are incredibly simple in design and have only two main components:

- Stator: Part that remains stationary.
- Rotor: Rotating piece housed inside the stator that rotates.

In the above diagram, the rotor and stator are shown in relation to their placement in an electric motor (Fig. 6.29).

The rotor of an electric motor contains the coils that become influenced by a magnetic field generated by windings on the stator of a motor. The magnetic flux that passes through the rotor's coils induces an opposing magnetic field causing a

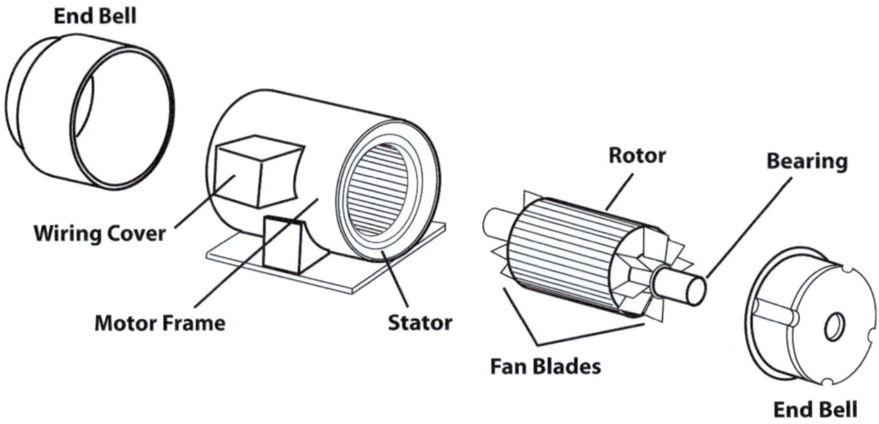

Fig. 6.29 Exploded diagram of AC motor

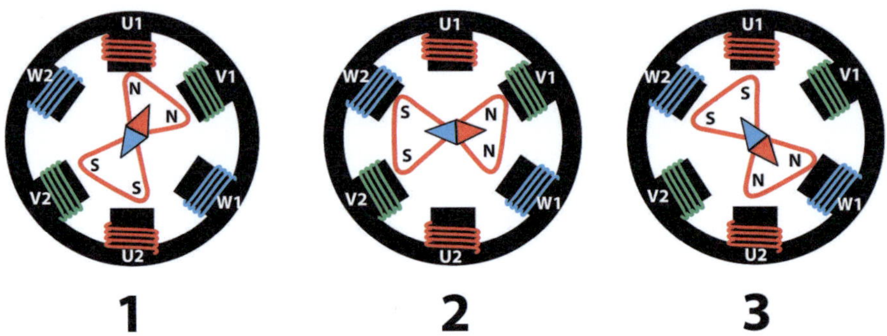

Fig. 6.30 Schematic of a rotating magnetic field in AC motor [11]

force to be applied to the rotor. This force manifests itself as a torque that causes the rotor to rotate as it tries to align its magnetic field with the one generated by the coils on the stator. The force generated by electromagnetic repulsion is also the reason why an AC motor is able to generate maximum torque from a standstill, as this force does not depend on the velocity of the motor's rotation (Fig. 6.30).

The three coils on the stator in green, red, and blue are wound in a specific geometry with angles 120 degrees apart and work together to create a rotating magnetic current. These coils are connected together with the power source in a specific pattern so that the magnetic field inside of the stator rotates as the polarity of the AC current reverses. The three phases of the AC current supplied must also be 120 degrees apart in time. This can be likened to how an IC engine has a specific ignition pattern that causes the crankshaft to rotate.

When selecting an AC machine for a task, it is important to understand the parameters that each one is designed for. Factors such as the current draw, torque output, and voltage requirements are all important when selecting an AC machine.

Rated Torque (T_r (rated)): The rated torque is the amount of torque produced by a motor at its rated power and rated speed; also known as the full-load torque.

Rated and Base Speeds (ω_r (rated)): The rated speed or nominal speed is the rotational speed at which the motor produces its rated torque. It can also be given in rotations per minute (N_r (rated)):

$$N_{r(rated)} = 60\frac{\omega_{r(rated)}}{2\pi}. \tag{6.82}$$

The base speed of a motor is the lowest speed where it can deliver full power.

Rated Power ($P_{r(rated)}$): This is the power output of the machine when it is operating at its rated torque and speed.

$$P_{r(rated)} = T_{r(rated)} x \omega_{r(rated)}. \tag{6.83}$$

Peak Operation: This is when an AC machine operates at its engineering limits. When this occurs, it produces either peak torque ($T_{r(peak)}$) or its maximum velocity while staying within its current and temperature limits. In particular, peak torque is usually only available for a short period of time before damage to the motor or to the electrical system may occur.

Starting Torque: The starting torque is also known as the stall torque and is the amount of torque available when the AC machine is stalled or starting up. It is important to note that while there will be no significant problems if it occurs on a machine buffered by an inverter such as in the case of an electric vehicle, there will be problems if the machine is connected directly to the grid.

In the below graph, the curves for rated torque T_r and power P_r are plotted to show their relationship with the rotational speed ω_r of the electric motor. Note how there are two distinct sections separated by the rated speed. Once the rated rotational speed has been exceeded, the power of the motor stays constant while the torque decreases as the rotational speed increases until the maximum rotational speed is reached (Fig. 6.31).

Constant Torque Mode is when the motor operates with a constant torque output in the region before its rated speed $\omega_r(rated)$ has been reached. The AC motor's ability to operate in this mode is what gives it the torque advantage over an IC engine. The stall torque is produced immediately before the motor starts rotating, and it is maintained as the power output increases linearly. The torque generated is limited by the input current and conduction loss. The torque, in this mode, is equivalent to the peak torque

$$T_r = T_{r(peak)}. \tag{6.84}$$

Constant-Power Mode is when the motor is operating faster than its rated speed with a constant power output. Once the rated speed has been exceeded, the power curve stays linear while the torque decreases inversely as motor speed continues to increase. This mode is also known as field-weakened mode and is used when a vehicle is being driven at high speeds. The motor will begin to reach its design

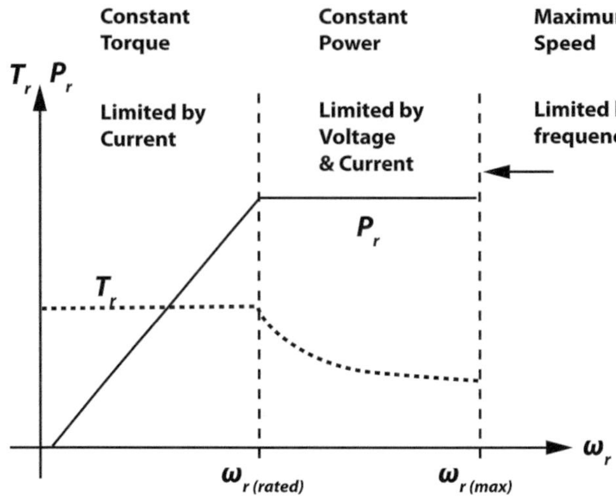

Fig. 6.31 Effect of speed on torque and power of a AC motor

limitations as its speed increases, and factors such as heat, saturation, and core loss will begin to affect its magnetic field while electrical limitations such as insulation thickness will affect performance. The power output of the motor will be limited by voltage and current input.

The inverse relationship for torque output may be calculated with this equation

$$T_{r(peak)} = \frac{P_{r(peak)}}{\omega_r}. \tag{6.85}$$

Note how the lower the rotational velocity is, the higher the torque will be.

Maximum-Speed Mode is when the motor operates at its maximum speed $\omega_{r(max)}$. This speed is limited by the inverter and the mechanical constraints of the motor's components.

Efficiency

AC machines operate at different levels of efficiency based on operating speed. Power losses such as heat buildup, and slip caused by the difference in rotor frequency from the synchronous frequency can be mitigated with different motor designs.

The efficiency of a motor can be graphically represented by plotting efficiency contours in relation to torque and speed on a graph. It is possible to identify the rated speed where the motor transitions from constant torque to constant power operation from the curve, and the color gradient of the contours underneath the line illustrates motor efficiency. From the data provided, it is possible to determine the motor is most efficient when operating at medium speed between 4000 RPM and 6000 RPM, which is the speed it typically operates at during normal driving. At lower speeds

Table 6.4 Example of engine characteristics

Engine speed (rpm)	500	1000	1750	2500	3000	3500	4000	4500	5000
Engine torque (N.m)	339	379.7	406.8	393.2	363.4	325.4	284.8	233.2	189.8

with higher torque output, the motor is less efficient with efficiency dropping below 70%.

The motor-inverter efficiency graph is also of interest because it illustrates the efficiency of the vehicle's inverter. This value is calculated as the product of the motor's efficiency and the efficiency of the electric drive system's inverter. Inverters are an important component of the vehicle's drivetrain efficiency because they supply power to the electric motor. They are typically very efficient with at least 90% efficiency, but can operate as high as 99% efficiency.

Lower efficiency areas can result in high temperatures for the motor and inverter that may lead to lower performance or damage. However, vehicles do not usually drive in these conditions for periods of time long enough to cause damage.

This data is important to an engineer because it can be used to develop control algorithms optimizing motor performance and efficiency while the vehicle is in operation. It is with these control algorithms that an HEV will decide how and when it will use its electric motor and IC engine based on driving conditions.

An engineer designing the electric drive of an HEV must use the engineering parameters of an AC machine to make an appropriate choice that meets design requirements. Then a suitable control algorithm for the systems microcontrollers may be developed based on an understanding of the relationships that govern a motor's performance. The designed controller for the system must strike a balance between efficiency, performance, and the limitations of the AC machine to meet the design objectives of the vehicle (Table 6.4).

Problems

1. A vehicle weighs 20.02 kN and has a wheelbase of 279.4 cm. The center of gravity is 127 cm behind the front axle and 50.8 cm above ground level. The frontal area of the vehicle is 2.32 m^2 and the aerodynamic drag coefficient is 0.45. The coefficient of rolling resistance is given by $R_r = 0.0136 + 0.4 \times 10^{-7} V^2$, where V is the speed of the vehicle in kilometers per hour. The rolling radius of the tires is 33 cm. The coefficient of road adhesion is 0.8. Estimate the possible maximum speed of the vehicle on level ground and on a grade of 25% as determined by the maximum tractive effort that the tire-road contact can support if the vehicle is (a) rear-wheel-

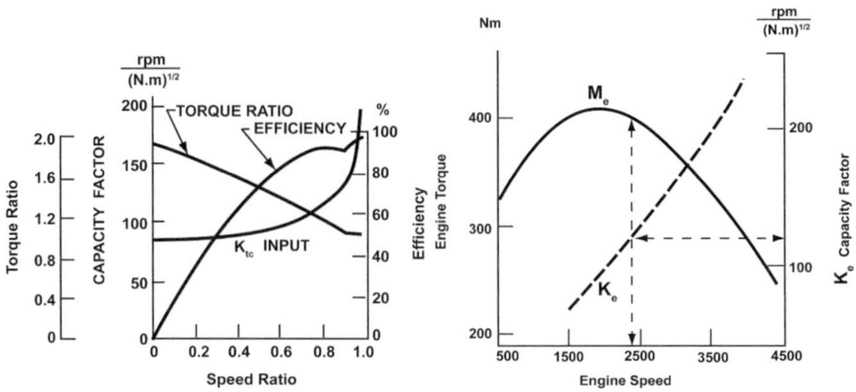

(a) Torque converter (b) Engine

Fig. 6.32 Characteristics of torque converter and engine

drive and (b) front-wheel-drive. Plot the resultant resistance versus vehicle speed, and show the maximum thrust of the vehicle with the two types of drive.

2. The vehicle described in Problem 2 is equipped with an engine having torque-speed characteristics as shown in the table below. The gear ratios of the gearbox are: first, 4.03; second, 2.16; third, 1.37; and fourth, 1.0. The gear ratio of the drive axle is 3.54. The transmission efficiency is 88%. Estimate the maximum speed of the vehicle on level ground and on a grade of 25%, as determined by the tractive effort that the engine torque with the given transmission can provide, if the vehicle is rear-wheel-drive. Plot the vehicle thrust in various gears versus vehicle speed.
3. A vehicle is equipped with an automatic transmission consisting of a torque converter and a three-speed gearbox. The torque converter and the engine characteristics are shown in Fig. 6.32a, b, respectively. The total gear reduction ratio of the gearbox and drive axle is 2.91 when the gear is engaged. The combined efficiency of the gearbox, propeller shaft, and the drive axle is 0.90. The rolling radius of the tire is 33.5 cm. Calculate the tractive effort and speed of the vehicle when the third gear is engaged and the engine is running at 2000 rpm with an engine torque of 407 N.m. Also, determine the overall efficiency of the transmission, including the torque converter.
4. An EV battery pack has 60kWh. The pack has 96 cells in series per string with 80 parallel strings. Each battery cell has a nominal voltage of 3.6 V, 3 A of maximum current per cell, and an internal resistance of 40 milliohms per cell.

   a. Determine the battery terminal voltage and maximum battery pack current.
   b. Determine the output power at maximum current.
   c. Determine the discharge time at maximum current (ignore the battery voltage change during discharge for simplicity)

5. Determine and explain the characteristics of each powertrain mode, (i.e., startup mode, acceleration mode, braking or deceleration mode, normal driving mode, battery charging mode) for the following powertrains. Characteristics may include: the device which supplies traction power to propel the vehicle in a particular mode, and how the battery is charged in a particular mode.
   a. Series hybrid electric vehicles with a gasoline engine.
   b. Parallel hybrid electric vehicles with a gasoline engine.

References

1. Wong JY (2008) Theory of ground vehicles. Wiley
2. Setz HL (1961) Computer predicts car acceleration. SAE Trans 69:351–360
3. Shigley JE (1960) The mechanics of walking vehicles. Technical report, ARMY TANK-AUTOMOTIVE CENTER WARREN MI
4. Bosch R, Peter (Tr.) Girling (1996) Automotive handbook. Society of Automotive Engineers, US
5. Toronto transit commission orders 150 series hybrid buses; agency will operate largest hybrid electric fleet in canada (2022) http://search.proquest.com.uproxy.library.dc-uoit.ca/wire-feeds/toronto-transit-commission-orders-150-series/docview/445444767/se-2. Accessed 2022-09-27
6. Office of energy efficiency and renewable energy (2022) https://www.fueleconomy.gov. Accessed 2022-09-27
7. Alternative fuels data center (2022) https://afdc.energy.gov/fuels/properties. Accessed 2022-09-27
8. Un-Noor F, Padmanaban S, Mihet-Popa L, Mollah MN, Hossain E (2017) A comprehensive study of key electric vehicle (EV) components, technologies, challenges, impacts, and future direction of development. Energies 10(8):1217
9. Jain S, Kumar L (2018) Fundamentals of power electronics controlled electric propulsion. In: Power electronics handbook. Elsevier, pp 1023–1065
10. Kurzweil P (2010) Gaston planté and his invention of the lead-acid battery–the genesis of the first practical rechargeable battery. J Power Sources 195(14):4424–4434
11. Sharma A, Zanotti P, Musunur LP (2019) Enabling the electric future of mobility: robotic automation for electric vehicle battery assembly. IEEE Access 7:170961–170991

Chapter 7
Road Vehicle Braking Performance

The general requirements that a vehicle braking system should meet are self-evident. The brakes should be capable of stopping the vehicle safely within the shortest distance possible on all types of road surfaces and in all motoring conditions. During braking the vehicle should not deviate to one side or the other, and should be completely under the control of the driver, and the brakes must not interact with other vehicle components.

Most cars can sustain approximately one-fourth of their curb weight as an added load and will transfer one-fifth of their load from the rear to the front wheels during a hard brake application. Conventional brake balancing will, therefore, provide optimum brake distribution for a given vehicle load at only one deceleration, below this deceleration, the rear wheels are doing less work than they should. Above this pre-established level, they provide excess torque, resulting in premature wheel slide, with subsequent loss of control or stability. There is no compensation for changes in static loading of the vehicle, or the dynamic weight transfer occurring during braking.

Load sensing brake proportioning is a means of optimizing the ratio of front-to-rear wheel retarding forces for the full range of vehicle loading and decelerations. A relatively simple system of providing two-axle vehicles with improved brake balance has been developed. Simulation studies of vehicle brake system requirements show the effects of a pressure-regulating device used to implement these requirements.

A brief survey of some basic issues related to brakes and braking performance will be given next.

1. Road-Type Adhesion

    - If the braking force at the road tire interface is greater than the functional force, that is, the adhesion between tire and road, the wheel will lock and any further increase in pedal effort will cause no further increase in deceleration. Also, the locked wheel will not develop a sufficient cornering force and so directional stability and/or controllability of the vehicle will be lost.
    - In many accidents with all classes of vehicles, loss of control occurred, caused by the braking force exceeding the adhesion force. The relating force exerted by a tire is developed as the wheel slips as well as rolls and the force builds up

M. El-Gindy and Z. El-Sayegh, *Road and Off-Road Vehicle Dynamics*,
https://doi.org/10.1007/978-3-031-36216-3_7

to peak value before decreasing to a lower value as the wheel locks in the fully sliding conditions. The ratio of the tangential force transmitted by the wheel to the normal load carried by the wheel is termed the coefficient of adhesion.
- When the tire is subjected to braking, the braking force is reduced at a larger percentage slip ratio. In the fully locked wheel condition, there is virtually no sideways force available for cornering.
- The main forces that affect road adhesion (μ_A) are the material and tread pattern of the tires, the nature of the road surface, and the water depth. Secondary effects arise from other factors such as size, inflation pressure, and loads.
- The tire treads material influences (μ_A) and a considerable increase in (μ_A) obtained by making the tread of high hysteresis rubber. Benefits are obtained if the surface deforms the rubber even in the presence of water film.
- In general, the brake designer must make provision for the vehicle to cope with widely varying road conditions. For maximum deceleration of a vehicle on a given surface, all wheels must be on the point of locking simultaneously and the maximum deceleration cannot exceed ($\mu_A \times g$). To ensure this condition, the braking force at each wheel must be in proportion to the load being carried.

2. One complication influencing braking performance is that the static loads carried on each axle vary depending on whether the vehicle is loaded or not. Another complication is load transfer which causes the wheel loads to vary with the degree of braking. Load transfer in turn can be divided into:

   - Steady State Load Transfer: Longitudinal braking forces in which (F_a) are independent of time;
   - Transient Load Transfer: Longitudinal braking forces are dependent on time.

7.1 Brake Mechanisms

Brakes are almost exclusively of the friction type and are classified as drum and disc. The torque developed for a given input depends on its design and particularly the coefficient of friction between lining and drum or disc.

In comparing brakes, we use the non-dimensional quantity S that is the shoe factor, defined as the total frictional force on a shoe or pad divided by the applied load actuating the shoe or pad.

$$S = \frac{fdN}{F} \tag{7.1}$$

Each shoe factor multiplied by the actuating force, F, and the drum radius or the mean radius for a pad, R, gives the brake shoe torque.

$$T = S \times R \times F \times \eta \tag{7.2}$$

where η is equal to 90% in particular, it is essential to ensure that not too great a shoe factor is used as this may give rise to excessive sensitivity S which can be defined as the percentage change in torque either for a given change in μ or for a percentage change in μ. Experience suggests that an S value below 2 gives a reasonably stable brake for most purposes, as will be discussed in detail in Sect. 7.4.3.

7.2 Brake Components

The components of the brakes, the drum, and the friction material also influence the stresses that a brake can support. Thermal expansion of the drum can alter the shoe factor, and therefore, the torque; the torque tends to increase when the drum is contracting upon cooling. The brake will, therefore, grab and give an erratic performance and may also be noisy.

Expansion, deflections, lining compression, and clearances also affect the shoe-center life, which is the radial displacement of the shoe at the center of the shoe and limits the effectiveness of the brake.

With discs, thermal expansion effects are smaller than with drums, as the expansion is toward the pads but pad compressibility can cause considerable lost motion in the brake system.

If large amounts of heat are dissipated over small areas of a cold metal member, the resulting thermal gradients are high and metallurgical transformations can take place at these areas giving rise to small-discolored hard “hot spots”. Hot spots may cause local cracking, and as they are slightly above the metal surface, they machine away the friction material. A cracked metal surface can also give rise to brake temperature.

Asbestos resin-based friction materials incorporating inorganic and other fillers continue to be developed to withstand higher temperatures and exhibit lower wear rates. These materials have the advantages of good performance and low wear over the wide range of temperatures experienced in service while still remaining cheaper than other types of friction materials.

7.3 Choice of Brakes

Weight and performance determine the type of drum brake to be fitted. On lighter cars of medium performance, drum brakes of two leading shoes or leading/trailing shoe brakes are fitted to the front wheels. On heavier cars, duo-servo brakes are used, and these are very suitable for the rear brakes to give a good hand brake performance without power assistance, and disc brakes are fitted to the front wheels.

On heavy vehicles, drum brakes have the advantage, as conventional discs are more exposed to road debris than drums and score badly and the pad life is less than the lining life, as ample power is available with most systems used on heavy vehicles.

Loading/trailing shoe brakes are normally fitted although two loading shoe brakes are also used for the front wheels with loading/trailing shoe brakes at the rear wheels. The shoes are generally forced apart by a cam or by a wedge mechanism.

7.4 Braking Systems for Road Vehicles

The outlines of the considerations involved in selecting a braking system for a rigid vehicle are basically any braking system that should be capable of being operated safely for a reasonable period of time without maintenance. To ensure an acceptable pedal effort and travel, bearing in mind the purpose for which the vehicle is to be used, it is relevant to note that the requirements of the countries in which the vehicle has to operate must also be met.

The braking system starts with the tire on the road and ends with the driver's foot on the brake pedal. The selection of the wheel brakes must be suitable to cope with the energy and the torque created at the tire/road contact area as a result of the force applied from the driver's foot.

7.4.1 Basic Limitations of a Fixed Ratio System

To achieve maximum deceleration or "Ideal braking", the front and rear wheels must approach the locking point simultaneously. At this point, the brake force requirements at any wheel are equivalent to the wheel load related to the tire to road adhesion, frequently referred to as the "dynamic load".

As the wheel load changes due to weight transfer, different braking force requirements will be required to achieve the ideal over the tire road adhesion range.

Some compromise is necessary and, in practice, a condition for optimum braking performance is selected and the system pressures are modulated, wherever possible, with valves to produce satisfactory braking over the range.

7.4.2 Selection of Wheel Brakes

The selection of brakes for a vehicle is influenced by the following:

1. **Vehicle Weight**: The ultimate strength requirement of a brake is primarily determined by the torque created by the load on the tire and the adhesion between it and the load. In this respect, safety factors are especially relevant as it is this developed torque which is transmitted to the brake parts and can be well in excess of that determined from nominal values of lining friction/brake factor and line pressure. For instance, failures of brake components have been experienced even when the ultimate strength has been in excess of twice the deceleration capability of the actuation/brake package based on nominal performance values. Spike testing is now typically a part of most brake proving programs. Weight is also the key factor in terms of energy input levels during hill descents or for that matter on any single brake application.
2. **Engine Power**: Ultimately, engine power determines two things, the time between successive brake applications and the maximum speed at the commencement

of the brake application. Brake usage on main roads bears a relationship with maximum speed and, for all practical purposes, a direct relationship to engine power on cross-country routes. In reality, until recently, the power to weight ratio was only significant in the case of trucks, with around 8 BHP/TON having their brake sizes influenced more by weight than engine power. It is evident that the increase of power and the reduction of weight, in future, will have to be reconciled with the effect of the engine power in any brake specification.

3. **Disc or Drum Brakes**:

   a. Passenger Cars: The drum brake obtained a bad reputation on medium/heavy cars of the late (1950s) because of the continual up-rating in power to weight ratios and improved road holding which reflected in excessive levels of duty at the brakes. Adopting the disc brake reconciled many of the problems of these vehicles but, it is relevant to note, this change resulted in an increase in the rubbed area of 30–40%.
   b. Trucks: This choice is somewhat limited because of the lack of availability of suitable disc brakes. This situation is a legacy left by the disappointments of the early 1960s. The whole problem facing disc brakes on heavy commercial vehicles is related to the difficulty of getting a sufficiently large disc into the wheel. Because of this restriction, it was necessary to utilize discs at a high specific energy level, which results in excessive pad/disc wear and even disc craving problems. Multiple discs are an obvious development but, because of the attendant mechanical sophistication necessary to provide adequate solutions, current market forces do not make them commercially attractive when compared with drum brakes. The drum brake maintains its precedence by the simple fact that wheels diametrically limit the use of the disc brakes. It is also relevant to note that drum brakes have the ability to self-limit, a feature which cannot be overemphasized and is one reason for their continued use in heavy vehicles despite the disc brake's superior fade and smoothness characteristics.

4. **Drum brakes**: Different drum brake designs result from considerations with regard to application, support, and adjustment of the brake shoes. Brakes are assessed in terms of the brake factor C^* which expresses the ratio of longitudinal force to tensioning force at the brake shoe. The brake factor takes into consideration the effect of the coefficient of friction and the internal transmission ratio of the brake.

   a. Simplex drum brakes: there are different types of simplex drum brakes as shown in Fig. 7.1, primarily depending upon the type of application (floating, fixed) and the type of support (rotating shoes, sliding shoes). Wheel brakes with floating brake application and rotating-shoes support, for example, are common. In the case of hydraulic braking force actuation, the brakes are applied by means of floating pressure pistons whose travel is not fixed, and which develop tensioning forces which are equal in both directions. One of the two shoes is the leading shoe, i.e., the frictional force between the brake lining and the brake drum act to support the application force, whereas the fric-

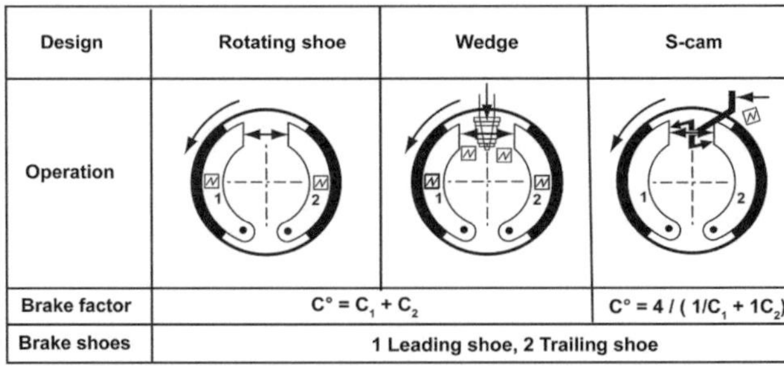

Fig. 7.1 Simplex drum brakes [1]

tional forces in the cases of the trailing shoe act to oppose the tensioning force.

In the case of the simplex brake, C^* is the sum of the values of the individual shoes and is approximately 2.0. A disadvantage of this design is the great difference in the braking effect between the two brake shoes and the resultant greatly increased wear on the leading shoe. For this reason, the trailing shoe is often equipped with a much thinner lining than the leading shoe.

A simplex drum brake can also be actuated by means of a wedge unit, a design which is being used more and more often in light-duty commercial vehicles.

The type of wheel brake used most often in heavy-duty commercial vehicles is the pneumatic S-cam simplex drum brake with fixed application. The advantages of the S-cam brake include the uniform lining wear on leading and trailing shoes as a result of fixed application and resulting long lining life. This type of drum brake has an application mechanism which is simpler, more reliable, and insensitive to temperature. It comprises diagram cylinder, automatic slack adjuster, brake shaft, and S-cam. The disadvantages include the high internal forces, and thus relatively heavy brake construction because unequal cam forces occur and lead to high free bearing forces. A relatively low brake factor results in a correspondingly large amount of application work during braking. Due to the roughly equivalent application travel of leading and trailing shoes, the application forces behave in the opposite way to that of the brake factors of the individual shoes. In general, the brake factor of this brake with the same coefficient of friction is somewhat lower than that of the above-mentioned simplex brake.

b. Duo-duplex drum brakes: The duo-duplex brake (e.g., with wedge-type actuator) as shown in Fig. 7.2 has two leading shoes. This type of drum brake is used on medium and heavy-duty trucks. This type of drum brake is characterized by floating application and the corresponding necessary sliding shoe support. The brake adjustment mechanism is an integral part of the wedge

Design	Duo-Duplex Wedge	Duo-Servo Constraintly-guided adjustment	Disc brake
Operation	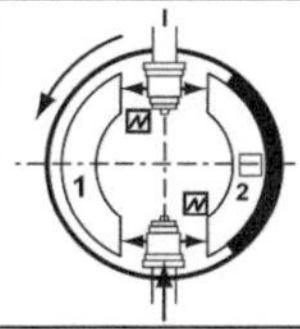	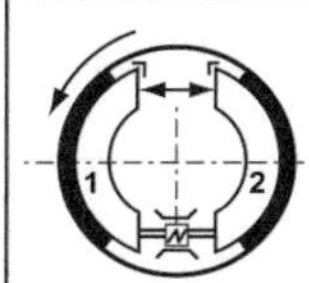	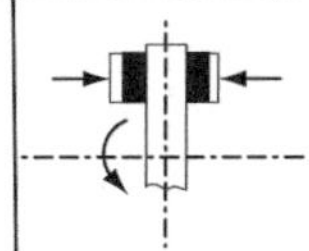
Brake factor	$C^{\circ} = C_1 + C_2$	$C^{\circ} = C_1 + C_2\,(k_1 + k_2 \cdot C_1)$	$C^2 = 2\mu$
Brake shoes	1 Leading shoe, 2 Trailing shoe		

Fig. 7.2 Duo-drum brakes and disc brake [1]

units. An advantage of this type of brake is the nearly equal brake lining wear on both shoes and the higher internal transmission ratio in comparison to simplex brakes. With two leafing shoes, a brake factor of approximately 3 is achieved, however, these figures cannot be held constant throughout a long period of braking due to the susceptibility to fading of this type of brake.

c. Duo-Servo drum brakes: Duo-servo brakes are widely used in light-duty commercial vehicles (particularly at the rear axles). They are characterized primarily by the fact that the support force of the primary shoe is used as the tensioning force for the secondary shoe both during forward and reverse vehicle motion. The brake factor is approximately 5.0. The reason for the popularity of the duo-servo brake is that the high brake factors allow even relatively heavy vans and light-duty trucks up to a weight of approximately 7.5 tons to be operated with a vacuum-assisted braking system. At the same time, the integral manually operated parking brake is able to generate a braking torque of considerable magnitude. Significant fluctuations in the brake factors occur, however, this limits the range of applications of this system of brake force apportioning which is well matched to the vehicle in question.

5. **Disc brakes**: Disc brakes as used in most passenger cars are slowly beginning to be used in commercial vehicles as well. Aside from their use in fast coaches, the current use of disc brakes in commercial vehicles is primarily limited to front axle brakes for commercial vehicles up to a weight of approximately 7.5 tons which are driven in a manner similar to passenger cars. Advantages over drum brakes include much better gradual braking efficiency, equal wear of the inboard and outboard brake pads if the proper degree of heat dissipation is provided, brake noise is better suppressed, and relatively constant brake factor performance with lower susceptibility of fading.
The disadvantages include the possible short brake pad life when used on heavy-duty commercial vehicles, and higher acquisition and operating costs as compared to drum brakes. The high degree of adaptive braking at high highway speeds is

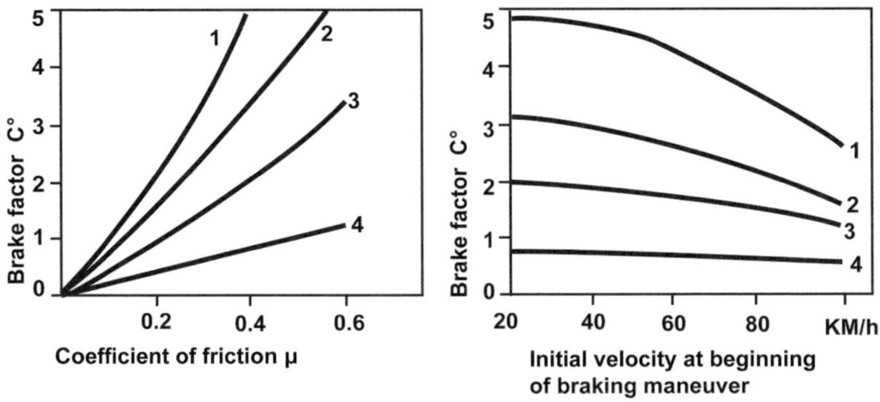

Fig. 7.3 Brake factor as a function of the coefficient of friction and initial road speed [1]

generally handled better by disc brakes with less fading and lower susceptibility to disc cracking than with drum brakes. The brake factor is approximately 0.75. Floating-caliper disc brakes continue to replace the fixed-caliper brakes generally used to date. This tendency has resulted from efforts to design lighter and cheaper brake assemblies which are more temperatureresistant. The caliper itself is not subjected to braking torque, and thus promotes gradual and constant brake action.

Other Operational Criteria include:

- "On/off" stop from maximum speed—no sensible heat loss—heat sink capability/temperature rise is a main limiting characteristic.
- Driving down a long hill descent—energy input is dependent on grade and gross vehicle weight—ability to reject heat at a rate sufficient to avoid excessive temperatures, main requirement.
- Driving on the level—energy input is governed by engine power—ability to reject heat at a rate sufficient to avoid excessive temperatures, main criteria.

The 1st and 2nd criteria can fairly readily be assessed but the 3rd one is more complex because the energy input level is most influenced by the driver (Fig. 7.3).

The Significance of Brake Types:

In any one torque range, there is normally a choice of brakes available including disc brakes and various types of drum brakes. Look at the brake stability characteristics to find the answer to the following question: Which brake type is suitable for a vehicle?

To answer this question, we have to analyze the fade characteristics. Fade sensitivity of a brake is influenced by the rate of change of brake factor with lining friction—which is a means of indicating the fade resistance capability of various brake designs—good fade resistance being synonymous with small rate of change in brake factor.

Table 7.1 Order of fade resistance for different brakes

Brake type	Order of fade resistance
Disc Brake	1 (Good)
S Cam Brake	2
L/T Brake	3
Two Leading Shoe	4
Duo Servo	5 (Poor)

On the basis of evaluation, it can be shown that the disc brakes and "S" cam are evidently the best as they show the least fade. Table 7.1 shows the brake types in descending order of fade resistance.

1. **Matching weight and brake distribution**: During braking, the braking forces do not act through the center of gravity of the vehicle, there is a load transfer from the rear to the front, which alters the load distribution at the wheels; the distribution of braking forces on modern cars are designed to take account of this.
2. **Locking of wheels**: When the front wheels locked first, the vehicle tended to move in a straight line, while when the rear wheels locked less than about half a second before the front wheels, the car also went straight. However, if the rear wheels were locked for more than half a second before the front wheels became locked, the car deviated from a straight path. It was noted that the two wheels on either the front or rear axle did not always lock simultaneously, and that, if the locking of one of the rear wheels was delayed, then no deviation occurred until both rear wheels became locked.
3. **Locked rear wheels**: The deviation (spin out) of a particular vehicle from a given path is consistent when the rear wheels only are locked and there is no braking at the front wheels, except at low speeds, when the car stopped before the deviation became large. It is often said that experienced drivers can steer out of a rear-wheel skid caused by locked rear wheels but detailed analysis of accident reports do not support this view.
4. **Locked front wheels**: Locking the front wheels, assuming that only these become locked or that they are locked before the rear wheels, results, of course, in the inability to steer. This may not be as serious as loss of stability due to locking the rear wheels, since steering can be regained by a release or partial release of the brakes. When the rear wheels are locked and the deviation of the car had exceeded a certain value of about 20° for most cars, directional control cannot be regained even by the complete release of the brakes and the most skillful driving.

7.4.2.1 Measures Designed to Prevent Loss of Control

Unstable behavior and loss of control during emergency braking have been shown to be factors in a large number of accidents. Some remedies will not be discussed. There are three main ways to prevent loss of control:

1. **Brake Distribution**: The most direct method of ensuring stability during braking is to design the brake distribution so that, for surfaces having the highest resistance to skidding, and for a lightly loaded vehicle, it matches the weight distribution, taking into account load transfer during braking. Brake distributions are used at the highest deceleration.
   There are several techniques and methods of altering the brake distribution ratio. One of these techniques is using a simple pendulum, which responds to vehicle deceleration, and by means of a beam and cam mechanism actuates a hydraulic piston and cylinder in the brake line in order to correct the brake distribution to suit the deceleration that is occurring.
2. **Pumping the Brakes**: In this method, the brakes are momentarily released and re-applied as rapidly as possible. It was found that this method enables the car to be kept on a straight path on a variety of surfaces, both wet and dry. At the same time, the braking distances achieved were generally slightly shorter than those achieved when braking with locked wheels.
3. **Open-loop and closed-loop control systems**: These systems help the driver to avoid wheel lock-up problems and will be discussed in another section.

7.4.3 Simplified Drum Brake Model

In Fig. 7.4 the moment about point O using the L-Shoe (leading shoe) is defined as

$$F_n.M = F_{nl}.Y - F_{fl}.x \tag{7.3}$$

$$F_{fl} = \mu.F_{nl} \tag{7.4}$$

$$F_n = \frac{F_{nl}(Y - \mu.x)}{M} \tag{7.5}$$

$$F_{nl} = \frac{F_n M}{(Y - \mu.x)} \tag{7.6}$$

$$F_{fl} = \frac{\mu F_n.M}{(Y - \mu.x)} \tag{7.7}$$

The moment using the T-shoes (trailing shoe) is defined as

Fig. 7.4 Simplified drum brake model

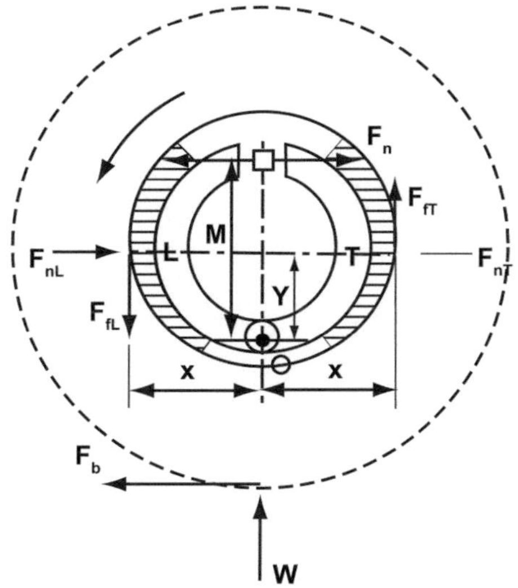

$$F_n.M = F_{nT}.Y + F_{fT}.x \tag{7.8}$$

$$F_{fT} = F_{nT}.\mu \tag{7.9}$$

$$F_{nT} = \frac{F_n.M}{(Y + \mu.x)} \tag{7.10}$$

$$F_{fT} = \frac{\mu F_n.M}{(Y + \mu.x)} \tag{7.11}$$

The shoe factor is defined for the leading edge as

$$\text{shoe factor} = \frac{F_{fl}}{F_n} \tag{7.12}$$

while for the trailing edge:

$$\text{shoe factor} = \frac{F_{fT}}{F_n} \tag{7.13}$$

The brake factor is defined as:

$$\text{Brake Factor} = \frac{F_{fl} + F_{fT}}{F_n} \tag{7.14}$$

It should be noted that the brake factor is an indication of the brake stability versus change of the coefficient of friction μ (between the friction material & the drum or the disc). The fade resistance, also known as the brake fade, is when the disc brakes are better than drum brakes because of the cooling rate.

It should be noted from Eqs. 7.7 and 7.11 that the friction force produced by the leading shoe, F_{fL}, is greater than the friction force produced by the trailing shoe, F_{fT}. This phenomenon is called the "self-energization effect", which is associated only with the leading-trailing drum brakes. The disc brakes do not have the self-energization effect, since both sides of the pads are producing the same friction forces.

7.5 Braking Characteristics of a Two-Axle Vehicle

The major external forces acting on a decelerating two-axle vehicle on a flat surface and neglecting the aerodynamic force are shown in Fig. 7.5.

The resulting retarding force is

$$F_{res} = F_b + f_r W \tag{7.15}$$

Furthermore, the normal loads on the front and rear axles, F_{zf} and F_{zr}, respectively, are

$$F_{zf} = \frac{1}{L}\left[Wl_2 + h\left(\frac{W}{g}a\right)\right] \tag{7.16}$$

$$F_{zr} = \frac{1}{L}\left[Wl_1 - h\left(\frac{W}{g}a\right)\right] \tag{7.17}$$

The equilibrium force is then defined as

$$F_b + f_r W = F_{bf} + F_{br} = f_r W = \frac{W}{g}a \tag{7.18}$$

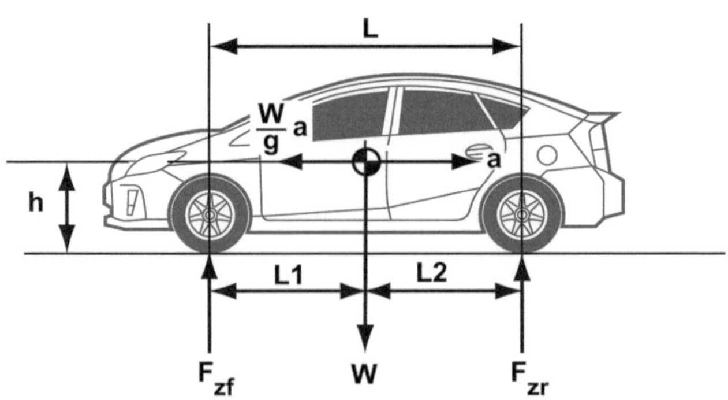

Fig. 7.5 Forces acting on a two-axle vehicle during braking

where $F_b = F_{bf} + F_{br}$. The normal loads on the axles become

$$F_{zf} = \frac{1}{L}\left[Wl_2 + h(F_b + f_r W)\right] \tag{7.19}$$

$$F_{zr} = \frac{1}{L}\left[Wl_1 - h(F_b + f_r W)\right] \tag{7.20}$$

The maximum braking forces on the front and rear axles

$$F_{bfmax} = \mu F_{zf} = \frac{\mu W\left[l_2 + h(\mu + f_r)\right]}{L} \tag{7.21}$$

$$F_{brmax} = \mu F_{zr} = \frac{\mu W\left[l_1 - h(\mu + f_r)\right]}{L} \tag{7.22}$$

Distribution of the braking forces

$$\frac{C_{bf}}{C_{br}} = \frac{F_{bfmax}}{F_{brmax}} = \frac{l_2 + h(\mu + f_r)}{l_1 - h(\mu + f_r)} \tag{7.23}$$

? Example 7.1

For a light truck with 68% of the static load on the rear axle, calculate the distribution of the braking force ratio. Given $l_2/L = 0.32, l_1/L = 0.68, h/L = 0.18, f_r = 0.01$, and the friction coefficient is 0.85.

Solution. The maximum braking forces of the front and rear tires that the tire-ground contact can support will be developed at the same time only if the braking force distribution between the front and rear brakes satisfies the following condition:

$$\frac{C_{bf}}{C_{br}} = \frac{0.32 + 0.18(0.85 + 0.01)}{0.68 - 0.18(0.85 + 0.01)} = 0.887 \tag{7.24}$$

Next is the lock-up sequence analysis

$$F_b + f_r W = F_{bf} + F_{br} + f_r W = \frac{W}{g}a \tag{7.25}$$

Now the normal loads can be rewritten as

$$F_{zf} = \frac{W}{L}\left(l_2 + \frac{a}{g}h\right) \tag{7.26}$$

$$F_{zr} = \frac{W}{L}\left(l_1 - \frac{a}{g}h\right) \tag{7.27}$$

The braking forces of the front and rear axles as determined by the brake system design are expressed by

$$F_{bf} = C_{bf} F_b = C_{bf} W \left(\frac{a}{g} - f_r \right) \tag{7.28}$$

and

$$F_{br} = C_{br} F_b = (1 - C_{bf}) F_b = (1 - C_{bf}) W \left(\frac{a}{g} - f_r \right) \tag{7.29}$$

The front tires approach lock-up when

$$F_{bf} = \mu F_{zf} \tag{7.30}$$

The vehicle deceleration rate (in g-units) associated with the impending lock-up of the front tires can be defined by

$$\left(\frac{a}{g} \right)_f = \frac{\mu l_2 / L + C_{bf} f_r}{C_{bf} - \mu h / L} \tag{7.31}$$

Similarly, it can be shown that the rear tires approach lock-up when the deceleration rate is

$$\left(\frac{a}{g} \right)_r = \frac{\mu l_1 / L + (1 - C_{bf}) f_r}{1 - C_{bf} + \mu h / L} \tag{7.32}$$

For a given vehicle with a particular braking force distribution on a given road surface, the front tires will lock first if

$$\left(\frac{a}{g} \right)_f < \left(\frac{a}{g} \right)_r \tag{7.33}$$

On the other hand, the rear tires will lock first if

$$\left(\frac{a}{g} \right)_r < \left(\frac{a}{g} \right)_f \tag{7.34}$$

? Example 7.1

A passenger car weighs 21.24 kN and has a wheelbase of 2.87 m. The center of gravity is 1.27 m behind the front axle and 0.508 m above ground level. The braking effort distribution on the front axle is 60%. The coefficient of rolling resistance is 0.02. Determine which set of tires will lock first on two road surfaces: one with a coefficient of adhesion 0.8, and the other with coefficient of adhesion 0.2 (Fig. 7.6).

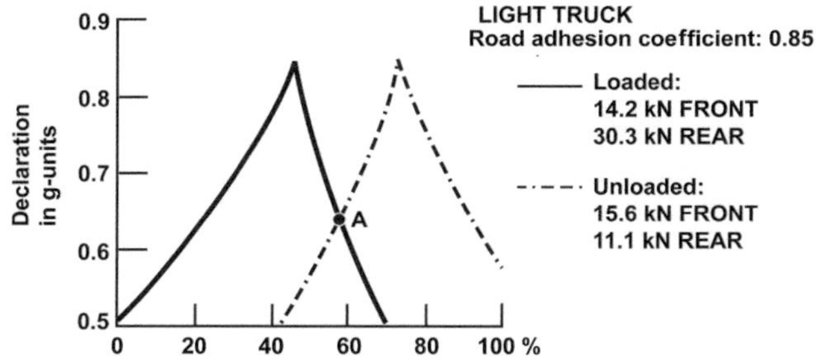

Fig. 7.6 Effect of braking effort distribution on the braking performance of a light truck [2]

Solution. On the road surface with $\mu = 0.8$, the vehicle deceleration associated with the impending lock-up of the front tires is determined as

$$\left(\frac{a}{g}\right)_f = \frac{\mu l_2/L + C_{bf} f_r}{C_{bf} - \mu h/L} = 1.0 \tag{7.35}$$

The vehicle deceleration associated with the impending lock-up of the rear tires is determined as

$$\left(\frac{a}{g}\right)_r = \frac{\mu l_1/L + (1 - C_{bf}) f_r}{1 - C_{bf} + \mu h/L} = 0.67 \tag{7.36}$$

Since $\left(\frac{a}{g}\right)_f > \left(\frac{a}{g}\right)_r$, then the rear tires will lock first on the road surface with $\mu = 0.8$.

On the road surface with a coefficient of adhesion of 0.2

$$\left(\frac{a}{g}\right)_f = \frac{\mu l_2/L + C_{bf} f_r}{C_{bf} - \mu h/L} = 0.219 \tag{7.37}$$

$$\left(\frac{a}{g}\right)_r = \frac{\mu l_1/L + (1 - C_{bf}) f_r}{1 - C_{bf} + \mu h/L} = 0.221 \tag{7.38}$$

Since $\left(\frac{a}{g}\right)_f < \left(\frac{a}{g}\right)_r$, then the front tires will lock first on the road surface with $\mu = 0.2$.

7.6 Adhesion Utilization

It is a measure of the interaction of adhesion and braking ratio η.

$$\eta = \frac{a}{\mu_A} g \tag{7.39}$$

where, η is the coefficient of adhesion utilization, a is the actual deceleration at which one axle locks, $\mu_A g$ is the maximum possible deceleration.

$$\eta = \frac{F_{zf}}{W\left(F_f - \mu_A h/d\right)} \tag{7.40}$$

$$\eta = \frac{F_{zr}}{W\left(F_r - \mu_A h/d\right)} \tag{7.41}$$

7.7 Transient Load Transfers

During braking, a pitching moment is applied to the vehicle and this imparts pitch and vertical motions to the vehicle body. The resulting effects may be critical as the vertical axle loads can vary, and in a critical situation cause axle locking and affect the vehicle stability.

The existence of transient load changes is thus a potential source of danger on a vehicle, as these can increase the probability of locking under critical adhesion conditions.

7.7.1 Improving Stability

Some means of adjusting the braking ratio to the road tire adhesion and loading on each wheel would help to give the shortest braking distance without loss of control. Variable braking ratios can be achieved by sensing, by the pipeline pressure or deceleration and providing accordingly restricted braking to the rear wheels, in such a way that, as deceleration is increased, more braking is applied to the front wheels and so reduces the risk of the rear wheels locking. The braking ratio can then be made to approximate to the ideal for a fixed static loading.

Load sensing proportioning values are practically essential for trucks, or buses, which can experience large changes in the load distribution (for static loading adds to deceleration). On articulated vehicles, two valves are used, one placed on the tractor rear axle and the other on the trailer axle, and this can cover most of the requirements of load and deceleration changes.

In practice, variations in brake torque for a given input limit the effectiveness of brake proportioning arrangements. For example, a designer might arrange his system so that the braking ratio is optimized at 65/35 for a particular set of conditions on a drum-braked car. If the brakes, however, should be heavily worked, the front brakes will become hotter than the rear. And if they should fade, the braking ratio may fall to, say 50/50, with the result that in an emergency stop the rear wheels instead of the front will be locked first and braking distance will be increased.

Of course, one way to maintain the ratio constant is to fit more stable brakes. Another and more general way to cope with changing conditions is to fit anti-lock devices. Anti-lock devices all work on the principle of determining the deceleration of the sensed wheel. If this is above a certain value, that is, if the wheel is on the point of locking, the brake actuating force is reduced. The wheel speeds up and the actuation force is re-applied. The cycle is then repeated as long as the wheel is on the point of locking. These devices give improved stability and controllability in emergencies. These systems would have a definite benefit in 4–7.5% of accidents.

7.8 Braking Control

1. **Ergonomics**: it is well known that most drivers can exert a force 2–3 times their own body weight for a short time and even higher forces in an emergency situation. However, there is poor correlation between maximum force and body weight as there is between age and maximum force. Females are capable of exerting a maximum force of about 55% of that of males. Pedal efforts of 356–400 N are reasonable for the weakest car drivers to produce a deceleration of 1.0 g. In the case of commercial vehicles, 445 N pedal effort develops approximately 0.6 g deceleration.
   Similarly, for handbrakes, a 222 N effort is satisfactory to hold the car on a 25% gradient.
   Lever movements should preferably not exceed about 60 and 90 mm for pedal and hand brake lever movements, respectively, for cars.
2. **Actuating Systems**: The actuating systems connect the pedal with the brake and provide the gain force to operate the brake at approximately 3 times the load on the wheel. This is achieved by, hydraulic, air or mixed systems.

   - Hydraulic systems have small light components, which can be readily adopted to incorporate a servo unit. They are used with and without a servo unit on vehicles weighing up to 500 kg. But on vehicles up to 10000 kg weight, a servo unit is essential. On vehicles up to 13,000 kg weight, air/hydraulics are used fitted with a hydro-vacuum servo.
   - Direct air systems can provide very large brake actuating forces and are used on heavier vehicles with spring brake actuators for the secondary and parking brake. Air brake systems on bigger vehicles have to be quite complex in order to meet safety requirements. Systems and their components have been highly developed and designed to minimize time delays.

The systems and size of components are selected together with the size and type of brake to give not only the required torque, but also to suit the performance and operating duties of the vehicles.

7.9 Open-Loop Brake Control

Trends in car usage and design favor the development of improved braking. Conditions, such as higher sustained cruising speeds now permitted on expressways and turnpikes, place increased demands on brake performances. Higher performance engines available in many models, front wheel drives, and the generally greater demand on car components as to performance, safety, and durability encourage improved designs.

For many years, and since the advent of hydraulic brakes, brake balance for two-axle vehicles has been achieved by dividing the braking force between front and rear wheels in accordance with some compromised percentage. In more recent years, this figure has been roughly changed to 60% of braking being accomplished by the front wheel and 40% by the rear wheels.

Most cars can sustain approximately one-fourth of their curb weight as an added load and will transfer one-fifth of their load from the rear to the front wheel during a hard brake application. Conventional brake balance will, therefore, provide optimum brake distribution for a given vehicle load at only one deceleration, below this deceleration, the rear wheels are doing less work than they should. Above this pre-established level, they provide excess torque, resulting in premature wheel slide, with subsequent loss of control. There is no compensation for changes in static loading of the vehicle, or the dynamic weight transfer occurring during braking. Dissimilar front-to-rear brake characteristics associated with hybrid brake combinations, such as front disc brakes and drum rear brakes, exaggerate the need for better brake balancing. A degree of improvement is provided by fixed brake proportioning, which has been used on European cars for some years. Dynamic weight transfer can be compensated for in part, but no correction for vehicle loading is provided. Since the one-half ton variation in the static load capacity of the vehicle requires approximately the same magnitude of redistribution of the brake balance, the implementation of the load sensing proportioning was a significant improvement.

With these considerations in mind, it was believed that a greatly improved, safer braking system could be developed. This system would adjust braking to the rear wheels in proportion to the weight carried by that axle with considerations given to the longitudinal load transfer during deceleration.

The load sensing brake proportioning is a means of optimizing the ratio of front-to-rear wheel retarding forces for the full range of vehicle loading and deceleration. A relatively simple system of providing two-axle vehicles with improved brake balance has been developed. Simulation studies of vehicle brake system requirements, effects of a pressure-regulating device used to implement these requirements, and the resulting improvements are compared in actual tests of vehicle hardware.

The performance of the new system provides nearly optimum vehicle braking. The areas of improvement available with load sensing proportioning are:

- Shorter stopping distance.
- Safer, more controllable stops.
- Lower pedal efforts.

- Reduction in pedal effort variation with vehicle load.
- Increased mileage between brake service.
- More nearly compatible performance of disc-front, duo-servo-rear brake combinations.

In order to achieve the shortest possible stopping distance without affecting the stability or the controllability of a vehicle, one would like to control the braking torque at each wheel at a level which would produce the greatest contribution to the total retarding force applied to the vehicle. This principle was recognized at least as early as 1907. However, its implementation is still being planned. Two principal approaches have been employed the first utilizes open loop control of the brake application force in response to one or more measured variables. The second is a closed loop controller, which senses wheel rotation and modulates braking to avoid wheel lock which is called ABS. Variable front-to-rear brake proportioning has been proposed to compensate for the forward load transfer during deceleration as well as for variations in vertical load distribution among the tées produced by loading of the vehicle. Compensation for deceleration load transfer has commonly utilized line pressure or deceleration as a signal to control valve, which decreases the ratio of rear brake pressure to front brake pressure as deceleration increased. The resulting brake distribution approximates the ideal value for a particular static loading. For vehicles with a relatively small ratio between full rated load and curb weight, this may be a satisfactory control. For vehicles such as trucks or buses, which may experience a large change in weight distribution with load, the load sensing proportioning value has been employed. Typically, these units sense load at the rear axle only by measuring suspension deflection. This signal then is utilized to operate a proportioning valve to modify the ratio of front-to-rear brake line pressure, since wheel loading is influenced by both deceleration and the static load, neither of the two proportioning principles just described is entirely adequate to achieve ideal brake distribution. Combination systems have been devised to respond both to static load and deceleration. In addition to rear-to-front load transfer, vehicles are also subjected to a lateral load transfer in a turn. In order to accommodate braking under these circumstances, inertial sensors have also been devised to modify side-to-side proportioning.

Since brake proportioning operates in an open loop fashion, it is subject to a number of errors. One of the first assumptions made is that the brake torque is proportional to the brake line pressure, since it is the brake line pressure, which controls brake sensitivity, particularly with fade induced by hard usage. The second assumption is that the tire to road friction coefficient is the same for all 4 tires. This may not be true due to local differences in the pavement surface or differences in the tires or operating conditions. For example, the usage of studded or snow tires on the rear can vastly change the desired brake torque distribution. Similarly, when running in deep water, the rear tires operate in the path cleared by the front tires and, therefore, have different friction characteristics. Engine drive torque with the car in gear also modifies the tractive force distribution, since the driveshaft driving or braking torque must be summed with the brake torque at the driving wheels. All of these factors limit the accuracy of open loop brake proportioning schemes.

7.10 Principle of a Load Sensing Brake Proportioning System

Some of the main characteristics of a load sensing brake proportioning system are:

$$q_1 = \frac{F_{bfmax}}{W} \tag{7.42}$$

$$q_2 = \frac{F_{brmax}}{W} \tag{7.43}$$

$$\left(\frac{a}{g}\right) = b = \mu \tag{7.44}$$

$$\frac{l_2}{L} = \Psi_1 \tag{7.45}$$

$$\frac{l_1}{L} = \Psi_2 \tag{7.46}$$

$$X = \frac{h}{L} \tag{7.47}$$

Furthermore, the following can be derived:

$$q_1 + q_2 = \mu\left(\frac{F_{zf} + F_{zr}}{W}\right) = \mu \tag{7.48}$$

The load on the front is defined as

$$F_{zf} = \frac{W}{L}\left(l_2 + \frac{a}{g}h\right) \tag{7.49}$$

$$= \frac{W}{L}(l_2 + \mu h) \tag{7.50}$$

while the load on the rear can be defined as

$$F_{zr} = \frac{W}{L}\left(l_1 - \frac{a}{g}h\right) \tag{7.51}$$

$$= \frac{W}{L}(l_1 - \mu h) \tag{7.52}$$

Thus the following can be expressed as

$$\frac{F_{bf}}{W} = q_1 = \mu\left(\frac{l_2}{L} + \mu\frac{h}{L}\right) \tag{7.53}$$

$$= (q_1 + q_2)\left(\frac{l_2}{L} + (q_1 + q_2)\frac{h}{L}\right) \tag{7.54}$$

$$= (q_1 + q_2)(\Psi_1 + (q_1 + q_2)X) \tag{7.55}$$

Equation 7.55 can be rewritten as

$$q_2^2 + q_2\left(2q_1 + \frac{\Psi_1}{X}\right) + \left(q_1\frac{\Psi_1 - 1}{X} + q_1^2\right) = 0 \tag{7.56}$$

where

$$A = 2q_1 + \frac{\Psi_1}{X} \tag{7.57}$$

$$C = q_1\frac{\Psi_1 - 1}{X} + q_1^2 \tag{7.58}$$

Thus, the quadratic equation can then be written as

$$x^2 + A.x + C = 0 \tag{7.59}$$

and the solution to the quadratic equation is defined as

$$x_{1,2} = -\frac{A}{2} \pm \sqrt{\frac{A^2}{4} - C} \tag{7.60}$$

Thus

$$q_2 = -\left(q_1 + \frac{\Psi_1}{2X}\right) \pm \sqrt{\frac{q_1}{X} + \left(\frac{\Psi_1}{2X}\right)^2} \tag{7.61}$$

This equation represents the relationship between the front & rear brakeforce distribution as a vehicle parameter change, and represents the optimal brake distribution (Fig. 7.7).

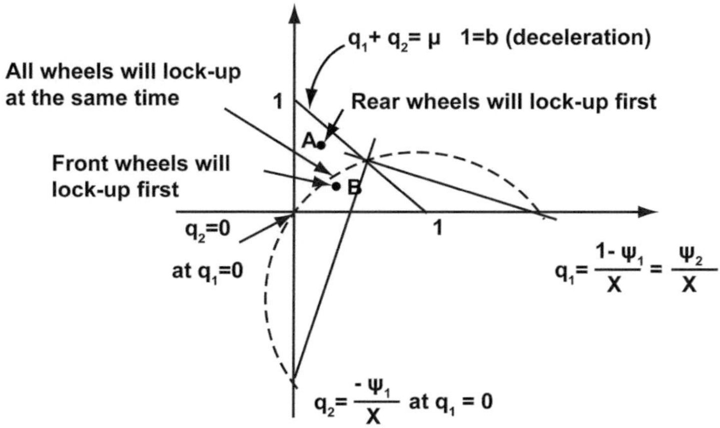

Fig. 7.7 Relationship between q_1 and q_2 over a wide range of road friction coefficient

7.11 Anti-Lock Braking System

When a driver brakes heavily in attempting to avoid an accident, the vehicle does not always behave in the way intended. The front wheels may lock, in which case the car cannot be steered; the rear wheels may lock, and the vehicle tends to slew round. If, however, all wheels lock, the car tends to rotate, provided that the rear wheels lock more than a certain time before the front wheels.

These consequences of heavy braking can be avoided or lessened in several ways, including the use of special braking techniques, alternatives to the brake distribution, or the fitting of anti-skid devices. Special mention is made of anti-skid devices which make use of the fact that the braking forces exerted by a wheel just before locking occurs are greater them those exerted by a locked wheel. Although the problem of loss of control during emergency braking can be partly solved by adopting appropriate brake distribution and by special braking techniques, the use of anti-locking devices offers prospects of much greater improvement both in directional control and in braking performance, and consequently in safety. The problem of emergency braking is either loss of controllability or stability or both.

- **Anti-Locking Devices**: Anti-locking devices operate by releasing the brake as soon as the wheel tends to lock, and by re-applying it when the wheel speeds up again. Since the brake is released for part of the braking time, one of the main concerns is to ensure that the braking distances are not increased. It is useful to consider the reasons why such braking distances may, in fact, be shorter than those with locked wheels, even though the brakes are released for some parts of the time.
- **Anti-locking devices applied to rear wheels only**: The Lockheed application uses an inertia-operated sensing device driven from the propeller shaft of a rear-wheel-drive car to control a servo unit. This system is no longer applicable to passenger cars but may be applied to heavy dump trucks.
- **Anti-locking devices controlling all wheels**: The first car with a braking system, which incorporates an anti-locking device controlling all four wheels, is that being developed by Harry Ferguson Research Co., Ltd. This firm has been working on the design of a car having four-wheel drive and a transmission and braking system in which a single anti-locking unit could easily be incorporated to control the braking on all wheels. These features were incorporated in a passenger vehicle. Recently, several ABS systems have been developed and controlling all brakes becomes almost a standard option in many automobile makes.

Figure 7.8 shows the forces and moments acting on a tire during braking. The following equations present the wheel model, from which we can predict the longitudinal slip and wheel deceleration.

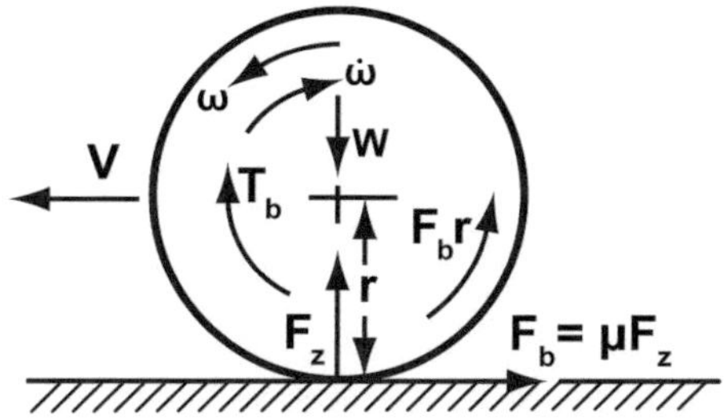

Fig. 7.8 Forces and moments acting on a tire during braking [2]

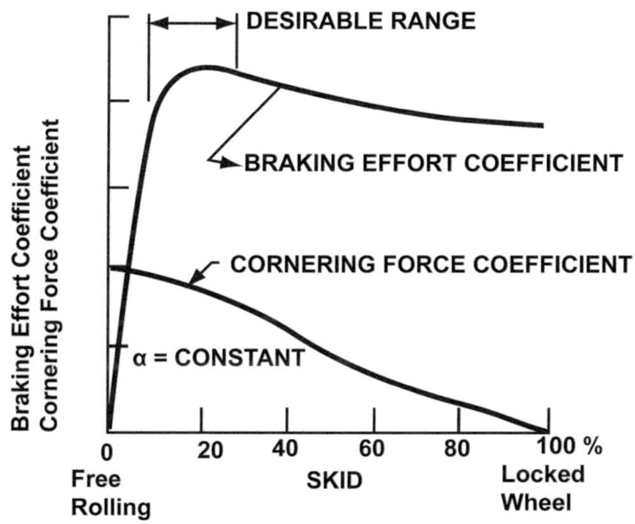

Fig. 7.9 Effect of skid on cornering force coefficient of a tire [2]

$$\dot{\omega} = \frac{M_t - M_b}{I_w} = \frac{F_b r - M_b}{I_w} \tag{7.62}$$

$$a_c = \frac{F_b}{W/g} \tag{7.63}$$

$$s = \left(1 - \frac{r\omega}{v}\right) \times 100\% \tag{7.64}$$

Figure 7.9 shows the effect of longitudinal slip (skid) on the braking effort coefficient and the cornering force coefficient. It is clearly shown that the cornering force is almost diminished when the wheel is locked ($s = 100\%$), that is why the ABS system is important to maintain the longitudinal slip within the range of 10–25% to obtain the maximum braking force and a reasonable level of cornering force.

Figure 7.10 shows the main components of an ABS system, which are as follows:

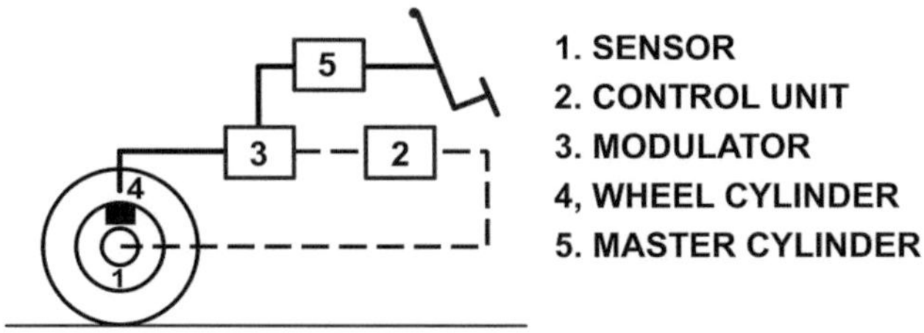

Fig. 7.10 Elements of an anti-lock braking system [2]

1. Wheel sensor, 1: shown in Fig. 7.10. It monitors the speed of wheel rotation. These sensors constantly monitor the speed of the wheel's rotation and send it to the ABS control unit 2. Since these have a functional role in control parameters, wheel speed sensors are an important automobile component that must be well-maintained.
2. Control Unit, 2: The ABS control unit 2 is a microprocessor that runs the stored prediction and reselection criteria checks of a vehicle's anti-lock braking system and processes information from wheel speed sensor 1 and sends the decision (ON/OFF) to operate the oil pump in the modulator 3. If the decision is "Prediction", the signal from the control unit will turn OFF the valve that delivers the oil pressure to the wheel that is predicted to achieve the lock-up situation. Contrarily, if the control unit decision is "Reselection", the signal sent to the modulator will turn the pipeline valve ON to reapply the oil pressure to the accelerating wheel to regain the brake torque level within the designed longitudinal slip range that will improve its braking efficiency and produce adequate braking force. The control unit is considered to be the brain of the ABS system, where the required calculations and vehicle parameters in addition to sensors' data are processed.
3. Modulator, 3: The modulator 3, receives the oil pressure from the master cylinder, 5, when the driver decides to brake it will activate the pump in the modulator at this point and if the vehicle speed is higher than approximately 5 km/h, the control unit will start to process the wheel speed signal and perform the Perdition and Reselection calculations to control the pump output pressure delivered to each wheel in case of having 4-channel ABS system.
4. Wheel cylinder, 4. The wheel cylinder for either disc or drum brakes. Actuates the pressure received from the modulator to the pads, in the case of disc brakes, or to the shoes, in the case of the drum brakes.
5. Master cylinder, 5: The master cylinder is connected to the brake booster that is connected to the brake pedal. The master cylinder produces the oil pressure signal that activates the pump in the modulator's pump (Fig. 7.11).

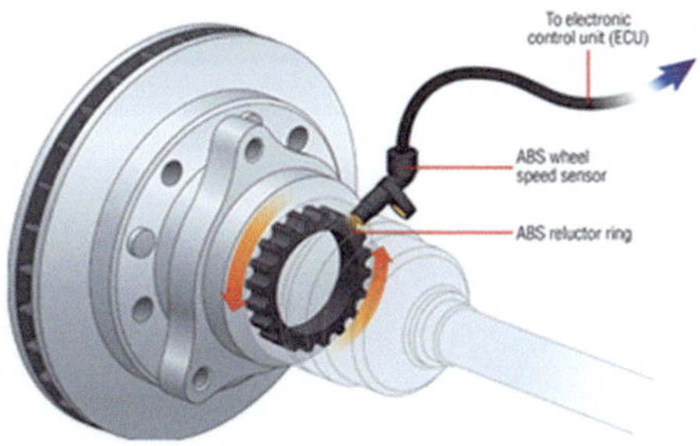

Fig. 7.11 Schematic of the anti-lock braking system wheel sensor (Image courtesy of www.mechanexpert.com/abs-sensors/)

7.12 Example of Different Logic Algorithms of ABS

Some prediction methods (Apply brake pressure)

1. Predict wheel lock if:
$$1\text{ g} < \dot{\omega} \leq 1.6\text{ g} \tag{7.65}$$
2. Using the track and hold circuit in the control unit, or if $a_y > 0.5$ g and $\omega < 15\%\omega_k$ also release brake pressure.
3. If $s > s_o$, (s_o predicted approximately 10%) s_o is estimated based on using several methods.

Some Reselection Methods (Reapply brake pressure)

- If any of the above criteria are no longer satisfied.
- In certain systems, a fixed time delay after the release of the brakes.
- or if $\dot{\omega}r > 2.2 \sim 3$ g

The following should also be noted :

- The cycle for reducing, holding, and restoring the brake pressure is repeated, typically 5–16 times/s.
- Usually, the ABS is deactivated if the speed ≅3–5 km/h.
  - "Select—Low" the CU will use the information from the slower of the two wheels, to apply the same pressure to both (good for rear tires on asymmetrical roads.)
  - "Select—High" the CU will use the information for the faster wheel.

7.12.1 Examples

1. Let $1.0\text{ g} < \dot{\omega}r \leq 1.6$ g.
2. We have two options:

- option 1: $-\dot{\omega}r \geq 1.6$ g and $\omega < 0.95\omega_k$ and $\frac{-a}{g} \leq 0.5$ g.
- option 2: $\frac{a}{g} > 0.5$ g and $\omega < 0.85\omega_k$.

3. If $s > i_{so}$ (i_{so} stored value $\cong$ 10%)

$$s = 1 - \frac{\omega r}{v} \tag{7.66}$$

.AND.

$$\dot{\omega}r = \dot{\omega}_o r(\dot{\omega}_o r \cong 1\text{–}1.6\text{ g}) \tag{7.67}$$

4. Another example of the prediction and reselection of an algorithm is as follows:
   - Prediction criteria

$$-\dot{\omega}_k r \geq 1.75\text{ g}$$
$$\text{.AND.}$$
$$\omega_k r - \omega r \geq 1.5\text{ (m/s)}$$

   - Reselection criteria

$$\dot{\omega}r \geq 2.2\text{–}3\text{ g}$$
$$s \leq 0.1$$

Figure 7.12 shows one cycle of the prediction and reselection of this algorithm. Where P_b is the applied pressure, ω_k is the cut-off angular velocity, ωr is the angular velocity, and $\dot{\omega}r$ is the angular acceleration. When the pressure is ON it means reselection and OFF means prediction of wheel lock.

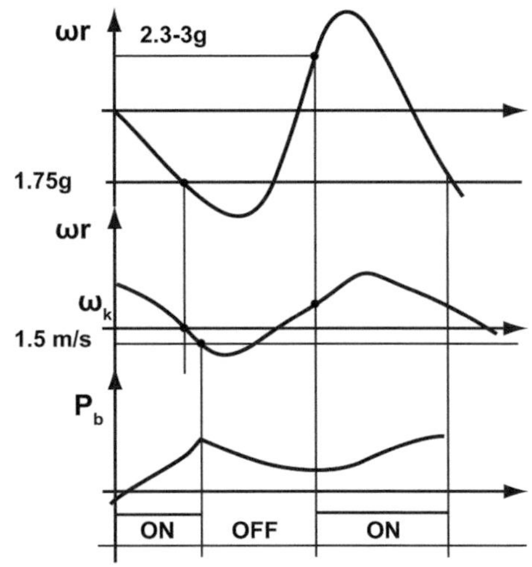

Fig. 7.12 ABS prediction and reselection based on angular velocity of a wheel

7.13 Boolean Algebra and Fluid Logic

This is the algebra of logic, although it was originally abstract mathematics.

7.13.1 The Language of Boolean Algebra

In all that follows, it is assumed that a variable can only have two states ON(1) or OFF(0).

The variables can then be written as

$$A = 1 \quad \text{or} \quad B = 1 \tag{7.68}$$

Three basic operations in Boolean algebra are
.NOT., .AND., & .OR.

7.13.1.1 The .NOT. Operation

$$\overline{A} = \text{ .NOT. } A \tag{7.69}$$

Thus, at any particular time

$$A = 1 \text{ then } \overline{A} = 0 \tag{7.70}$$

7.13.1.2 The .AND. Operation

.AND. gate is a device which hasseveral inputs and is designed so that there is an output only when a certain definite set of input conditions are met. The .AND. gate only has output (S) when all of its inputs are present (Fig. 7.13).

$$S = ABCD\ldots\ldots \tag{7.71}$$

Meaning that S is (one), only if the inputs A and B and C and 0, etc., is each one. If $S = AB$, the gate can be represented as (Fig. 7.14).

7.13.1.3 The .OR. Operation

The .OR. gate gives an output if any one of the inputs are present. This is written

$$S = A + B + C + D + \ldots. \tag{7.72}$$

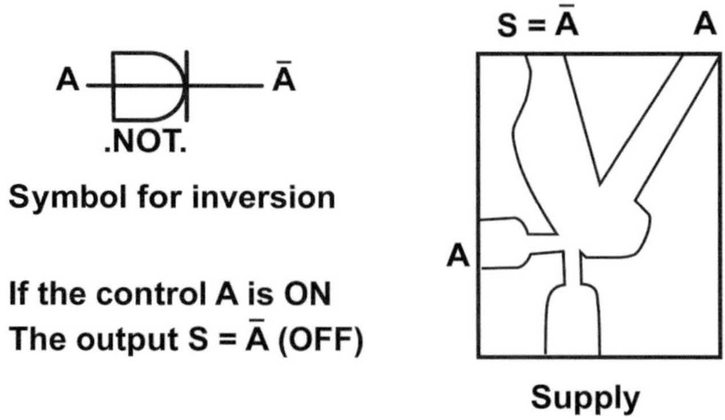

Fig. 7.13 .NOT. gate control unit

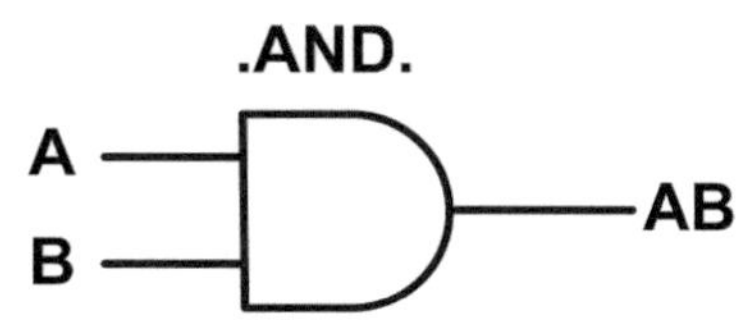

Truth Table

Inputs		Output
A	B	S
0	0	0
0	1	0
1	0	0
1	1	1

Fig. 7.14 .AND. gate control unit

The truth table for two inputs .OR. gate (Fig. 7.15).

7.13.1.4 Some Further Logic Gates

1. The .NOR. operation is the inversion of .OR. and written as

$$S = \overline{A + B + C + \ldots\ldots}$$

   If $S = AB$, the gate can be represented as
2. The .OR. gate gives an output if any one of the inputs is present. This is written $S = A + B + C + D + \ldots.$ The truth for two inputs .OR. gate (Fig. 7.16).
3. The .NAND. operation is the inverse of .AND. and is written as

$$S = \overline{ABC\ldots..}$$

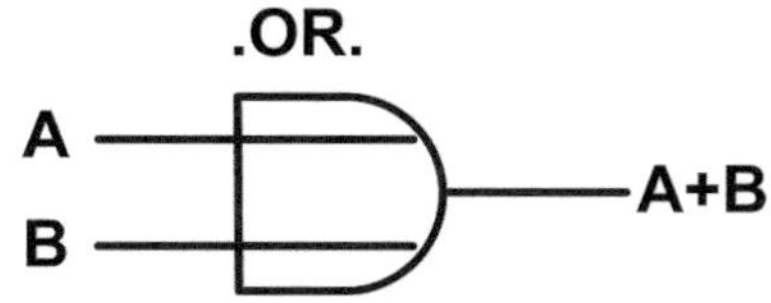

Truth Table

Inputs		Output
A	B	S
0	0	0
0	1	1
1	0	1
1	1	1

Fig. 7.15 .OR. gate control unit

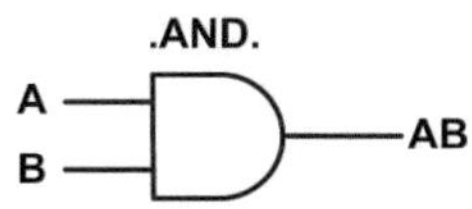

Truth Table

Inputs		Output
A	B	S
0	0	0
0	1	0
1	0	0
1	1	1

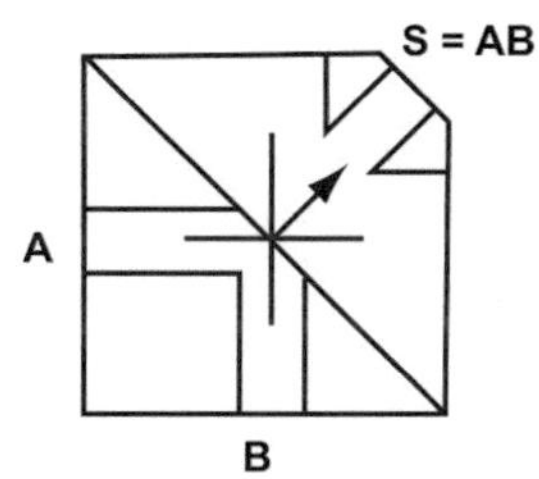

Fig. 7.16 .NOR. gate control unit

The truth table of NAND is

4. Flip-Flop or Bistable

If the cost input was A, output S_2 will be ON. If it was B, then output S_1, will be ON. The memory inherent in these devices results in the state of the output being affected by not only its present inputs, but also its past inputs (Figs. 7.17 and 7.18).

The De Morgan theorems

1. $\overline{ABC} = \overline{A} + \overline{B} + \overline{C}$

   $\overline{A + B + C} = \overline{A}\,\overline{B}\,\overline{C}$ The inverse of an expression is obtained by inverting all the variables and then changing the .AND. to .OR. or the .OR. to .AND. (Tables 7.2 and 7.3).
2. $A + AB = A$
3. $A + \overline{A} = 1$
4. $A\overline{A} = 0$

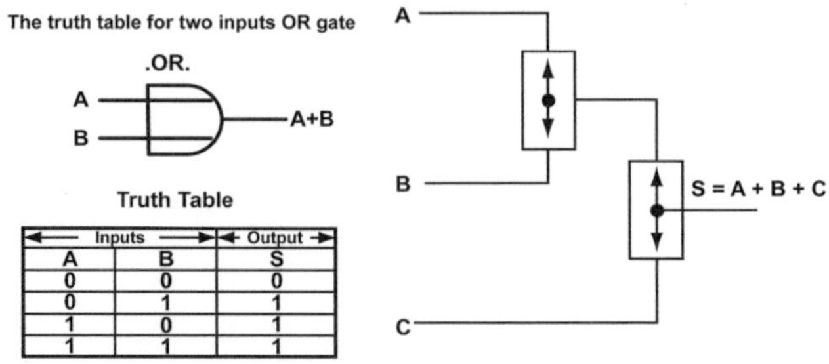

Fig. 7.17 Two input .OR. gate control unit

Fig. 7.18 Flip-flop operation using two .NOR. gates

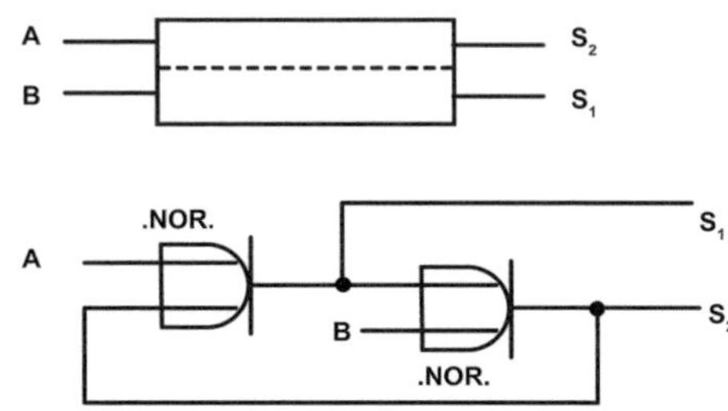

Table 7.2 Truth table for NAND gate

Input	Input	Output
A	*B*	*S*
0	0	1
0	1	1
1	0	1
1	1	0

5. $A + 1 = 1$
6. .AND. logic can be obtained for .NOT. and .NOR. logic.
   $S = AB$ (.AND.) gate
   $\overline{S} = \overline{A} + \overline{B}$ (.OR.) gate
   by inverting
   $S = \overline{\overline{S}} = \overline{\overline{A} + \overline{B}}$
   $S =$ (.NOT. A) .NOR. (.NOT. B)

7.13.1.5 Example of Anti-Locking Braking System Logic Circuit

Figure 7.19 shows a suggested simulation model for an ABS system using prediction and reselection algorithms, using the information received from a vehicle model or actual vehicle sensors.

Table 7.3 Multiplication of logic expressions

A	B	C	AB	AC	$AB + AC$	$B + C$	$A(B + C)$
0	0	0	0	0	0	0	0
0	0	1	0	0	0	1	0
0	1	0	0	0	0	1	0
0	1	1	0	0	0	1	0
1	0	0	0	0	0	0	0
1	0	1	0	1	1	1	1
1	1	0	1	0	1	1	1
1	1	1	1	1	1	1	1

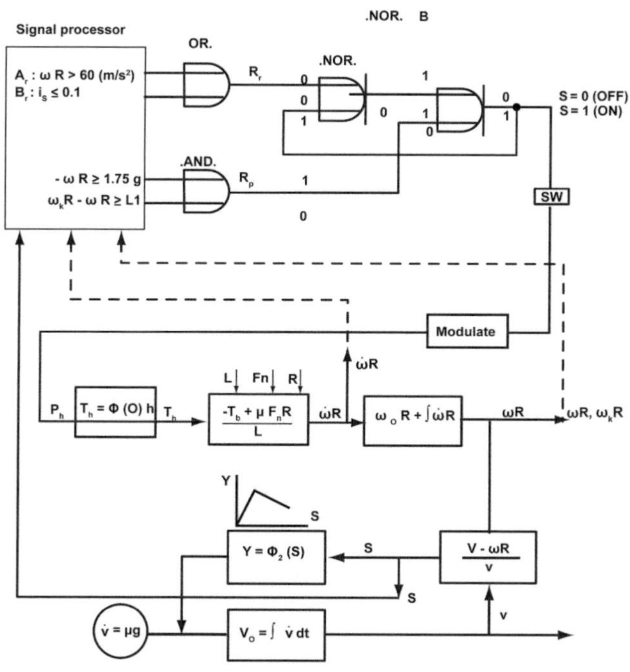

Fig. 7.19 Diagram of suggested ABS system model

Figure 7.20 shows laboratory test results of a commercial anti-lock brake system. The application of the pressure and reselection (pressure is off) was repeated as far the control unit is satisfying the prediction and reselection criteria. The number of cycles in this system are low (about 3 cycles per second) [2].

Figure 7.21 shows various popular ABS systems as follows (from top to bottom): 1- 4-channel ABS with 4 sensors that send the angular velocity of each wheel to the control unit then the signals are processed and compared with the installed prediction and reselection criteria for each wheel. This system is considered to be the most

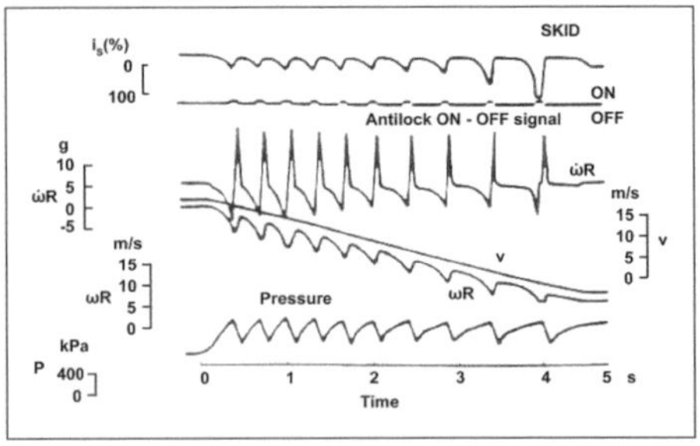

Fig. 7.20 Operating characteristics of an anti-lock system for heavy commercial vehicles with pneumatic braking systems [2]

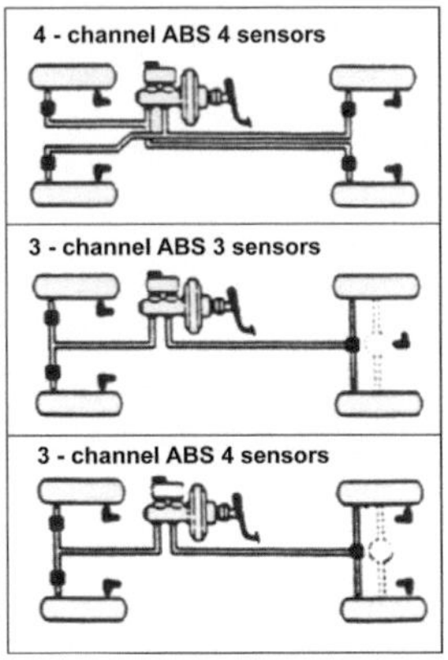

Fig. 7.21 Various layouts of anti-lock systems for passenger cars. (Reproduced with permission of the Society of Automotive Engineers from reference [3])

reliable ABS system. 2- 3-channel ABS with 3 sensors. This system is processing the signals from the front right and left wheels separately and one signal from the rear axle average angular velocity of the left and right wheel. Based on processing the signal of the rear axle, the prediction and reselection criteria will be applied to both rear wheels.

3-channel and 4 sensors. This system is more accurate than the previous system and applies the concept of SELECT LOW on the rear axles. In case the vehicle is in turn and if the prediction criteria is met by slower rear wheels, the pressure on both wheels will be off. This may result in less braking efficiency, but locking the rear wheels can be prevented to maintain the vehicle stability.

7.14 Evaluation of Vehicle With Anti-Lock Braking System

Evaluation of the braking performance of vehicles equipped with anti-lock systems is different from that required for vehicles equipped with conventional brake systems. Usually, ABS-equipped vehicles are evaluated from the standpoint of stability and control that the vehicle maintains during braking on a special road surface arrangement. These tests and their method of evaluation are summarized in the next sections.

1. Adhesion utilization: Adhesion utilization is the ratio of (a) a vehicle's deceleration rate from 40 to 20 km/h, when full brake application snubs are made from 50 km/h on a dry surface with a peak coefficient of friction of about 0.8 and on a wet surface with a peak coefficient of friction of about 0.4; to (b) the deceleration theoretically possible using the peak coefficient of friction available at each braked wheel (under comparable dry and wet conditions). This ratio gives the percentage of adhesion utilization. This braking performance requirement for vehicles equipped with anti-lock braking systems is contained in the Economic Commission for Europe (ECE) regulations.
   In these tests, the first snub is made with a pressure less than the lock-up pressure. Successive snubs are made, increasing the pressure for each snub until lock-up occurs. The time to decelerate from 40 km to 20 km/h is measured, and used to calculate the braking force which, in turn, is used to calculate the dynamic weight on the axle. The ratio of the braking force to the dynamic weight is defined as the peak coefficient of friction.
2. Stability and controllability capability: In the ECE regulations, there are additional tests to ensure that the anti-lock systems provide stability and controllability and do not allow the wheels to lock more than momentarily on various surfaces under various maneuvers. The most important tests and their performance measures are presented next; those which are excluded are less important, or do not add significantly valuable information.

   a. Uniform high friction tests: Full, rapid, and straight-line brake applications from 80% of a vehicle's maximum allowable speed, on a uniform high friction surface (say, dry asphalt of 80 SN) in a 3.66 m width lane. In this test, all the same variables needed for evaluating the stopping distance of vehicles equipped with conventional braking system must be measured. Tests should be performed at least at an initial speed of 60 km/h.
   b. Uniform low friction tests: Full, rapid, and straight-line brake applications from 40 km/h on uniform low friction surface (say, wet jennite of 20 SN) in a 3.66 m width lane. In this test, all variables measured are similar to those needed for evaluating the stopping distance of vehicles equipped with conventional braking systems. The operative criterion in this case is that the vehicle must stop under full driver control without leaving the lane.
   c. Split friction tests: Full, rapid, and straight-line brake applications on a surface with split coefficients of friction, i.e., the wheels on one side run on high adhesion (80 SN or more) and the other side on low adhesion (where the lower coefficient of friction should be less than half of the high one) at an initial speed of 50 ± 2 km/h. The requirements of the performance measures are:
      i. The mean deceleration rate must not be less than 50% of the mean value of the decelerations obtained from the braking tests on a uniform high friction surface and the low adhesion surface with the anti-lock system in operation.

ii. It must be possible to keep the vehicle within a 3.66 m wide lane, then the dividing line between the two surfaces is in the lane's center. Steering corrections (if any) must thereby not exceed ±180°.

d. Changing friction tests: Full rapid brake application stops should be made on a surface where the peak coefficient of friction changes abruptly from high (80 SN) to low (20 SN), and conversely, from low to high. Tests must be conducted with a fully laden vehicle.
In the test for changing friction from low to high, and from high to low, the initial speed for these stops must be at least 50 km/h when the first axle enters the high friction track, and at most 40 km/h when the first axle enters the low friction track, respectively. In the test for changing friction from low to high, the performance measure is defined as the ability of the anti-lock device to regain full braking force within 2 s after its wheel(s) have passed from the low to the high friction surface. In the test for changing from a high to low friction surface, the measure is defined as the ability of the anti-lock device to make the transition without significant wheel lock-up.

e. Braking-in-a-turn tests: These tests should be conducted on a uniform low-friction turn. There is no standard test procedure for braking-in-a-turn maneuvers, and therefore, no standard radius of curvature for such tests. A common radius, which NHTSA uses [4] is 152 m (500 ft). The speed prior to braking is dependent on the test vehicle's maximum allowable speed in such a turn, however, 80% of the maximum speed could be used. In this test, there are no limits set as to the amount of lock-up. The performance measure is defined as the ability of a vehicle to make as short a stop as possible with full controllability without leaving the 3.66 m lane.
Other brake test maneuvers have been recommended and performed in other countries, such as braking in a J-turn which is used by the Swedish Road and Traffic Research Institute [5], and braking-in-a-turn-change—maneuver which is used by the American National Highway Traffic Safety Administration (NHTSA) and the University of Michigan Transportation Research Institute (UMTRI) [6], and by Canadian researchers. The experience of the researchers who have conducted uniform lane-change, J-turn, and braking-in-a-turn tests [7] believe that it is not necessary to conduct all of these tests because there is some redundancy among them. These researchers also prefer the braking-in-a-turn test because it takes less driver skill and is not affected by the driver's ability to initiate braking at precisely the same point for each test since the curve has a uniform radius.

Problems

1. A vehicle has a loaded weight of 20.02 kN with a wheelbase of 279.4 cm. The center of gravity is 127 cm behind the front axle and 50.8 cm above the ground. The vehicle operates over a variety of surfaces with coefficients of adhesion

ranging from 0.2 to 0.8. What would you recommend regarding the brake effort distribution between the front and rear axles with the objective of avoiding the loss of directional stability on slippery surfaces under emergency braking conditions?

2. A pick-up truck has an empty weight of 4800 lb and the front axle carries 65% of the weight when stationary on a level ground. When fully loaded, it weighs 7800 lb and the front axle supports 40% of the weight when stationary on a level road. The wheelbase of the vehicle is 135 in, and its center of gravity is 48 and 32 in above ground level under loaded and empty conditions, respectively. The coefficient of rolling resistance is 0.018.

    - The empty pick-up truck is traveling at a speed of 50 mph on a down slope of 2% with a coefficient of road adhesion of 0.3. Calculate the possible minimum distance required to bring the vehicle to rest, if: 1. The brake system had a fixed braking effort distribution of 60% on the front axle, and, 2. All tires are allowed to lock-up at the same time during braking.
    - To improve the braking performance of the pick-up truck under loaded conditions; a load-apportioning valve is proposed to be incorporated into the brake system. What should be the optimum characteristics (i.e., q_1 vs q_2) of the load-apportioning valve when the vehicle operates over a variety of road surfaces with coefficients of road adhesion ranging from 0.15 to 0.85? Plot q_1 vs q_2 within this adhesion range and determine their values at the coefficient of adhesion of 0.15 and 0.85.

References

1. Bosch R, Peter (Tr.) Girling (1996) Automotive handbook. Society of Automotive Engineers, US
2. Wong JY (2008) Theory of ground vehicles. Wiley
3. Leiber H, Czinczel A (1979) Antiskid system for passenger cars with a digital electronic control unit. SAE Trans 1694–1700
4. Radlinski RW, Flick MA (1986) Tractor and trailer brake system compatibility. SAE Trans 872–895
5. Sandberg U (1987) Road traffic noise–the influence of the road surface and its characterization. Appl Acoust 21(2):97–118
6. Ehlbeck JM, Murphy RW (1975) An evaluation of antilock system performance on heavy duty air braked commercial vehicles. In: Proceedings of a symposium on commercial vehicle braking and handling, p 193
7. Strandberg L (1989) Braking characteristics of 400 heavy trailer combinations from Denmark, Finland, Norway and Sweden. Statens Väg-och Trafikinstitut., VTI särtryck 135

Chapter 8
Multi-wheel Combat Vehicle Dynamics and Control

This chapter presents extensive discussions of multi-wheeled compact vehicles, in particular 8 × 8 configurations. Off-road vehicle behavior depends not only on the total power provided by the engine but also on the power distribution among the drive axles/wheels. In turn, this distribution is primarily regulated by the drivetrain layout and the torque distribution devices. A number of simulation studies, during longitudinal and cornering maneuvers, are conducted to investigate the contribution of typical significant parameters. In addition, the influences of different drivetrain arrangements are presented. The obtained results defined that both traction and cornering responses of multi-wheeled off-road vehicles are highly affected by the driving torque distributed between axles/wheels.

8.1 Combat Vehicle Technology

Wheeled combat vehicles have become extremely popular for military use by offering maneuverable, portable, and fuel-efficient qualities when compared to a tracked vehicle. Wheeled combat vehicles are capable of maximum speeds of 100–110 km/h [1] with light armor and modular configurations for troop and/or infantry transportation. They are designed to withstand typical combat threats such as ballistics, mines, improvised explosive devices, and rocket propelled grenades. They are also extremely maneuverable on various terrains due to the multi-axle configurations that offer more efficient weight distribution. The added axles also allow for a higher maximum Gross Vehicle Weight Rating (GVWR) than smaller vehicles. With three or four axles, the wheelbase of these wheeled combat vehicles has resulted in a vehicle with a large turning radius. Many manufacturers of these vehicles have introduced rearaxle steering to reduce the turning radius.

The MOWAG Piranha shown in Fig. 8.1a reduces its turning radius by steering the rear axle [2]. The FNSS PARS III 8 × 8 shown in Fig. 8.1b steers all axles with

M. El-Gindy and Z. El-Sayegh, *Road and Off-Road Vehicle Dynamics*,
https://doi.org/10.1007/978-3-031-36216-3_8

(a) MOWAG Piranha v (gdels.com)

(b) TFNSS PARS III 8x8 (fnss.com.tr)

(c) Patria AMV (patria.fi)

(d) Hagglunds SEP (Military-today.com)

Fig. 8.1 Production and pre-production RAS combat vehicles

a gradual decrease of steer by wire output and locking over a certain speed [3]. The Patria AMV (Fig. 8.1c offers rear axle steering as an optional method for decreasing the turning radius [4]. The Hagglunds SEP shown in Fig. 8.1d was a proposed electric combat vehicle including rear axle steering. This project was canceled due to lack of international support [5].

Most of these vehicles offer only rear axle steering as this provides significant improvement on the turning radius, which is the initial intention of adding this feature.

Multi-wheeled vehicles that are used mainly for military or for special purposes have to fulfill several main requirements. One of the concerns of these requirements is the off-road vehicle mobility, which is the ability of the vehicle to cope with challenging cross-country terrains. Off-road terrains are characterized by deformable irregular surfaces with abrupt slopes and obstacles of a distinctive nature. The interaction between wheeled vehicles and soft terrain is complex and strongly dominated by the terrain's mechanical properties. Furthermore, some soils can behave excessively in terms of sinkage and slippage according to the applied vertical load and driving moment on the wheel.

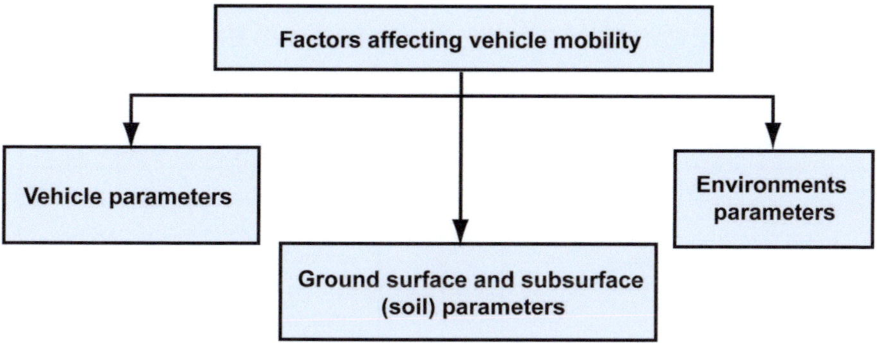

Fig. 8.2 Factors affecting vehicle mobility [6]

8.2 Off-Road Vehicle Mobility

Wheeled vehicles that are used in the military or for special purposes have to satisfy several requirements, and mobility is one of the most important concerns.

The available publications related to off-road vehicle mobility evaluation show significant and useful efforts in this area. These efforts brought to light some methods and techniques that can be used in vehicle mobility evaluation.

The mobility of the vehicle is influenced by many parameters, (see Fig. 8.2) which make the evaluation process complicated, the main factors affecting vehicle mobility are

- Vehicle design and construction parameters.
- Soil parameters.
- Environmental parameters.

In the present work, climate conditions and the driver's skill are assumed to be satisfactory. Hereafter, only vehicle and soil parameters are to be considered when studying the parameters influencing the off-road vehicle mobility evaluation.

8.2.1 Vehicle Parameters Affecting Vehicle Mobility

The vehicle parameters have considerable influence on vehicle mobility. Figure 8.3 shows the vehicle parameters affecting vehicle mobility that include vehicle performance, geometric configuration, vehicle construction, and economy of operation [6].

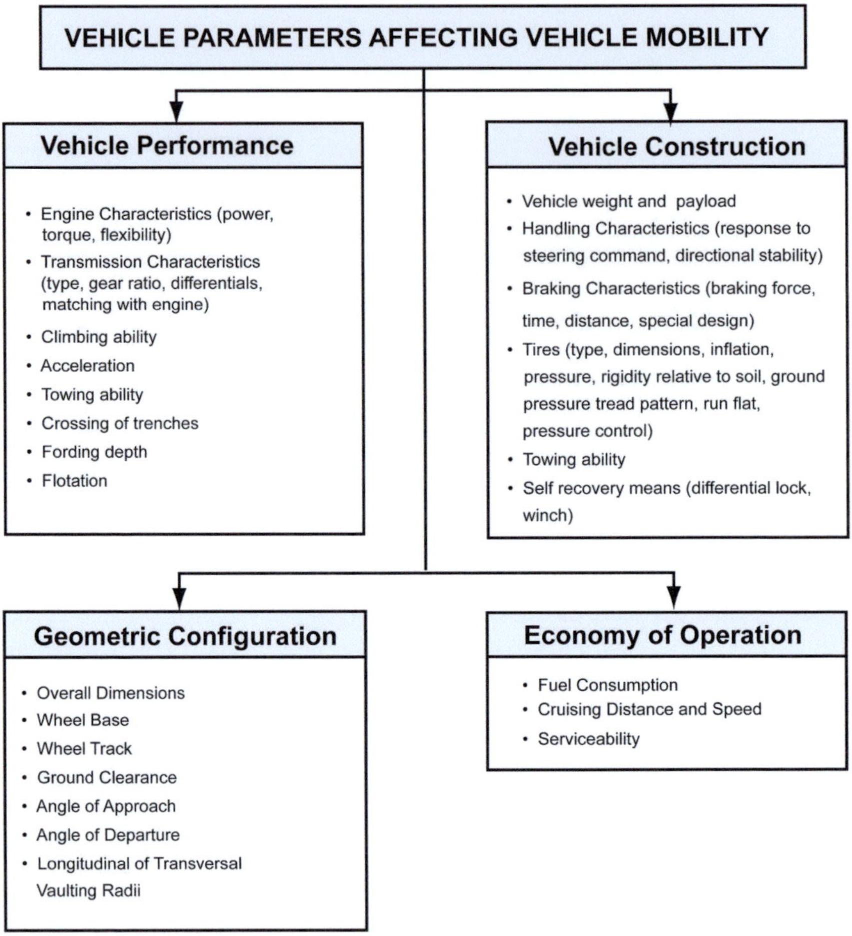

Fig. 8.3 Vehicle parameters affecting vehicle mobility [6]

8.2.1.1 Vehicle Performance

The vehicle performance can be evaluated based on the study of engine characteristics, transmission characteristics, climbing ability, acceleration, towing ability, crossing of obstacles, crossing of trenches, and flotation. The transmission may be divided into two groups such as axled and H-shaped as shown in Figs. 8.4 and 8.5.

Axle designs are used with dependent and independent suspensions as well; the primary transmitters and inter-wheel differentials are located on the axles. Power distribution between the axles is affected by either one or more distributor cases while H-shaped transmissions are usually used on vehicles with high off-the-road mobility with an independent suspension. The use of H-shaped transmission provides greater road clearance and better utilization of the inner volume of the body [7].

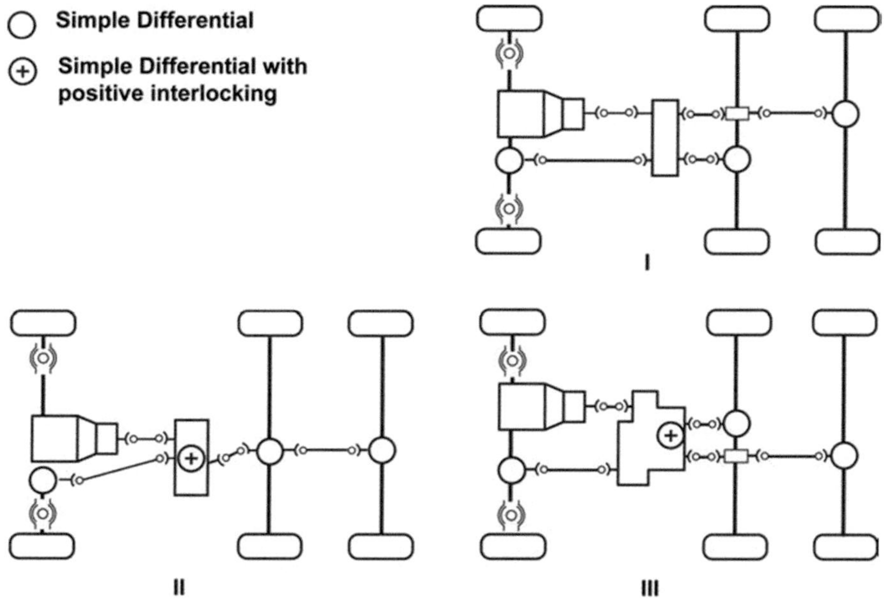

Fig. 8.4 Axle designs of transmission [7]

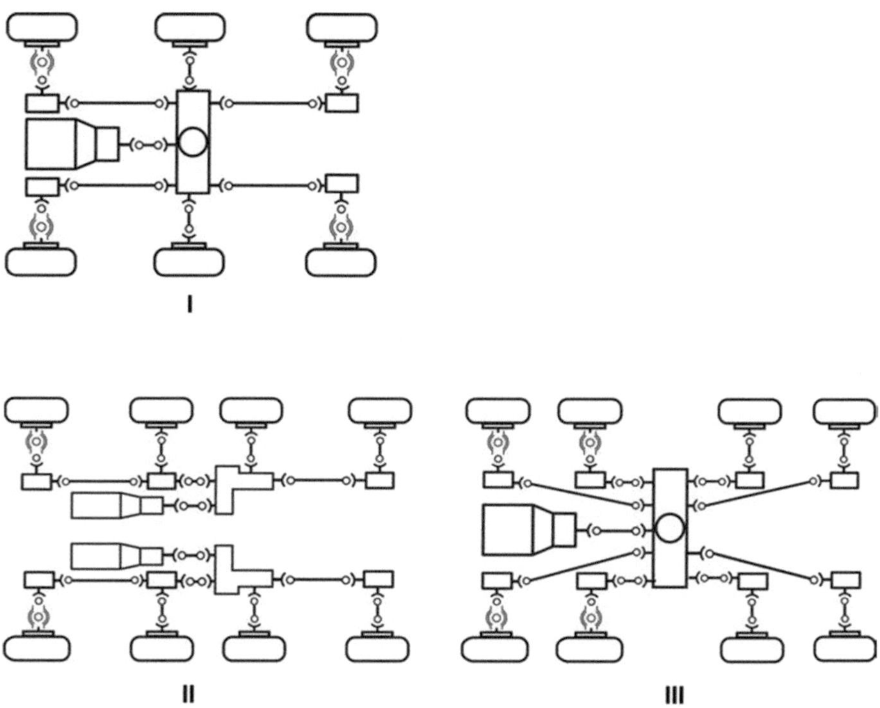

Fig. 8.5 H-shaped and combined designs of transmission [7]

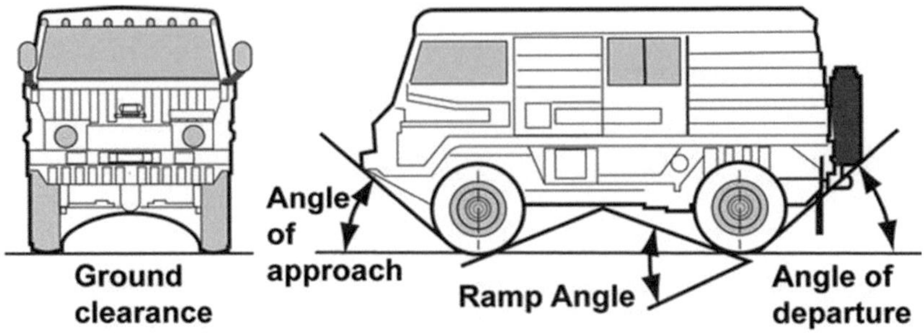

Fig. 8.6 Geometrical properties of a wheeled off-road vehicle [6]

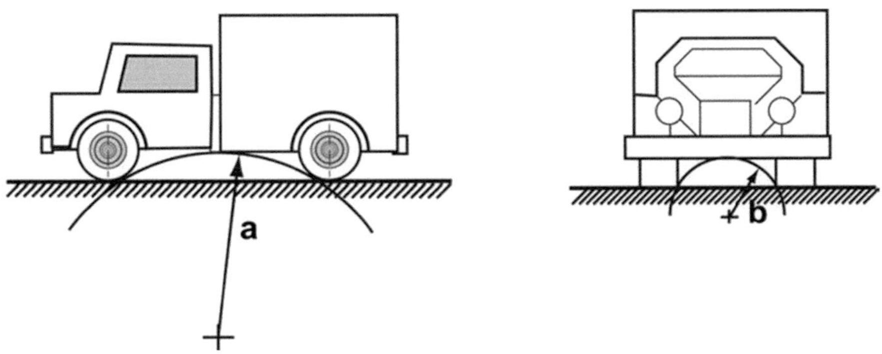

Fig. 8.7 Vaulting radii, **a** Longitudinal and **b** Transversal [6]

8.2.1.2 Vehicle Geometric Configuration

Vehicle geometric configuration refers mainly to the vehicle shape and dimensions including vehicle overall height, width and length, wheelbase, ground clearance, angle of approach, angle of departure, longitudinal vaulting radius, and transversal vaulting radius as shown in Figs. 8.6 and 8.7 [6].

8.2.1.3 Vehicle Construction

Vehicle construction deals with some design parameters of the vehicle such as vehicle weight and payload, handling characteristics, tire forces, and self-recovery.

1. Vehicle weight and payload: The ability of a low-weight vehicle to carry greater loads indicates higher vehicle performance. On soft terrain, the optimum load carrying capacity varies with the mechanical properties of the soil. Rolling resistance increases with increasing vehicle weight due to increased soil sinkage [6].
2. Tires: The primary functions of tires are to support the weight of the vehicle, to cushion the vehicle over surface irregularities, to provide sufficient traction to

drive and brake, and to provide adequate steering control and directional stability. [6].

Vehicle mobility performance depends on several tire parameters. The following items are to be investigated: tire types, inflation pressure and rigidity with relative to the soil, ground pressure, tire tread pattern, and tire pressure control.

1. Tire types: According to the construction, there are two main types of tires that are commonly used: bias-ply tires and radial-ply tires. Radial-ply tires show the following advantages over the bias-ply tires [6]:

   - Less slippage.
   - Increased drawbar pull.
   - Less tread wear.
   - Better distribution of torque.
   - Less rolling resistance.
   - Excellent upholding during cornering.

   Dkebowski et al. [8] investigated the performance of five different agricultural tractor tires on thirty-two different terrain conditions to compare the obtained results with a predictive approach valid for a range of tire sizes, loads, and soil conditions.
   Aubel et al. [9] studied the effect of dual wheel configuration on both rolling resistance and sinkage of towed rigid wheels on sand. The conducted study stated that using dual tires instead of single one reduces both sinkage and rolling resistance.
2. Inflation pressure: The increase of tire inflation pressure increases the tire stiffness and reduces the contact area. Bekker [10] found that the drawbar pull increases with a reduction of inflation pressure. Figures 8.8 and 8.9 show the drawbar pull-slip curves for fine and coarse sand respectively.
3. Specific ground pressure: Specific ground pressure is known as the weight per unit contact area between tire and ground. In addition, low specific ground pressure, especially for soft soils, is recommended for higher mobility performance.
4. Tire tread pattern: It is the appropriate arrangement of ribs, grooves, lugs, and sips in the tread. Road grip, wear, and driving noise are dependent on the type of tread pattern and its condition. The pattern itself is chosen according to the tire application. All wheels of a vehicle should be equipped with tires of the same tread pattern [11].
   Tread configuration, as shown in Fig. 8.10, affects the performance of off-road tires. In soft soils, the lugs will increase the operative tire radius, as it will be clogged with soil. While on rigid terrain, smooth tires will provide the same drawbar pull. In the case of high moisture terrains, traction aids will not provide sufficient traction [12].
5. Tire pressure control: Adjusting the inflation pressure according to the kind of soil is necessary to improve the tire-soil interaction. Vehicles equipped with pressure control systems have an increased off-road performance, as the tire pressure can be adjusted according to load and terrain conditions even during

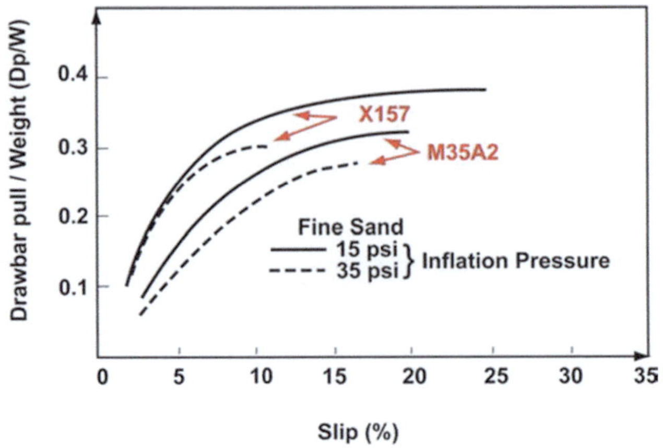

Fig. 8.8 (Drawbar pull/weight)—slip curves in fine sand [10]

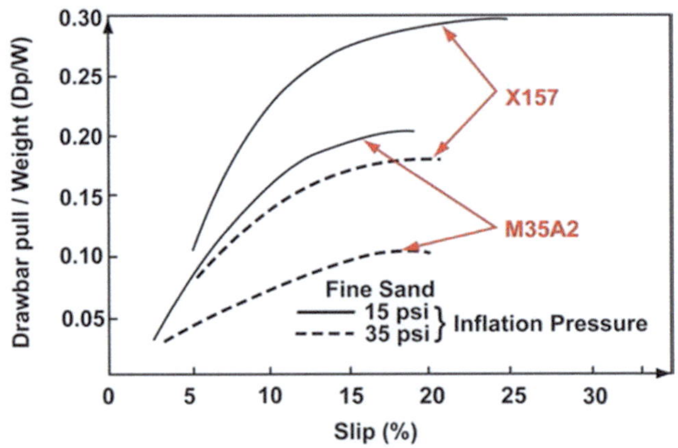

Fig. 8.9 (Drawbar pull/weight)—slip curves in coarse sand [10]

vehicle motion. This system is suitable for vehicles operating on a wide range of terrain types [6].

8.2.2 Soil Parameters Affecting Vehicle Mobility

The word "soil" is widely known as the surface layer of earth that supports our plant life. This definition is incomplete from the point of view of researchers and specialists such as terrain-vehicle engineers who design off-road vehicles capable of negotiating different kinds of soils. The soil parameters affecting vehicle mobility

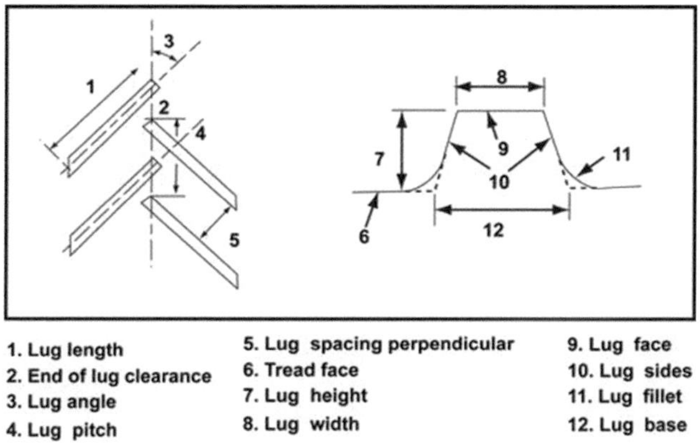

Fig. 8.10 Tread configuration [12]

could be permanent or transient parameters and soil behavior under loading as shown in Fig. 8.11.

Soil is a natural or artificial assembly of a specified range of solid particles, fluids, and gases in a three-phase system. Numerous classification systems have been developed; however, the majority of soil identification systems classify them according to

- Particle size and shape.
- Mineral composition.
- Specific gravity.

There are a number of systems for identification of soils according to their particle size such as

- (USDA) United States Department of Agriculture.
- (ASTM) American Society for Testing Material.
- (ISSS) International Society of Soil Science.
- (DIN) German Industrial Norms Classification.
- (BS) British Classification.
- (WES) Waterways Experimental Station.
- (MIT) Massachusetts Institute of Technology.

8.2.2.1 Permanent Parameters of Soils

a. Particle size composition of soils:

Particle size distribution in soil and its density influences the soil strength and compressibility, both of which are necessary for the consideration of flotation for vehicle mobility. Therefore, the grain size distribution of the soil influences mechanical,

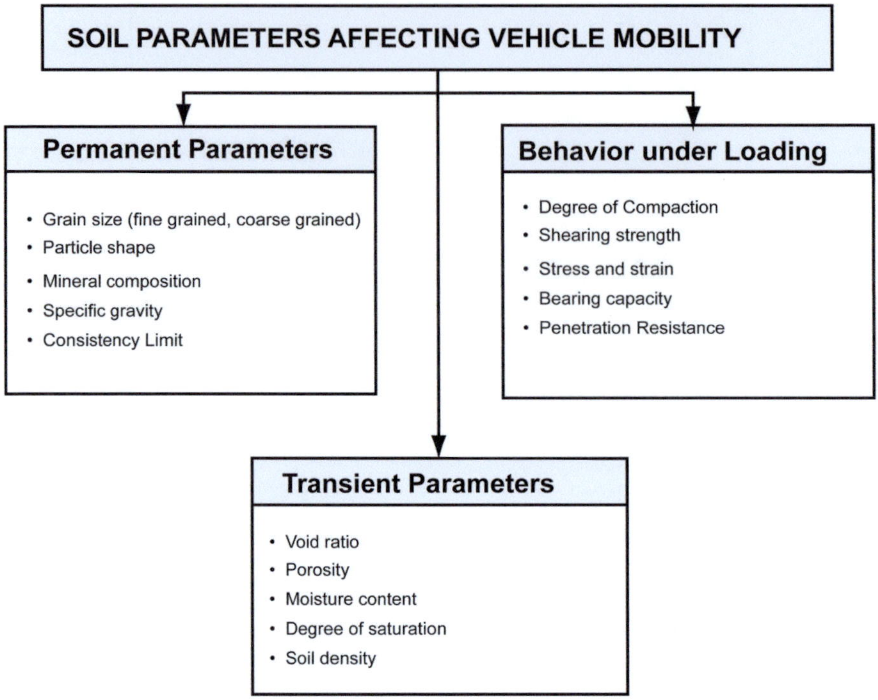

Fig. 8.11 Soil parameters affecting vehicle mobility [6]

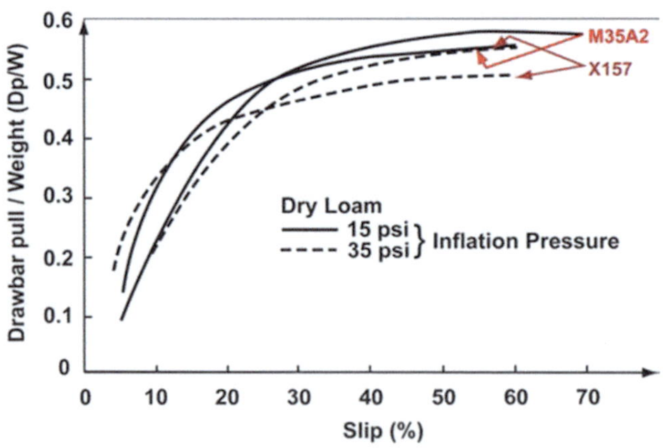

Fig. 8.12 Drawbar pull/weight versus slip curves in hard loam [10]

physical, and biological properties of soils. The effect of grain size distribution on the output drawbar pull of the tested vehicles in different soil types like loam, fine sand, and coarse sand is investigated by Czako [10] as shown in Fig. 8.12.

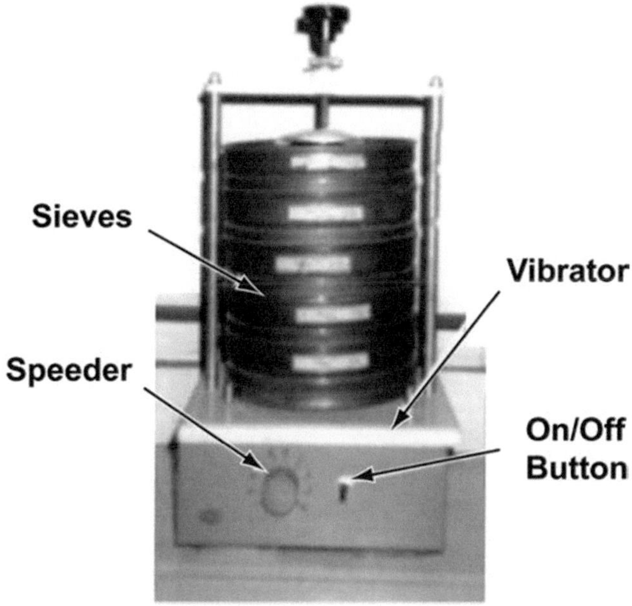

Fig. 8.13 A typical stack of sieves in a mechanical sieve shaker

Particle size is defined as "the effective diameter of the soil particle", and its distribution is affected by chemical, physical, and biological properties. Identification of soils according to grain size is generally accomplished by one of these three methods:

1. Sieve analysis.
2. Hydrometer analysis.
3. Combined analysis.

Sieve Analysis:

- This method consists of shaking a representative soil sample (500 g) through a nest of wire screens of known size openings arranged in a stack with the coarsest on top and finest on bottom as shown in Fig. 8.13.
- After the sample has been shaken sufficiently the weights retained by each sieve are expressed as a percentage of the total sample weight and the results plotted semi-logarithmically as shown in Fig. 8.14.

Hydrometer Analysis:

The hydrometer analysis is used for soil particle size that is too small to be suspended by sieving. This method is based on Stokes' equation for setting velocity of spheres falling freely through a viscous fluid. This terminal velocity (v) is directly related to the diameters of setting spheres (d). The relation between Stokes' equation and the

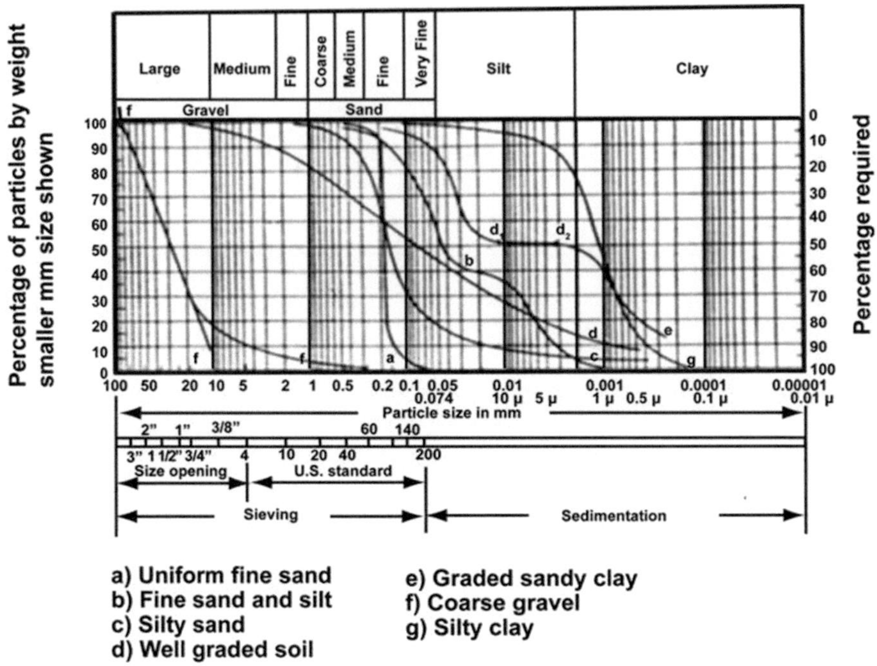

Fig. 8.14 Gradation curves for various types of soils

particle diameter is made through the time rate of change of density of the fluid as the soil particles settle out of suspension.

$$v = \frac{G_S - G_L}{18.\eta} d^2 \tag{8.1}$$

where v is the terminal velocity, G_S is the specific gravity of soil, G_L is the specific gravity of liquid in which soil particles are falling, η is the viscosity of the liquid, and d is the diameter of the particle.

Combined Analysis:

This method employs both above methods used to cover the full range of particle size. Knowledge of the gradation characteristics of surface soil layers is of interest in trafficability studies as it may provide the first indication of potential mobility problems. In general, soils with well-graded curves of particle size distribution provide better trafficability especially after repeated passes of vehicles. This is referred to their better particles redistribution due to load application which in turn increases their density and bearing capacity.

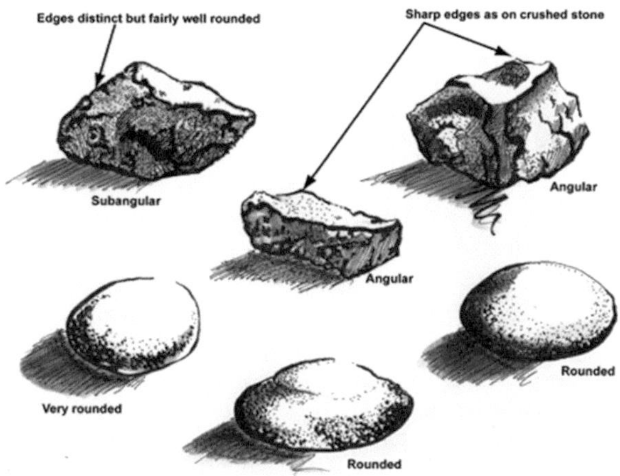

Fig. 8.15 Particle shape of soils in the silt range or coarser

b. Particle shape:

- The shape of the particle refers to the relative proportions of its length, width, and thickness relative to the surface geometry. The shape of soil particles plays an important role in the mechanical behavior of the soil.
- Most of the particles in the silt range and coarser are virtually equidimensional, their surface geometry designations are shown in Fig. 8.15.
- On the other hand, most particles in the clay size range are irregularly shaped, either elongated, flat, flaky, rod like or lathe-shaped, thus a lot of water may be held as adsorbed water within a clay mass as shown by Fig. 8.16.

$$\text{Specific Surface} = \frac{\text{Surface Area}}{\text{Weight}} = \frac{\text{Surface Area}}{\rho \text{Volume}} \tag{8.2}$$

Sand grains are close to cubes or spheres in shape, and have specific surfaces near the minimum value. Clay particles are flaky and have much greater specific surface values.

c. Mineral Composition:

- Minerals are produced mainly from the chemical weathering and decomposition of feldspars. Mineralogical composition of the particles in the sand and gravel size range has little effect on soil properties. The strength of the particles is not critical for behavior of the soil in the range of stresses encountered in off-road locomotion (Table 8.1).
- On the other hand, mineralogical composition of the fine particles greatly affects the behavior of the soils even if they constitute a small percentage of the total mass. The reason is that the gravitational forces of fine particles (dimension $< 1\ \mu$) are approximately equal to their bond forces, and the magnitude of bond forces is

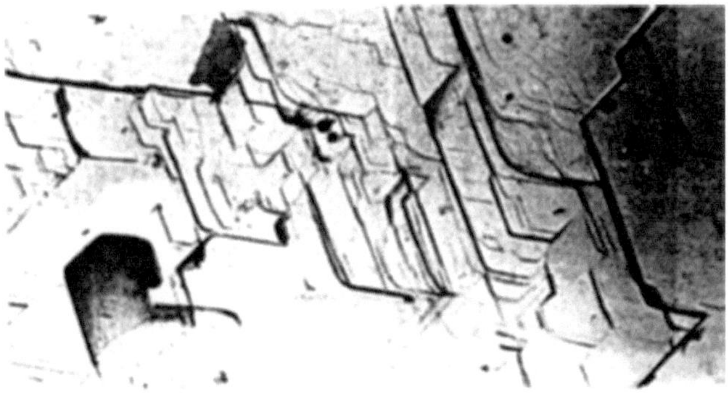

Fig. 8.16 Particle shape of fine soils in the clay range

Table 8.1 Mechanical properties of clay minerals

Mineral	Specific surface, m^2/g	Particle shape	Particle size, μ
Kaolonite	10–20	Platy	$d = 0.3$ to 3 Thickness $= 1/3$ to $1/10$
Illite	80–100	Platy	$d = 1$ to 2 Thickness $= 1/10d$
Nontronite	800	Lath length	$d = 0.4$ to 2 Thickness $= 1/100d$

directly related to the mineralogical composition. Table 8.2 shows some parameters of some common clay minerals.

- The mineralogical composition is determined either instrumentally or chemically. Some examples are X-ray diffraction, infrared adsorption, and electron microscopy.

d. Specific Gravity:

- Specific gravity of a soil, G_S, is defined as the ratio between the unit weight of the soil, γ_s, and the unit weight of distilled water at 4 °C (γ_w) and is written as

$$G_S = \frac{\gamma_s}{\gamma_w} \tag{8.3}$$

- Measurement: The jar is weighed empty (M_1). A quantity of dry soil is placed in the jar and the jar weighed (M_2). The jar is filled with water, air removed by stirring, and weighed again (M_3). The jar is emptied, cleaned, and refilled with water—and weighed again (M_4).

Table 8.2 Parameters of common clay minerals

Mineral	G_S
Calcite	2.71–3.72
Chlorite	2.60–3.00
Gypsum	2.20–2.40
Kaolonite	2.50–2.66
Hematite	4.30–5.30
Quartz	2.65
Talcum	2.60–2.70

$$G_S = \frac{\text{Mass of soil}}{\text{Mass of water displaced by soil}} \tag{8.4}$$

$$= \frac{M_2 - M_1}{(M_4 - M_1) - (M_3 - M_2)} \tag{8.5}$$

e. Consistency limits:

Consistency limits (or Atterberg limits) measure the water content at which remolded soils pass from one state to another. States of the soil are dry, solid, semisolid, plastic, and liquid.

- Liquid Limit: is determined experimentally by letting a groove in the soil paste, made by a special cutting tool, flow together under the repeated impact of dropping the cup containing the soil paste from a specified height, Fig. 8.17.
- Plastic Limit: is defined as the lowest water content at which the soil can be rolled into threads 3 mm in diameter without the threads breaking into pieces or crumbling, Fig. 8.18. Soils exhibiting water content below the plastic limit are likely to be trafficable by most of the off-road vehicles.
- Shrinkage Limit: the water content at which it ceases to decrease its volume upon reduction of its water content. The shrinkage limit is determined by measuring the volume of an oven-dried soil pat and calculating the water content necessary to fill the voids completely.
- Saturation Limit: is defined as the water content at which a drop of water at the soil surface ceases to infiltrate into the soil. Its value is roughly half of the liquid limit.
- The Plasticity Index: is the difference between the liquid limit and the plastic limit. It is actually a measure of the cohesive properties of soils. The index has practical application to vehicle mobility since materials that have a high index tend to soften in wet weather and become slippery.

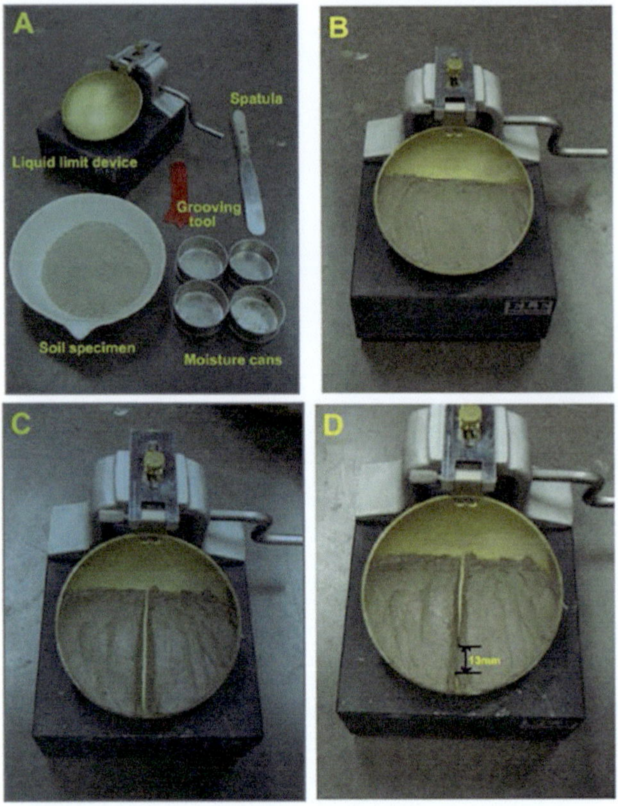

Fig. 8.17 Experimental determination of liquid limit according to Atterberg

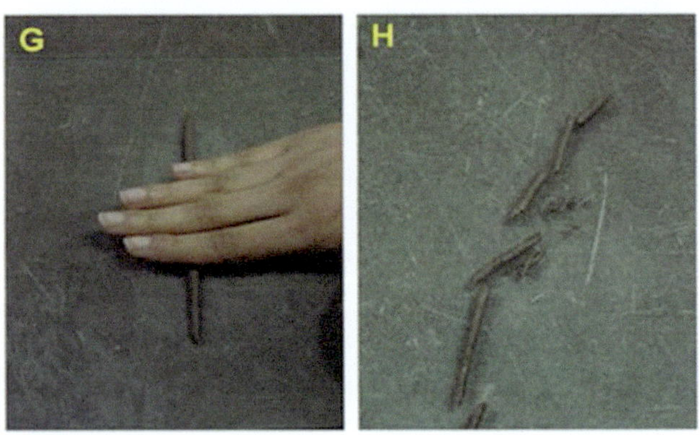

Fig. 8.18 Experimental determination of plastic limit according to Atterberg

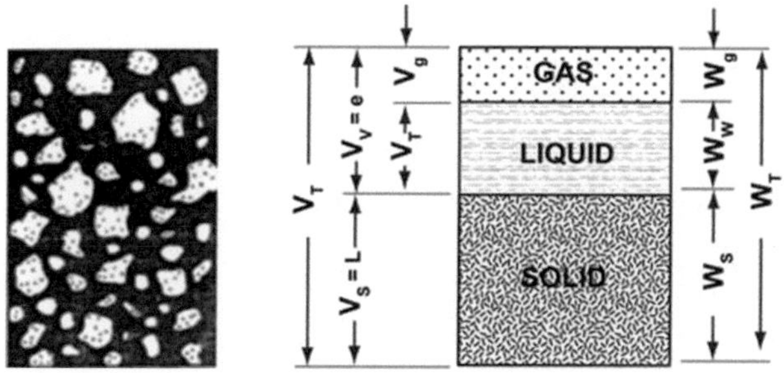

Fig. 8.19 Elements of natural soil

8.2.2.2 Transient Parameters

The soil state in the in-situ conditions is described by the so-called transient characteristics. Unlike engineering materials, soil is a multiphase medium that contains three distinct phases: solid, liquid, and gas, Fig. 8.19.

a. Void Ratio and Porosity:

Void Ratio (e) is the ratio between the volume of voids (V_v) and the volume of solids (V_s). The voids can be completely filled with air, completely filled with water, or partially filled with each.

$$e = \frac{V_v}{V_s} \tag{8.6}$$

Porosity of a soil (n) is defined as the ratio between the volume of voids (V_v) and the total volume (V_T).

$$n = \frac{v_v}{v_T} = \frac{e}{1+e} \tag{8.7}$$

b. Moisture content:

Moisture content (W) is defined as a percentage between the weight of water (W_W) in a given soil sample and the corresponding weight of dry solids (W_S), and is written as $\left(W = \frac{W_W}{W_S}\right)$. In nature, moisture content varies with depth and the greatest variations usually take place at the surface since it is closest to daily and seasonal fluctuation of the weather.

The moisture content has a significant effect on the wheeled vehicles traction and resistance coefficient as shown in Figs. 8.20 and 8.21 [1]. Crolla et al. [2] deduced the soil friction resistance per unit area and the moisture content relationship which presented that the soil has a single peak close to the plastic limit as shown in Fig. 8.22.

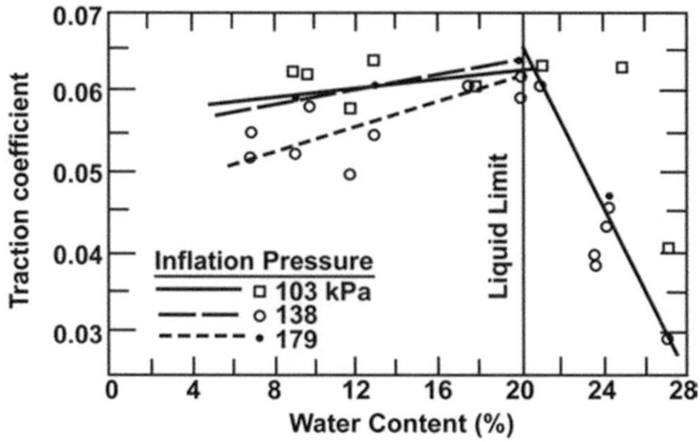

Fig. 8.20 Net traction coefficient-water content at different inflation pressures [1]

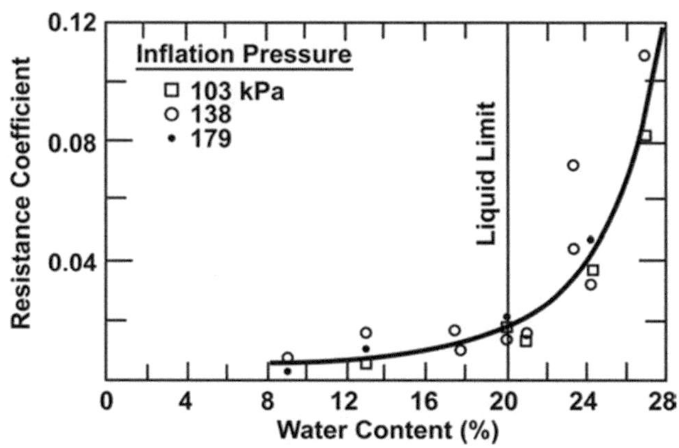

Fig. 8.21 Resistance coefficient-water content at different inflation pressures [1]

c. Bulk, Dry, and Relative Densities:

- Bulk density (γ_T) or total unit weight is defined as the ratio of the total weight of the soil (W_T) to the total volume (v_T), and is written as

$$\gamma_T = \frac{W_T}{v_T} = \frac{W_S + W_W}{v_S + v_v} = \gamma_s \frac{1 + W}{1 + e} \tag{8.8}$$

- Dry density (γ_d) or dry unit weight is defined as the ratio between the weight of solids (W_S) and the total volume of the soil sample (v_T), and is written as

$$\gamma_d = \frac{W_S}{v_T} = v_S \gamma_S (v_S + v_v) = \frac{\gamma_S}{1 + e} = \frac{\gamma_T}{1 + W} \tag{8.9}$$

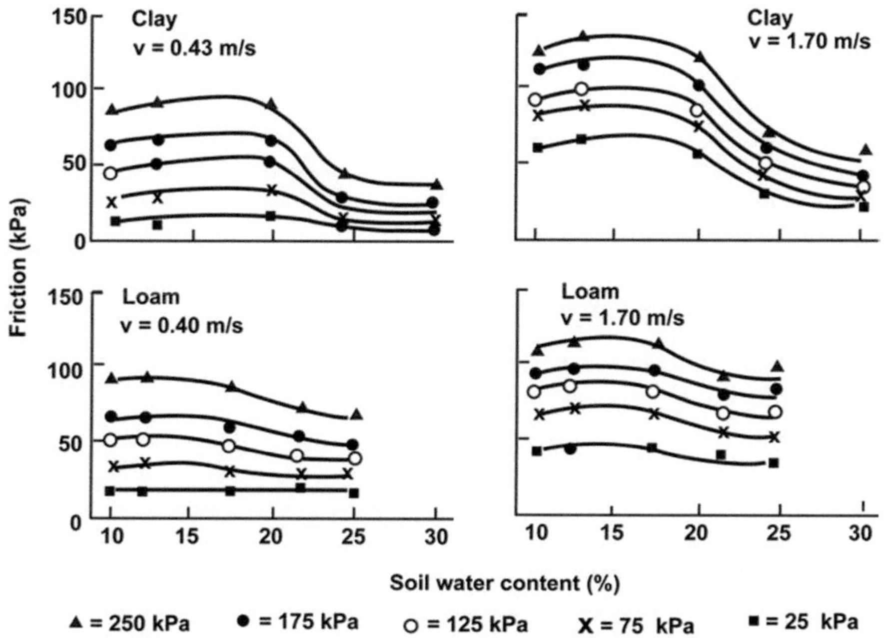

Fig. 8.22 Experimental relation between friction and soil water content [2]

Table 8.3 Description of the soils with different relative density

Relative density, %	Descriptive term
0–15	Very loose
15–35	Loose
35–65	Medium
65–85	Dense
85–100	Very dense

- Relative Density, D_r: can be expressed by

$$D_r = \frac{e_{max} - e}{e_{max} - e_{min}} = \frac{\gamma_{dmax}}{\gamma_d} = \frac{\gamma_d - \gamma_{dmin}}{\gamma_{dmax} - \gamma_{dmin}} \tag{8.10}$$

where γ_{dmax} is the densest dry unit weight, γ_{dmin} is the loosest dry unit weight, e_{max} is the loosest void ratio, is the densest void ratio. Table 8.3 gives indicative values of relative density for granular soils. A well-graded soil (that has a wide range of particle sizes) is generally more dense than uniformly graded soil having a predominant grain size in the same range. The reason is that the voids among the large particles can be filled more easily with smaller particles.

d. Moisture-Density Relationships:

- In general, the denser a soil becomes, the greater its shear strength. A mass of soil that has a very loose arrangement contains a large number of voids and is less stable under load than a very dense soil. Soil density can be increased by compaction. Compaction is a process in which the soil particles are artificially arranged and packed together into a closer state of contact by mechanical means.
- It has been demonstrated that the moisture content of the soil at the time of compaction is most critical, especially for fine grained soils.
- Compaction properties of a certain soil can be experimentally evaluated using the so-called moisture-density relationship. That relationship is determined in the laboratory by dropping a weight of 2.5 kg 25 times from a distance of 30.5 cm onto a soil mass (with certain moisture content) encountered in a standard test mold, Fig. 8.23.
- The process of soil compaction has to be repeated several times changing the moisture content and the corresponding dry density and the moisture content values are plotted.
- The degree of compaction is a function of two variables, the moisture content and the compaction effort. For a specific amount of compaction effort or energy there is one moisture content, the "optimum", at which a given soil attains its maximum dry density.
- Cohesionless soils such as sand do not respond noticeably to variations in compacting moisture content and compaction effort in comparison to cohesive soils such as clays.
- In general, well-graded soils (those having a wide range of particle sizes) exhibit better compaction behavior than uniformly graded soils having a predominant grain size in the same range. The reason is that the voids among the large particles can be filled more easily with smaller particles by the action of applied compaction effort.

8.2.2.3 Soil Behavior Under Loading

The main soil parameters depending on vehicle loading and affecting vehicle mobility may be summarized as follows:

- Shear strength.
- Bearing capacity.

a. Soil shear strength:

It can be defined as the soil maximum resistance to shearing stresses and depends on moisture content, soil type, and grain size distribution of the soil. The soil shear strength can be determined using Eq. 8.11 depending on two parameters, soil cohesion (c), and internal friction angle (ϕ). The two parameters are obtained based on

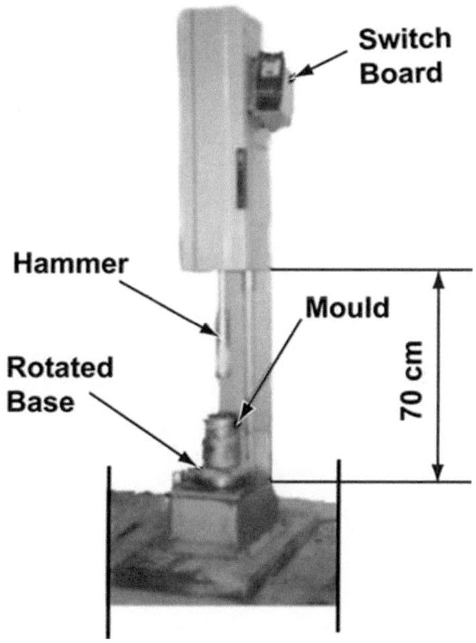

Fig. 8.23 Compaction test equipment

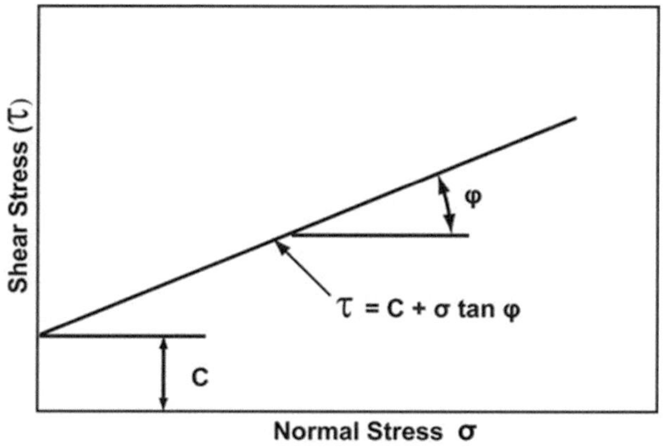

Fig. 8.24 The Mohr-coulomb relationship [3]

the Mohr-coulomb failure criterion as shown in Fig. 8.24:

$$\tau_m = c + \sigma \tan \phi \tag{8.11}$$

where τ_m is the maximum shear stress, σ is the normal stress, c is the soil cohesion, and ϕ is the angle of internal friction.

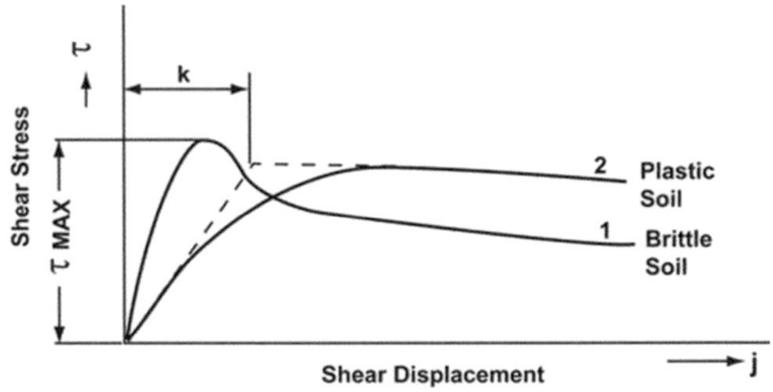

Fig. 8.25 Shear stress-displacement curves [3]

There are two types of shear stress curves; the first one presents the maximum shear stress, τ_{max} and a part of residual shear stress, τ_r after yielding as shown by curve 1 in Fig. 8.25. The second one is the shear stress-displacement curve as shown by curve 2 in Fig. 8.25 [3].

b. Soil bearing capacity:

The bearing capacity is the required average load per unit area on the contact area to reach the supporting soil mass failure [4]. The bearing capacity theory estimates the maximum load that the vehicle can exert on the terrain without failure. The pressure sinkage relationship of terrain, assuming homogeneous characteristics, can be determined using Eq. 8.12 [3].

$$P = \left(\frac{k_c}{b} + k_\phi\right) z^n \tag{8.12}$$

where P is the ground pressure, b is the width of contact area, z is the sinkage, n is the exponent of deformation, and k_c, k_ϕ are the terrain constants.

Terzaghi's bearing capacity formula is given by the following Eq. [5].

$$P_f = \alpha C N_c + q' N_q + \frac{1}{2}\gamma b_f N_\gamma \tag{8.13}$$

where P_f is the bearing capacity, α is the shape factor, C is the cohesion, q' is the effective surcharge, γ is the unit weight, b_f is the footing width, and N_c, N_q, N_γ are the bearing capacity factors.

8.3 Torque Management Devices Implemented in AWD Vehicles

Off-road vehicles have different running abilities, i.e., higher traction, tractive efficiency, and improved mobility. These depend not only on the total tractive effort available by the power plant but also on its distribution between the driving wheels, which can be determined by actuating vehicle systems and characteristics of the power dividing mechanisms, e.g., inter-wheel, inter-axle reduction gear and transfer cases. The locking features of these mechanisms control the force distribution between driving wheels. Consequently, they can control both vehicle longitudinal performance and handling characteristics [13]. Mohan and Williams [14] organized different AWD traction control systems, including passive and active devices, by using general principles and their strategies as shown in Fig. 8.26.

Lanzer [16] suggested a torque split factor to evaluate the impact of tractive force on drivability, handling, ease of operation, cost, and compatibility with the ABS

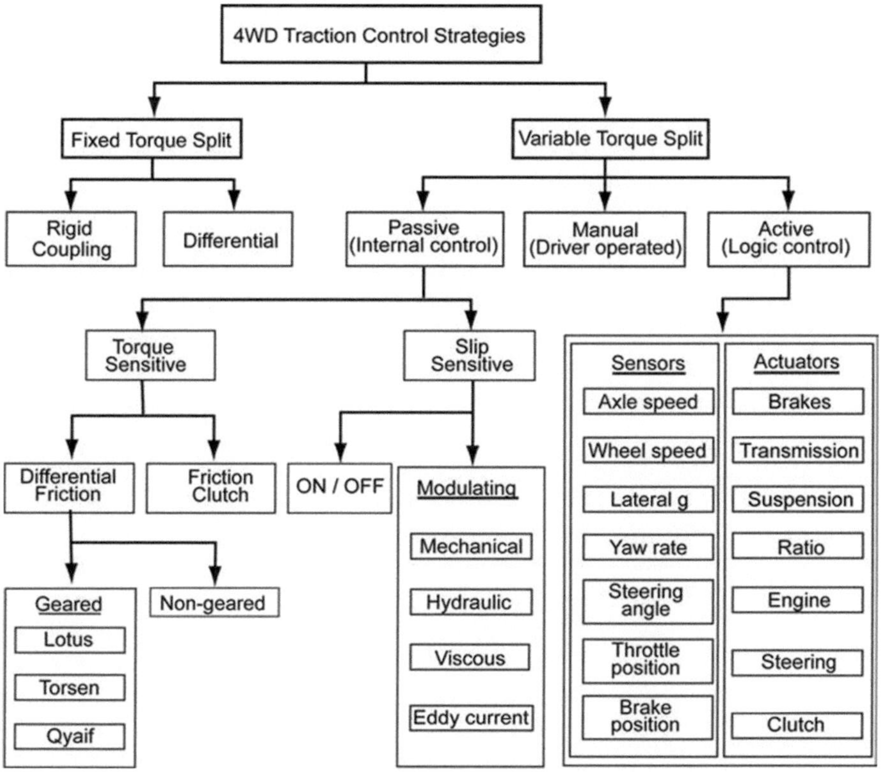

Fig. 8.26 4WD Traction control strategies [15]

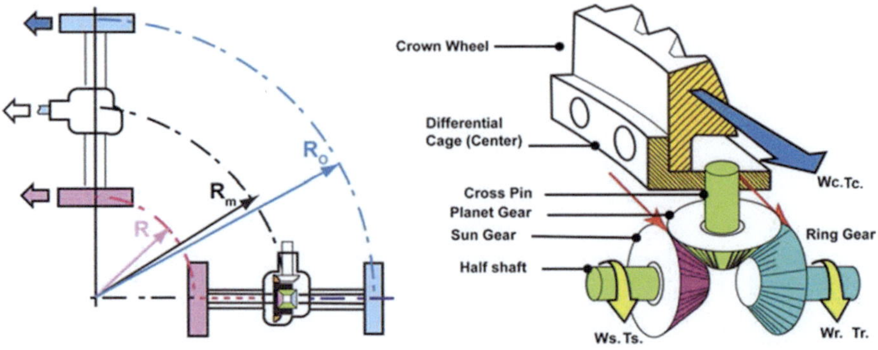

Fig. 8.27 Principles of open differential gearing [17]

system for different 4WD systems based on the performed comparison between permanent and part time 4WD systems.

8.3.1 Mechanical Differential (Open and Locked)

The conventional open differential has been the standard device for an automotive powertrain for a long time. This device is simple and effective in providing the necessary speed differential between the driving wheels during vehicle turning, Fig. 8.27.

However, it cannot take full advantage of the available traction at the driving wheels on roads with different levels of adhesion. Consequently, the vehicle's maximum driving power is limited to twice the torque at the low friction side of the driving wheels which means that any increase in the engine throttle causes the low friction side wheels to spin more, which would increase the slip sinkage in the case of driving on an off-road terrain [18].

The ordinary bevel-gear differential can be presented as a set of planetary gears, the gear attached to the left half-axle can be considered as the sun gear with angular speed (ω_s), the other gear attached to the right half-axle can be considered as the ring gear with an angular speed (ω_r). The crown wheel is considered as the planet carrier with an angular velocity (ω_c) [19]. In addition, the driving speed and torque along the lateral axis can be calculated as shown in Eq. 8.14:

$$\omega_c = \frac{\omega_r + \omega_s}{2} \text{ and } T_s = T_r = \frac{T_c}{2} \tag{8.14}$$

where T_s is the sun gear torque, T_r is the ring gear torque, T_c is the carrier gear torque. The locked differential has the ability to lock the two outputs together using an electric, pneumatic and hydraulic or frictional mechanism. This mechanism can

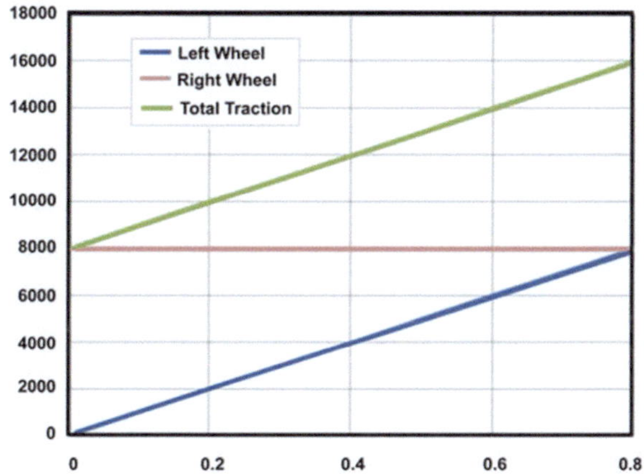

Fig. 8.28 Variation of traction force with adhesion coefficient (Locked differential)

be selected manually, and when the differential is locked, the wheels will have the same speed as shown in Eq. 8.15.

$$\omega_c = \omega_r = \omega_s \text{ and } T_c = T_s + T_r \tag{8.15}$$

? Example 8.1

For a vehicle with rear drive, consider the weight on the rear axle is 20000 N, the adhesion coefficient on the right wheel is 0.8 and the adhesion on the left wheel varies from 0 to 0.8. For locked differential calculate the traction force and draw the variation of traction force with adhesion coefficient.

Solution. The variation of traction force with adhesion coefficient for locked differential is shown in Fig. 8.28. If there is a big difference between Max. Min. T and T then the differential lock increases the available traction force. This is critical in the case of low adhesion coefficient necessary for off-road vehicles and military vehicles (Table 8.4).

8.3.2 Clutch-Type LSD

Torque bias can be introduced only by adding a friction clutch to the system as shown in Fig. 8.29. The clutch type Limited Slip Differential (LSD) has the same mechanical parts used in the open differential, but it has a set of clutches and springs.

Table 8.4 Total traction force as calculated for several adhesion coefficients

Adhesion coefficient, ϕ_L	Adhesion coefficient, ϕ_R	Traction force P_{TL}	Traction force P_{TR}	Total traction force $P_T = P_{TL} + P_{TR}$
0	0.8	0	8000	8000
0.1	0.8	1000	8000	9000
0.2	0.8	2000	8000	10000
0.3	0.8	3000	8000	11000
0.4	0.8	4000	8000	12000
0.5	0.8	5000	8000	13000
0.6	0.8	6000	8000	14000
0.7	0.8	7000	8000	15000
0.8	0.8	8000	8000	16000

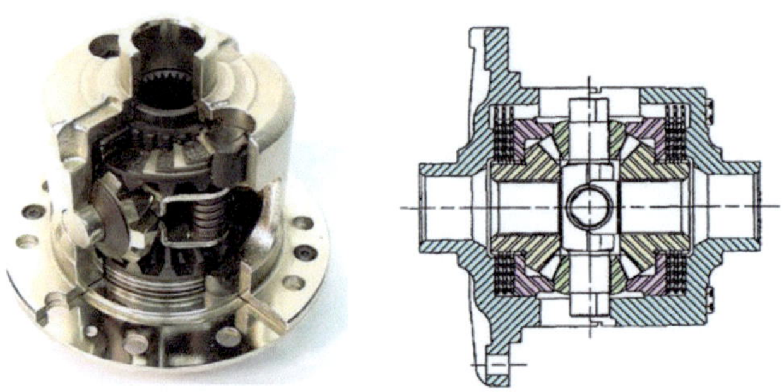

Fig. 8.29 Clutch type limited slip differential [3]

The clutch's objective is to keep both wheels at the same rotating speed. The spring's stiffness combined with the clutch friction regulates how much torque is required to overcome the clutch resistance. The main disadvantage is the frictional wear on the clutch, which results in a deterioration of the performance of the differential. The biased torque based on the applied force in the friction disc is given by Eq. 8.16.

$$C_f = nfN\frac{R_1 + R_2}{2} sgn\Delta\omega \tag{8.16}$$

where n is the number of slipping surfaces, f is the clutch dynamic coefficient of friction, N is the normal load applied on the clutch disc, R_1, R_2 are the outer and inner clutch disc radii, $\Delta\omega$ is the differential angular speed of the rotating discs.

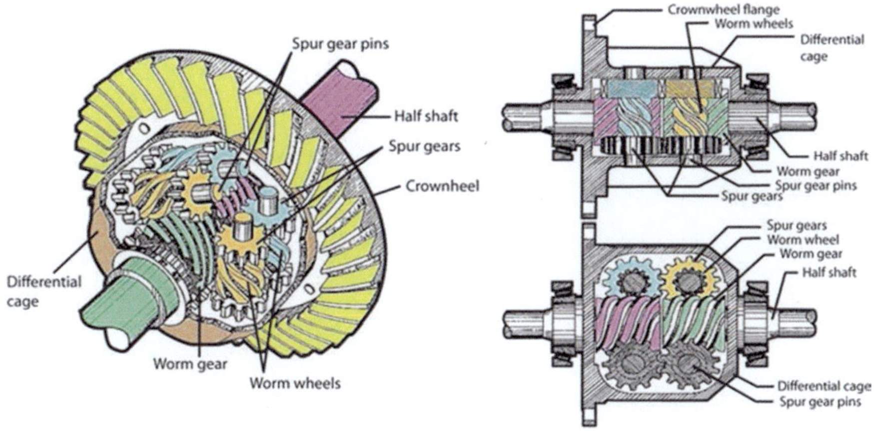

Fig. 8.30 Torsen limited slip differentials [20]

8.3.3 Torsen LSD

Torsen differentials have been involved in the powertrain driveline since 1983, and they are frequently used in high-performance AWD vehicles. The Torsen (Torque sensing) differential is a purely mechanical device that performs as an open differential in the case of having the same driving torque for both wheels as shown in Fig. 8.30. While in the case of losing traction of one of the wheels, the differential gears will use torque difference between the wheels to bind them together.

Harnisch [20] studied and compared the operating principles and performance of the Torsen differentials with open differentials. In addition, Shih and Bowerman [21] compared the torque bias ratio and the efficiency of friction clutch-based LSD, Torsen differentials, and lockable differential devices. It should be stated that the LSD differential biases torque is based on the available torque at the slipping wheel. Several differentials are designed with a preload to ensure that there will be some torque available to the wheel with good traction. In addition, this preload must be limited to prevent opposing handling effects in the vehicle, [22].

8.3.4 Viscous-Lock Devices

Viscous coupling consists of a sealed housing and a splined hub. A set of thin plates are alternately connected to the housing and the hub. The intervening space between the plates and the housing is partially filled with high viscosity silicone oil as shown in Fig. 8.31. If one set of wheels attempts to spin faster, the adjacent plates will rotate faster in comparison with the others. The fluid follows the faster plates and drags

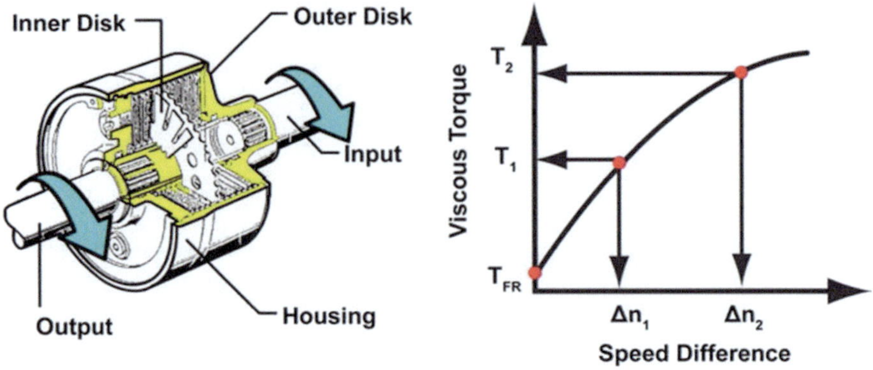

Fig. 8.31 Viscous coupling characteristics [23]

the slower plates with it. This action will add additional torque to the slower set of wheels.

Harnisch [23] introduced several applications of viscous coupling in all-wheel drive vehicles. In addition, they developed a simple empirical equation to calculate the transmitted viscous torque (T) based on the speed difference (Δn) and the friction torque (T_{FR}) as shown in Eq. 8.17:

$$T = T_{FR} + \alpha \Delta n^{b} \tag{8.17}$$

Their method of calculation has been supported by several experimental measurements to predict the empirical constants (a, b) as shown below

$$a = \frac{log\left(\frac{T_2 - T_{FR}}{T_1 - T_{FR}}\right)}{log\left(\frac{\Delta n_2}{\Delta n_1}\right)} \tag{8.18}$$

$$b = \frac{T_2 - T_{FR}}{\Delta n^{a}} \tag{8.19}$$

Harnisch [24] developed a theory to define the conditions necessary for initiating and sustaining Static Timing Analysis (STA) in rotary viscous couplings. In addition, he verified the processes that produce STA by proposing a sequence of events that are qualitatively viable and consistent with one another.

8.3.5 Electronically Controlled LSD

The ordinary controlled limited slip differential has limited capabilities due to its design while both traction and handling can be directly optimized by electronically controlling the differential's output. In addition, if the vehicle is equipped with one of the advanced traction or braking control systems, the differential can resist by

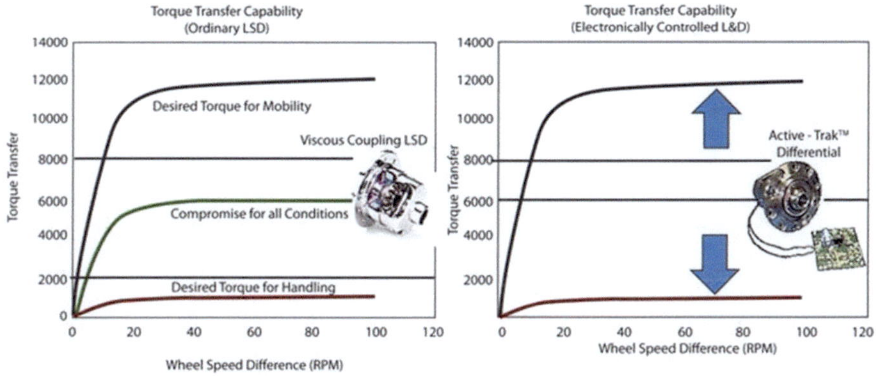

Fig. 8.32 Passive versus electronically controlled LSD [25]

applying a torque to the wheel that is slowing down. This reduces the effectiveness of both the differential and the control systems. Optimal mobility and handling can easily be achieved by programming the differential to react differently to specific external conditions. Figure 8.32 shows the torque transfer range of an electronically controllable differential compared with an ordinary viscous coupling LSD [25].

A Proportional-Integral-Differential (PID) controller is used to calculate the engagement force based on using various inputs to determine the vehicle operating condition. Inputs include individual wheel speeds, steering angle, throttle position, vehicle speed, brake status, transfer case mode, and temperature. The controller determines how much correction is needed based on the difference between the actual and theoretical wheel speeds.

Xia et al. [26] investigated different coupling solutions by employing a magnetic particle clutch, coupled, in a quasi-static torque split arrangement with a planetary gear system. The proposed arrangement increases the torque capacity of the coupling by directing only a fraction of the torque through the magnetic particle clutch.

The term "Torque vectoring" is defined as a driveline device capable of controlling both the magnitude and direction of torque to influence traction and vehicle dynamics. Such devices may be applied between wheels of the same axle or between axles in AWD applications, as torque vectoring can deliver power to any wheel instantly without using either the brakes or engine management. Torque vectoring depends on using advanced differentials that can distribute power to the wheels that have traction, which means that wheels do not need to be stopped.

8.3.6 Control Architecture

Mohan et al. [27] presented a novel torque vectoring called "Differential System Dynamic Trak", which can be applied to both the inter-axle and the inter-wheel differential systems. The "Dynamic Trak" has three multi-plate clutches as shown in Fig. 8.33. The main clutch either offers a limited slip or complete lock-up ability based

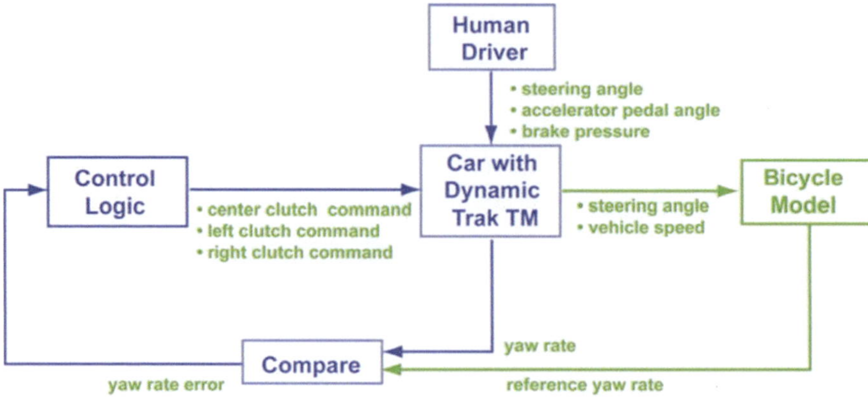

Fig. 8.33 Torque vectoring differential [27]

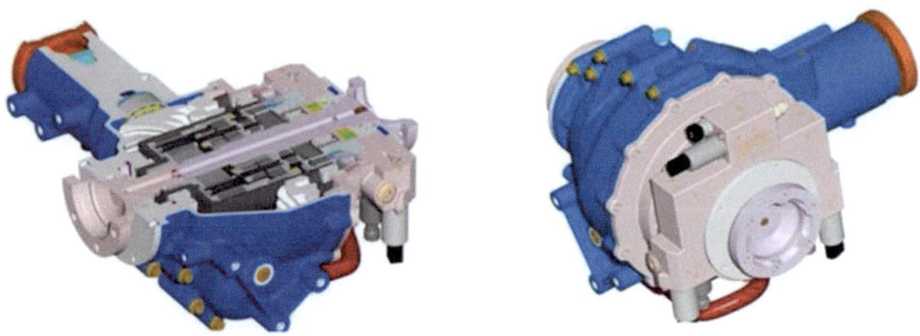

Fig. 8.34 Ricardo's cross-axle torque vectoring

on the driving conditions. The two exterior clutches regulate the torque delivered to the left or right shafts/wheels. An electronic control unit actively controls the three clutches to manage the torque delivered to the two output shafts/wheels. The "Dynamic Trak" can provide a maximum of 100% torque bias.

Mitsubishi Super All Wheel Control (S-AWC) integrates its Active Center Differential (ACD), Active Stability Control (ASC), Active Yaw Control (AYC), and ABS control as shown in Fig. 8.33. The feedback control depends on a direct yaw moment control strategy that affects left-right torque vectoring and controls cornering maneuvers based on the desired yaw rate during different vehicle driving states. S-AWC succeeded in enhancing vehicle stability performance at different driving situations.

Ricardo's Torque Vectoring technology used in Audi A6 4.2l V8 Quattro Avant allows the driving torque to be redistributed based on vehicle speed and road conditions is shown in Fig. 8.34. In addition, Mohan et al. [28] developed a simplified model of center differential control containing: the equations, which describe the vehicle, the model structure, important values, and parameters for the simulation.

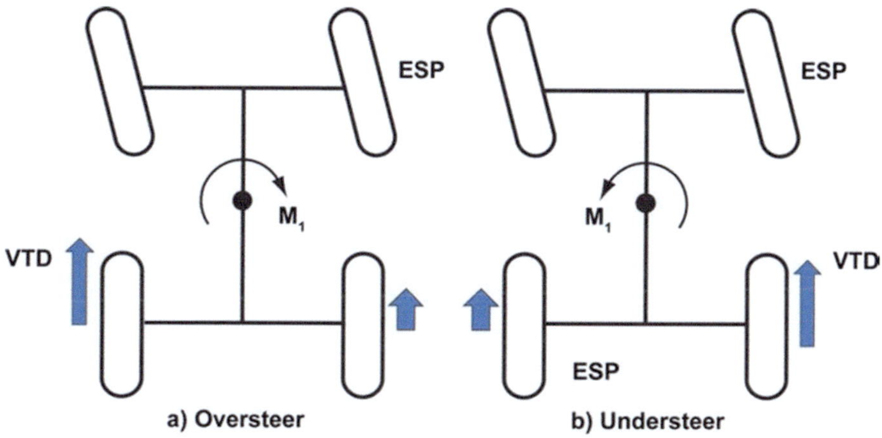

Fig. 8.35 Integrated control of VTD and ESP [29]

In addition, the authors described a concept of a simplified torque of split control system.

Park et al. [29] introduced a control system to enhance vehicle stability and controllability performance based on two control systems such as Electronic Stability Program (ESP) and Variable Torque Distribution (VTD). The control strategy depends on identifying the driving situations based on the vehicle slip angle as shown in Fig. 8.35.

In the case of steady-state conditions, the VTD system is used, while the ESP controller is primarily used for emergency maneuvers. To solve this difference, an individual subsystem should be activated depending on operating conditions as shown in Fig. 8.36.

Perumpral et al. [30] developed a torque vectoring control strategy based on using a 2-DOF linear Parameter Varying (LPV) control to enhance the vehicle performance as shown in Fig. 8.37.

Liu et al. [31] developed a torque vectoring control strategy using a PID and LQR controllers for longitudinal and lateral dynamics respectively for hybrid electric vehicle as shown in Fig. 8.38. Simulation results presented enhancements in the vehicle performance.

8.4 Vehicle Modeling and Validation

The actual vehicle configuration and simulation model of a multi-wheeled combat vehicle are shown in Fig. 8.39. The vehicle is equipped with four axles, which can be operated in either 4WD or 2WD. The front two axles are steering axles (δ_1 and δ_2). The vehicle is equipped with independent suspensions. The vehicle model consists

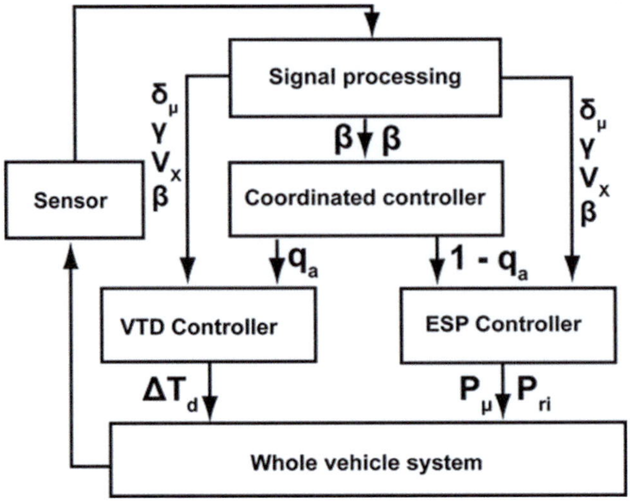

Fig. 8.36 Block diagram of integrated control [29]

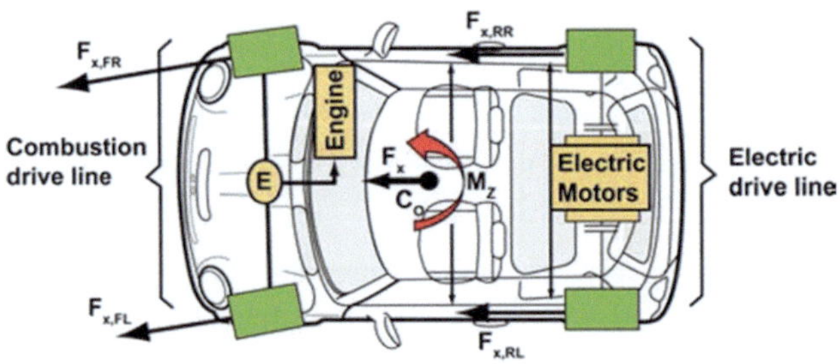

Fig. 8.37 Basic design of a TtR-HEV [30]

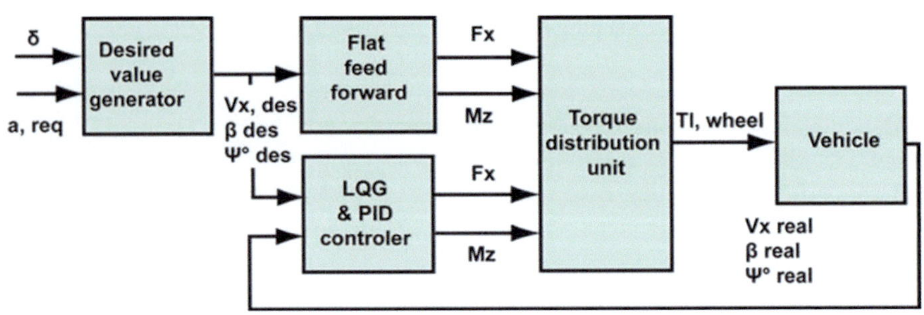

Fig. 8.38 Torque vectoring control structure [31]

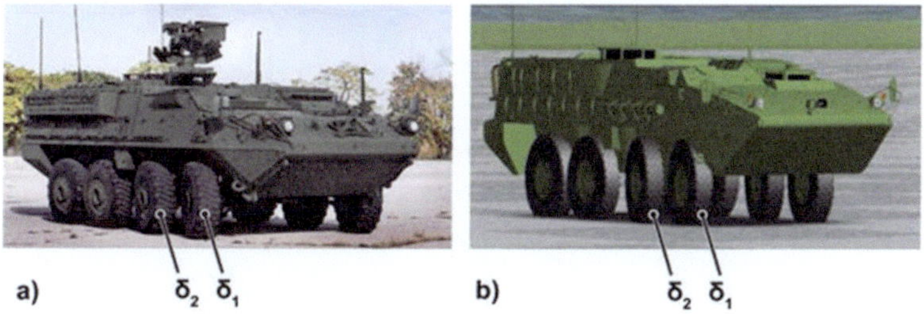

Fig. 8.39 Actual vehicle configuration (**a**) and the simulation model (**b**)

of 22° of freedom, namely, pitch, yaw, and roll of the vehicle sprung mass and spin and vertical motions of each wheel of the eight wheels.

8.4.1 Vehicle Modeling

The vehicle model has been developed using TruckSim and is based on the actual vehicle configuration for multi-wheeled combat vehicle design parameters, including non-linear tire/terrain interaction characteristics in the form of look-up tables for both rigid and soft terrain. The tire/soft terrain characteristics were obtained from FEA off-road tire models developed using PAM-CRASH.

As can be seen in Fig. 8.40, the vehicle is equipped with two front steering axles. The individual steering angle according to Ackerman's conditions, for a specific turning radius, can be determined by plotting perpendicular lines on the four steering wheels and the rear two axles at their geometric center.

The average inner and outer steering angles (δ_i and δ_0, respectively) for the first and second axles have been approximated and calculated using Eq. 8.20.

$$\cot \delta_0 - \cot \delta_i = \frac{B}{L} \tag{8.20}$$

Figure 8.41 shows the relationship between gearbox output and the steering angle at the ground of each road wheel of the first and second axle, at the nominal suspension position and in the absence of tire forces, without accounting for speed effects.

The developed combat vehicle model is used to study vehicle maneuverability on rigid and soft terrain at different speeds and powertrain configurations (8 × 4 and 8 × 8). The predictions of the vehicle handling characteristics and transient response during a lane change on rigid road at different vehicle speeds were compared with field tests results. Measured and predicted results are compared based on vehicle steering, yaw rates, and accelerations. Published validation criteria have been used to validate the simulation results. At each measurement location, the model predicted Root Mean Square (RMS) value should agree with the measured RMS acceleration

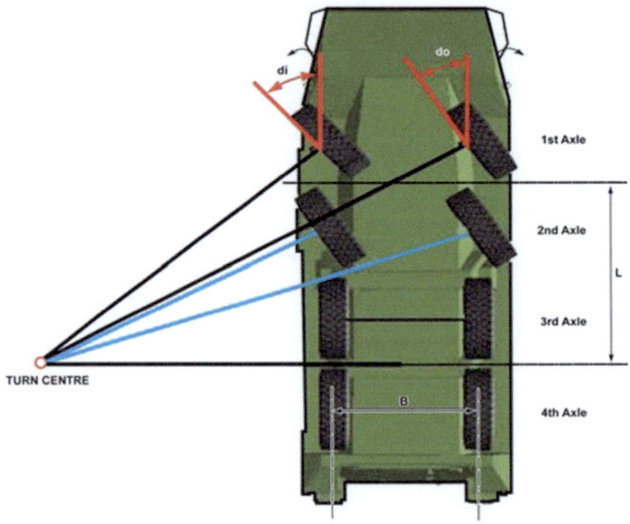

Fig. 8.40 Ackerman steering of eight-wheel vehicle with multi-axle steering

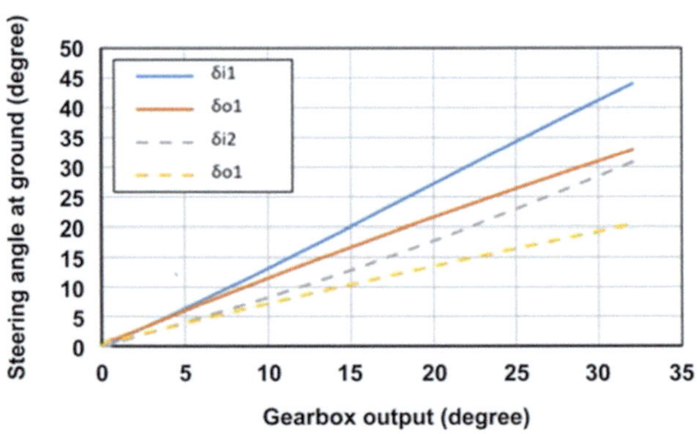

Fig. 8.41 First and second axles steering angle versus gearbox output

within ±10%. The model time domain data and measured time domain data skewness, and kurtosis values should agree within ±50% of the measured data values to provide a comparison on wave shape in the time domain. The kurtosis, skewness, and RMS are defined as follows:

Kurtosis measures the peaks of the random data and was chosen as a statistical parameter because it is an excellent indicator of extreme values and how they relate to the general data. It is extremely useful in picking out wild points.

$$\text{Kurtosis} = \frac{\sum (x_i - \mu)^4}{N\sigma^4} - 3 \tag{8.21}$$

Table 8.5 Test courses matrix

No.	Test course	Vehicle speed	Additional test data
1	Double lane change (NATO AVTP-1 03-160 W)	40, 53, 72, 81 km/h and maximum	
2	Constant step slalom (NATO AVTP-1 03-30)	40, 53, 60 km/h and maximum	30 m cone spacing
3	J-Turn (75ft radius)	30, 35, 40, 45, 50 km/h	
4	Turning circle (4 × 8 & 8 × 8)	Crawling	Maximum cramping angle = 34 deg

where x_i is the ith value, μ is the mean, N is the number of data points, σ is the sample standard deviation.

Skewness is a measure of the probability distribution of random variables, skewness is a measure of one-sidedness.

$$\text{Skewness} = \frac{\mu^3}{\sigma^3} \tag{8.22}$$

Root Mean Square (RMS) is the magnitude of varying quantity of data. It is relatively insensitive to wild points, and it does not provide an indication of variation about the mean.

$$RMS = \sqrt{\frac{1}{N}\sum_{i=1}^{N} x^2} \tag{8.23}$$

8.4.2 Vehicle Model Validation

The vehicle model was tested in four different test courses, Double Lane Change, Constant Step Slalom, J-Turn with 8 × 4 powertrain drive, and Turning circle test with two different powertrain configurations (8 × 4 and 8 × 8). All tests have been conducted on rigid roads with tire inflation pressure of 0.72 MPa. Table 8.5 shows the test courses and vehicle speeds used to validate the vehicle model. In the following sections, a sample of the performed validation tests of each test course will be demonstrated.

8.4.2.1 Double Lane Change (NATO AVTP-1 03-160W)

This maneuver is designed to examine the vehicle's transient response. The vehicle was tested during the lane-change maneuver at different speeds; Fig. 8.42 shows a schematic drawing of the lane change test course.

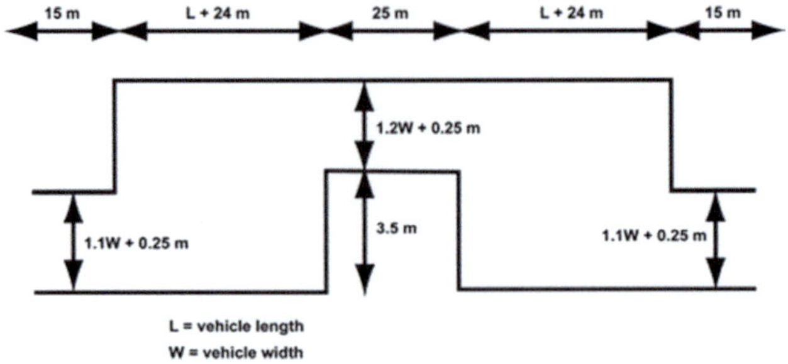

Fig. 8.42 NATO (AVTP 03-160) lane change test course [32]

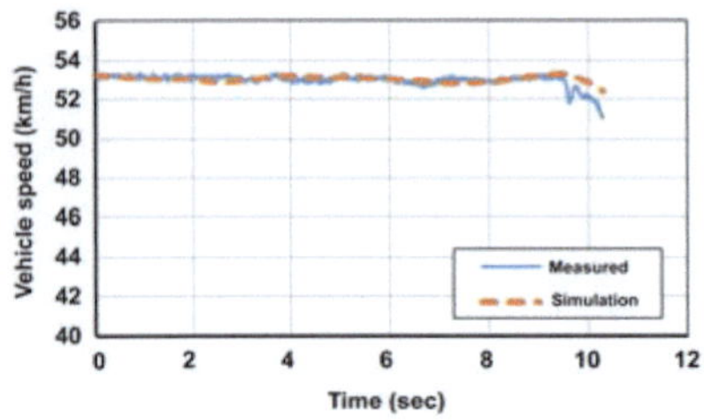

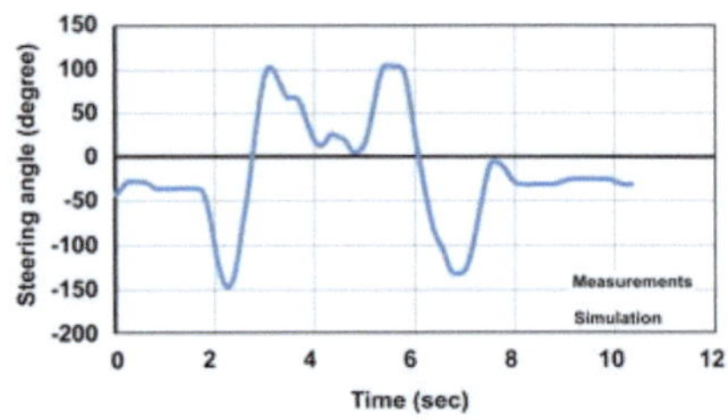

(a) Vehicle speed

(b) Vehicle steering angle

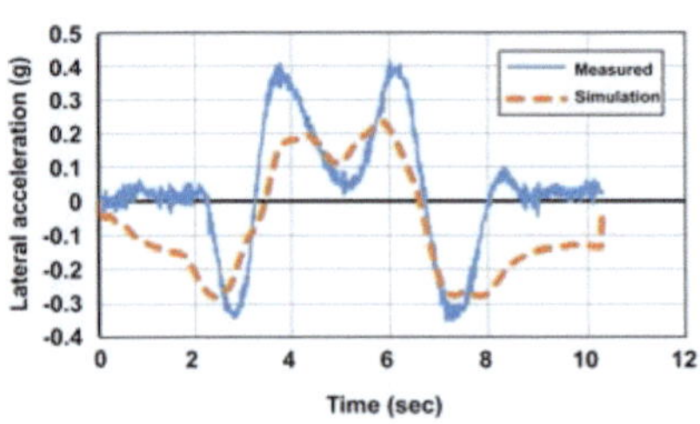

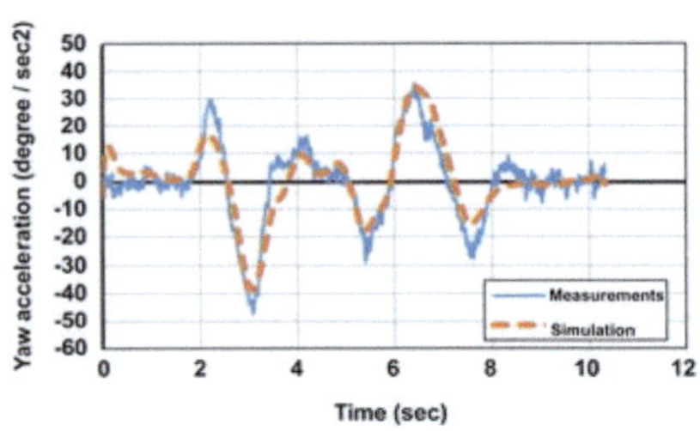

(c) Vehicle lateral acceleration

(d) Vehicle yaw acceleration

Fig. 8.43 Vehicle time history at a speed of 53 km/h

a. NATO Lane Change—53 km/h

This test was performed using the simulation speed as shown in Fig. 8.43a which is simulated to replicate what was measured during the experimental testing. As can be seen, the simulation speed and measured speed are constants with the approximate value of 53 km/h. The steering wheel input used in the simulation was obtained from the measurements as shown in Fig. 8.43b. The vehicle lateral acceleration and yaw acceleration are given in Fig. 8.43c, d.

Validation criteria have been used to validate the simulation results of this test. Table 8.6 shows that the lateral acceleration validation criteria are found to be within

Table 8.6 Validation results for left lane change at 53 km/h using validation criteria

	Measured	Simulation	Min.	Max.
Kurtosis	3.081	2.426	1.541	4.622
Skewness	1.208	0.747	0.604	1.812
RMS	0.369	0.031	0.027	0.033
Yaw acceleration				
	Measured	Simulation	Min.	Max.
Kurtosis	11.265	8.006	5.632	16.897
Skewness	2.715	2.418	1.358	4.0725
RMS	199.168	172.325	179.251	219.085

the recommended range (minimum and maximum values) of the kurtosis, skewness, and RMS. In the case of the yaw acceleration, the skewness and kurtosis values are found to be within the recommended range, while the predicted RMS is value found to be outside the recommended range, due to the high noise level of the supplied measured data.

b. NATO Lane Change—85 km/h

This test was performed using the simulation speed as shown in Fig. 8.44a which is simulated to replicate what was measured during the experimental testing. As can be seen, the simulation speed and measured speed are constants with the approximate value of 85 km/h. The steering wheel input used in the simulation was obtained from the measurements as shown in Fig. 8.44b. The vehicle lateral acceleration and yaw acceleration are given in Fig. 8.44b, c. As can be seen there is a good agreement between the measurement and simulation in both shape and peaks' locations.

Validation criteria have been used for validation. Table 8.7 shows that the lateral acceleration validation criteria are found to be within the recommended range of the kurtosis and skewness, while the predicted RMS value is found to be outside the recommended range due to the high noise level of the supplied measured data. In the case of the yaw acceleration, the kurtosis and skewness values are found to be within the recommended range, while the predicted RMS value is found to be outside the recommended range. In addition, the simulation results are compared with eight additional different tests. The calculated skewness and kurtosis values are found to be within the recommended range. While the model prediction of RMS values of the lateral acceleration and yaw acceleration did not agree with some of the measured ones within $\pm 10\%$ due to the high noise level of the measured lateral acceleration and yaw acceleration data.

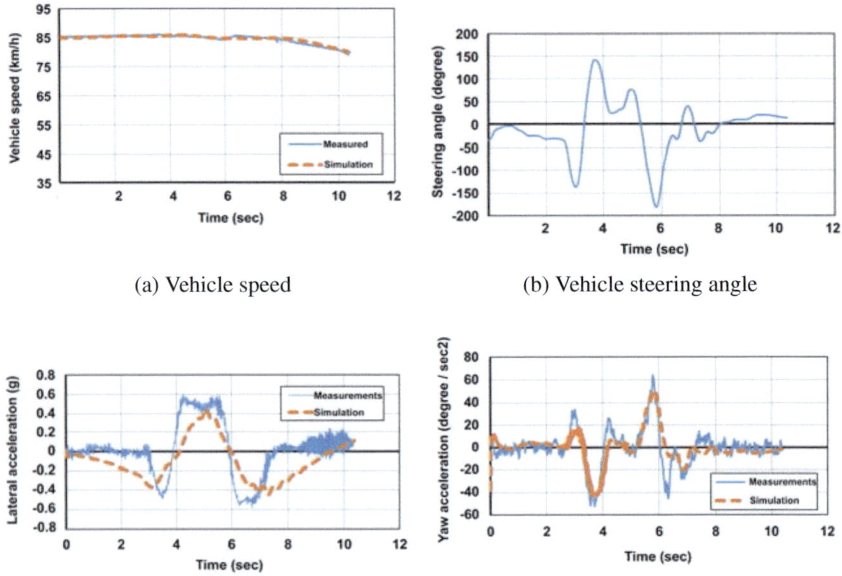

(a) Vehicle speed

(b) Vehicle steering angle

(c) Vehicle lateral acceleration

(d) Vehicle yaw acceleration

Fig. 8.44 Vehicle time history at a speed of 85 km/h

Table 8.7 Validation results for left lane change at 85 km/h validation criteria

	Measured	Simulation	Min.	Max.
Kurtosis	2.480	2.474	1.240	3.72
Skewness	1.045	0.842	0.523	1.567
RMS	0.076	0.053	0.068	0.083
Yaw acceleration				
	Measured	Simulation	Min.	Max.
Kurtosis	13.348	11.116	6.674	20.022
Skewness	3.074	2.960	1.537	4.611
RMS	283.030	204.939	254.727	311.333

8.4.2.2 Constant Step Slalom (NATO AVTP-1 03-30)

This maneuver is designed to examine the vehicle's transient response. The vehicle was tested during a constant step slalom maneuver at different speeds. Figure 8.45 shows a schematic drawing of the constant step slalom test course.

a. 30 m slalom 40 km/h

This test was performed using the simulation speed as shown in Fig. 8.46a, which is simulated to replicate what was measured during the experimental testing. As can be

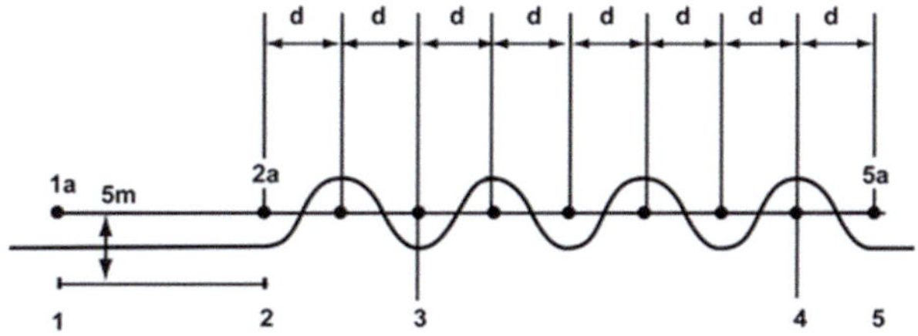

Fig. 8.45 NATO (AVTP-1 03-30) constant step slalom test course [32]

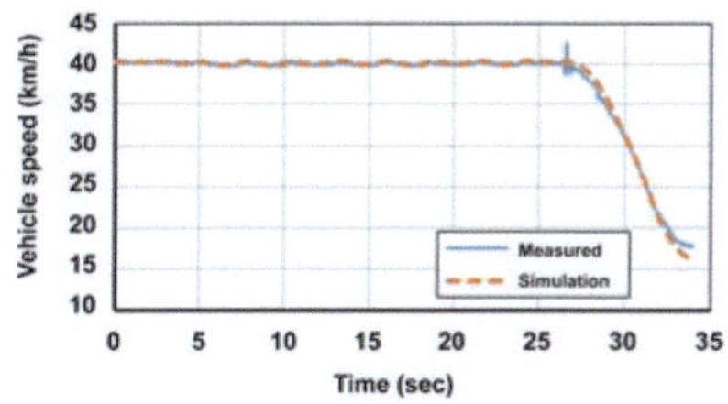

(a) Vehicle speed

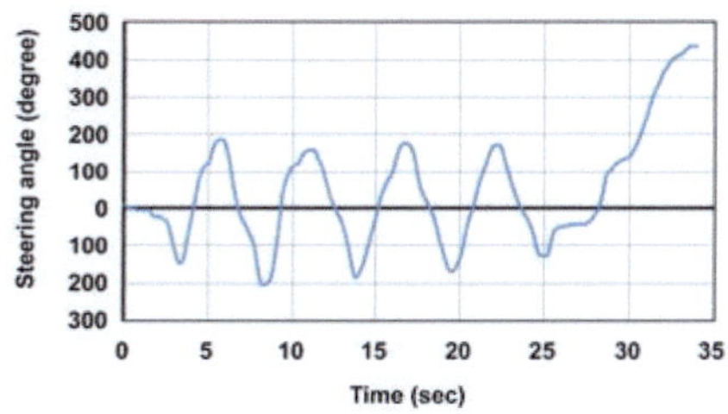

(b) Vehicle steering angle

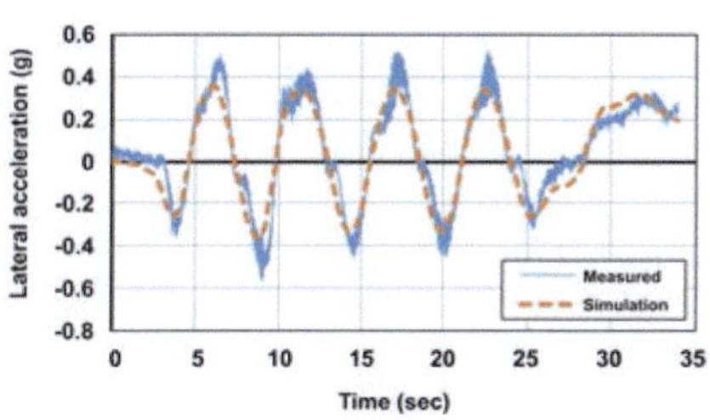

(c) Vehicle lateral acceleration

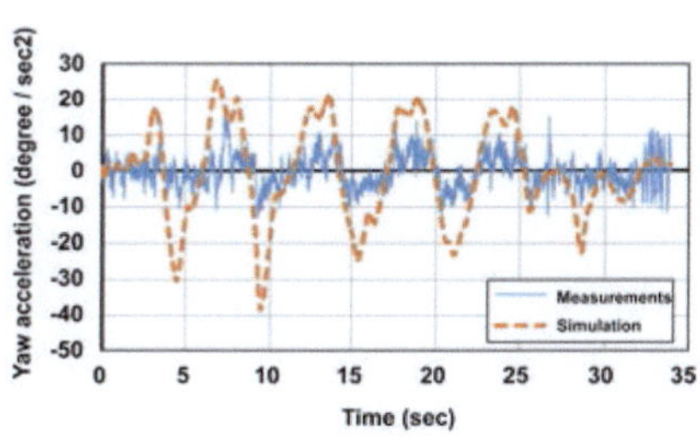

(d) Vehicle yaw acceleration

Fig. 8.46 Vehicle time history at a speed of 40 km/h

seen, the simulation speed and measured speed are constants with the approximate value of 40 km/h. The steering wheel input used in the simulation was obtained from the measurements as shown in Fig. 8.46b. The vehicle lateral acceleration and yaw acceleration are given in Fig. 8.46c, d.

Validation criteria have been used for validation. Table 8.8 shows that the lateral acceleration validation criteria are found to be within the recommended range of the kurtosis and RMS while the predicted skewness are found to be outside the recommended range. In the case of the yaw acceleration, the kurtosis and skewness values are found to be within the recommended range, while the predicted RMS values are found to be outside the recommended range.

Table 8.8 Validation results for constant step slalom at 40 km/h validation criteria

Lateral acceleration				
	Measured	Simulation	Min.	Max.
Kurtosis	4.017	1.739	2.008	6.025
Skewness	1.213	0.242	0.606	1.819
RMS	0.057	0.052	0.051	0.062
Yaw acceleration				
	Measured	Simulation	Min.	Max.
Kurtosis	16.496	11.516	8.248	24.745
Skewness	3.031	2.417	1.515	4.546
RMS	21.722	173.463	19.550	23.894

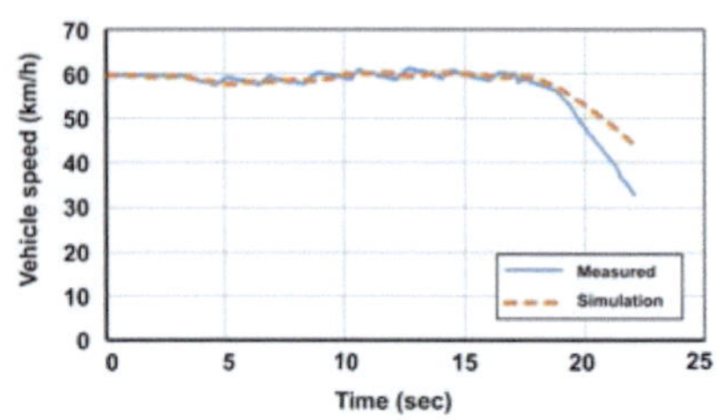

(a) Vehicle speed

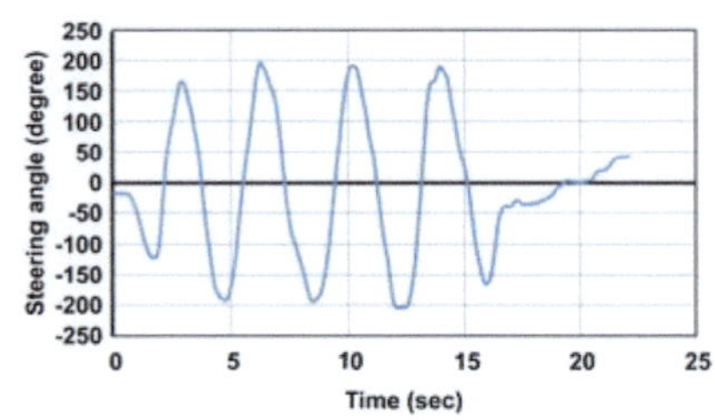

(b) Vehicle steering angle

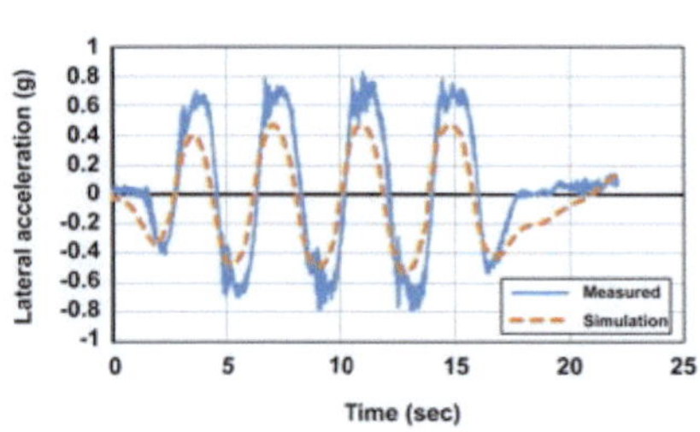

(c) Vehicle lateral acceleration

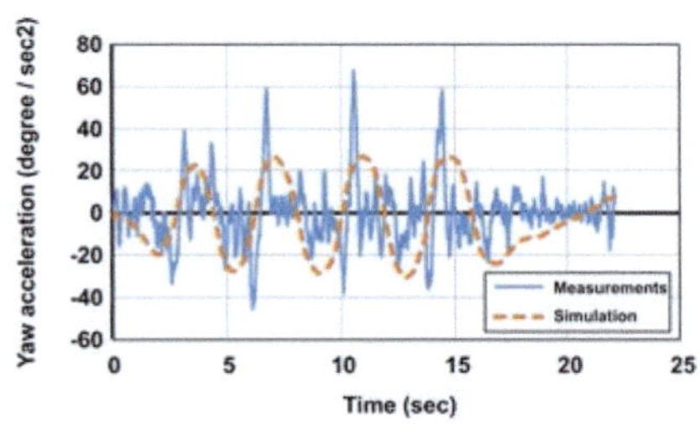

(d) Vehicle yaw acceleration

Fig. 8.47 Vehicle time history at a speed of 60 km/h

b. 30 m slalom 60 km/h

This test was performed using the simulation speed as shown in Fig. 8.47a, which is simulated to replicate what was measured during the experimental testing. As can be seen, the simulation speed and measured speed are constants with the approximate value of 60 km/h.

Table 8.9 Validation results for constant step slalom at 40 km/h using validation criteria

Lateral acceleration				
	Measured	Simulation	Min.	Max.
Kurtosis	1.766	1.889	0.883	2.649
Skewness	0.466	0.472	0.233	0.699
RMS	0.187	0.092	0.169	0.206
Yaw acceleration				
	Measured	Simulation	Min.	Max.
Kurtosis	27.132	1.889	13.566	40.698
Skewness	4.521	0.472	2.260	6.781
RMS	218.464	302.704	196.618	240.310

The steering wheel input used in the simulation was obtained from the measurements as shown in Fig. 8.47b. The vehicle lateral acceleration and yaw acceleration are given in Fig. 8.47c, d.

Validation criteria have been used for validation. Table 8.9 shows that the lateral acceleration validation criteria are found to be within the recommended range of the kurtosis and skewness, while the predicted RMS is found to be outside the recommended range, but still very close to it. In the case of the yaw acceleration, the kurtosis, dkewness, and RMS values are found to be outside the recommended range.

8.4.2.3 J-Turn (22 m radius)

a. 75 ft J turn—25 km/h

This test was performed using the simulation speed as shown in Fig. 8.48a which is simulated to replicate what was measured during the experimental testing. As can be seen, the simulation speed and measured speed are constants with the approximate value of 25 km/h.

The steering wheel input used in the simulation was obtained from the measurements as shown in Fig. 8.48b. The vehicle lateral acceleration and yaw acceleration are given in Fig. 8.48c, d.

Validation criteria have been used for validation. Table 8.10 shows that the lateral acceleration validation criteria are found to be within the recommended range of the kurtosis and RMS while the predicted skewness value is found to be outside the recommended range. In the case of the yaw acceleration, the skewness and RMS are found to be within the recommended range, while the predicted kurtosis value is found to be outside the recommended range but area still very close to it.

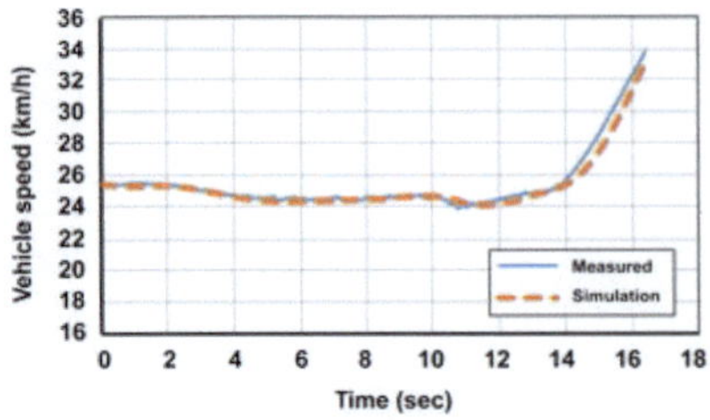

(a) Vehicle speed

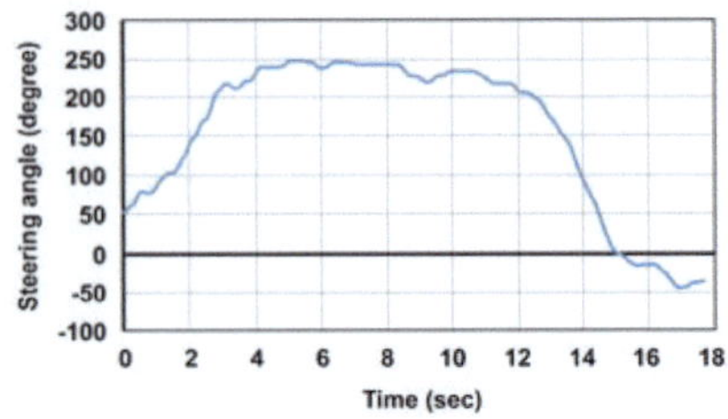

(b) Vehicle steering angle

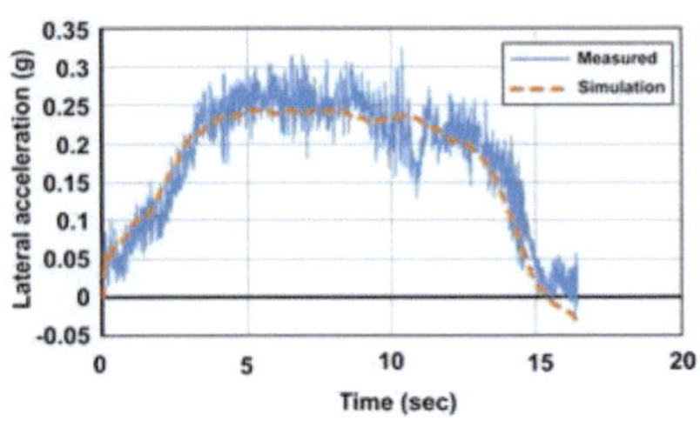

(c) Vehicle lateral acceleration

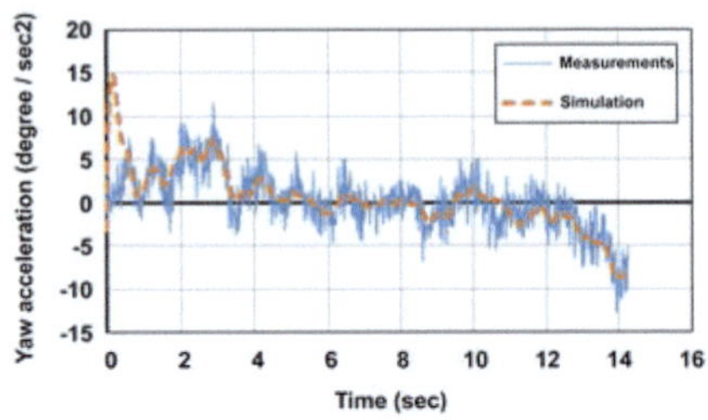

(d) Vehicle yaw acceleration

Fig. 8.48 Vehicle time history at a speed of 25 km/h

Table 8.10 Validation results for constant step slalom at 25 km/h using validation criteria

Lateral acceleration				
	Measured	Simulation	Min.	Max.
Kurtosis	1.705	1.450	0.852	2.557
Skewness	−0.008	−0.401	−0.004	−0.011
RMS	0.037	0.035	0.033	0.041
Yaw acceleration				
	Measured	Simulation	Min.	Max.
Kurtosis	12.930	6.171	6.465	19.394
Skewness	2.824	1.997	1.412	4.236
RMS	12.444	11.926	11.200	13.688

b. 75 ft J turn—45 km/h

This test was performed using the simulation speed as shown in Fig. 8.49a which is simulated to replicate what was measured during the experimental testing. As can be seen, the simulation speed and measured speed are constants with the approximate value of 45 km/h.

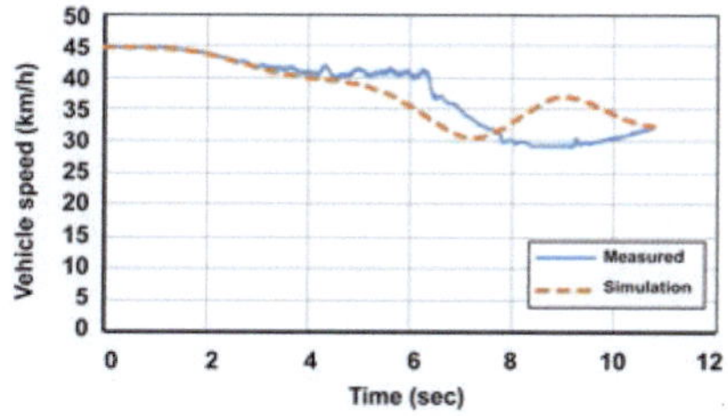

(a) Vehicle speed

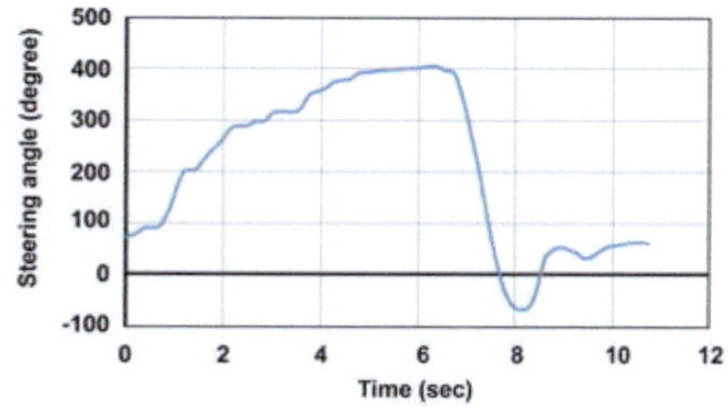

(b) Vehicle steering angle

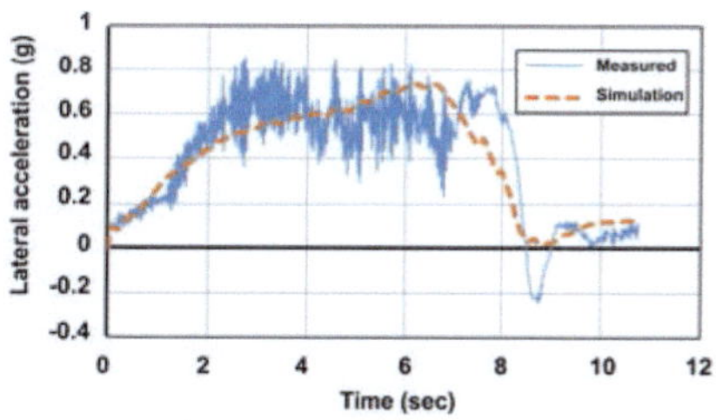

(c) Vehicle lateral acceleration

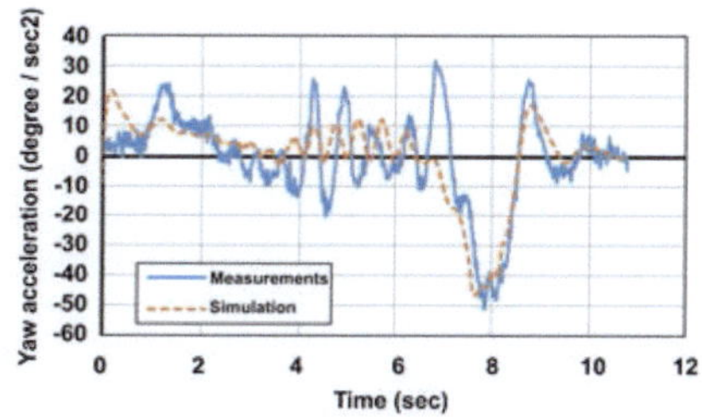

(d) Vehicle yaw acceleration

Fig. 8.49 Vehicle time history at a speed of 45 km/h

Table 8.11 Validation results for constant step slalom at 45 km/h using validation criteria

Lateral acceleration				
	Measured	Simulation	Min.	Max.
Kurtosis	1.924	1.651	0.962	2.886
Skewness	0.286	0.215	0.143	0.429
RMS	0.236	0.224	0.213	0.260
Yaw acceleration				
	Measured	Simulation	Min.	Max.
Kurtosis	11.997	12.238	5.999	17.996
Skewness	2.997	3.157	1.498	4.495
RMS	227.727	195.675	204.954	250.499

The steering wheel input used in the simulation was obtained from the measurements as shown in Fig. 8.49b. The vehicle lateral acceleration and yaw acceleration are given in Fig. 8.49c, d.

Validation criteria have been used for validation. Table 8.11 shows that the lateral acceleration validation criteria are found to be within the recommended range of the kurtosis, skewness, and RMS. In the case of the yaw acceleration, the kurtosis and skewness are found to be within the recommended range, while the predicted RMS value is found to be outside the recommended range but still very close to it. In

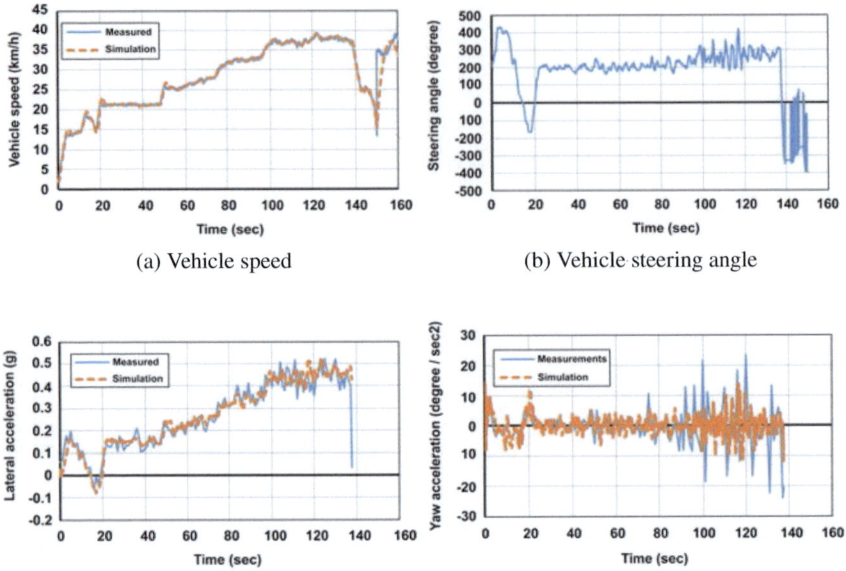

(a) Vehicle speed (b) Vehicle steering angle

(c) Vehicle lateral acceleration (d) Vehicle yaw acceleration

Fig. 8.50 Vehicle time history

addition to the demonstrated results, the simulation results are compared with eight additional different tests. The calculated skewness and kurtosis values are found to be within the recommended range. The model prediction of RMS values of the lateral acceleration and yaw acceleration did not agree with some of the measured ones within ±10% due to the high noise level of the measured lateral acceleration and yaw acceleration data.

8.4.2.4 Turning Circle (8 × 8 & 8 × 4)

a. Turning Circle (8 × 4) Right

This test was performed using the simulation speed as shown in Fig. 8.50a which is simulated to replicate what was measured during the experimental testing (crawling speed). The steering wheel input used in the simulation was obtained from the measurements as shown in Fig. 8.50b. The vehicle lateral acceleration and yaw acceleration are given in Fig. 8.50c, d.

Validation criteria have been used for validation. Table 8.12 shows that the lateral acceleration validation criteria are found to be within the recommended range of the kurtosis, skewness, and RMS. In the case of the yaw acceleration, the kurtosis and skewness are found to be within the recommended range while the predicted RMS value is found to be outside the recommended range.

Table 8.12 Validation results for turning circle (8 × 4) right using validation criteria

Lateral acceleration				
	Measured	Simulation	Min.	Max.
Kurtosis	1.932	1.832	0.966	2.899
Skewness	0.546	0.535	0.273	0.818
RMS	0.091	0.094	0.082	0.101
Yaw acceleration				
	Measured	Simulation	Min.	Max.
Kurtosis	30.746	21.334	15.373	46.119
Skewness	4.890	3.887	2.445	7.335
RMS	23.239	10.847	20.915	25.563

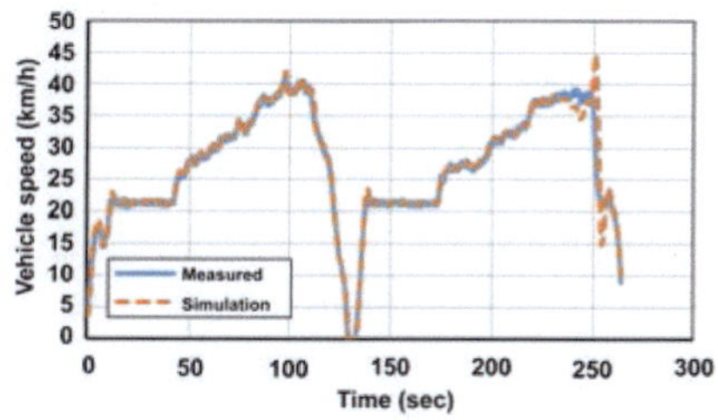

(a) Vehicle speed

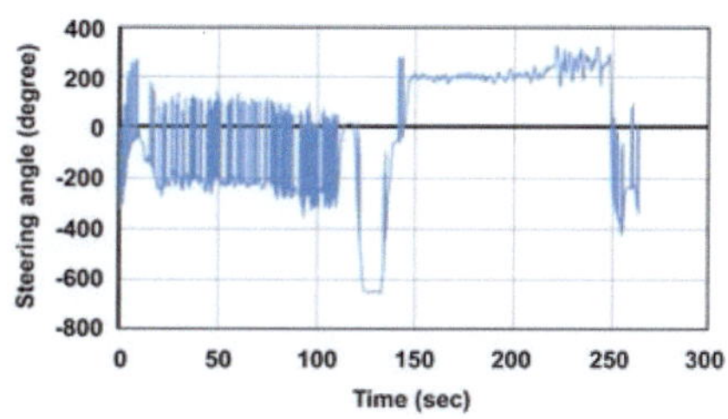

(b) Vehicle steering angle

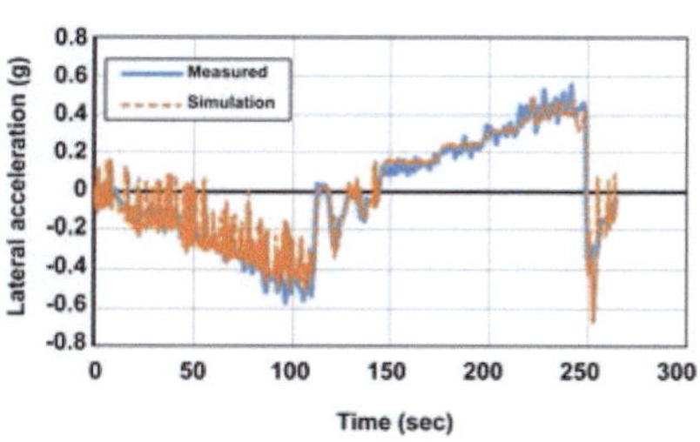

(c) Vehicle lateral acceleration

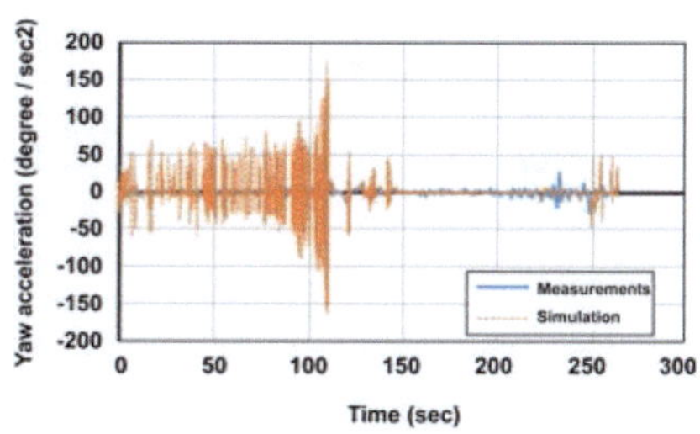

(d) Vehicle yaw acceleration

Fig. 8.51 Vehicle time history

b. Turning Circle (8 × 8) Left and Right

This test was performed using the simulation speed as shown in Fig. 8.51a which is simulated to replicate what was measured during the experimental testing (crawling speed). The steering wheel input used in the simulation was obtained from the measurements as shown in Fig. 8.51b. The vehicle lateral acceleration and yaw acceleration are given in Fig. 8.51c, d.

Table 8.13 Validation results for turning circle (8 × 4) right

Lateral acceleration				
	Measured	Simulation	Min.	Max.
Kurtosis	2.505	2.464	1.253	3.758
Skewness	0.861	0.719	0.431	1.292
RMS	0.082	0.074	0.074	0.090
Yaw acceleration				
	Measured	Simulation	Min.	Max.
Kurtosis	94.777	501.404	47.389	142.166
Skewness	8.442	17.864	4.221	12.663
RMS	15.339	146.568	13.805	16.873

Validation criteria have been used for validation. Table 8.13 shows that the lateral acceleration validation criteria are found to be within the recommended range of the kurtosis, skewness, and RMS. In the case of the yaw acceleration, the skewness, kurtosis, and RMS are found to be outside the recommended range.

In addition to the demonstrated results, the simulation results were compared with two additional different tests. The calculated skewness and kurtosis values were found to be within the recommended range. The model prediction of RMS values of the lateral acceleration and yaw acceleration did not agree with some of the measured ones within ±10% due to the high noise level of the measured lateral acceleration and yaw acceleration data.

8.5 Active Torque Distribution Control System

In passenger vehicles, the rapidly increasing applications of all-wheel drive (AWD) require the development of vehicles not only with higher traction capability but also with better maneuverability. Although improving traction performance is of prime concern for off-road vehicle applications, handling behavior is an important aspect of new vehicles, which requires the capability to undergo high lateral accelerations, while maintaining a proper level of directional stability. The desired increase in mobility must be reached without making any compromises regarding safety or ease of operation or driver comfort. It is expected that the performance of off-road vehicles depends not only on the total tractive effort available by the power plant, but also on its distribution between the driving wheels.

One advancement in the field of road vehicles is the use of active torque distribution control systems to fulfill the function of torque split and transfer among all the driving wheels. The primary objective of this section is to develop an active torque distribution control strategy for a multi-wheeled combat vehicle with (8 × 4) powertrain configuration. The developed vehicle model is used to investigate different control strategies for torque distribution on rigid roads at different operating conditions. An active torque distribution control strategy will be presented in the following

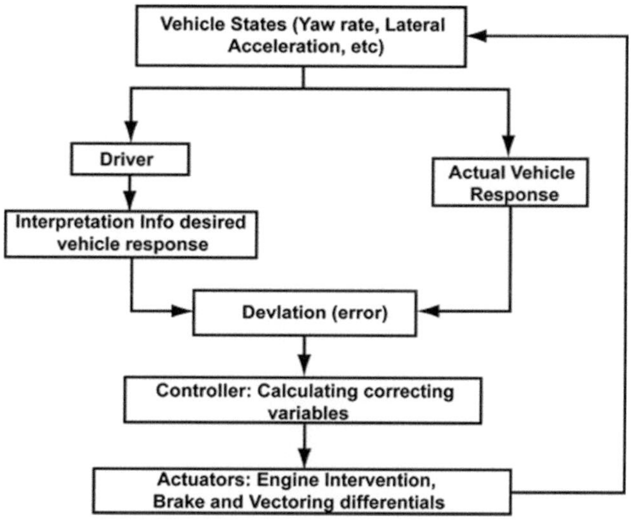

Fig. 8.52 Flow diagram of the vehicle dynamics control system [33]

sections, and comparison between the vehicle directional stability and performance with and without the developed control strategy will be performed and discussed.

8.5.1 Vehicle Dynamics Control

The primary objective of the Vehicle Dynamics Control (VDC) system is to enhance vehicle directional stability based on limiting the deviation of the vehicle's states from its desired states by utilizing different types of actuators such as engine management, braking system, and vectoring differentials as shown in Fig. 8.52.

8.5.2 Actual Vehicle Responses

The actual vehicle responses can be obtained based on real-time measurements using different sensors for wheel speed, yaw rate, steering angle, and lateral acceleration. The developed and validated non-linear vehicle model is utilized to generate the actual vehicle responses required for the proposed control strategy (Fig. 8.53).

8.5.3 Desired Vehicle Responses

A simplified vehicle model could be used to obtain the desired vehicle responses based on the driver responses: steering input, torque, and braking inputs [34]. In this research, the desired responses are obtained from a developed four-axle vehicle bicycle model and the considered vehicle states are the yaw rate and lateral acceleration. The primary goal of the proposed control system is to minimize the driver's required action in difficult driving situations. Accordingly, the driver has been excluded from all analyses of the control systems. The state space representation of the bicycle model is used to generate desired or target responses as given by [35, 36]. In most cases, the desired responses of the state variables are chosen from steady-state values of the bicycle model. For a given road wheel steering angle δ, the desired states are defined as follows: The slip angles:

First axle:

$$\alpha_1 = \delta_1 - \tan^{-1}\left(\frac{v + ar}{u}\right) \tag{8.24}$$

Assume small slip angles: $\tan\alpha = \alpha$, and $\cos\alpha = 1$

$$\alpha_1 = \delta_1 - \left(\frac{v + ar}{u}\right) \tag{8.25}$$

The second axle similarly becomes

$$\alpha_2 = \delta_2 - \left(\frac{v + br}{u}\right) \tag{8.26}$$

For simplification (δ_a, α_a) will be used to represent the first and second axle as follows:

$$\delta_a = \frac{\delta_1 + \delta_2}{2} \tag{8.27}$$

$$\alpha_a = \frac{\alpha_1 + \alpha_2}{2} \tag{8.28}$$

$$\alpha_a = \delta_a - \left(\frac{v + a_a r}{u}\right) \tag{8.29}$$

The third axle:

$$\alpha_3 = -\left(\frac{v - cr}{u}\right) \tag{8.30}$$

and for the fourth axle:

$$\alpha_4 = -\left(\frac{v - dr}{u}\right) \tag{8.31}$$

The cornering forces calculations:

$$F_{y_\alpha} = C_{\alpha_a}\alpha_a \tag{8.32}$$

where $C_{\alpha_a} = \frac{C_{\alpha_1} + C_{\alpha_2}}{2}$

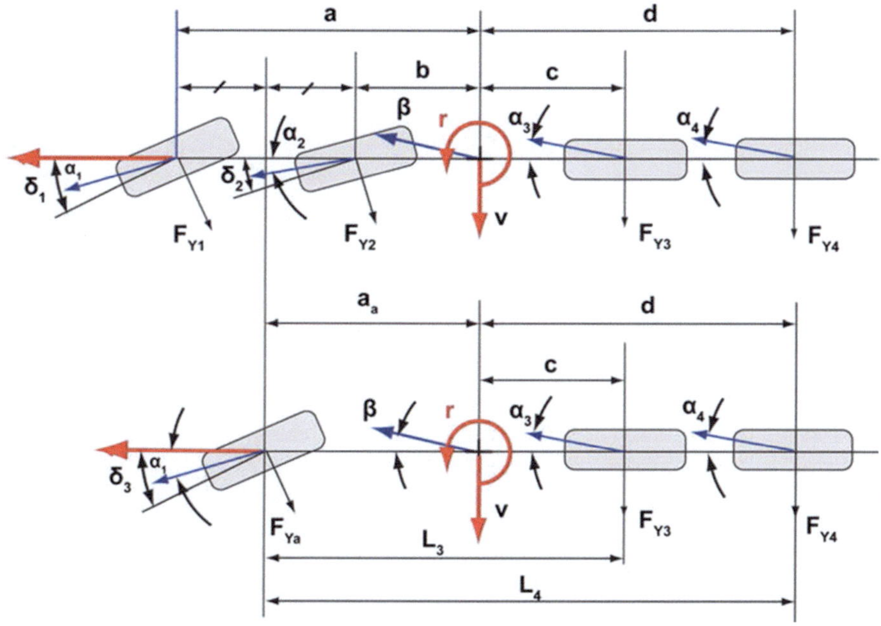

Fig. 8.53 **a** Four-axle vehicle bicycle model and **b** bicycle model with combined front axles

In addition, for third and fourth axle:

$$F_{y3} = C_{\alpha_3}\alpha_3 \tag{8.33}$$

$$F_{y4} = C_{\alpha_4}\alpha_4 \tag{8.34}$$

Equation of motion for the model:

The lateral and yaw equations of motion can be expressed as follows:

$$m\left(\dot{v} + ur\right) = F_{y_\alpha} + F_{y_3} + F_{y_4} \tag{8.35}$$

$$I\dot{r} = a_a F_{y_\alpha} - cF_{y_3} - dF_{y_4} \tag{8.36}$$

Substituting cornering forces in the equation of motion:

$$m\left(\dot{v} + ur\right) = C_{\alpha_a}\delta_a - C_{\alpha_a}\left(\frac{v + a_a r}{u}\right) - C_{\alpha 3}\left(\frac{v - cr}{u}\right) - C_{\alpha 4}\left(\frac{v - dr}{u}\right) \tag{8.37}$$

$$m\left(\dot{v} + ur\right) = -\left(C_{\alpha_a} + C_{\alpha_3} + C_{\alpha 4}\right)\frac{v}{u} - \left(aC_{\alpha_a} - cC_{\alpha_3} - dC_{\alpha 3}\right)\frac{r}{u} + C_{\alpha_a}\delta_a \tag{8.38}$$

$$I\dot{r} = -\left(aC_{\alpha_a} - cC_{\alpha_3} - dC_{\alpha_4}\right)\frac{v}{u} - \left(a_a^2 C_{\alpha_a} + c^2 C_{\alpha_3} + d^2 C_{\alpha_4}\right)\frac{r}{u} \\ = a_a C_{\alpha_a}\delta_a \tag{8.39}$$

Stability criteria include

$$\dot{v} = Pv \tag{8.40}$$

$$\dot{r} = Pr \tag{8.41}$$

$$P = \frac{d}{dt} \tag{8.42}$$

$$\left(mp + \frac{C_{\alpha_a} + C_{\alpha_3} + C_{\alpha_4}}{u}\right)v + \left(mu + \frac{a_a C_{\alpha_a} - cC_{\alpha_3} - dC_{\alpha_4}}{u}\right)r = C_{\alpha_a}\delta_a \tag{8.43}$$

$$\left(mu + \frac{a_a C_{\alpha_a} - cC_{\alpha_3} - dC_{\alpha_4}}{u}\right)v + \left(Ip + \frac{a_a^2 C_{\alpha_a} + c^2 C_{\alpha_3} + d^2 C_{\alpha_4}}{u}\right)r = a_a C_{\alpha_a}\delta_a \tag{8.44}$$

$$\begin{bmatrix} A_1 & B_1 \\ A_2 & B_2 \end{bmatrix}\begin{bmatrix} \frac{v}{\delta_a} \\ \frac{r}{\delta_a} \end{bmatrix} = \begin{bmatrix} C_{\alpha_a} \\ a_a C_{\alpha_a} \end{bmatrix} \tag{8.45}$$

For steady-state response, $P = 0$ and $\dot{v} = \dot{r} = 0$, thus:

$$\begin{bmatrix} A_{s_1} & B_{s_1} \\ A_{s_2} & B_{s_2} \end{bmatrix}\begin{bmatrix} \frac{v}{\delta_a} \\ \frac{r}{\delta_a} \end{bmatrix} = \begin{bmatrix} F_1 \\ T_1 \end{bmatrix} \tag{8.46}$$

$$\left[\frac{v}{\delta_a}\right]_{ss} = \frac{\begin{vmatrix} F_1 & B_{s_1} \\ T_1 & B_{s_2} \end{vmatrix}}{|A|} \tag{8.47}$$

$$\left[\frac{r}{\delta_a}\right]_{ss} = \frac{\begin{vmatrix} B_{s_1} & F_1 \\ B_{s_2} & T_1 \end{vmatrix}}{|A|} \tag{8.48}$$

where

$$A_{s_1} = \frac{C_{\alpha_a} + C_{\alpha_3} + C_{\alpha_4}}{u} \tag{8.49}$$

$$A_{s_2} = \frac{a_a C_{\alpha_a} - cC_{\alpha_3} - dC_{\alpha_4}}{u} \tag{8.50}$$

$$B_{s_1} = mu + \frac{a_a C_{\alpha_a} - cC_{\alpha_3} - dC_{\alpha_4}}{u} \tag{8.51}$$

$$B_{s_2} = \frac{a_a^2 C_{\alpha_a} + c^2 C_{\alpha_3} + d^2 C_{\alpha_4}}{u} \tag{8.52}$$

And the determinate of A can be written as

$$|A| = A_{s_1} B_{s_2} - A_{s_2} B_{s_1} \tag{8.53}$$

$$|A| = \frac{C_{\alpha_a} C_{\alpha_4} L_4^2 + C_{\alpha_a} C_{\alpha_3} (c + a_a)^2 + C_{\alpha_3} C_{\alpha_4} (d - c)^2}{u^2} + \frac{mu^2 \left(cC_{\alpha_3} + dC_{\alpha_4} - a_a C_{\alpha_a}\right)}{u^2} \tag{8.54}$$

where $L_4 = a_a + d$ and $L_3 = a_a + c$

$$\left[\frac{v}{\delta_a}\right]_{ss} = \frac{u \left(L_4 d C_{\alpha_4} C_{\alpha_a} + L_3 c C_{\alpha_3} C_{\alpha_a} - mu^2 a_a C_{\alpha_a}\right)}{C_{\alpha_a} C_{\alpha_4} L_4^2 + C_{\alpha_a} C_{\alpha_3} L_3^2 + C_{\alpha_3} C_{\alpha_4} (d - c)^2 + mu^2 \left(cC_{\alpha_3} + dC_{\alpha_4} - a_a C_{\alpha_a}\right)} \tag{8.55}$$

$$\left[\frac{r}{\delta_a}\right]_{ss} = \frac{u \left(L_4 C_{\alpha_4} C_{\alpha_a} + L_3 C_{\alpha_3} C_{\alpha_a}\right)}{C_{\alpha_a} C_{\alpha_4} L_4^2 + C_{\alpha_a} C_{\alpha_3} L_3^2 + C_{\alpha_3} C_{\alpha_4} (d - c)^2 + mu^2 \left(cC_{\alpha_3} + dC_{\alpha_4} - a_a C_{\alpha_a}\right)} \tag{8.56}$$

The steady-state acceleration and the curvature response will be as follows:

$$\left[\frac{A_y}{\delta_a}\right]_{ss} = \left[\frac{r}{\delta_a}\right]_{ss} u \tag{8.57}$$

$$\left[\frac{1/R}{\delta_a}\right]_{ss} = \frac{\left[\frac{r}{\delta_a}\right]_{ss}}{u} \tag{8.58}$$

The Ackerman steering angle at $u = 0$ is defined as

$$\left[\frac{1/R}{\delta_a}\right]_{ss_{u=0}} = \frac{L_4 C_{\alpha_4} C_{\alpha_a} + L_3 C_{\alpha_3} C_{\alpha_a}}{C_{\alpha_a} C_{\alpha_4} L_4^2 + C_{\alpha_a} C_{\alpha_3} L_3^2 + C_{\alpha_3} C_{\alpha_4} (d - c)^2} \tag{8.59}$$

And the fourth axle steering is

$$\delta_4 = \frac{C_{\alpha_a} C_{\alpha_4} L_4^2 + C_{\alpha_a} C_{\alpha_3} L_3^2 + C_{\alpha_3} C_{\alpha_4} (d - c)^2}{R \left(L_4 C_{\alpha_4} C_{\alpha_a} + L_3 C_{\alpha_3} C_{\alpha_a}\right)} \tag{8.60}$$

Let:

$$L_a = \frac{C_{\alpha_a} C_{\alpha_4} L_4^2 + C_{\alpha_a} C_{\alpha_3} L_3^2 + C_{\alpha_3} C_{\alpha_4} (d - c)^2}{L_4 C_{\alpha_4} C_{\alpha_a} + L_3 C_{\alpha_3} C_{\alpha_a}} \tag{8.61}$$

Thus $\delta_a = \frac{L_a}{R}$ and

$$K_{us} = -\left(\frac{L_a/R - \delta_a}{A_y/g}\right) \tag{8.62}$$

$$\delta_a = \frac{L_a}{R} + K_{us}\frac{A_y}{g} \tag{8.63}$$

The desired yaw rate r_d can be written as

$$r_d = \frac{u\delta_a\left(L_4 C_{\alpha_4} C_{\alpha_a} + L_3 C_{\alpha_3} C_{\alpha_a}\right)}{C_{\alpha_a} C_{\alpha_4} L_4^2 + C_{\alpha_a} C_{\alpha_3} L_3^2 + C_{\alpha_3} C_{\alpha_4}(d-c)^2 + mu^2\left(cC_{\alpha_3} + dC_{\alpha_4} - a_a C_{\alpha_a}\right)} \tag{8.64}$$

The desired lateral acceleration, A_{y_d}, can be written as

$$A_{y_d} = \frac{u^2\delta_a\left(L_4 C_{\alpha_4} C_{\alpha_a} + L_3 C_{\alpha_3} C_{\alpha_a}\right)}{C_{\alpha_a} C_{\alpha_4} L_4^2 + C_{\alpha_a} C_{\alpha_3} L_3^2 + C_{\alpha_3} C_{\alpha_4}(d-c)^2 + mu^2\left(cC_{\alpha_3} + dC_{\alpha_4} - a_a C_{\alpha_a}\right)} \tag{8.65}$$

The desired yaw rate and lateral acceleration for the combat vehicle used in the study are evaluated using the following vehicle dimensions:

$$a = d = 1930 \text{ mm} \tag{8.66}$$

$$b = c = 710 \text{ mm} \tag{8.67}$$

$$a_a = 1320 \text{ mm} \tag{8.68}$$

$$L_3 = 2030 \text{ mm} \tag{8.69}$$

$$L_4 = 3250 \text{ mm} \tag{8.70}$$

For a rigid road, the cornering stiffness of all axles is given as $C_{\alpha_a} = C_{\alpha_3} = C_{\alpha_4} =$ 7.68 kN/degree. The desired yaw rate and acceleration on hard roads can then be described as

$$r_d = \frac{\delta_a u}{3.063 + K_{us}u^2} \text{ rad/s} \tag{8.71}$$

$$A_{y_d} = \frac{\delta_a u^2}{3.063 + K_{us}u^2} \text{ m/s}^2 \tag{8.72}$$

For a soft soil such as clayey soil the stiffness of the axles is defined as

$$C_{\alpha_1} = 2.902 \text{ kN/degree} \tag{8.73}$$

$$C_{\alpha_2} = C_{\alpha_3} = C_{\alpha_4} = 3.116 \text{ kN/degree} \tag{8.74}$$

$$C_{\alpha_a} = 3.01 \text{ kN/degree} \tag{8.75}$$

Thus, the desired yaw rate and acceleration on hard roads can then be described as

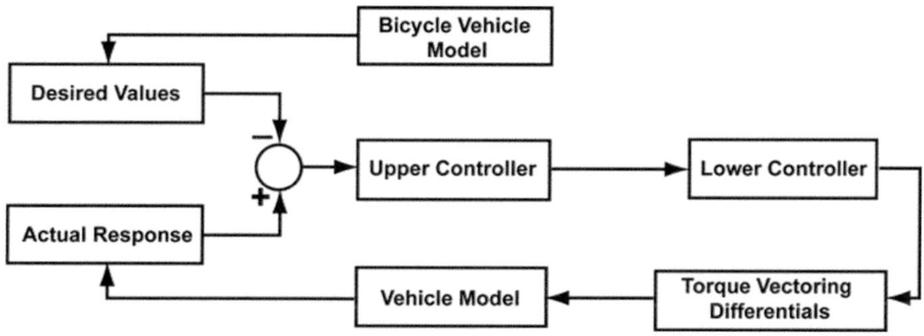

Fig. 8.54 Schematic of control architecture

$$r_d = \frac{\delta_a u}{3.075 + K_{us}u^2} \text{ rad/s} \tag{8.76}$$

$$A_{y_d} = \frac{\delta_a u^2}{3.075 + K_{us}u^2} \text{ m/s}^2 \tag{8.77}$$

where δ_a is in rad and u is in m/s. The respective errors in some desired variables are defined as follows. The lateral acceleration error is

$$e_{a_y} = A_y - A_{y_d} \tag{8.78}$$

and the yaw rate error:

$$e_r = r - r_d \tag{8.79}$$

A_y and r are the actual values of the corresponding vehicle states (lateral acceleration and yaw rate respectively) obtained from actual vehicle model. The lateral acceleration error e_{a_y} and yaw rate error e_r are the feedback variables used in the controller design as will be detailed in the following sections.

8.5.4 Architecture of the Proposed Control

This subsection describes the control structure adopted as shown in Fig. 8.54.

8.5.4.1 Development of the Upper Controller

The upper controller utilizes the developed four-axle bicycle model, and the actual vehicle response: yaw rate, lateral acceleration, longitudinal speed to prepare the desired vehicle responses as a first step in the upper controller. Then, three PID controllers are used to develop the needed corrective yaw moment based on the

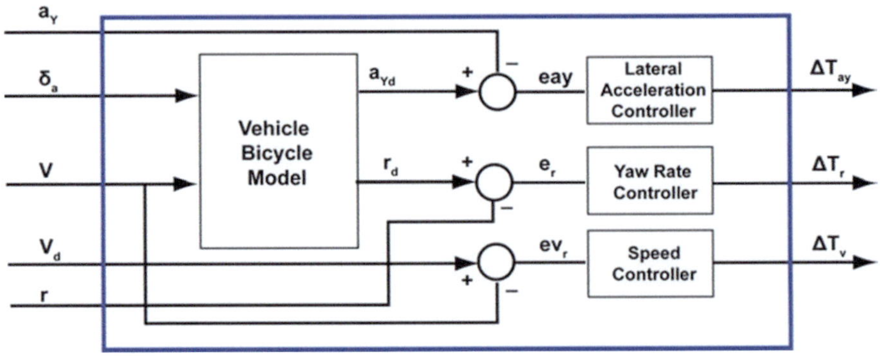

Fig. 8.55 Block diagram of the upper controller

differences between the actual and desired vehicle responses to enhance vehicle directional stability. The corrective yaw moment is then passed to the management system (the lower controller) as shown in Fig. 8.55.

8.5.4.2 Development of the Lower Controller

Generally, the lower controller objective is to produce the needed action to generate the required corrective yaw moment by the upper controller by means of either braking, driving or steering effort. In the proposed control system strategy, the lower controller is the torque distribution management system (torque vectoring differentials) that manages the torque distribution between all wheels independently to achieve the desired yaw moment. In addition, the physics description of the yaw moment control through torque distribution as achieved by vectoring differentials is described as follows.

a. Inter-axle torque distribution

More torque transfer to the front axle wheels will increase longitudinal slip of the front axle wheels while the rear axle wheels will drop and decrease the lateral forces generated by the front axle wheels compared to the rear ones. Accordingly, torque transfer from the rear to the front wheels induces an understeering effect.

b. Left to Right torque distribution

Reducing the driving torque delivered to the outer wheel in comparison to the inner one generates a yaw moment in the opposite direction of the turn that will induce an understeering effect on the vehicle. The differences in longitudinal forces produce a significant yaw moment while the differences in lateral forces, being partially compensating, lead to the generation of small positive yaw moments. Thus, a net positive yaw moment in the opposite direction of motion is generated, leading to understeer.

Active torque distribution systems utilize the physics described above for yaw moment control by varying the torques on individual wheels. In this research, yaw moment control is based on a left to right torque distribution strategy and various torque distribution approaches are considered and analyzed as follows.

c. Torque ratios variations approach

Osborn et al. [37] introduced a torque distribution strategy based on two torque ratios; the front-rear ratio and left-right ratio. The front-rear ratio, r_{fr}, is determined based on the calculated yaw rate error, while the left-right ratio, r_{lr}, is determined based on the calculated lateral acceleration error. The front-rear torque ratio can be defined as the ratio of the front left wheel torque to the sum of the front left and rear left wheel torques. In addition, the left-right torque ratio can be defined as the ratio of the front left wheel torque to the sum of the front left and front right wheel torques. These ratios are expressed mathematically as shown in the following equations:

$$r_{fr} = \frac{T_{fl}}{T_{fl} + T_{rl}} = \frac{T_{fr}}{T_{fr} + T_{rr}} \tag{8.80}$$

$$r_{lr} = \frac{T_{fl}}{T_{fl} + T_{fl}} = \frac{T_{rl}}{T_{rl} + T_{rr}} \tag{8.81}$$

Given a total driveline torque T, using the above definitions of torque distribution ratios, the four individual torques on the wheels can be evaluated from the following equations:

$$T_{fl} = T r_{fr} r_{lr} \tag{8.82}$$

$$T_{fr} = T r_{fr}(1 - r_{lr}) \tag{8.83}$$

$$T_{rl} = T(1 - r_{fr}) r_{lr} \tag{8.84}$$

$$T_{rr} = T(1 - r_{fr})(1 - r_{lr}) \tag{8.85}$$

The presented simulation response based on using the “torque-ratio” approach in [38] shows promise in achieving an adequate stability control system. The torque distribution ratios are constrained by the two ratios and the total torque on the vehicle always remains constant. Consequently, this approach reduced the control variables from four (each of four individual wheels) to two (two torque ratios) which reduces the torque distribution independence by limiting the total torque.

d. Differential torque distribution approach

This approach utilizes differential torque distribution by either addition or subtraction of corrective torque already produced by the upper controller. In addition, this approach does not limit the total torque as in the torque ratio variations approach

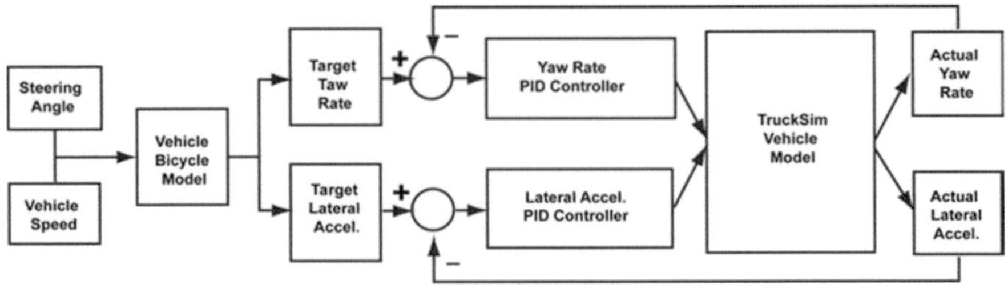

Fig. 8.56 Schematic of the proposed controllers interfaced with vehicle model

which allows independent torque control of each wheel. In this research, this approach is implemented in simulations based on the selected control variables, yaw rate, and lateral acceleration.

The torque distribution strategies are analyzed and implemented with and without controlling vehicle speed. Therefore, different standard maneuvers are performed at constant or nearly constant speed. Consequently, speed control is introduced as a PID speed controller. The speed error, e_v, is defined as the difference between the actual forward velocity, v_x, and the desired (test) forward velocity of the vehicle, v_{xd}.

$$e_v = v_x - v_{xd} \tag{8.86}$$

In all the performed simulations at constant speed, the total torque ΔT_v is considered to be equally distributed between all wheels. Therefore, the speed control torque is added to the corrective torques of each wheel. On the other hand, in the case of no speed control, constant torques "base torques" are delivered to each wheel and added to the corrective torques of each wheel. The total base torques on the left and right sides of the vehicle are given as follows:

$$T_L = T_{fl} + T_{rl} \tag{8.87}$$

$$T_R = T_{fr} + T_{rr} \tag{8.88}$$

where T_{fl}, T_{rl}, T_{fr}, and T_{rr} are the individual base torques acting on the individual wheels. The proposed control strategy used in this research was interfaced with the developed vehicle model in TruckSim as shown in Fig. 8.56.

1-Yaw rate control:

A proper controller can be developed to generate the necessary corrective yaw moment based on the yaw rate differences between the actual and desired values. The necessary corrective torque, ΔT_r, that will be added or subtracted to the base torques (in case of no speed control) or speed control torques of the individual wheels for generating the desired yaw moment is evaluated using a PID controller. In this research, half the corrective torques are added to the left wheels and half of them are subtracted from the right wheels for both the driving axles.

$$T_{l3_{new}} = T_{l3} + \frac{\Delta T_r}{2} \tag{8.89}$$

$$T_{l4_{new}} = T_{l4} + \frac{\Delta T_r}{2} \tag{8.90}$$

$$T_{r3_{new}} = T_{r3} - \frac{\Delta T_r}{2} \tag{8.91}$$

$$T_{r4_{new}} = T_{r4} - \frac{\Delta T_r}{2} \tag{8.92}$$

2-Lateral acceleration control:

For the lateral acceleration as a feedback variable, the required differential torque, ΔT_{a_y} can be evaluated from the PID controller based on the lateral acceleration error in a similar way as was done for yaw rate control.

$$T_{l3_{new}} = T_{l3} + \frac{\Delta T_{a_y}}{2} \tag{8.93}$$

$$T_{l4_{new}} = T_{l4} + \frac{\Delta T_{a_y}}{2} \tag{8.94}$$

$$T_{r3_{new}} = T_{r3} - \frac{\Delta T_{a_y}}{2} \tag{8.95}$$

$$T_{r4_{new}} = T_{r4} - \frac{\Delta T_{a_y}}{2} \tag{8.96}$$

3-Combined lateral acceleration and yaw rate control:

This approach combines the corrective torques being added to left wheels and subtracted from right wheels based on yaw rate and lateral acceleration errors. The final wheel driving torques on the individual wheels are calculated by the following equations:

$$T_{l3_{new}} = T_{l3} + \frac{\Delta T_r}{2} + \frac{\Delta T_{a_y}}{2} + \frac{\Delta T_v}{2} \tag{8.97}$$

$$T_{l4_{new}} = T_{l4} + \frac{\Delta T_r}{2} + \frac{\Delta T_{a_y}}{2} + \frac{\Delta T_v}{2} \tag{8.98}$$

$$T_{r3_{new}} = T_{r3} - \frac{\Delta T_r}{2} - \frac{\Delta T_{a_y}}{2} + \frac{\Delta T_v}{2} \tag{8.99}$$

$$T_{r4_{new}} = T_{r4} - \frac{\Delta T_r}{2} - \frac{\Delta T_{a_y}}{2} + \frac{\Delta T_v}{2} \tag{8.100}$$

where ΔT is the corrective differential torque to be transferred according to the error function for yaw rate, lateral acceleration, and longitudinal vehicle speed as follows:

$$\Delta T_r = K_{p_r} e_r + K_{i_r} \int e_r dt + K_{d_r} \frac{d}{dt} e_r \tag{8.101}$$

$$\Delta T_{a_y} = K_{p_{a_y}} e_{a_y} + K_{i_{a_y}} \int e_{a_y} dt + K_{d_{a_y}} \frac{d}{dt} e_{a_y} \tag{8.102}$$

$$\Delta T_v = K_{p_v} e_v + K_{i_v} \int e_v dt + K_{d_v} \frac{d}{dt} e_v \tag{8.103}$$

8.5.4.3 MATLAB/Simulink—TruckSim Co-Simulator

A co-simulator that consists of the TruckSim combat vehicle model and MATLAB/Simulink controller was developed to verify the proposed control strategy as shown in Fig. 8.57. The vehicle model in Matlab/Simulink represents the vehicle as specified in the TruckSim software and fits with the signal requirements of the Simulink control model.

Problems

1. A moist soil sample weighs 346 g. After drying at 105 °C its weight is 284 g. The specific gravity of the mass and of the solids is 1.86 and 2.70, respectively. Determine the following assuming that air is weightless:
   - The water content.
   - The void ratio.
   - The degree of saturation.
   - The porosity.
2. A test of the density of the soil in place was performed by digging a small hole in the soil, weighing the extracted soil, and measuring the volume of the hole. The moist soil weighed 895 g, and the volume of the hole was 426 cm^3. After drying, the sample weighed 779 g. Of the dried soil, 400 g was poured into a vessel in a very loose state. Its volume was subsequently determined to be 276 cm^3. That same 400 g was then compacted (vibrated and tamped) to a volume of 212 cm^3. Given that the specific gravity, $G = 2.71$ and the unit weight of water, $\gamma_w = 1$ g/cm^3, calculate the relative density D_r.
3. The water content of a soil sampled in the field can be calculated from the difference between the weight of a wet and an oven-dried soil sample. Suppose then that the initial weight of a moist soil sample plus container was measured to be 58.2 g. Also, suppose that the total weight of the sample plus container—after oven drying for a period of 24 h—was 50.6 g. If the self-weight of the container was 35.3 g, calculate the water content of the soil sample.
4. For a given terrain, explain with the aid of drawing the following characteristics:
   - Terrain mechanical properties: Shear strength-Bearing capacity.
   - Geometric characteristics.

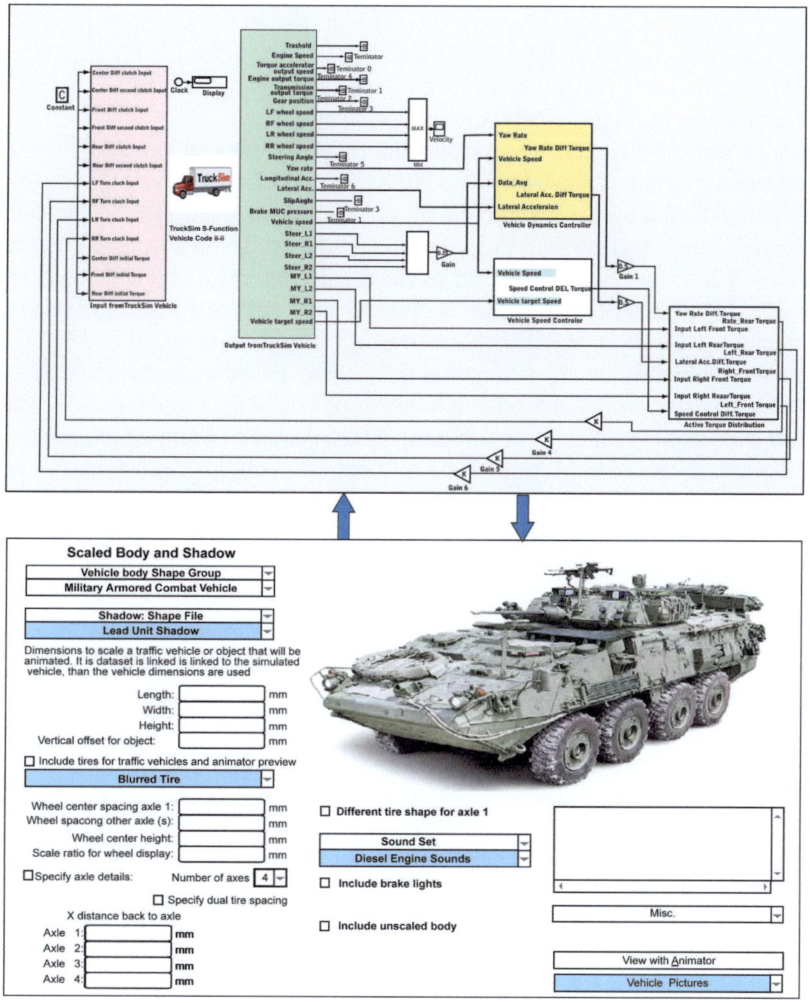

Fig. 8.57 MATLAB/Simulink—TruckSim co-simulator

5. What is meant by permanent and transient parameters of soils? Explain how to conduct the sieve analysis for determining the soil grain size distribution. Then draw typical soil particles gradation curves for well-graded soil and uniformly graded soil.
6. Describe with the aid of sketches the Atterberg device.
7. For a vehicle with rear drive, consider the weight on the rear axle is 20000 N, the adhesion coefficient on the right wheel is 0.8, and the adhesion on the left wheel varies from 0 to 0.8. For free differential calculate the traction force and draw the variation of traction force with adhesion coefficient.

8. For a vehicle with rear drive, consider the weight on the rear axle is 35000 N, the adhesion coefficient on the right wheel is 0.7, and adhesion on the left wheel varies from 0 to 0.7. For locked differential calculate the traction force and draw the variation of traction force with adhesion coefficient.
9. For a vehicle with rear drive, consider the weight on the rear axle is 30000 N, the adhesion coefficient on right wheel is 0.85, and adhesion on the left wheel varies from 0 to 0.85. For limited slip differential with bias ratio, B.R = 3, calculate the traction force and draw the variation of traction force with adhesion coefficient.
10. A 4 × 4 lorry has the following data: weight on front axle = 20 and 40 kN on the rear axle, coefficient of overloading on rear axle mb2 = 1.2, engine maximum moment = 300 N.m at 1700 r.p.m. and engine maximum speed = 3800 r.p.m., main gear box ratios are 4.2, 3.32, 1.9, 1, 0.81. Transfer case ratios are 2.4, 1. The final drive ratio is 4.5 (single reduction). Wheel dynamic radius = 0.50 m, and maximum coefficient of adhesion is 0.8. Based on tooth bending strength; calculate: If the axle has limited slip differentials with bias ratio 4; calculate and draw the traction force with adhesion on the rear axles when crossing a road with split adhesion condition. The coefficient of adhesion on the left tire varies from 0.1 to 0.8 and that on the right one is 0.8.

References

1. Crolla DA (1983) The steering behaviour of articulated body steer vehicles. In: Road Vehicle Handling, I Mech E Conference Publications 1983-5. Sponsored by Automobile Division of the Institution of Mechanical Engineers under patronage of Federation Internationale des Societies d'Ingenieurs des Techniques de l'Automobile (FISITA) he, number C123/83
2. Crolla DA, El-Razaz ASA, Alstead CJ, Hockley C (1987) A model to predict the combined lateral and longitudinal forces on an off-road tyre. In: Proceedings of the ninth international conference, pp 362–372
3. Ragheb H, El-Gindy M, Kishawy HA (2013) On the multi-wheeled off-road vehicle performance and control. Int J Veh Syst Model Testing 8(3):260–281
4. Danesin D, Girardin C, Sorniotti A, Morgando A, Velardocchia M (2004) Driveline layout influence on four wheel drive dynamics. SAE Trans 534–541
5. Dick WM (1995) All-wheel and four-wheel-drive vehicle systems. Technical report, SAE Technical Paper
6. Lessem A, Mason G, Ahlvin R (1996) Stochastic vehicle mobility forecasts using the nato reference mobility model. J Terramech 33(6):273–280
7. GbR AESCO (2005) Matlab/simulink module as2tm user's guide version 1.12
8. Dębowski A, Żardecki D (2011) Modelling of centre differential control. J KONES 18:135–142
9. Aubel Th (1994) The interaction between the rolling tyre and the soft soil–fem simulation by venus and validation. In: Proceedings of 6th European ISTVS conference, Vienna, Austria, pp 169–188
10. Bekker MG (1956) Theory of land locomotion university of Michigan press. Ann Arbor, p 522
11. Bekker MG (1960) Off-the-road locomotion university of Michigan press. Ann Arbor, pp 27–29
12. Bekker MG (1969) Introduction to terrain-vehicle systems. part i: The terrain. part ii: The vehicle. Technical report, Michigan University Ann Arbor

13. Abd El-Gawwad KA, Crolla DA, Soliman AMA, El-Sayed FM (1999) Off-road tyre modelling i: the multi-spoke tyre model modified to include the effect of straight lugs. J Terramech 36(1):3–24
14. Abd El-Gawwad KA, Crolla DA, Soliman AMA, El-Sayed FM (1999) Off-road tyre modelling ii: effect of camber on tyre performance. J Terramech 36(1):25–38
15. Abd El-Gawwad KA, Crolla DA, Soliman AMA, El-Sayed FM (1999) Off-road tyre modelling iii: effect of angled lugs on tyre performance. J Terramech 36(2):63–75
16. Abd El-Gawwad KA, Crolla DA, Soliman AMA, El-Sayed FM (1999) Off-road tyre modelling iv: extended treatment of tyre-terrain interaction for the multi-spoke model. J Terramech 36(2):77–90
17. Anbar ES (1993) Technical Evaluation of off-road Vehicle Mobility. PhD thesis, MSC thesis, Alazhar University, Engineering College, Cairo
18. Kaiser G, Holzmann F, Chretien B, Korte M, Werner H (2011) Torque vectoring with a feedback and feed forward controller-applied to a through the road hybrid electric vehicle. In: 2011 IEEE intelligent vehicles symposium (IV), pp 448–453. IEEE
19. Yanjin G, Guoqun Z, Gang C (2011) 3-dimensional non-linear fem modeling and analysis of steady-rolling of radial tires. J Reinf Plast Compos 30(3):229–240
20. Harnisch C, Lach B (2002) Off road drive of wheeled vehicles in dynamic realtime simulation. In: NATO SCI panel symposium. Berlin
21. Harnisch C, Lach B (2002) Off road vehicles in a dynamic three-dimensional realtime simulation. In: Proceedings of the 14th international conference of the international society for terrain-vehicle systems, pp 20–24
22. Harnisch C, Lach B, Jakobs R (2007) Orsis-news and further developments. J Terramech 44(1):35–42
23. Harnisch C, Lach B, Jakobs R, Troulis M, Nehls O (2005) A new tyre-soil interaction model for vehicle simulation on deformable ground. Veh Syst Dyn 43(sup1):384–394
24. Heisler H (2002) Advanced vehicle technology. Elsevier
25. Guo J, Chu L, Zhou F, Cao L (2011) Integrated control of variable torque distribution and electronic stability program based on slip angle phase. In: Proceedings of 2011 international conference on electronic & mechanical engineering and information technology, vol 7. IEEE, pp 3777–3780
26. Xia K (2011) Finite element modeling of tire/terrain interaction: application to predicting soil compaction and tire mobility. J Terramech 48(2):113–123
27. Mohan SK, Ramarao BV (2003) A comprehensive study of self-induced torque amplification in rotary viscous couplings. J Trib 125(1):110–120
28. Mohan SK (2004) Comprehensive theory of viscous coupling operation. SAE Trans 594–609
29. Park J, Kroppe WJ (2004) Dana torque vectoring differential dynamic trak™. SAE Trans 1057–1062
30. Perumpral JV, Liljedahl JB, Perloff WH (1971) A numerical method for predicting the stress distribution and soil deformation under a tractor wheel. J Terramech 8(1):9–22
31. Liu Q, Kaiser G, Boonto S, Werner H, Holzmann F, Chretien B, Korte M (2011) Two-degree-of-freedom lpv control for a through-the-road hybrid electric vehicle via torque vectoring. In: 2011 50th IEEE conference on decision and control and European control conference, pp 1274–1279. IEEE
32. Wong JY, Huang W (2006) An investigation into the effects of initial track tension on soft ground mobility of tracked vehicles using an advanced computer simulation model. Proc Inst Mech Eng Part D: J Automobile Eng 220(6):695–711
33. Karogal IS (2008) Independent torque distribution management systems for vehicle stability control. Clemson University
34. Ghoneim YA, Lin WC, Sidlosky DM, Chen HH, Chin Y-K (2000) Integrated chassis control system to enhance vehicle stability. Int J Veh Des 23(1-2):124–144
35. Wong JY (2001) Theory of ground vehicles. Wiley, New York

36. Genta G (1997) Motor vehicle dynamics: modeling and simulation, vol 43. World Scientific
37. Osborn RP, Shim T (2006) Independent control of all-wheel-drive torque distribution. Veh Syst Dyn 44(7):529–546
38. Wu Y-C (2000) Handling of multiaxle, all-wheel-drive off-road vehicles. PhD thesis, Carleton University

Chapter 9
Suspension Characteristics

Street-driven cars and trucks use a suspension system to keep the tires on the road and to provide acceptable riding comfort. A vehicle with a solid suspension or no suspension would bounce off the ground when the tires hit a bump. The purpose of the suspension system is to provide the vehicle with

1. A smooth ride;
2. accurate steering;
3. responsive handling;
4. support for the weight of the vehicle;
5. maintenance of acceptable tire wear.

The angular positions and the vertical force acting on a tire and the vehicle body depend on the suspension system, which locates the wheel relative to the vehicle body. Therefore, suspension system design plays an important role in the cornering and handling characteristics. It should be noted that

- The suspension system locates the wheels relative to the vehicle body.
- Angular position and vertical loads depend upon the suspension system.
- In a practical suspension system, the wheels are connected to the body through various links; these permit an approximately vertical motion of the wheel relative to the body.

9.1 Frame Construction and Platform

The frame construction usually consists of channel-shaped steel beams welded and/or fastened together. A full frame is defined as a frame that supports all the running gear of the vehicle and is completed in such a way that most vehicles can be driven without

M. El-Gindy and Z. El-Sayegh, *Road and Off-Road Vehicle Dynamics*,
https://doi.org/10.1007/978-3-031-36216-3_9

the body. The full frame is most common among tracks and larger rear-wheel-drive cars. Other types of frames include

1. Ladder frame: is a perimeter frame where the laterals connecting the members are straight across.
2. Perimeter frame: consists of welded frame members around the entire perimeter of the body.
3. Sub-type frame: is a partial frame often used on unit-body vehicles to support the power train and suspension.
4. Unit-body frame: is a design that combines the body with the structure of the frame.
5. Space frame: consists of formed sheet steel used to construct a frame for the entire vehicle.

The platform of any vehicle is the basic size and shape. A platform of a unit-body vehicle includes all major sheet components that form the load-bearing structure of the vehicle. Examples of common platforms include the following:

- Chevrolet Impala and Pontiac Grand Prix.
- Toyota Camry and Lexus ES 350.
- Buick Lucerne and Cadillac DTS.

9.2 Straight Motion Mechanics

The mobility analysis uses degrees-of-freedom (mobility) analysis, which is helpful in categorizing the wide range of real suspension designs (Figs. 9.1, 9.2 and 9.3).

It is very important to provide a straight motion path for some point of a suspension member. In real suspensions, the straight lines need precision design, which is a trade-off against the cost, and other factors. Some of the straight-line mechanisms can be listed as follows:

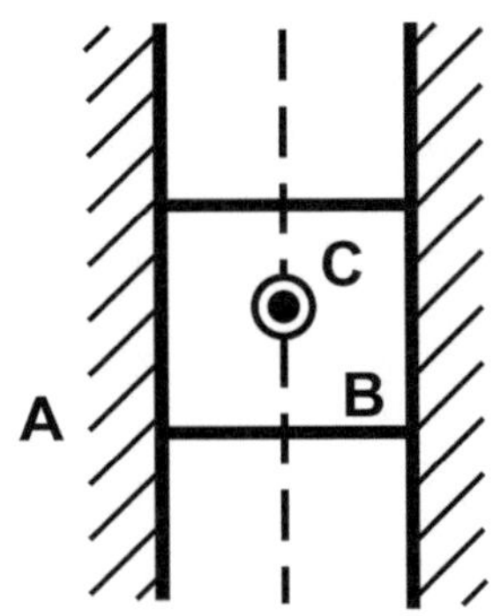

Fig. 9.1 Diagram of a slider

Fig. 9.2 Diagram of a Panhard rod

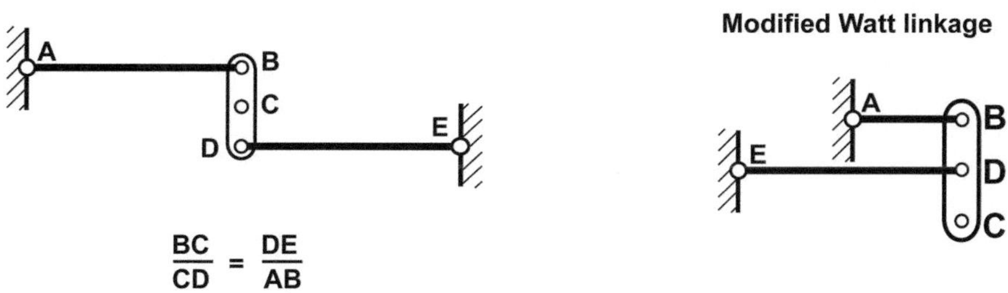

Fig. 9.3 Diagram of a Watt's linkage

1. Slider: It is a direct solution and a simple slider has a perfect straight-line motion and only one degree of freedom.
2. Panhard Rod (Radius Rod): Approximation of straight-line motion, widely used for axle locations and small displacements.
3. Watt's linkage: This is perceived as a logical development of the radius rod, introducing compensating errors. There is also what is called Modified Watt's linkage.
4. Others will be discussed, such as Roberts, Tcebichef, Evans, ... etc.

9.3 Types of Suspensions

The two main common types of suspensions are the independent and the dependent suspension systems.

9.3.1 Independent Systems

In the full three dimensions, an unconstrained object has six degrees of freedom: three translations and three of rotation. Relative to a chassis, a wheel has two degrees of freedom for a given steering position. One is the rotation and the other is the vertical motion. The wheel carrier has only one degree of freedom when the steering is fixed. Since an entirely free object has six degrees of freedom, it is necessary to provide constraints to remove five degrees of freedom from the carrier, including the constraint of the steering.

9.3.2 *Dependent Systems*

Axles are rigid and can be divided into three kinds:

1. The dead axle which is used at the rear with front drive, having non-driven wheels.
2. The live axle which has driven wheels and carries the differential.
3. The de Dion axle which has driven wheels but does not carry differential.

9.3.3 *Axle Design Requirements*

1. Rigidity to avoid displacements or large amplitude axle vibration when tractive forces are applied.
2. Must have two degrees of freedom (heave and roll), i.e., four degrees of freedom constraint. This requirement can be achieved by using 4 radius rods.

9.4 Suspension Principles

The suspension system uses various links, arms, and joints to allow the tires to move freely vertically. The front suspensions also have to allow the front tires to turn as well. All suspensions must provide the following supports:

1. Transverse tires support: the suspension system must accommodate the vertical movements of the tires and still keep the tires from moving away from the vehicle.
2. Longitudinal tires support: the suspension system must allow the tires to move vertically and still keep the tires from moving backwards whenever a bump is hit.

The design of the suspension and the location of the suspension mounting points on the frame or body are critical to proposer vehicle handing. Two important factors are called the anti-squat and anti-dive.

1. Anti-squat: refers to the reaction of the body of a vehicle during acceleration.
2. Anti-dive refers to the force that causes the front of the vehicle to drop down while braking.

9.5 Front Axle Suspension Systems Design

The following are examples of some common front axle designs and features:

Fig. 9.4 The A-arm front suspension. (Photo courtesy of Ford motor company)

9.5.1 Double Wishbone (A-arm) Axle

The advantage of this design is that it is less expensive, while the disadvantage is that the vehicle front structure becomes more complicated if the axle is driven. Design a suitable ratio of the upper and lower arms to minimize change of track.

The most common design for the front suspension of American cars following World War II used two lateral control arms to hold the wheel as shown in Fig. 9.4. The upper and lower control arms are usually of unequal length from which the acronym SLA (short-long arm) gets its name. The arms are often called "A-arms" in the United States and "wishbones" in Britain. This layout sometimes appears with the upper A-arm replaced by a simple lateral link, or the lower arm replaced by a lateral link and an angled tension strut, but the suspensions are functionally similar.

9.5.2 Spring (MacPherson) Strut Axle

Earle S. MacPherson developed a suspension with geometry similar to the unequal-arm front suspensions using a strut configuration as shown in Fig. 9.5. The strut is a telescopic member incorporating damping with the wheel rigidly attached at its lower end, such that the strut maintains the wheel in the camber direction.

The upper end is fixed to the body shell or chassis, and the lower end is located by linkages which pick up the lateral and longitudinal forces. Because of the need to offset the strut inboard of the wheel, the wheel loads the strut with an overturning moment which adds to friction in the strut. This is often counteracted by mounting the coil spring at an angle on the strut.

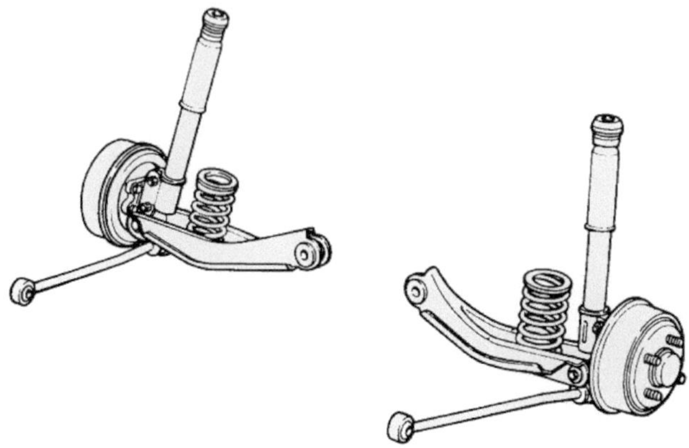

Fig. 9.5 The MacPherson strut suspension. (Photo courtesy of Ford motor company)

It should be noted that this suspension might also be used, the advantages of this system include

- Longer spring, thus soft, large stroke suspension.
- Large distance between mountings ensure low mounting load thus permitting use of soft supports.
- Allows favorable front structure design.

While the disadvantages of such a design include

- Friction forces in the shock absorber guides increase with tire width.
- High installed height.
- Anti-dive action is usually limited.

9.6 Rear Axle Suspension Systems Design

Optimum rear wheel suspension system parameters are

- Slightly negative camber when cornering, together with minimum camber change.
- Track changes as small as possible to ensure good driving safety and low tire wear.
- Protection against unfavorable steering effects which can arise from elastic deformation due to braking and lateral forces.

Examples of common rear axle design include the rigid axle, semi-trailing arm axle, de Dion axle, and multi-link axle.

9.6.1 Rigid Axle

The rigid axle offers the following advantages:

- No change in track and camber.
- Trend to tramp on rough road.

The most familiar form of the rigid drive axle is the Hotchkiss drive [1]. The axle is located by semi-elliptic leaf springs as shown in Fig. 9.6, and is driven through a longitudinal driveshaft with universal joints at the transmission and axle. The springs, mounted longitudinally, connect to the chassis at their ends with the axle attached near the midpoint.

In response to the shortcomings of leaf spring suspensions, the four-link rear suspension, shown in Fig. 9.7a, evolved as the suspension of choice in recent decades for larger passenger cars with solid rear-drive axles. The lower control arms provide longitudinal control of the axle while the upper arms absorb braking/driving torques and lateral forces. Occasionally, the two upper arms will be replaced by a single, triangular arm, but it remains functionally similar to the four-link. The ability to use coil springs (or air springs) in lieu of leaf springs provides a better ride and NVH by the elimination of the coulomb friction characteristic of leaf springs.

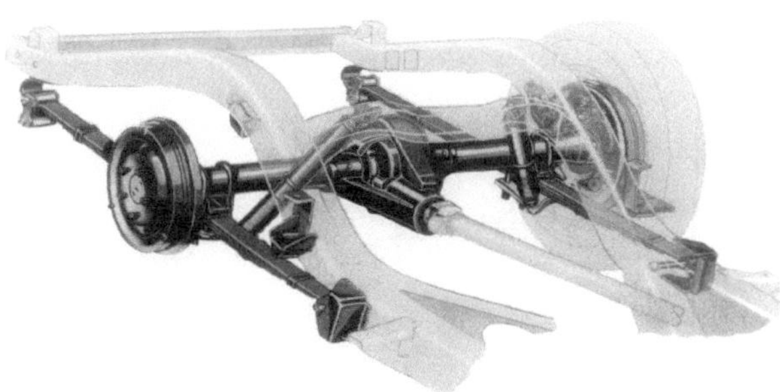

Fig. 9.6 The Hotchkiss rear suspension. (Photo courtesy of Ford motor company)

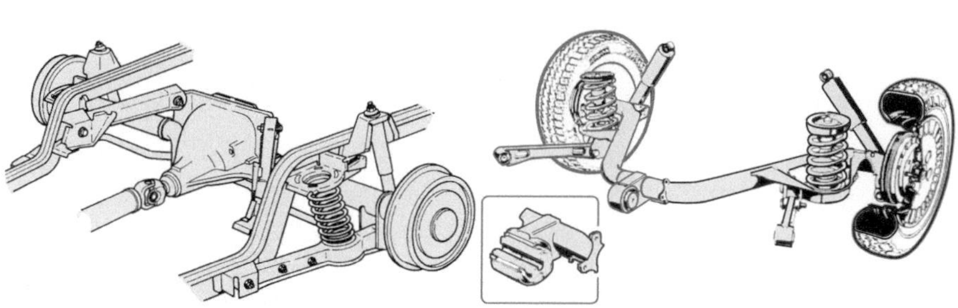

(a) The four-link rear suspension (b) 'Omega' rear wheel suspension

Fig. 9.7 Different rear suspensions

Figure 9.7 shows the "Omega" rear wheel suspension on the Lancia Y 10 and Fiat Panda, a drawbar axle with a U-shaped tube, drum brakes, inclined shock absorbers, and additional elastomer springs seated inside the low-positioned coil springs. The rubber element in the shaft axle bearing point, shown separately, has cut-outs similar to the front bearings of the two longitudinal trailing links and achieves the longitudinal elasticity necessary for radial tire dynamic rolling hardness. The middle bearing point is also the body pitch pole. The body roll center is located in the center of the axle but is determined by the level of the three mounting points on the body.

9.6.2 Semi-Trailing Arm Axle

- Used extensively for driven axles.
- Good cornering stability provided by negative camber.
- Slight change in track and camber.

The semi-trailing arm rear suspension was popularized by BMW and Mercedes Benz. This design, as shown in Fig. 9.8, gives rear wheel camber something between that of a pure trailing arm (no camber change relative to the body) and a swing axle.

Its pivot axis is usually about 25 degrees to a line running across the car. The semi-trailing arm produces a steering effect as the wheels move in jounce and rebound. The steer/camber combination on the outside wheel acts against the direction of cornering, thus generating roll understeer on the rear axle, but lateral force compliance steer will contribute oversteer if not controlled.

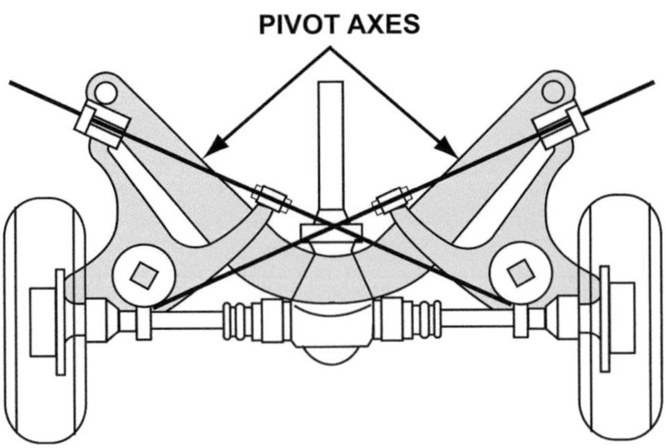

Fig. 9.8 The semi-trailing arm rear axle

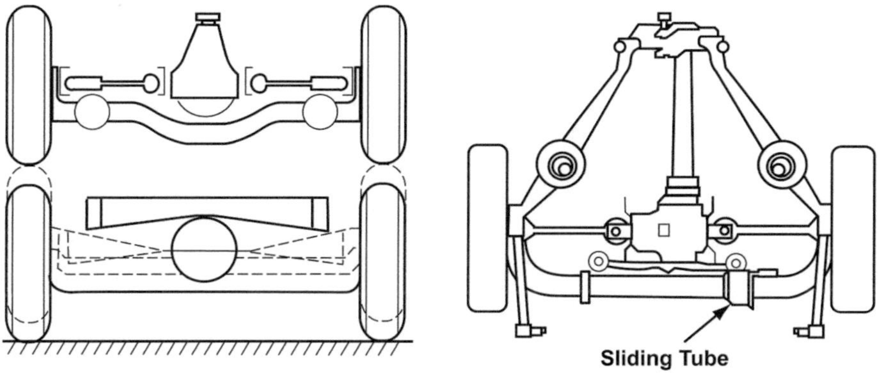

Fig. 9.9 The De Dion rear axle

9.6.3 De Dion Axle

A cross between the solid axle and an independent suspension is the classic, but little used, de Dion system (patented in 1894 by Count de Dion and George Bouton), shown in Fig. 9.9. It consists of a cross tube between the two driving wheels with a chassis-mounted differential and half shafts. Like a solid axle, the de Dion keeps the wheels upright while the unsprung weight is reduced by virtue of the differential being removed from the axle.

Axle control is provided by any of a variety of linkages from leaf springs to trailing arms. The design also has advantages for interior space because there is no need to provide differential clearance. One of the main disadvantages of the de Dion is the need to have a sliding tube or splined half shafts, which can add friction to the system. Some of the most common advantages of the de Dion axle are

- Combine benefits of independent suspension.
- Negligible unsprung masses.
- No change of track or camber.
- Elaborate and expensive design.

9.6.4 Multi-Link Axle

Some of the main advantages of the multi-link axle shown in Fig. 9.10 are

- Favorable camber characteristics (camber angle increases as the spring deflects and reduces when the suspension is in the design position).
- Slight track and toe-in change.

Figure 9.11 shows various methods for using Panhard (radius rods) to constrain four degrees of freedom (yaw, longitudinal, lateral, and angular motion along the

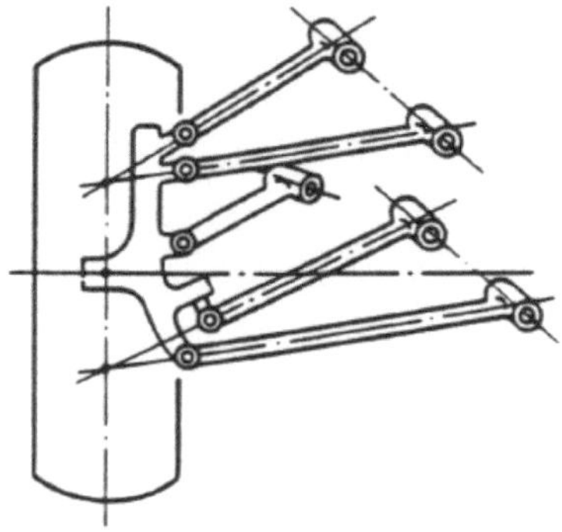

Fig. 9.10 Multi-link rear suspension of the Ford Taurus and Sable. (Photo courtesy of Ford motor company)

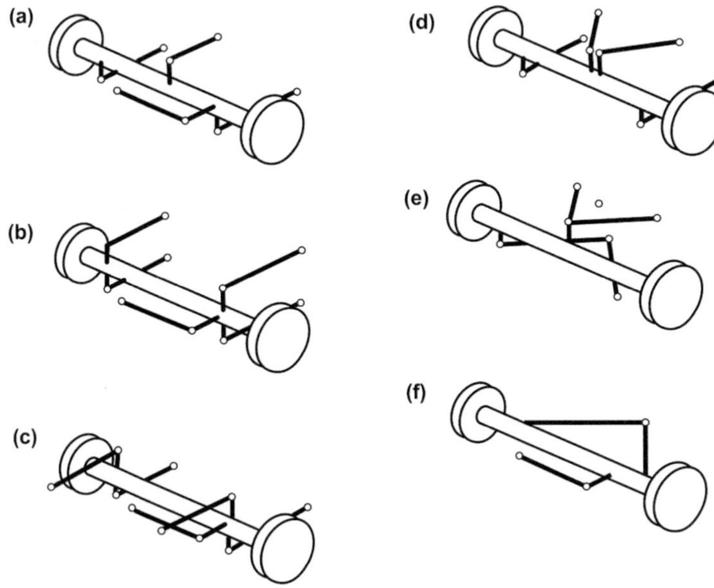

Fig. 9.11 Axle location systems

axle axis) of a vehicle rigid axle and allow only two degrees of freedom which are the roll and vertical motions.

Figure 9.10 shows that used on the Ford Taurus/Sable cars. The multi-link is characterized by ball-joint connections at the ends of the linkages so that they do not experience bending moments. Generally speaking, four links are required to provide longitudinal and lateral control of the wheels and to react brake torques.

Occasionally five links are used, as in the Mercedes Benz rear suspensions. The additional link over-constrains the wheel, but capitalizes on compliances in the bushings to allow more accurate control of toe angles in cornering. The use of linkages provides flexibility for the designer to achieve the wheel motions desired.

9.7 Roll Center

A Bumb is defined as the upward displacement of a wheel relative to the vehicle body. The heave is the vertical upward motion of the vehicle body. The roll center is a point in the center plane and in the vertical transverse plane of the wheel centers, at roll center height. While the roll center height is the height at which the lateral forces may be applied to the sprung mass without producing suspension roll. The lever arm is the distance between the vehicle center of gravity and the roll center of a suspension.

9.7.1 Independent Suspension Roll Centers

The procedure for finding the roll center of a symmetrical independent suspension is as follows:

1. Find the virtual reaction point of the suspension links (point A).
2. Draw a line from the tire-ground contact patch to the virtual reaction point.
3. The point where this line crosses the centerline of the body is the roll center (R).

Note that this procedure can be used for determining the roll center when the body is rolled; however, the suspensions are no longer symmetrical so both sets must be analyzed.

9.7.1.1 Swing Arm Geometry

The virtual reaction point of the upper and lower links is first obtained as shown in Fig. 9.12a. A line is drawn from the tire contact patch to the reaction point. The roll center is established where the line crosses the centerline of the vehicle. This suspension geometry is referred to as the "positive swing arm" because the roll center is located above the ground.

Negative swing arm geometry is shown in Fig. 9.12b. The virtual reaction point of the links is first obtained and connected to the tire contact patch as shown. The line is then projected downward to the car centerline below the ground. The roll center is negative; hence, the name "negative swing arm" geometry.

9.7.1.2 Parallel Horizontal Links

A suspension with parallel links that are horizontal (at design load) is shown in Fig. 9.13a. The virtual reaction point of the two links is therefore at infinity. Drawing a line from the tire contact patch toward infinity places the roll center in the ground plane.

Fig. 9.12 Independent suspension and swing axle [2]

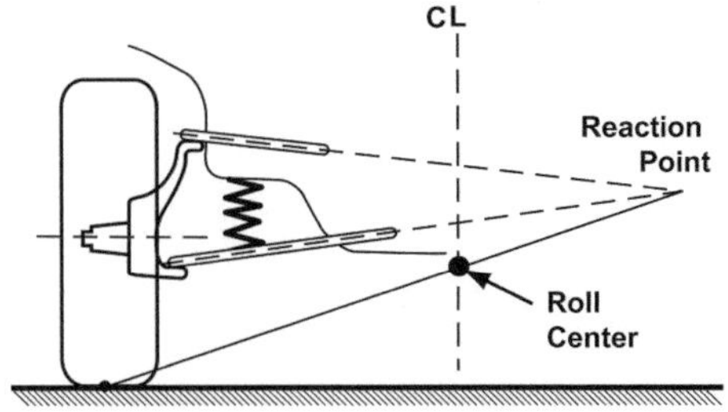

(a) Positive swing arm independent suspension

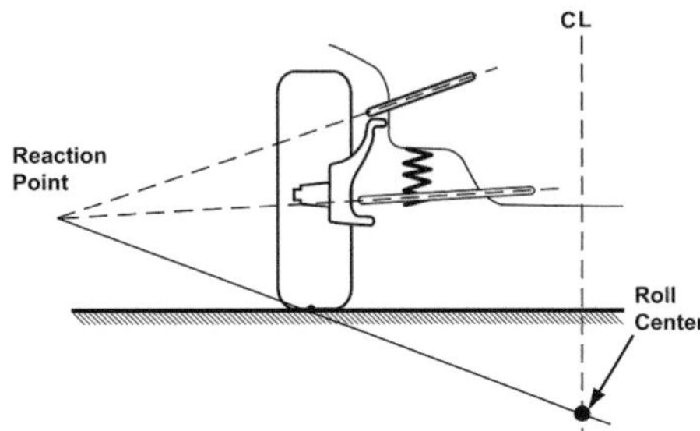

(b) Negative swing arm independent suspension

Fig. 9.13 Parallel link independent suspension [2]

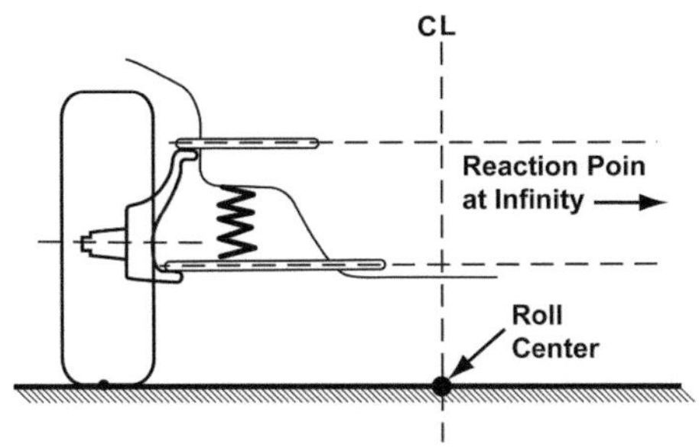

(a) Parallel links

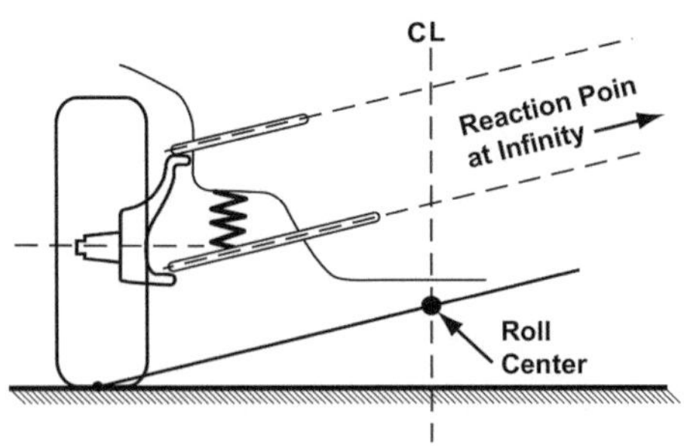

(b) Inclined parallel links

Another possibility is the use of parallel links, which are not horizontal at design load as shown in Fig. 9.13b. The virtual reaction point is at infinity. The line from the tire contact patch to the roll center is inclined at the same angle as the control arms. The roll center is elevated above the ground at the car centerline as shown. In this geometry, the roll center moves on the centerline of the car during rolling because the wheels camber with respect to the body. If the links of the suspension are equal, there will be no camber change with respect to the body and the roll center will remain stationary.

9.7.1.3 MacPherson Strut and Swing Axle

The MacPherson strut is a combination of a strut with a lower control arm as shown in Fig. 9.14a. The virtual reaction point must lie at the intersection of the axis of the lower control arm and a line perpendicular to the strut. The roll center is located on the centerline of the vehicle at the intersection with the line from the center of tire contact to the virtual reaction point.

A rear suspension swing axle is generically equivalent to that shown in Fig. 9.14b. The location of the roll center is easily obtained for this configuration because the virtual reaction point is the actual pivot of the axle. The line from the tire contact passes through the pivot and the roll center is located above the wheel center on the vehicle centerline.

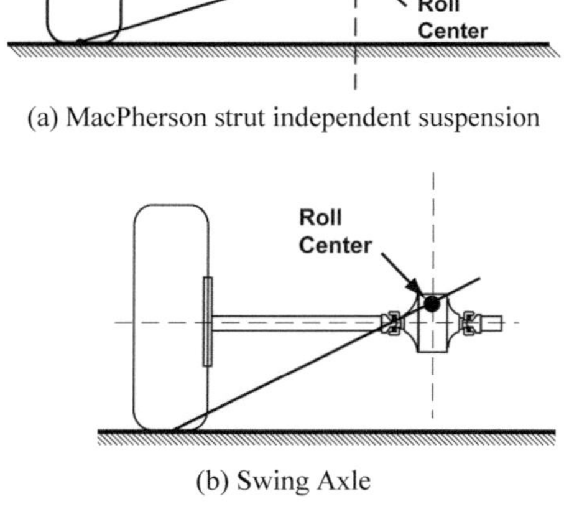

Fig. 9.14 Independent suspension and swing axle [2]

9.7.2 Dependent Suspension Roll Centers

The suspension roll axis and roll center can be determined from the layouts of the suspension geometry in the plan and elevation views. For the analysis, we draw again on the concept of a "virtual reaction point." (The virtual reaction point is analogous to the "instant center" used in kinematic analysis of linkages, but that term is not used here because of the implication that it defines a center of motion, when in fact, it does not). Physically, the virtual reaction point is the intersection of the axes of any pair of suspension control arms. Mechanistically, it is the point where the compression/tension forces in the control arms can be resolved into a single lateral force.

9.7.2.1 Multi-link Rear Suspension

Figure 9.15a shows a three-link suspension consisting of a track bar and two lower control arms. Because the track bar picks up lateral force directly, point A is established at the location where the track bar crosses the centerline of the vehicle. Point B is established as the virtual reaction point for the two lower control arms. Note that the upper link which reacts to the axle windup torque does not react lateral forces and is therefore ignored in the analysis.

Consider the case of a four-link suspension with a solid axle, as shown in Fig. 9.15b. The lateral force acting on the wheel in the top view must react as tension and compression forces in the control arms. The two long arms establish a virtual reaction point ahead of the axle at B, while the two short arms have a virtual reaction

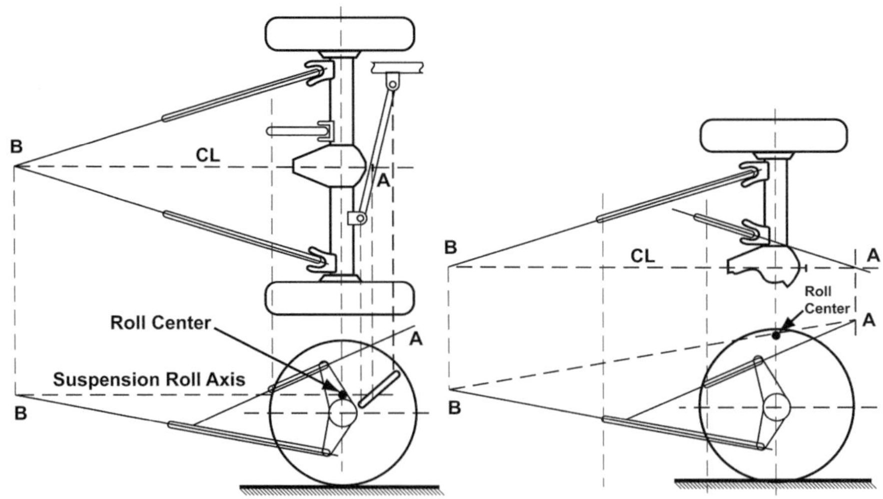

(a) Three-link rear suspension (b) Four-link rear suspension

Fig. 9.15 Three and four-link rear suspensions [2]

point behind the axle at A. In effect, each pair of arms acts like a triangular member pivoting at their respective virtual reaction points with these points establishing the suspension roll axis. Consequently, the lateral force will be distributed between the two points in inverse proportion to the length of the arms in order to achieve moment equilibrium on the axle (i.e., a large force at A and a small force at B).

The two forces at A and B must add up to F_y acting in the transverse vertical plane through the wheel centers. Given that points A and B are at different heights above the ground, their resultant at the axle centerline must be on the line connecting the two. This is the roll center for the axle. A general procedure for finding roll centers then is as follows:

1. In a plan view of the suspension find the linkages that take the side forces acting on the suspension. Determine the reaction points A and B on the centerline of the vehicle for forces in the links. In the case of paired control arms, this is a virtual reaction point.
2. Locate the points A and B in the side elevation view, thereby identifying the suspension roll axis.
3. The roll center is the point in the side view where the roll axis crosses the vertical centerline of the wheels.

9.7.2.2 Rear Leaf Spring Suspension

The design of this suspension is quite different from those discussed previously, but the general rules for determining the roll axis and center still apply. Referring to Fig. 9.16, it is seen that the leaf springs are the members that react to the side thrust. Because they are parallel to the centerline of the vehicle in the top view, the points A and B lie on the centerline of the car, both at infinity.

The lateral forces are applied to the body at the front spring eye and the rear shackle attachment point on the frame. The roll axis of the suspension is established by these points and the roll center is found on the line connecting the points. Although this analysis may seem less obvious than those discussed previously, it should be clear that a side force applied at this point will not roll the body, which is the essential definition of the roll center. Experimental measurements of leaf spring suspensions have generally confirmed the validity of this method in establishing the roll center.

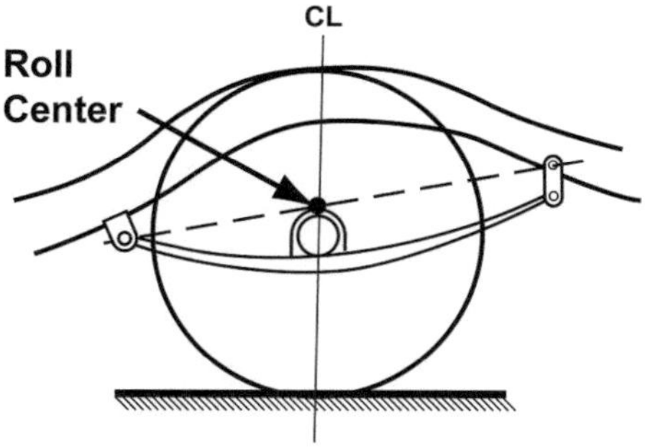

Fig. 9.16 rear leaf spring suspension [2]

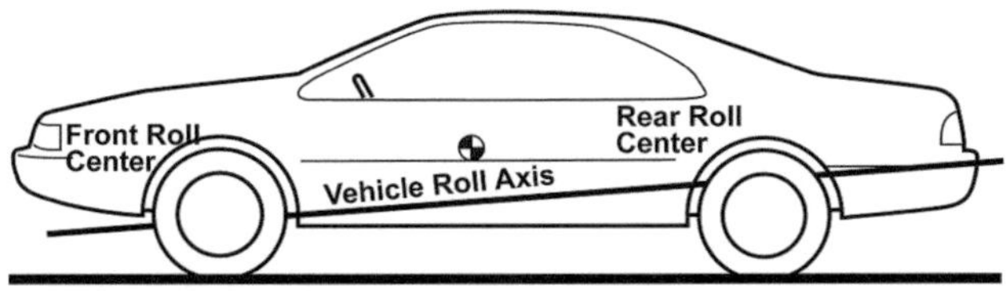

Fig. 9.17 Definitions of suspension roll center and roll axis

9.8 Roll Axis

Each suspension has a suspension roll center, defined as the point in the transverse vertical plane through the wheel centers at which lateral forces may be applied to the sprung mass without producing suspension roll [3]. It derives from the fact that all suspensions possess a roll axis, which is the instantaneous axis about which the unsprung mass rotates with respect to the sprung mass when a pure couple is applied to the unsprung mass.

The roll center is the intersection of the suspension roll axis with the vertical plane through the centers of the two wheels. These definitions are illustrated in Fig. 9.17. The roll center height is the distance from the ground to the roll center. Once the front and rear suspension roll centers are located, the vehicle roll axis is defined by the line connecting the centers. This axis is the instantaneous axis about which the total vehicle rolls with respect to the ground.

9.9 Locations of Main Inertia Axis and Roll Axis in the Longitudinal Direction

1. Roll axis (RA) displaced upward, parallel to and in the direction of horizontal main inertia axis (MA).

   - Fewer steering vibrations.
   - Increased steering stability.
   - Reduced fluctuation in dynamic wheel load.
   - Small roll change.

Fig. 9.18 Roll axis (RA) displaced upward, parallel to and in the direction of horizontal main inertia axis (MA)

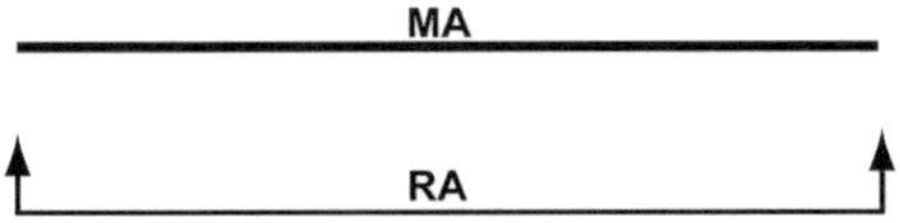

2. Roll axis (RA) displaced upward, parallel to and above horizontal main inertia axis (MA).

- Increased steering vibrations.
- Reduced steering stability.
- Increased dynamic wheel load on outside wheels and reduced dynamic load.
- on inside wheels.
- Smaller roll angle.

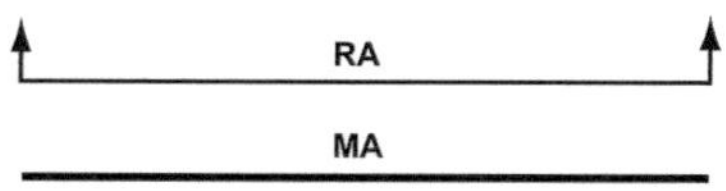

Fig. 9.19 Roll axis (RA) displaced upward, parallel to and above horizontal main inertia axis (MA)

3. Main inertia axis (MA) displaced upward, parallel to and in the direction of horizontal roll axis (RA).

   - Fewer steering vibrations.
   - Increased steering stability.
   - Increased dynamic wheel loads and wheel load fluctuation.
   - Greater Roll angles.

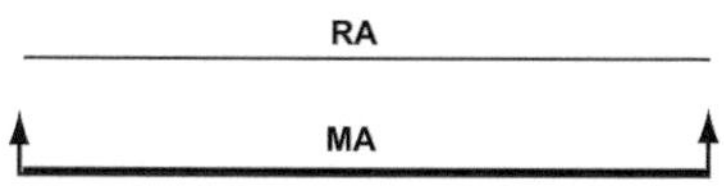

Fig. 9.20 Main inertia axis (MA) displaced upward, parallel to and in the direction of horizontal roll axis (RA)

4. Main inertia axis (IA) displaced upward, parallel to and above horizontal roll axis (RA).

   - Increased steering vibrations.
   - Reduced steering stability.
   - Increased dynamic wheel loads and wheel load fluctuation.
   - Greater roll angle.

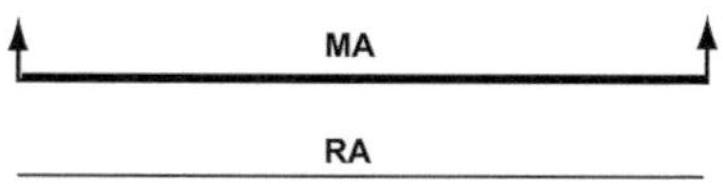

Fig. 9.21 Main inertia axis (IA) displaced upward, parallel to and above horizontal roll axis (RA)

5. Roll axis (RA) rotates forward from arbitrary position, main inertia axis (IA) remains horizontal.

   - Increased steering vibrations.
   - Reduced steering stability.
   - Increased and reduced dynamic loads at rear and front wheels, respectively.
   - Increased wheel load fluctuations.
   - Tendency to oversteer.

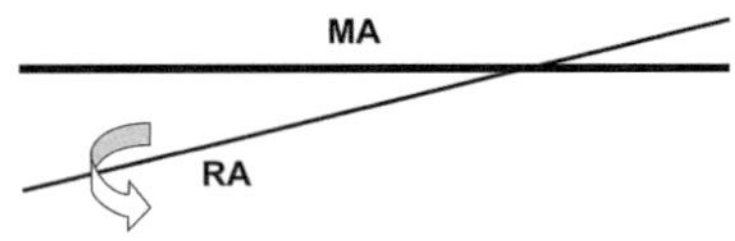

Fig. 9.22 Roll axis (RA) rotates forward from arbitrary position, main inertia axis (IA) remains horizontal

6. Main inertia axis (IA) rotates forward from arbitrary position, roll axis (RA) remains horizontal.
   - Fewer steering vibrations.
   - Increased steering ability.
   - Reduced dynamic wheel fluctuations.
   - Smaller Roll Angles.

Fig. 9.23 Main inertia axis (IA) rotate forward from arbitrary position, roll axis (RA) remains horizontal

9.10 Anti-Dive and Anti-Squat

The longitudinal load transfer incidental to braking acts to pitch the vehicle forward producing "brake dive". Just as a suspension can be designed to resist acceleration squat, the same principles apply to generation of anti-dive forces during braking. Because virtually all brakes are mounted on the suspended wheel (the only exception is in-board brakes on independent suspensions), the brake torque acts on the suspension and by proper design can create forces which resist dive.

The anti-dive is defined as the longitudinal load transfer due to braking results in pitch motion of the vehicle sprung mass. The suspension of the front and rear axles should be designed in a way to resist this pitch motion "dive". Theoretically, we can obtain 100% anti-dive during braking. Practically, maybe 50% anti-dive or less can be obtained. It is also not desired to get 100% (full) anti-dive [2].

9.11 Design and Damping Characteristics of Shock Absorbers

Automotive shock absorbers are, invariably, hydraulic dampers that dissipate vibration energy by modulating the flow through restrictions. The primary function of the shock absorber is to dissipate energy associated with the vertical motion of body or wheels, arising from control inputs or winds or road roughness. A damping or dissipative mechanism within a vehicle suspension is extremely vital to reduce the vehicle response overshoots and to minimize the influence of unavoidable resonances. Although friction dampers were extensively used in road vehicles in the past, telescopic hydraulic dampers represent the current standard.

The primary components of a shock absorber include a hydraulic cylinder, piston, orifice restrictions and valves, rod and piston seals, mountings, bump stops, hydraulic fluid and inert gas. While many configurations of dampers exist, current shock absorber designs can be grouped into three categories based upon their design and performance:

- Single-tube emulsion type.
- Double-tube type.
- Single-tube anti-emulsion type.

Figure 9.24 illustrates the schematics of single-tube emulsion type and double-tube hydraulic dampers. All the designs include a number of constant orifice passages and, as a minimum, two control valves. The control valves are tuned to achieve different damping characteristics in compression and rebound. The relative motions across the damper, caused by the body and wheel motion, give rise to the difference in pressure of fluid across the piston. The pressure drop caused by the fluid flow through the orifice restrictions represents the dissipation of pressure energy within the fluid. Adequate heat dissipation properties are thus vital near the orifice restrictions. Using the typical values of specific gravity of damper oil (0.8) and specific heat (0.4), it has been established that a pressure drop of 1 MPa (145 psi) causes a temperature rise of 0.075°C. With the maximum pressure likely to be in the range 3–8 MPa, it can be shown that the temperature increase per pass is in the 0.225–0.6°C.

Single-Tube: The single-tube emulsion type damper comprises a piston and rod in a fixed volume cylinder filled with an emulsion of oil and gas (usually nitrogen). Under static conditions, the gas separates from the liquid and re-emulsification is achieved very quickly under dynamic motions. The piston comprises two valves: compression and rebound control valves, which control the pressure differential across the piston during compression and rebound, respectively. A reduction in cylinder volume due to the rod in chamber II necessitates the presence of certain gases within the damper. While the hydraulic fluid is incompressible due to its high bulk modulus, the emulsion is highly compressible due to entrapped gas. The two valves are thus calibrated for the pressure due to liquid-gas emulsion, not the pure hydraulic oil. During compression, the increase in volume due to the piston rod contained in the cylinder causes a rise in the pressure of emulsion in chamber I. The pressure difference across the piston exerts a force termed as the damping force. While a single-tube emulsion type damper offers superior heat dissipation properties, its performance is severely limited due to the emulsion and cavitation effects.

Single-tube anti-emulsion type dampers employ an additional reservoir with a secondary piston separating the gas from oil. The additional chamber may be either built-in or remote, as illustrated in Figs. 9.24 and 9.25. This design yields high performance by eliminating the gas-oil emulsion and utilizing the high heat dissipation properties of the single tube. High performance seals, however, are required to avoid

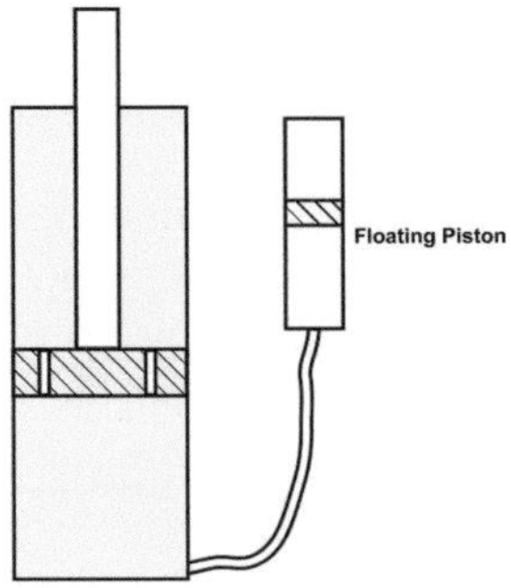

Fig. 9.24 Schematic of a single-tube shock absorber with remote reservoir

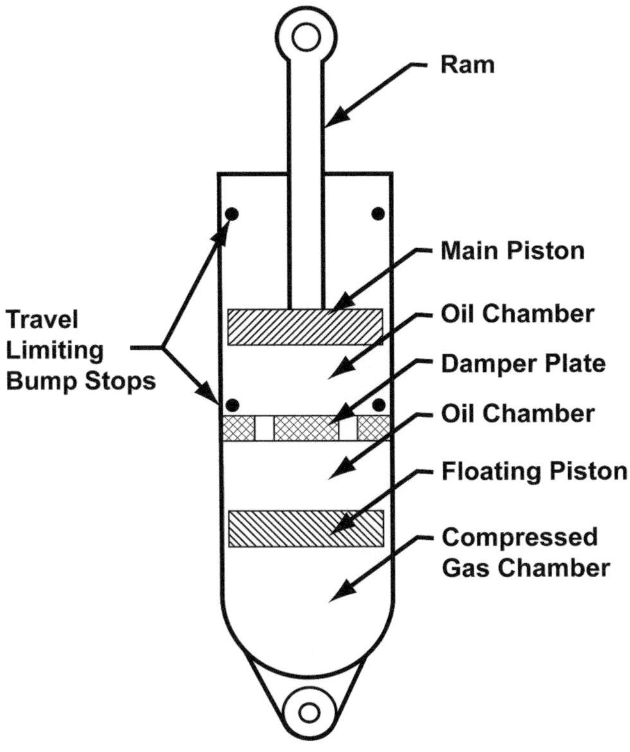

Fig. 9.25 Schematic of a single-tube shock absorber with built in reservoir

the leakage flows through the floating piston. Flow control valves are introduced within the orifices in the piston or the damper plate.

Double-Tube: Double-tube shock absorbers are most commonly used in road vehicles. Although many variations exist in view of the flow paths and control valves,

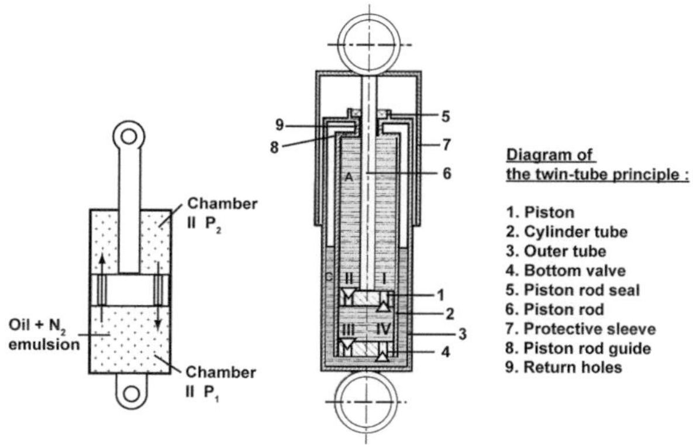

Fig. 9.26 Single-and double-tube emulsified

this design primarily comprises two annular cylinders, piston, and control valves as shown in Fig. 9.26.

Chambers I and II are filled with oil, while chamber III is partly filled with gas. In the compression stroke the volume of fluid in chamber II displaced by the piston is greater than that received by chamber I. The fluid thus flows from II to I and II to III. In the extension stroke, the fluid flows from I to II and III to II. Compression/extension of the gas in chamber III gives rise to a spring force.

Although the double-tube design tends to reduce the emulsion of gas in the oil, considerable cavitation occurs at high speeds. The double-tube design, due to its construction, yields poor heat dissipation properties.

9.11.1 Force-Velocity Relationship

The damping properties of a shock absorber are often described by their force-velocity relationship, which may be derived from the pressure and flow equations. The methodology can be perhaps best demonstrated through analysis of a symmetric double-tube shock absorber with constant orifices as shown in Fig. 9.27.

Force Equation: Dynamic force developed by the damper is related to the pressure differential across the piston, and the cross-section areas due to piston and the rod:

$$f_D = p_2\left(a_p + a_R\right) - p_1 a_p \tag{9.1}$$

where p_1 and p_2 are the instantaneous pressures of fluids in chambers I and II. f_D is the damping force, and a_R and a_p are the rod area and net piston area on the chamber I side. The force equation may be written as

$$f_D = a_p\left(p_2 - p_1\right) + a_R\left(p_2 - p_3\right) + p_3 a_R \tag{9.2}$$

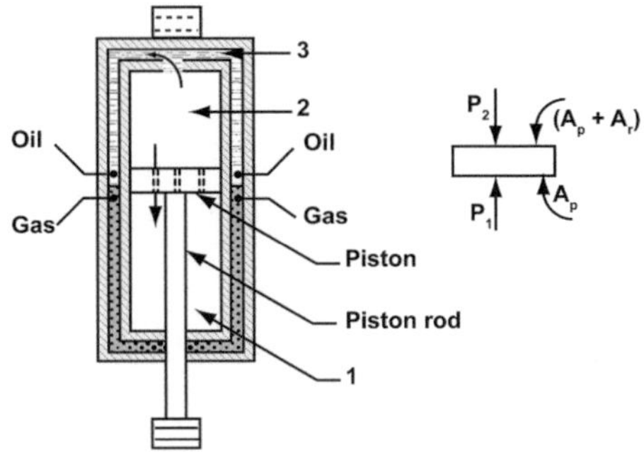

Fig. 9.27 Schematic of the Bostrom shock absorber

Flow Equations: Since the flow is symmetric in compression and rebound, the compression flow path alone may be considered. Let z be the piston motion relative to the cylinder and z be the piston velocity. The fluid volume displaced by chamber II is $(a_p + a_R)z$, and that received by chamber I is $a_p z$. The difference of fluid volume, $a_R z$, flows through the cylinder orifice to chamber III. Since the orifice areas are very small, the fluid flow quickly approaches turbulent conditions. The flow from chamber II to I can thus be described as

$$a_p\dot{z} = nC_{D1}a_1\sqrt{\frac{2\,(p_2 - p_1)}{\rho}} \tag{9.3}$$

Similarly, the flow from II to III can be expressed as

$$a_R\dot{z} = C_{D2}a_2\sqrt{\frac{2\,(p_2 - p_3)}{\rho}} \tag{9.4}$$

where C_{D1} and C_{D2} are flow discharge coefficients determined from the geometry of the orifices, piston and cylinder a_1 and a_2 are the areas due to piston and cylinder orifices, and ρ is the fluid density, n is the number of orifices in the piston.

Pressure Equations: Equations 9.3 and 9.4 yield the following pressure differentials:

$$p_2 - p_1 = \frac{\rho}{2n^2C_{D1}^2}\left[\frac{a_p}{a_1}\right]^2\dot{z}^2 \tag{9.5}$$

and,

$$p_2 - p_3 = \frac{\rho}{2n^2C_{D2}^2}\left[\frac{a_R}{a_2}\right]^2\dot{z}^2 \tag{9.6}$$

Assuming a polytropic process for the gas, the gas pressure can be expressed as

$$p_3v_3^\gamma = p_0v_0^\gamma \tag{9.7}$$

where p_3 is the instantaneous absolute gas pressure and $v_3 = v_0 - a_R z$ is the corresponding volume. The gas pressure can thus be expressed as

$$p_3 = \frac{p_0 v_0^\gamma}{(v_0 - a_R z)} - p \tag{9.8}$$

The damping force due to constant specific double-tube damper is obtained as

$$f_d = \frac{\rho}{2}\left[a_p\left(\frac{a_p}{a_1 n C_{D1}}\right)^2 + a_R\left(\frac{a_R}{a_2 n C_{D2}}\right)^2\right]\dot{z}^2 + a_R\left[\frac{p_0 v_0^\gamma}{(v_0 - a_R z)^\gamma} - p_{atm}\right] \tag{9.9}$$

The first term in Eq. 9.9 represents the damping force related to the square of the relative velocity, while the second term describes the force due to the air spring. The damping force can thus be expressed as

$$f_d = C_2 \dot{z}^2 \tag{9.10}$$

The damping force versus velocity will be increasing sharply according to Eq. 9.10. This kind of behavior is undesirable in the case of automotive shock absorbers and will result in a hard ride on uneven surfaces.

? Example 9.1

Using Eqs. 9.9 and 9.10, calculate and plot the damping force versus the relative velocity range 0 to 10 m/s as the piston extends from 0 cm to 11 cm. Assuming the following parameters:

$$\begin{aligned}
\rho &= 860\,\text{kg/m}^3;\\
a_p &= 9\,\text{m}^2;\\
a_R &= 1\,\text{m}^2;\\
a_1 &= 0.05\,\text{m}^2;\ a_2 = 0.05\,\text{m}^2;\\
n &= 4;\\
C_{D1} &= 1.7;\ C_{D2} = 0.62
\end{aligned}$$

Solution The complete damping force equation f_d is

$$f_d = \frac{\rho}{2}\left[a_p\left(\frac{a_p}{a_1 n C_{D1}}\right)^2 + a_R\left(\frac{a_R}{a_2 n C_{D2}}\right)^2\right]\dot{z}^2 + a_R\left[\frac{p_0 v_0^\gamma}{(v_0 - a_R z)^\gamma} - p_{atm}\right]$$

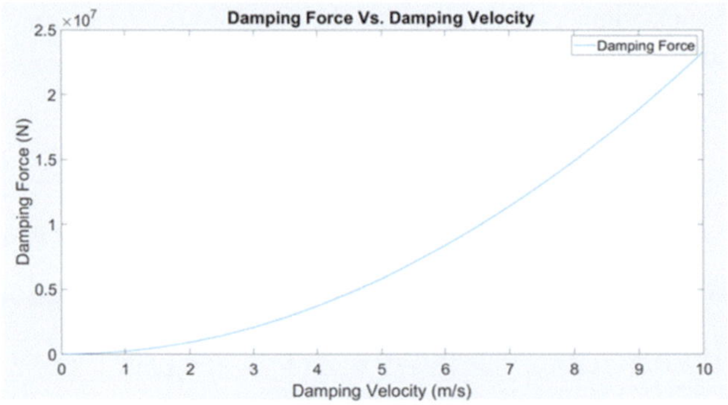

Fig. 9.28 Damping force as a function of damping velocity

The first term in Eq. 9.9 represents the damping force related to the square of the relative velocity, and the second term describes the spring force of the system. The equation of the damping force with respect to velocity f_d is

$$f_d = \frac{\rho}{2}\left[a_p\left(\frac{a_p}{a_1 n C_{D1}}\right)^2 + a_R\left(\frac{a_R}{a_2 n C_{D2}}\right)^2\right]\dot{z}^2$$

Using MATLAB coding, one can plot the damping force as a function of damping velocity as shown in Fig. 9.28.

9.11.2 *Design Considerations Treatment of Damping in Vehicle Dynamics Studies*

Dynamic analysis of vehicles and suspension systems largely uses the concept of a linear damper, with force proportional to the velocity, in order to utilize convenient and efficient linear analytical tools. Although Eq. 9.10 clearly illustrates a highly nonlinear damping force, the modern hydraulic dampers exhibit nearly linear behavior for low piston velocities. Modern designs use complex valving to achieve the force-velocity characteristics considered desirable for vehicle ride and control performance. The constant orifice velocity squared damper is undesirable due to its many limitations:

- Damper yields excessively high fluid pressure and thus high stresses.
- High pressure further poses severe design constraints for piston and rod seats.
- The damping force, increasing proportional to the square of the relative velocity, yields to a poor ride.
- High pressure differential causes excessive heat build-up and thus cavitation.

Modern dampers are designed to operate near a selected design pressure through variable area orifices. Variable area orifices are achieved by introducing spring-loaded or pressure control valves. For low velocity across the damper these valves remain closed. The flow occurs through constant area restrictions, and the fluid pressure and the damping force increases as a square of the velocity, as shown in Fig. 9.29. With increasing velocity, as the fluid pressure approaches the selected design pressure the valves open gradually, thereby, increasing the effective orifice area. The rate of increase of fluid pressure and the damping force thus reduce considerably, as shown in Fig. 9.29.

The damping characteristics illustrated in Fig. 9.29 yield high damping coefficient (slope of the curve) near low velocities and a low damping coefficient at high velocities. Such damping characteristics are thus considered highly desirable. Many different designs of blow-off and pressure control valves are currently being used in shock absorbers. Figure 9.30 illustrates one of the designs of spring-loaded compression and rebound valves.

Figure 9.30 shows the MacPherson strut of the Fiat Panda manufactured by Monroe: the spring seat 2 for taking the coil spring, the tab 3 (for fixing the steering arm) and the bracket parts 4 and 5 to which the steering knuckles are bolted to the

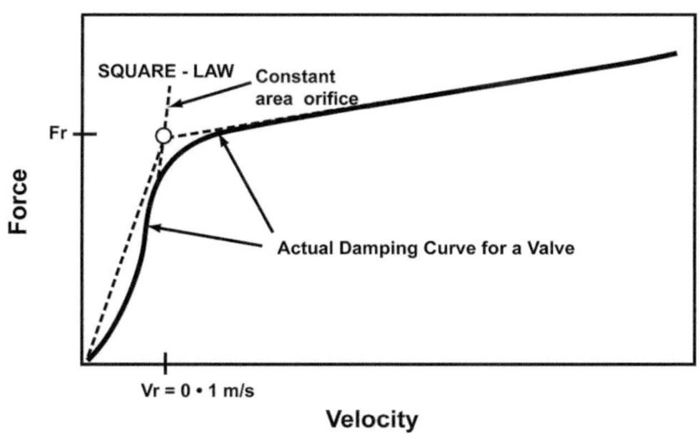

Fig. 9.29 Force-velocity relationship

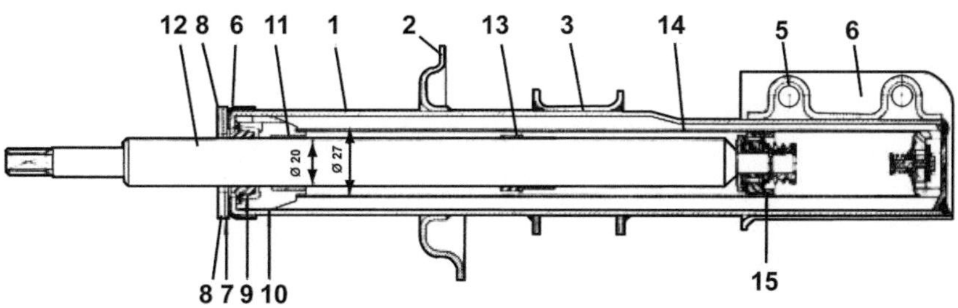

Fig. 9.30 MacPherson strut

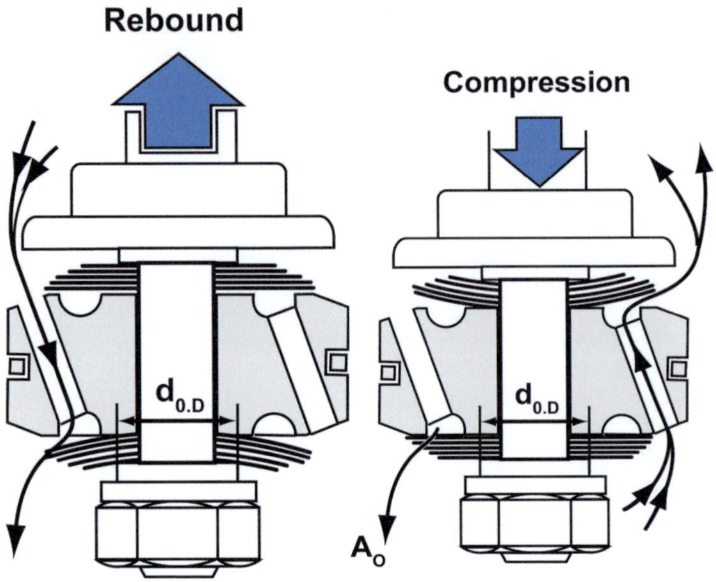

(a) Rebound state of the piston valve (b) Compression state of the piston valve

Fig. 9.31 MacPherson strut variable orifices during rebound and compression

outer tube. The stop disc 7 is supported on the rolled edge 6 of the outer tube, 12 and its two transverse grooves, 8 ensure that the supplementary spring cannot create overpressure in 6 the interior; this would press dirt and deposits into the seal 9. The bush 11 is pressed into the sintering iron rod guide 10 from the bottom and its surface conditioned to reduce friction (to the piston rod 12). The rod is 20 mm in diameter and, in the midrange, carries the rebound stop 13; when the wheel is fully extended, the minimum bearing span (center bush 11 to center piston) is 120 mm. The rod 12 is drawn in at the bottom to provide space for the rebound stage and check valve 13. The low-friction ring 15 provides the seal between the piston, which is 27 mm in diameter, and the cylinder tube 14.

Figure 9.31a, b shows the MacPherson strut orifices in rebound and compression, respectively. The orifice area during rebound and compression is controlled by a set of springs that deflects more when the speed of the piston is increasing to increase the orifice area which results in reduced damping force. This means that the increase of orifice area will result in a reduction of damping coefficient relative to the low speed of the piston.

Modern dampers are also designed to yield asymmetric damping characteristics in compression and rebound. The rebound damping of these dampers is considerably higher than the compression damping, as shown in Fig. 9.32. Different rebound and compression damping is achieved by selecting different orifice areas and valve settings in compression and rebound.

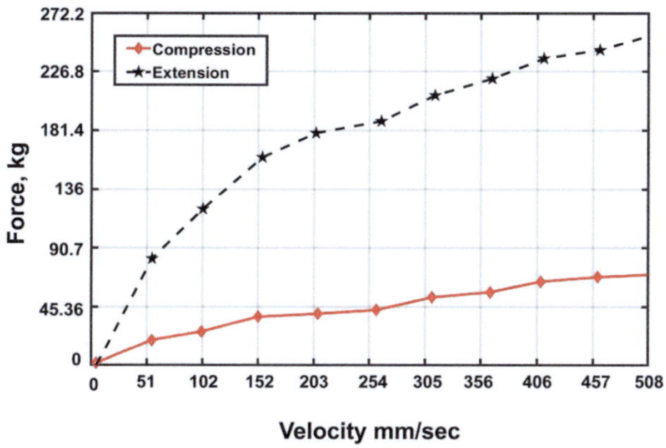

Fig. 9.32 Compression and extension characteristics

Fig. 9.33 Tire entering and exiting a bump

Figure 9.32 shows a typical measurement of an automotive shock absorber for both compression and extension. The slope keeps reducing as the velocity keeps increasing, this is due to the increase in the Orpheus area of the shock absorber piston. The variable Orpheus area increase as the relative velocity increases and that results in a reduction of the damping coefficient in order to avoid a harsh ride on rough surfaces and at high speeds.

When the tire enters a bump as shown in Fig. 9.33a, the shock absorber will be under compression. In this case, the damping coefficient (slope of damping force versus relative vertical velocity) must be reduced to avoid a high force transmitted to the chassis that may cause discomfort to the driver and passengers. This explains why the damping force variation in Fig. 9.33b has fewer values in the case of extension. While the tire is departing the bump, the shock absorber will be in a state of extension, therefore the damping force becomes higher to avoid homering the road surface by slowing down the tire motion while leaving the bump.

9.12 Human Response to Vibration

In general, passenger ride comfort (or discomfort) boundaries are difficult to determine because of the variations in individual sensitivity to vibration and of a lack of a generally accepted method of approach to the assessment of human response to vibration. Considerable research has been conducted by a number of investigators in an attempt to define ride comfort limits. A variety of methods for assessing human tolerance to vibration have been developed over the years [4, 5]. They include the following:

1. Subjective Ride Measurements. The traditional technique for comparing vehicle ride quality in the automotive industry in the past was to use a trained jury to rate the ride comfort, on a relative basis, of different vehicles driven over a range of road surfaces. With a large enough jury and a well-designed evaluation scheme, this method could provide a meaningful comparison of the ride quality of different vehicles. The degree of difference in ride quality, however, cannot be quantitatively determined by this type of subjective evaluation.
2. Shake Table Tests. In an attempt to quantitatively study human response to vibration, a large number of shake table experiments have been performed over the years. Most of this research pertains to the human response to sinusoidal excitation. It is intended to identify zones of comfort (or discomfort) for humans in terms of vibration amplitude, velocity, or acceleration in a given direction (such as foot-to-head, side-to-side, or back-to-chest) over a specific frequency range.
3. Ride Simulator Tests. In these tests, ride simulators are used to replicate the vibration of the vehicle traveling over different road surfaces. In some facilities, an actual vehicle body is mounted on hydraulic actuators, which reproduce vehicle motions in pitch, roll, and bounce (or heave). Road inputs are fed into the actuators. Using the simulator, it is possible to establish a human tolerance limit in terms of vibration parameters.
4. Ride Measurements in Vehicles. Shake table tests and ride simulator tests described above are conducted under laboratory conditions. They do not necessarily provide the same vibration environments to which the passenger is subjected to while driving on the road. Therefore, on-the-road ride measurements, particularly for passenger cars, have been performed. This test method attempts to correlate the response of test subjects in qualitative terms, such as "unpleasant" or "intolerable", with vibration parameters measured at the location where the test subject is situated under actual driving conditions.

The assessment of human response to vibration is complex in that the results are influenced by the variations in individual sensitivity, and by the test methods used by different investigators. Over the years, numerous ride comfort criteria have been proposed. Figure 9.34 shows one of such criteria for vertical vibration described in the Ride and Vibration Data Manual J6a of the Society of Automotive Engineers [6]. The recommended limits shown in the figure are also referred to as Janeway's comfort criterion. It defines the acceptable amplitude of vibration as a function of frequency.

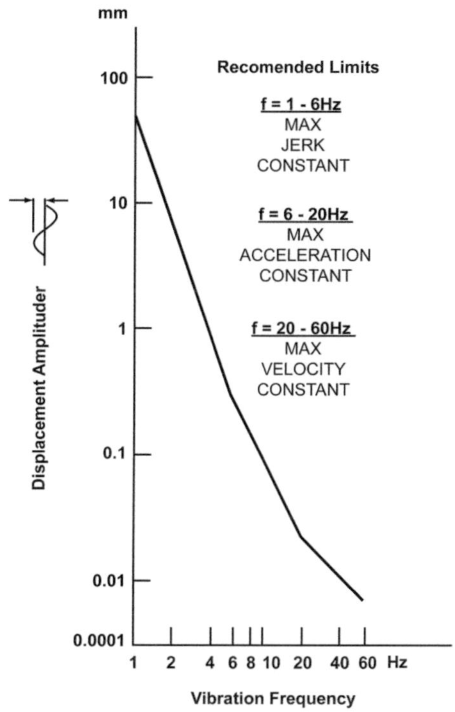

Fig. 9.34 Vertical vibration limits for passenger comfort proposed by Janeway

It can be seen that as the frequency increases, the allowable amplitude decreases considerably. The Janeway comfort criterion consists of three simple relationships, each of which covers a specific frequency range. In the frequency range 1–6 Hz, the peak value of jerk, which is the product of the amplitude and the cube of the circular frequency, should not exceed $12.6\,\text{m/s}^3$ ($496\,\text{in/s}^3$).

For instance, at 1 Hz (2π rad/s), the recommended limit for amplitude is $\frac{12.6\,\text{m.s}^{-3}}{(2\pi\ \text{s}^{-1})^3} = 0.0508\,\text{m}(2\,\text{in})$. In the frequency range 6–20 Hz, the peak value of acceleration, which is the product of the amplitude and the square of the circular frequency, should be less than $0.33\,\text{m/s}^2$ ($13\,\text{in/s}^2$), whereas in the range 20–60 Hz, the peak value of velocity, which is the product of the amplitude and the circular frequency, should not exceed 2.7 mm/s (0.105 in/s). Janeway's comfort criterion is based on data for vertical sinusoidal vibration of a single frequency. When two or more components of different frequencies are present, there is no established basis on which to evaluate the resultant effect. It is probable, however, that the component that taken alone represents the highest sensation level will govern the sensation as a whole. Furthermore, all of the data used to establish the ride comfort boundaries were obtained with test subjects standing or sitting on a hard seat.

9.13 Vehicle Ride Models

To study the ride quality of ground vehicles, various ride models have been developed. For a passenger car with independent front suspensions, a seven degrees of freedom model may be used. In this model, the pitch, bounce, and roll of the vehicle body, as well as the bounce of the two front wheels and the bounce and roll (tramp) of the solid rear axle are taken into consideration. The mass of the vehicle body is usually referred to as the "sprung mass", whereas the mass of the running gear together with the associated components is referred to as the "unsprung mass".

9.13.1 Multi-Wheeled Combat Vehicle Ride Dynamics

For a cross-country military vehicle shown in Fig. 9.35, a twenty-three degrees of freedom model may be used, which includes the pitch, bounce, and roll of the vehicle body and the bounce of each road wheel of the 8 wheels.

To evaluate the vertical dynamics of an 8 × 8 vehicle, the vehicle was tested on the sinusoidal 6-inch Washboard, 10-inch Half-Round, and Belgian Block road profiles. Figure 9.36 shows the 6-inch Washboard road elevation profile used for the simulation runs.

The 8 × 8 vehicle model was simulated on the 6-inch Washboard road to predict the dynamic performances at the speeds of 4 and 10 mph and at tire pressures of 47 and 81 psi. Figure 9.35 shows the 8 × 8 vehicle simulation model running on the 6-inch Washboard. The measured and simulation data were compared at a low-speed range and various inflation pressures.

The vertical accelerations of the vehicle's center of gravity (CG), pitch rate at the CG, longitudinal speed, and the road side (RS) strut displacement at all four axles were selected for comparison with the measurement data. The overlaying plots of the measured and predicted data are shown in Figs. 9.37, 9.38, 9.39, 9.40, 9.41, 9.42 and 9.43. For vertical acceleration skewness and RMS and strut displacement for

Fig. 9.35 A ride model for a military muti-wheeled vehicle

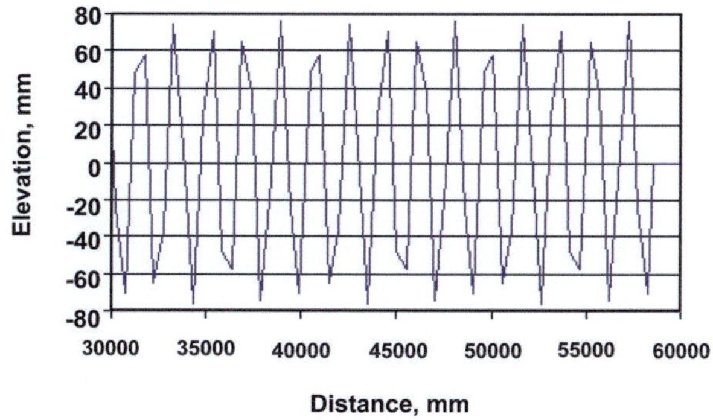

Fig. 9.36 6-inch washboard road profile

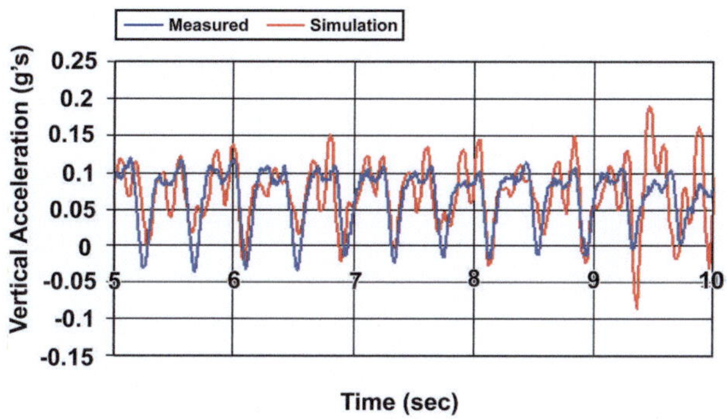

Fig. 9.37 Time history comparison of vertical acceleration

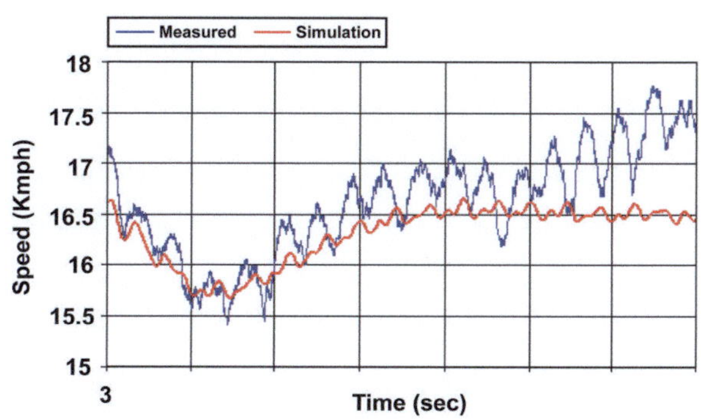

Fig. 9.38 Time history comparison of speed

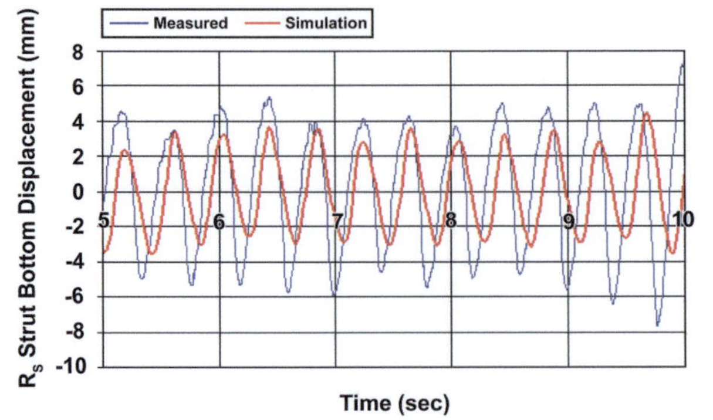

Fig. 9.39 Time history comparison of pitch rate

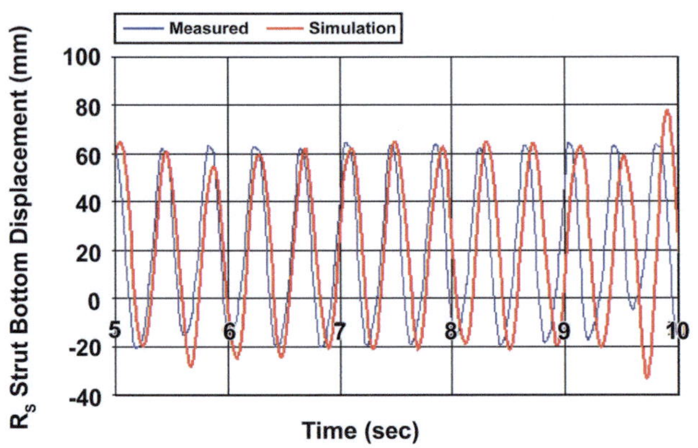

Fig. 9.40 Time history comparison of RS strut displacement axle 1

Axle 3 and Axle 4, the predicted values are not within the ATC validation criteria. The poor statistical correlation is attributed to the vehicle speed variation. As can be seen in the vehicle speed plot, the vehicle speed amplitudes of the measurement and simulation are not equivalent. This is because it is very hard to model the effect of a driver's braking and deceleration during an actual test event.

9.13.2 Simplified Two Axle Vehicle Ride Model

To study the vibrational characteristics of the vehicle, equations of motion based on Newton's second law for each mass have to be formulated. Natural frequencies and amplitude ratios can be determined by considering the principal modes (normal modes) of vibration (or the free vibration) of the system. When the excitation of the system is known, the response can, in principle, be determined by solving the

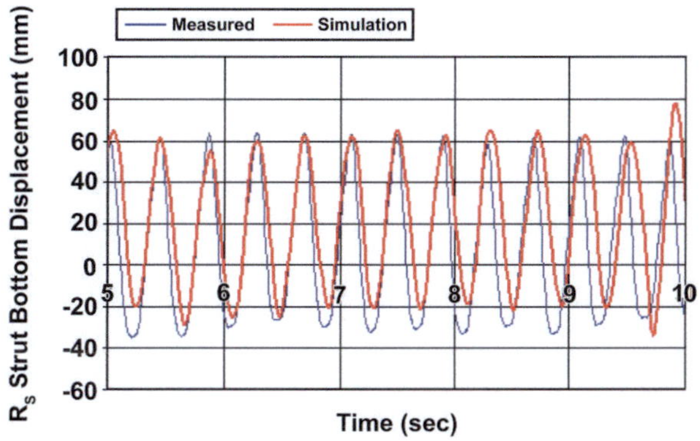

Fig. 9.41 Time history comparison of RS strut displacement axle 2

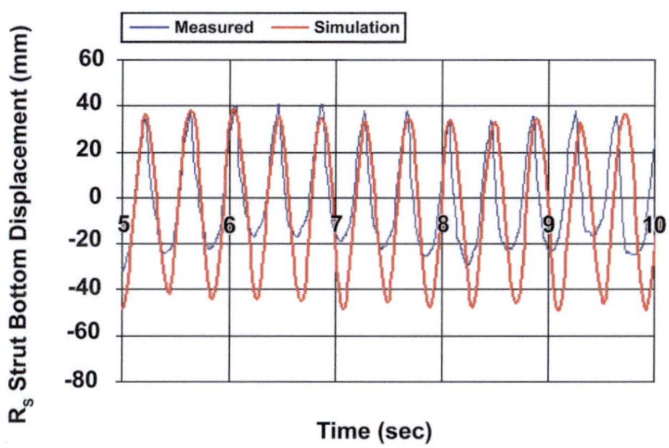

Fig. 9.42 Time history comparison of RS strut displacement axle 3

equations of motion. However, as the degrees of freedom of the system increase, the analysis becomes increasingly complex. Digital computer simulations are usually employed.

A vehicle represents a complex vibration system with many degrees of freedom. It is possible, however, to simplify the system by considering only some of its major motions. For instance, to obtain a qualitative insight into the functions of the suspension, particularly the effects of the sprung and unsprung mass, spring stiffness, and damping on vehicle vibrations, a linear model with two degrees of freedom, as shown in Figs. 9.44 and 9.45, may be used. On the other hand, to reach a better understanding of the pitch and bounce vibration of the vehicle body, a two-degrees-of-freedom model, as shown in Fig. 9.46, may be employed.

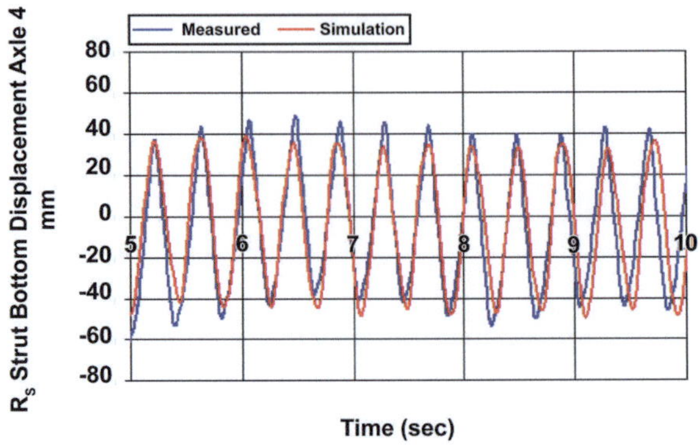

Fig. 9.43 Time history comparison of RS strut displacement axle 4

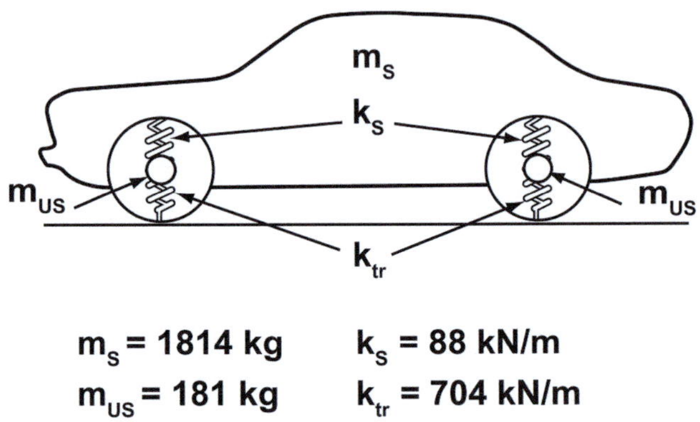

Fig. 9.44 A two-degrees-of-freedom ride model for the sprung and unsprung mass

Fig. 9.45 A quarter-car model

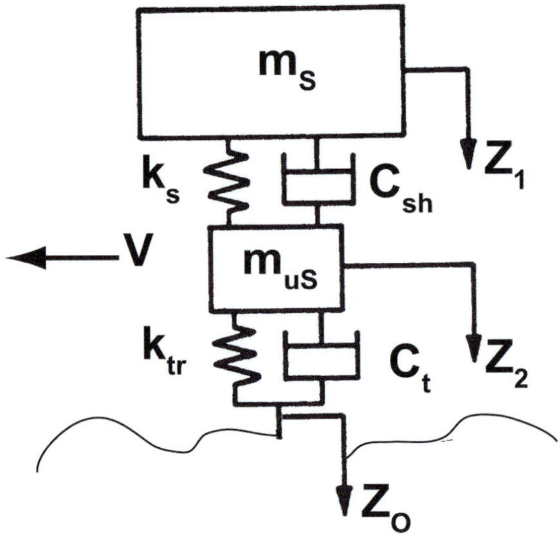

9.13.3 Two-Degrees-of-Freedom Vehicle Model for Sprung and Unsprung Mass

The two-degrees-of-freedom model shown in Figs. 9.44 and 9.45 includes an unsprung mass representing the wheels and associated components and a sprung mass representing the vehicle body. Their motions in the vertical direction can be described by two coordinates, z_1 and z_2 as shown in Fig. 9.45, with origins at the static equilibrium positions of the sprung and unsprung mass, respectively. This model can be used to represent a quarter of a car. As a result, it is often referred to as the "quarter-car" model. By applying Newton's second law to the sprung and unsprung mass separately, the equations of motion of the system can be obtained.

For vibrations excited by surface undulation, the equations of motion are as follows:

For the sprung mass:

$$m_s \ddot{z_1} + c_{sh}(\dot{z_1} - \dot{z_2}) + k_s(z_1 - z_2) = 0 \tag{9.11}$$

For the unsprung mass:

$$m_{us} \ddot{z_2} + c_{sh}(\dot{z_2} - \dot{z_1}) + k_s(z_2 - z_1) + c_t \dot{z_2} + k_{tr} z_2 = F(t) = c_t \dot{z_0} + k_{tr} z_0 \tag{9.12}$$

where m_s is the sprung mass, m_{us} is the unsprung mass, c_{sh} is the damping coefficient of the shock absorber, c_t is the damping coefficient of the tire, k_s is the stiffness of the suspension spring, k_{tr} is the equivalent spring stiffness of the tire, and $F(t)$ is the excitation acting on the wheels and induced by surface irregularities. If z_0 is the elevation of the surface profile and $\dot{z_0}$ represents the vertical velocity of the tire at the ground contact point, which is the slope of the profile multiplied by the forward speed of the vehicle, then the excitation due to surface undulation may be expressed by $c_t \dot{z_0} + k_{tr} z_0$, as shown in Eq. 9.12. Excitations due to aerodynamic forces and to vibrations of the engine and driveline are applied to the sprung mass, while those due to non-uniformities of the tire/wheel assembly are applied to the unsprung mass. If the excitation of the system is known, then, in principle, the resulting vibrations of the sprung and unsprung mass can be determined by solving Eqs. 9.11 and 9.12.

To determine the natural frequencies of the two-degrees-of-freedom system shown in Fig. 9.46, the free vibration of the system is considered (or the principal modes of vibration are considered). The equations of motion for free vibration are obtained by setting the right-hand sides of both Eqs. 9.11 and 9.12 to zero. For an undamped system, from Eqs. 9.11 and 9.12, the equations of motion for free vibration are as follows:

$$m_s \ddot{z_1} + k_s z_1 - k_s z_2 = 0 \tag{9.13}$$

$$m_{us} \ddot{z_2} + k_s z_2 - k_s z_1 + k_{tr} z_2 = 0 \tag{9.14}$$

The solutions to the above differential equations can be assumed to be in the following form:

$$z_1 = Z_1 \cos \omega_n t \tag{9.15}$$

$$z_2 = Z_2 \cos \omega_n t \tag{9.16}$$

where ω_n is the undamped circular natural frequency, and Z_1 and Z_2 are the amplitudes of the sprung and unsprung mass, respectively. Substituting the assumed solutions into Eqs. 9.13 and 9.14, one obtains the following amplitude equations:

$$\left(-m_s\omega_n^2 + k_s\right) Z_1 - k_s Z_2 = 0 \tag{9.17}$$

$$-k_s Z_1 + \left(-m_{us}\omega_n^2 + k_s + k_{tr}\right) Z_2 = 0 \tag{9.18}$$

These equations are satisfied for any Z_1 and Z_2 if the following determinant is zero:

$$\begin{bmatrix} \left(-m_s\omega_n^2 + k_s\right) & -k_2 \\ -k_2 & \left(-m_{us}\omega_n^2 + k_s + k_{tr}\right) \end{bmatrix} = 0 \tag{9.19}$$

Expanding the determinant leads to the characteristic equation of the system:

$$\omega_n^4 m_s m_{us} + \omega_n \left(-m_s k_s - m_s k_{tr} - m_{us} k_s\right) + k_s k_{tr} = 0 \tag{9.20}$$

The solution of the characteristic equation yields two undamped natural frequencies of the system, ω_{n1}^2 and ω_{n2}^2:

$$\omega_{n1}^2 = \frac{B_1 - \sqrt{B_1^2 - 4A_1C_1}}{2A_1} \tag{9.21}$$

$$\omega_{n2}^2 = \frac{B_1 + \sqrt{B_1^2 - 4A_1C_1}}{2A_1} \tag{9.22}$$

where

$$A_1 = m_s m_{us}$$
$$B_1 = m_s k_s + m_s k_{tr} + m_{us} k_s$$
$$C_1 = k_s k_{tr}$$

Although each of these leads to frequencies $\pm\omega_{n1}$ and $\pm\omega_{n2}$, the negative values are discarded as being of no physical significance. The corresponding natural frequencies in Hz (cycles/s) are expressed by

$$f_{n1} = \frac{1}{2\pi}\omega_{n1} \tag{9.23}$$

$$f_{n2} = \frac{1}{2\pi}\omega_{n2} \tag{9.24}$$

For a typical passenger car, the sprung mass m_s is an order of magnitude higher than the unsprung mass m_{us}, while the stiffness of the suspension spring k_s is an order of magnitude lower than the equivalent spring stiffness of the tire k_{tr}, as shown in Fig. 9.45. In view of this, an approximate method may be used to determine the two natural frequencies of the system. The approximate values of the undamped natural frequencies in Hz of the sprung and unsprung mass, f_{n-s} and f_{n-us}, can be expressed by

$$f_{n-s} = \frac{1}{2\pi}\sqrt{\frac{k_s k_{tr}/(k_s + k_{tr})}{m_s}} \tag{9.25}$$

$$f_{n-us} = \frac{1}{2\pi}\sqrt{\frac{k_s + k_{tr}}{m_{us}}} \tag{9.26}$$

With the values of m_s, m_{us}, k_s, and k_{tr} shown in Fig. 9.45, the two natural frequencies calculated using Eqs. 9.23 and 9.24 are 1.04 and 10.5 Hz, respectively, which are found to be practically identical to those obtained using Eqs. 9.25 and 9.26. The natural frequency of the unsprung mass is an order of magnitude higher than that of the sprung mass. For passenger cars, the damping ratio provided by shock absorbers is usually in the range of 0.2–0.4, and thus the damping of the tire is relatively insignificant. Consequently, there is little difference between the undamped and damped natural frequencies, and undamped natural frequencies are commonly used to characterize the system.

9.13.4 Two-Degrees-of-Freedom Vehicle Model for Pitch and Bounce

Because of the wide separation of the natural frequencies of the sprung and unsprung mass, the up and down linear motion (bounce) and the angular motion (pitch) of the vehicle body and the motion of the wheels may be considered to exist almost independently. The bounce and pitch of the vehicle body can therefore be studied using the model shown in Fig. 9.46. In this model, damping is neglected.

By applying Newton's second law and using the static equilibrium position as the origin for both the linear displacement of the center of gravity z and the angular displacement of the vehicle body θ, the equations of motion for the system can be formulated. For free vibration, the equation of motion for bounce is

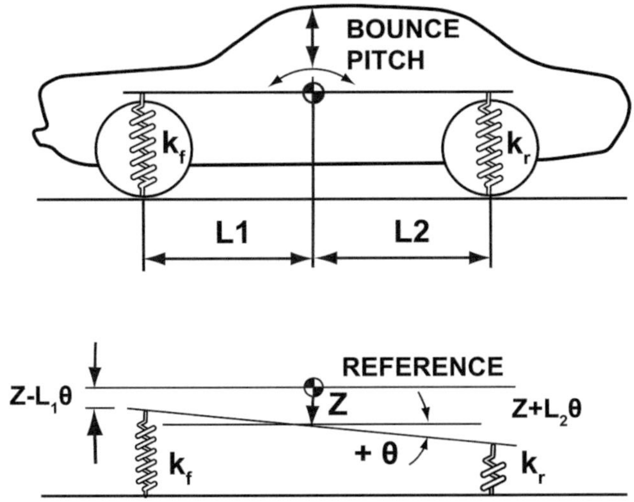

Fig. 9.46 A two-degrees-of-freedom ride model for bounce and pitch of the sprung mass

$$m_s\ddot{z} + k_f(z - l_1\theta) + k_r(Z + l_2\theta) = 0 \tag{9.27}$$

and the equation of motion for pitch is

$$I_y\ddot{\theta} - k_f l_1(z - l_1\theta) + kk_r l_2(z + l_2\theta) = 0 \tag{9.28}$$

where k_f is the front spring stiffness, k_r is the rear spring stiffness, and I_y and r_y are the mass moment of inertia and radius of gyration of the vehicle body about the y-axis, respectively. By letting:

$$D_1 = \frac{1}{m_s}(k_f + k_r)$$
$$D_2 = \frac{1}{n_s}(k_r l_2 - k_f l_1)$$
$$D_3 = \frac{1}{I_y}(k_f l_1^2 + k_r l_2^2)$$

Equations 9.27 and 9.28 can be written as

$$\ddot{z} + D_1 z + D_2\theta = 0 \tag{9.29}$$
$$\ddot{\theta} + D_3\theta + \frac{D_2}{r_y^2}z = 0 \tag{9.30}$$

It is evident that D_2 is the coupling coefficient for the bounce and pitch motions, and that these motions uncouple when $k_f l_1 = k_r l_2$. With $k_f l_1 = k_r l_2$, a force applied to the center of gravity induces only bounce motion, while a moment applied to the body produces only pitch motion. In this case, the natural frequencies for the uncoupled bounce and pitch motions are

$$\omega_{nz} = \sqrt{D_1} \tag{9.31}$$

$$\omega_{n\theta} = \sqrt{D_3} \tag{9.32}$$

It is found that this would result in a poor ride. In general, the pitch and bounce motions are coupled and an impulse at the front or rear wheel excites both motions. To obtain the natural frequencies for the coupled bounce and pitch motions, the free vibration of the system is considered (or the principal modes of vibration are considered). The solutions to the equations of motion can be expressed in the form of

$$z = Z \cos \omega_n t \tag{9.33}$$

$$\theta = \Theta \cos \omega_n t \tag{9.34}$$

where ω_n is the undamped circular natural frequency, and Z and Θ are the amplitudes of bounce and pitch, respectively. Substituting the above equations into Eqs. 9.29 and 9.30, one obtains the following amplitude equations:

$$(D_1 - \omega_n^2)Z + D_2\Theta = 0 \tag{9.35}$$

$$\left(\frac{D_2}{r_y^2}\right) Z + (D_3 - \omega_n^2)\Theta = 0 \tag{9.36}$$

The characteristic equation for the system can be found as

$$\omega_n^4 - (D_1 + D_3)\omega_n^2 + \left(D_1 D_3 - \frac{D_2^2}{r_y^2}\right) = 0 \tag{9.37}$$

From Eq. 9.37, two undamped natural frequencies ω_{n1} and ω_{n2} can be obtained:

$$\omega_{n1}^2 = \frac{1}{2}(D_1 + D_3) - \sqrt{\frac{1}{4}(D_1 - D_3)^2 + \frac{D_2^2}{r_y^2}} \tag{9.38}$$

$$\omega_{n2}^2 = \frac{1}{2}(D_1 + D_3) + \sqrt{\frac{1}{4}(D_1 - D_3)^2 + \frac{D_2^2}{r_y^2}} \tag{9.39}$$

These frequencies for coupled motions, ω_{n1} and ω_{n2}, always lie outside the frequencies for uncoupled motions, ω_{nz} and $\omega_{n\theta}$.

From Eqs. 9.35 and 9.36, the amplitude ratios of the bounce and pitch oscillations for the two natural frequencies ω_{n1} and ω_{n2} can be determined. For ω_{n1},

$$\frac{Z}{\Theta}\Big|_{\omega_{n1}} = \frac{D_2}{\omega_{n1}^2 - D_1} \tag{9.40}$$

and for ω_{n2},

$$\frac{Z}{\Theta}\Big|_{\omega_{n2}} = \frac{D_2}{\omega_{n2}^2 - D_1} \tag{9.41}$$

It can be shown that the two amplitude ratios will have opposite signs.

To further illustrate the characteristics of the bounce and pitch modes of oscillation, the concept of oscillation center is introduced. The location of the oscillation center is denoted by l_0 measured from the center of gravity, and it can be determined from the amplitude ratios. Thus, one center is associated with ω_{n1}, and the other with ω_{n2}. For ω_{n1},

$$l_{01} = \frac{D_2}{\omega_{n1}^2 - D_1} \tag{9.42}$$

and for ω_{n2}

$$l_{02} = \frac{D_2}{\omega_{n2}^2 - D_1} \tag{9.43}$$

When the value of the amplitude ratio is negative, the oscillation center will be located to the right of the center of gravity of the vehicle body, in accordance with the sign conventions for z and θ. On the other hand, when the value of the amplitude ratio is positive, the oscillation center will be located to the left of the center of gravity. In general, a road input at the front or rear wheel will cause a moment about each oscillation center, and therefore will excite both bounce and pitch oscillations. In other words, the body motion will be the sum of the oscillations about the two centers as shown in Fig. 9.47.

Usually, the oscillation center that lies outside the wheelbase is called the bounce center, and the associated natural frequency is called the bounce frequency. On the other hand, the oscillation center that lies inside the wheelbase is called the pitch center, and the associated natural frequency is called the pitch frequency.

? Example 9.1

Determine the pitch and bounce frequencies and the locations of oscillation centers of an automobile with the following data:

- Sprung mass $m_s = 2120$ kg.
- Radius of gyration $r_y = 1.33$ m.
- Distance between the front axle and center of gravity $l_1 = 1.267$ m.
- Distance between the rear axle and center of gravity $l_2 = 1.548$.
- Front spring stiffness $k_f = 35$ kN/m.
- Rear spring stiffness $k_r = 38$ kN/m.

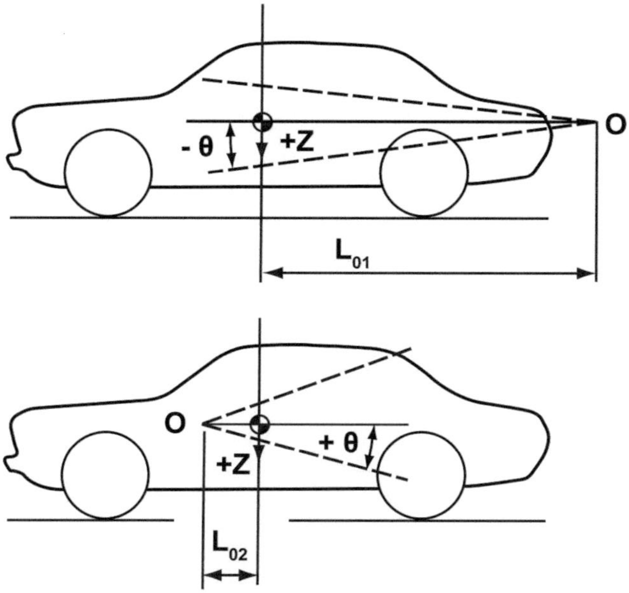

Fig. 9.47 Oscillation centers for bounce and pitch of sprung mass

Solution. The constants D_1, D_2, and D_3 are first calculated as follows:

$$D_1 = \frac{k_f + k_r}{m_s} = \frac{3500 + 38000}{2120} = 34.43\,\text{s}^{-2}$$

$$D_2 = \frac{k_r l_2 - k_f l_1}{m_s} = \frac{3800 \times 1.548 - 3500 \times 1.267}{2120} = 6.83\,\text{m.s}^{-2}$$

$$D_3 = \frac{k_f l_1^2 + k_r l_2^2}{m_s r_y^2} = \frac{35000 \times 1.267^2 + 3800 \times 1.584^2}{2120 \times 1.33^2} = 39.26\,\text{s}^{-2}$$

$$\left(\frac{D_2}{r_y}\right)^2 = 26.37\,\text{s}^{-4}$$

$$D_3 + D_1 = 73.69\,\text{s}^{-2}$$

$$D_3 - D_1 = 4.83\,\text{s}^{-2}$$

$$\omega_{n1}^2 = \frac{1}{2}(D_1 + D_3) - \sqrt{\frac{1}{4}(D_1 - D_3)^2 + \left(\frac{D_2}{r_y}\right)^2}$$

$$= 31.17\,\text{s}^{-2}.$$

Thus $\omega_{n1} = 5.58\,\mathrm{s}^{-1}$ or $f_{n1} = 0.89\,\mathrm{Hz}$

$$\omega_{n2}^2 = \frac{1}{2}(D_1 + D_3) + \sqrt{\frac{1}{4}(D_1 - D_3)^2 + \left(\frac{D_2}{r_y}\right)^2}$$
$$= 42.52\,\mathrm{s}^{-2}.$$

Thus $\omega_{n2} = 6.52\,\mathrm{s}^{-1}$ or $f_{n2} = 1.04\,\mathrm{Hz}$. The locations of the oscillation centers can be determined using Eqs. 9.42 and 9.43.

For ω_{n1},

$$l_{01} = \frac{D_2}{\omega_{n1}^2 - D_1} = 2.09\,\mathrm{m}$$

and for ω_{n2}

$$l_{02} = \frac{D_2}{\omega_{n2}^2 - D_1} = 0.84\,\mathrm{m}.$$

This indicates that one oscillation center is situated at a distance of 2.09 m (82 in) to the right of the center of gravity, and the other is located at a distance of 0.84 m (33 in) to the left of the center of gravity, as shown in Fig. 9.47.

For most passenger cars, the natural frequency for bounce is in the range of 1.0–1.5 Hz, and the natural frequency for pitch is slightly higher than that for bounce. For cars with coupled front-rear suspension systems, the natural frequency for pitch may be lower than that for bounce. In roll, the natural frequency is usually higher than those for bounce and pitch primarily because of the effect of anti-roll bars. The natural frequency for roll usually varies in the range of 1.5–2.0 Hz for cars.

The locations of the oscillation centers have practical significance to ride behavior. One case of interest is that when the motions of bounce and pitch are uncoupled (i.e., $k_f l_1 = k_r l_2$). In this case, one oscillation center will be at the center of gravity, and the other will be at an infinite distance from the center of gravity. The other case of interest is that when $r_2 = l_1 l_2$. In this case, one oscillation center will be located at the point of attachment of the front spring to the vehicle body (or its equivalent), and the other at the point of attachment of the rear spring to the body. This can be verified by setting $l_{01} = l_2$ and $l_{02} = l_1$ in Eqs. 9.42 and 9.43, respectively. Under these circumstances, the two-degrees-of-freedom model for pitch and bounce shown in Fig. 9.46 can be represented by a dynamically equivalent system with two concentrated masses at the front and rear spring attachment points (or their equivalents), as shown in Fig. 9.48. The equivalent concentrated mass at the front will be $m_s l_2/(l_1 + l_2)$ and that at the rear will be $m_s l_1/(l_1 + l_2)$.

The equivalent system is, in fact, two single-degree-of-freedom systems with natural frequency $\omega_{nf} = \sqrt{\frac{k_f(l_1+l_2)}{m_s l_f}}$ for the front, and natural frequency $\omega_{nr} = \sqrt{\frac{k_r(l_1+l_2)}{m_s l_1}}$ for the rear. Thus, there is no interaction between the front and rear suspensions, and input at one end (front or rear) causes no motion of the other. This is a desirable condition for a good ride. For practical vehicles, however, this condition often cannot be satisfied. Currently, the ratios of $r_y^2/l_1 l_2$ vary from approximately 0.8 for

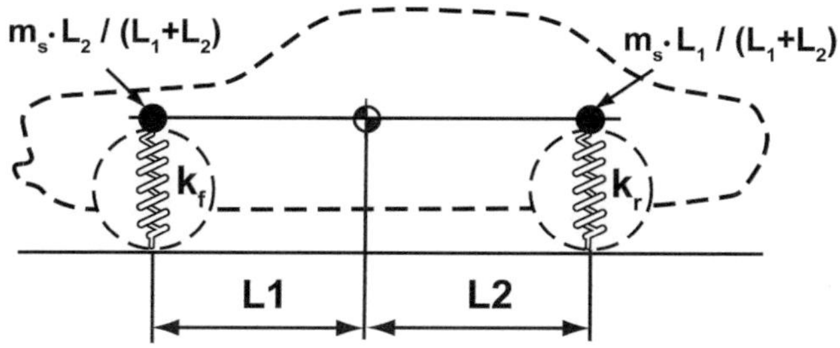

Fig. 9.48 Equivalent system having two concentrated masses for the vehicle body

sports cars through 0.9–1.0 for conventional passenger cars to 1.2 and above for some front-wheel-drive cars.

In considering the natural frequencies for the front and rear ends, it should be noted that excitation from the road to a moving vehicle will affect the front wheels first and the rear wheels later. Consequently, there is a time lag between the excitation at the front and that at the rear. This results in a pitching motion of the vehicle body. To minimize this pitching motion, the equivalent spring rate and the natural frequency of the front end should be slightly less than those of the rear end. In other words, the period for the front end ($2\pi/\omega_{nf}$) should be greater than that for the rear end ($2\pi/\omega_{nr}$). This ensures that both ends of the vehicle will move in phase (i.e., the vehicle body is merely bouncing) within a short time after the front end is excited. From the point of view of passenger ride comfort, pitching is more annoying than bouncing. The desirable ratio of the natural frequency of the front end to that of the rear end depends on the wheelbase of the vehicle, the average driving speed, and the wavelengths of the road profile.

As noted previously, a variety of multibody dynamics software packages, such as MSC ADAMS, DADS have become commercially available in recent years. They can be used to simulate the vibrations of ground vehicles in detail.

Problems

1. A passenger car of sprung mass, $m_s = 1800\,\text{kg}$, combined unsprung mass (front plus rear), $m_{us} = 190\,\text{kg}$, combined suspension spring stiffness (front and rear) $K_s = 99\,\text{kN/m}$ and combined tire stiffness, $K_{tr} = 800\,\text{kN/m}$. Assuming this 2-DOF model has a point contact and is moving at a linear velocity of 20 m/s on a perfect sinusoidal road profile of wavelength of 10 m and amplitude of 6 cm, determine the natural frequencies of the sprung and unsprung masses.
2. A passenger car of sprung part weight is 18 kN, its center of gravity is 110 cm behind the front axle, and the wheelbase is 3 m. The combined stiffness of the front and rear suspensions is, 27 kN/m and 22 kN/m, respectively. The sprung

mass pitch moment of inertia is 1894.8 kg m^2. Calculate the undamped pitch and bounce natural frequencies.

3. For the vehicle described in problem 9.2, calculate the locations of the oscillation centers.
4. If the pitch moment of inertia is varied in problem 9.2, determine the value of the moment of inertia that will lead to locating the oscillation centers at the centers of the front and rear unsprung masses (axles).

References

1. Gerd HZ, Edward BM (1959) Short time human tolerance to sinusoidal vibrations. Technical report, Air Force Aerospace Medical Research Lab Wright-Patterson AFB OH
2. Thomas DG (1992) Fundamentals of vehicle dynamics. Technical report, SAE Technical Paper
3. Max H (1955) Tractor seat suspensions for easy riding. SAE Trans 452–470
4. Bruce DVD (1968) Human response to vehicle vibration. SAE Trans 328–370
5. Richard AL, Fred P (1968) Analytical analysis of human vibration. SAE Trans 346–370
6. Society of Automotive Engineers (1965) Vehicle dynamics committee. In: Ride and Vibration Data Manual-SAE J6a: Report of Riding Comfort Research Committee Approved July 1946 and Last Revised by the Vehicle Dynamics Committee, October 1965. SAE

Chapter 10
Underride Protection Devices

Improvements to highway safety are in continual demand. One of the most severe instances of vehicle collision occurs as a result of vehicle weight and sizing mismatch. The fitment of Underride Protection Devices (UPDs) on trucks and trailers is studied as a method to improve crash compatibility between passenger vehicles and trucks and/or trailers involved in head-on, rear end, or side highway crashes. While some countries require the use of UPDs, no such regulation exists in North America. North America's use of Conventional Tractors also presents a variation to Cab-over Engine Tractors popular in Europe. In this chapter, an introduction to the following underride protection devices will be made:

1. Front Underside Protection Device (FUPD).
2. Rear Underride Protection Device (RUPD).
3. Side Underride Protection Device (SUPD).

10.1 Front Underside Protection Device (FUPD)

Passenger vehicle safety and occupant protection has improved by leaps and bounds in a relatively short period of time. These improvements are a direct result of the transient nature of safety tests and requirements demanded by governing bodies. There exists no point of rest in the search for improved safety, only relative performance increases. The state-of-possible philosophy summarizes the relentless demand for research in this area. It suggests vehicle safety improves continuously based upon technological and economic conditions [1]. The combination of these variables has provided interest and funding toward resolving issues presented in collisions involving severe vehicle mismatches. Specifically, the vehicles selected for study are chosen from both ends of the size spectrum. Instances of head-on type collisions between

M. El-Gindy and Z. El-Sayegh, *Road and Off-Road Vehicle Dynamics*,
https://doi.org/10.1007/978-3-031-36216-3_10

tractor-trailers and small passenger vehicles have proven to be of high severity. The use of Front Underride Protection Devices (FUPD) is a proposed method of decreasing the severity of these collisions. In general terms, the FUPD mounts to the front of the tractor and provides the colliding passenger vehicle with a structure against which it may react. The term "underride" describes the sliding of the passenger car beneath the tractor-trailer's structural members. In the case of severe underride, large values of intrusion are introduced to the passenger vehicle's occupant compartment. There are regulations in place in Europe [2], Australia [3], Japan [4], and India [5] necessitating the use of FUPDs. No such regulation currently exists in North America. Criticism can be found directed at the ineffectiveness of underride protection devices in regulated regions. The type of tractor utilized in North America is also distinct. A demand for understanding the possible benefits and design strategies of FUPDs for use in North America is therefore present.

Whether FUPDs do or do not become a requirement in North America, the demand for improved safety is ever-present. The consideration of small passenger vehicles continues to be of importance, as cars make up 55.4% of all light vehicles on Canadian roads. On these same roads tractors-trailers record the majority of total travel kilometers [6]. The instances of interaction between the two vehicle types are high as a result. Transport Canada's National Collision Database averages indicate 12.4% of all road fatalities involve tractor-trailers. Similarly, 18.3% of all fatalities involve collisions with heavy trucks. The term "heavy truck" groups together data consisting of straight trucks and tractor-trailers. Of the recorded fatalities, 73.6% were occupants of the vehicle colliding with the heavy truck. Further statistical analysis indicates 30.8% of the collisions mentioned may be classified as involving two vehicles in a head-on orientation [7]. The total economic cost resultant from motor vehicle collisions throughout the year 2000 in the United States was an estimated $230.6 billion. This expense includes the loss of 41,821 lives, 5.3 million injuries, and 28 million damaged vehicles [8]. The Insurance Institute for Highway Safety (IIHS) in association with the Fatality Analysis Reporting Systems (FARS) reported 3,413 fatalities involving heavy trucks in 2010 on American roads. This accounted for 9% of all vehicle collisions at the time. 97% of these accidents resulted in fatalities to the occupant(s) of the passenger vehicles. Of the heavy trucks involved, 75% were recorded as tractor-trailers with the remaining 25% being classified as straight trucks [9]. The Large Truck Crash Causation Study in the United States was published to further stress the significance of the vehicle underride problem [10]. There is a belief that a properly designed FUPD would aid in reducing North American highway fatality and injury rates. The process of properly designing an FUPD to accomplish this goal has many unknowns. The European-based industry driven Vehicle Crash Compatibility Project (VC-Compat) has highlighted the need for increased understanding of vehicle-to-vehicle structural interactions in collisions as well [11]. Hence with demands from public and industry sectors, the FUPD project was undertaken.

10.1.1 FUPD Regulations

Global regulation of Front Underride Protection Devices was pioneered by the Economic Commission for Europe (ECE). The initial fitment and compliance regulation set forth by this organization may be found within document ECE R93 [2]. The demand for application of ECE R93 is directed by requirement 2000/40/EC, as of the year 2000 [9]. Direct adoptions of this regulation ensued with India as AIS-069 in 2006 [5], Japan in 2007 [4] and Australia as ADR84/00 in 2009 [3]. There remains no such regulation requiring the use of FUPDs in North America.

ECE R93 details a number of geometric and load compliance requirements for the FUPD. Geometrically, the FUPD must have a maximum ground clearance of 400 mm, and a minimum frontal cross-section height of 120 mm. Quasi-static load testing is performed to ensure stiffness compliance. In such testing cases, the FUPD may be mounted upon a tractor or in a pre-equipped state on a test bench. Three load points (denoted as P_1, P_2, and P_3) are tested sequentially. These test points are as indicated in Fig. 10.1. A ram is to apply the designated load amount in the tractor's longitudinal direction for a minimum 0.2 s. Load points P_1 and P_3 are assigned a value of 80 kN, while point P_2 is assigned 160 kN. Upon completion of loading, the device is not to have exceeded 400 mm of deformation measured from the front of the tractor [2].

An attempt was made to determine the effectiveness of mandatory FUPD fitment in Europe. Unfortunately, due to a lack of properly classified statistical data, the results were found to be inconclusive [13]. Previous work from industry sponsored academic sources has provided commentary on the effectiveness of FUPDs designed to meet ECE R93 specifications. The resultant findings suggested setting a minimal ground clearance height in addition to demanding higher stiffness out of the devices [14]. A secondary work group has suggested implementing increased point load magnitudes, in an effort to improve the reliability of the device. Under such conditions, it was advised to upgrade P_1, P_2 and P_3 to 400 kN, 300 kN and 200 kN respectively [12]. Clearly there exists some concern regarding the insufficiencies of ECE R93 and the resultant poor FUPD performance. The regulation's inadequacies have also been addressed internally by industry leaders such as Volvo, Mercedes-Benz, and Scania. An external review of FUPDs set forth by these companies for testing indicated measured deformation values well shy of the maximum allowable 400 mm. In most cases, the maximum deformation experienced by the loaded devices lies within the 50 mm to 150 mm range [15].

The variations in tractor designs found in either Europe or North America are a result of the contrast in vehicle length measurement standards. The European style of tractor, named cab-over engine, was conceived as an attempt to shorten the overall length of the tractor-trailer combination. This was encouraged through directive 96/53/EC restricting the maximum combined tractor-trailer length to 16.5 m, while restricting the trailer length to a maximum length of 13.6 m. Effectively, the available design space for the tractor becomes restricted to 2.5 m [16]. Conversely in North America, the conventional style tractor remains the popular choice. The Surface

Fig. 10.1 ECE R93 geometry and test points [12]

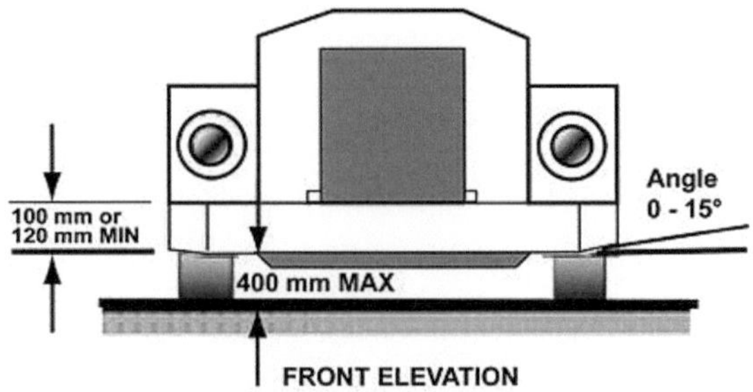

(a) Front View

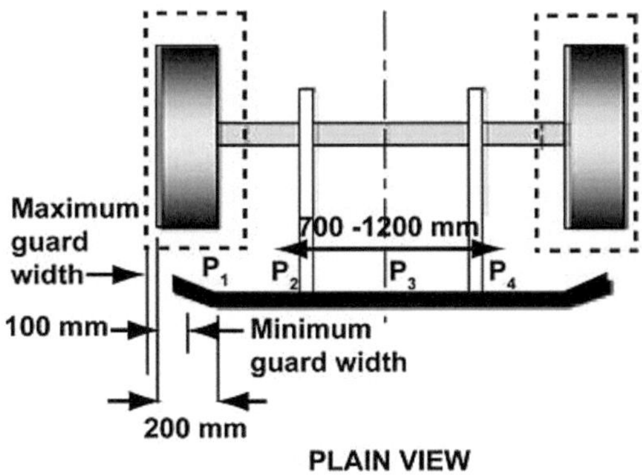

(b) Plain View

Transportation Assistance Act implemented by the United States in 1982 limits only the length of the trailer [17]. The present body of work will investigate Front Underride Protection Devices from a North American perspective, and only conventional style tractors are to be considered.

10.1.2 Variations in Tractor Design

There is a very distinct variation of tractor design used in North America compared to the designs used in Europe. The conventional tractor (Fig. 10.2a) dominates the North American trucking fleets with Volvo trucks, Mack trucks, etc. The European style of tractors (Fig. 10.2b), called cab-over engine tractors, are primarily used in Europe due to the contrast in vehicle length measurement standards, in an attempt to shorten the overall length of the tractor-trailer combination (Fig. 10.3). Directive 96/53/EC conforms the tractor-trailer combination to limiting the total length to a maximum of 16.5 m, and the trailer length to a maximum of 13.6 m. Consequently, this only allows for the tractor to be 2.5 m in length for the maximum trailer capacity

(a) Conventional tractor (b) EU Cab-over engine tractor

Fig. 10.2 Tractor-trailer variants

Fig. 10.3 Conventional tractor (North America) and cab-over engine tractor (EU) measurement standards [18]

(FKA, 2011). The North American standards only limits the maximum length of the trailer allowing for a range of styles of conventional tractors [17]. Due to the work's focus on North America, only FUPDs for conventional tractors would be considered.

With the flexibility of design, North American conventional trucks are designed with various differences. Figure 10.4 shows the various front axle positions in conventional trucks with either Axle Forward as shown in Fig. 10.4a which is closer to the front of the tractor, or Axle Back as shown in Fig. 10.4b which is closer to the rear axle. From a vehicle dynamic prospective, the most important difference between set forward or set back axles is the allowable payload that can be hauled. Farther apart truck axles (overall wheel base) increase the allowable carrying payload. However, due to limitations of payload capacities from bridge laws this configuration is limited. Setback configurations allow for better turning radius, better visibility, and increased fuel economy due to the allowable design space to slope the hood. Both styles are important in the design of FUPDs, however the axle back configuration will allow for more intrusion when impacted.

Another geometric design difference in conventional tractors is the placement height of the front bumper (Fig. 10.5); classified as above axle as shown in Fig. 10.5a, below axle Fig. 10.5b and center of axle Fig. 10.5c. As depicted, if the front

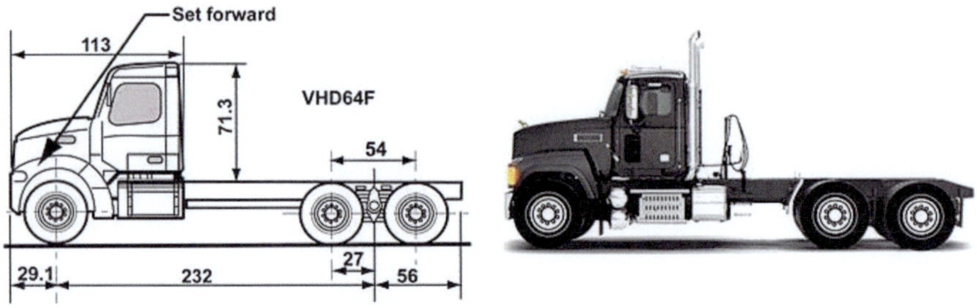

(a) Set Forward Front Axle Tractor Configuration

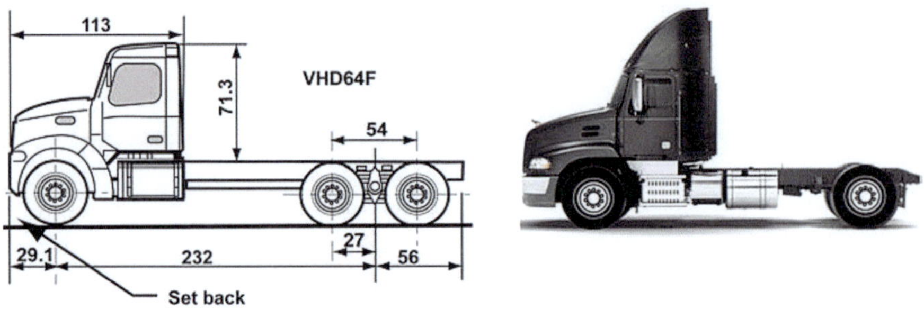

(b) Set Back Front Axle Tractor Configuration

Fig. 10.4 Axle position variants in conventional tractors (National highway traffic safety administration, heavy-vehicle crash data collection and analysis to characterize rear and side underride and front override in fatal truck crashes, 2013; Mack trucks, 2015)

(a) above axle (b) center of axle (c) below axle

Fig. 10.5 Front bumper height classification (National highway traffic safety administration, heavy-vehicle crash data collection and analysis to characterize rear and side underride and front override in fatal truck crashes, 2013) bumper height on the tractor is above the center of the axle it is deemed above axle. This also contributes to the available design space of a FUPD.

As the work is supported by Volvo Group Trucks Technology, the work only focuses on the development of a set back axle and center of axle bumper height configuration; specifically, a Volvo VNL series tractor.

10.1.3 Frontal Crash Testing

Passenger vehicles are regulated for frontal collision testing ensuring crash worthiness and occupant safety before being allowed on the roads. These tests are governed differently according to country. Primarily the vehicle is given an impacting forward speed into a rigid wall/barrier or deformable barrier. A rigid wall or barrier is an immovable and non-deformable structure which absorbs all applied energies, while allowing only the impacting object to deform. Fixed rigid barrier testing simulates a severe automotive collision (Fixed Rigid Barrier Collision Tests [19]. In addition, some regulations require the impact of the vehicle at full width (100% overlap) or impacting only a percentage of a barrier causing a smaller overlap. Overlap impact is the percentage that the barrier covers the vehicle. The United States Department of Transportation's National Highway Traffic Safety Administration (NHTSA) regulates all automotive crash performance under Federal Motor Vehicle Safety Standards (FMVSS). Canada's Department of Transportation regulates vehicle standards similarly to NHTSA under Canadian Motor Vehicle Safety Standards (CMVSS). North American frontal crash standard (FMVSS 208/CMVSS 208) requires automotive manufacturers to perform full wrap frontal collision tests at 56 km/h against a rigid barrier and reviews only occupant injury/safety using a 50th percentile adult male test dummy. The vehicle is given an initial impact speed of 56 km/h impacting at a rigid wall at 100% overlap, Fig. 10.6a (Canadian Motor Vehicle Safety Standards; National Highway Traffic Safety Administration, Occupant Crash Protection (FMVSS 208)). The forces of a single vehicle impacting the rigid wall are similar to the impact of two vehicles of the same weight just under the impact speed [20].

FMVSS/CMVSS regulations are a good step in the right direction for collision testing, however other occupant and vehicle safety organizations have criticized the regulations for being insufficient. Insurance Institute for Highway Safety (IIHS) is an independent, non-profit scientific and educational organization in the United States dedicated to a modern, scientific approach to identifying a full range of options for improving collision safety. The IIHS set a guideline for frontal testing with a different and stricter approach than NHTSA by evaluating at different overlap conditions and higher speed. The vehicle test impacts a rigid barrier with a deformable aluminum honeycomb at 64 km/h at a moderate overlap (40%) and small overlap (25%) configuration. The small overlap test simulates the impact of another vehicle or an object like a tree or utility pole when colliding with the front corner of a vehicle. IIHS's rating system evaluates the occupant injury metrics from Hybrid III dummies and structural performance of the vehicle's structure/safety cage (Insurance

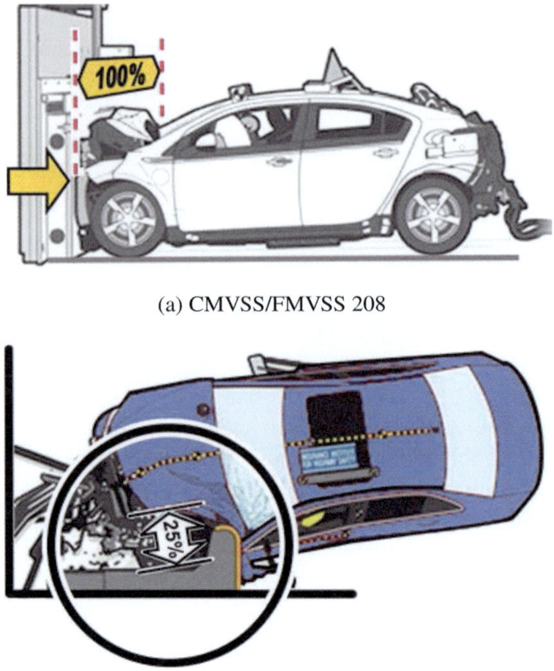

(a) CMVSS/FMVSS 208

(b) IIHS Small-Overlap Frontal Impact (PHILPOT, 2012)

Fig. 10.6 Frontal crash testing

Institute for Highway Safety, Frontal Crash Test, 2016). This standard is utilized by European Unions under Directive 96/79/EC under ECE R94. South Korea's Ministry of Land, Transport and Maritime Affairs (MOLIT) adopts both NHTSA and IIHS testing methods and regulated that both forms of testing be passed; full wrap frontal collision at 56 km/h and an offset frontal collision at 64 km/h (South Korea's Ministry of Land, Transport, and Maritime Affairs (MOLIT).

10.1.4 Rigid and Energy Absorbing Underride Protection

There are two main design concepts in the development of underride protection. The rigid FUPD concept consists of stiff structural components. Sufficiently stiff FUPD components provide the inherit crashworthiness features of the passenger vehicle with a reaction interface. With the design of an ideally rigid FUPD, all energy absorbed throughout the collision would be experienced by the components of the passenger vehicle. In reality the rigid FUPD will experience some form of deformation and may still be termed as rigid on a relative basis. An alternative approach taken by energy absorbing FUPDs (eaFUPD), attempts to further reduce crash severity by allowing both the passenger vehicle and tractor to absorb energy in

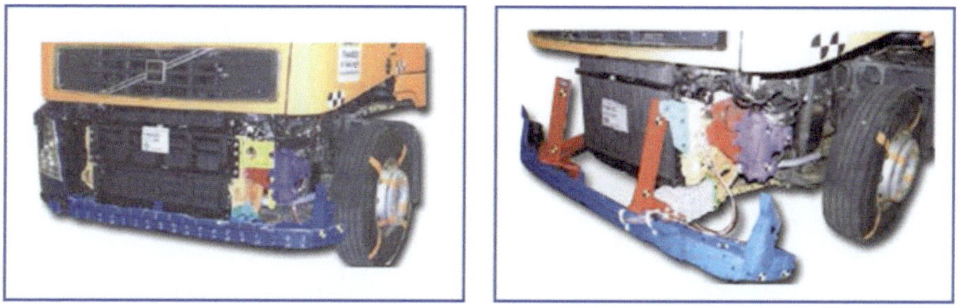

Fig. 10.7 VC-compat's volvo eaFUPD (left) and special eaFUPD (right) [22]

the collision. One analysis of such energy absorbing systems has suggested a design with the ability to absorb 130 kJ may provide protection at vehicle closing speeds elevated by 25–30% over a rigid application [21].

Additional research performed within the VC-Compat work groups further analyzed the functionality of both rigid and energy absorbing FUPDs. The database analysis suggested the implementation of theoretical eaFUPDs in place of rigid FUPDs could result in the saving of 160 lives, while mitigating 1200 serious injuries per year. In an effort to put theory into practice, physical studies were also conducted. The conceived energy absorption was to result from deformation experienced by metallic foam components built into the FUPD. Unfortunately, the device showed minimal performance gains in physical testing. The poor performance was attributed to difficulty in proper activation of energy absorbing elements [22]. In effect, even under ideal test conditions and vehicle alignment, the device did not perform as designed. Actual highway collisions experience a long list of unknowns including vehicle-to-vehicle alignment and closing speed. In such instances, the robustness of an eaFUPD would become even more important (Fig. 10.7).

10.1.5 Working Foundation

There was some investigatory work published previous to the undertaking of this project. Castellanos et al previously employed Finite Element Analysis (FEA) software LS-DYNA as a method of testing a proposed FUPD in a virtual environment. In this work, a 900 kg Geo Metro was collided with a stationary tractor fitted with an FUPD. From this work, a number of recommendations were made for further progress in the field, listed as follows [23]:

1. FUPD should absorb approximately 100 kJ of energy during static tests.
2. Ridedown acceleration (y-direction average) shall not exceed 20 g at 64 km/h with 900 kg Geo Metro (NCHRP350 Limit).
3. Minimum occupant compartment intrusion and vehicle underride

4. Shall not generate deceleration (50ms longitudinal-direction average) greater than 30 g at 56 km/h.
5. Accident Severity Index (ASI) should not exceed 3 at 64 km/h.

Krusper and Thomson proposed their own virtual FUPD investigation during the same time period. Within their work, LS-DYNA was selected as the method of analysis. Energy absorbing components, modeled by springs, were attached to a rigid bar. The rigid bar was set up to interact with a colliding passenger vehicle. A Ford Taurus model represented the passenger vehicle in this study. The work summarized the possibilities of accident mitigation through to the tuning of energy absorption stiffness parameters [24]. More imaginative work suggested the addition of an energy absorbing honeycomb structure beyond the front of a tractor. Tractor length limitations combined with aerodynamic requirements diffused the practicality of such designs [25]. An adaptive frontal structure, equipped with a full braking system, was also conceived. Hydraulic brakes activated by cylinders equipped with controllable flow valves were suggested as part of this theoretical approach. The system would replace plastic deformation with hydraulic dissipation to absorb energy during the collision. Feasibility for application became an issue not to be overcome by this concept [26]. The idea of designing underride guards to generate deflection in the passenger vehicle away from the tractor was also investigated. Large device overhangs coupled with excessive design space requirements plagued this concept. The post-deflection case was also considered, where it was proposed an out-of-control passenger vehicle heading toward traffic would not be an appropriate solution to the underride problem.

10.1.5.1 Testing Setup

The testing of underride protection devices undertaken within this work relies solely and completely on finite element analysis. Highly detailed publicly available open source vehicle models provide the source of testing validation. The methods undertaken to provide relatively inexpensive physical FUPD compliance testing involve the use of point loading as described by ECE R93. Alternatively, the device could be subjected to costly full vehicle collisions. Similar computational expense savings are experienced when using point load evaluation in a simulation environment. The comprehension of these two distinct types of FUPD testing procedures highlight the design strategy proposed.

In order to simplify the design process and boost project outcome, an effective design strategy must be clearly outlined. In the work presented, this strategy takes advantage of both sequential and simultaneous parameter isolation. The procedure is often repeated building upon that which has come before. With each level of parameter isolation, the design scope is reduced according to the simplification made. The reduction in design scope within this case is determined through three sequential experiments. Each experiment refines FUPD design search space with a goal of converging on ultimate performance. The selection of experiment type must therefore

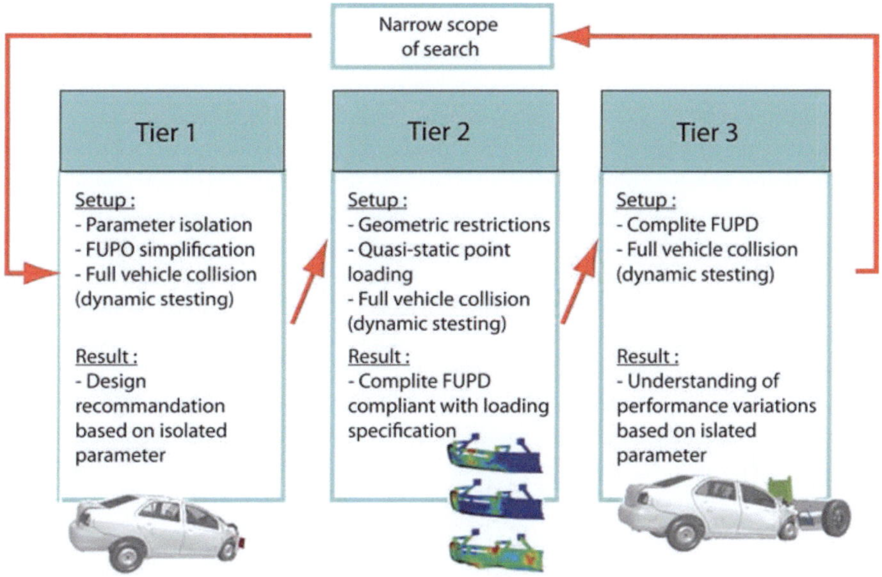

Fig. 10.8 Three tier design strategy

be initially broad enough to investigate a global perspective, and transition to a narrower region in the later phases. Figure 10.8 provides a visual of the sequential experimental approach in designing FUPDs for maximum performance.

The overall simplification process for design parameter isolation is presented as a three Tier approach. Tier 1 investigates collisions between full passenger vehicle models and a simplified FUPD. The simplified FUPD is initially represented by a rigid and fixed member, undergoing some form of variation to the isolated parameter. From this initial investigation, conclusions may be drawn to guide the development of a complete FUPD. The complete FUPD is a more realistic representation, with support member, frame contact, section thickness designations, and non-rigid material definitions. The complete FUPD development is accomplished within Tier 2. Specifics such as available component geometric spacing are considered through the use of Topology Optimization. Section thickness and support component shape are selected through the use of LS-OPT. These two sub-steps taking advantage of optimization employ relatively inexpensive quasi-static point load testing with implicit solutions. An FUPD design is developed as a result, which complies with point load specifications (either ECE R93 or above). Minimizing physical mass and deflection of the FUPD ensures the delivered guard is not over-designed. The FUPD is then presented to Tier 3. Tier 3 subjects the underride protection device to full vehicle collisions. This yields final commentary on the initially isolated variable and associated performance relationship. The search scope is refined, and a new design variable is selected for isolation reinitializing the process.

In order to understand Tier 2 more thoroughly, the following description is offered. Tier 2 implements quasi-static loading to ensure FUPDs are compliant with deformation constraints under three case point loads. The stage is composed of Topology Optimization as well as Shape and Size Optimization evaluated sequentially. A component design envelope is first defined with geometry modeled in CAD. Introducing the geometry to LS-DYNA allows meshing of the design envelope. Appropriate boundary conditions such as loading and constraints are also applied. The design envelope is then subjected to mass reduction through Topology Optimization. The Topology Optimization procedure yields insight regarding favorable locations of component load paths. These load paths are used as a guide in defining rough section geometry within CAD once again. Loading and boundary conditions are then applied to the refined component. Section thickness and cross-section shape are parameterized. This allows Shape and Size Optimization to be conducted simultaneously. The minimization of total FUPD mass and deformation under loading define the objective function. An array of feasible FUPD designs results from this process, of which one or more may be selected to advance to Tier 3 of testing.

10.1.5.2 Performance Evaluation Metric

The selection of appropriate evaluation criteria is paramount to the process of defining FUPD design variable relationships. A number of quantitative metrics for testing roadside safety hardware subjected to passenger vehicle collisions are outlined as part of NCHRP 350. Occupant Impact Velocity (OIV), Occupant Compartment Deformation Index (OCDI), and Theoretical Head Impact Velocity (THIV) are just a few of the suggested evaluation models [1]. For initial selection of FUPD performance comparison methods, one must outline the desired outcome. In other words, establish a definition related to FUPD performance increase. The common form of passenger vehicle regulation testing (as mentioned within the validation process) deals with a collision between the vehicle and either a rigid or deformable wall. The wall provides a large surface area with which the inherent crashworthiness features of the vehicle may react. While colliding head-on with a wall might be a rarity, these tests provide an optimistic view of passenger protection. Once the wall is replaced by a vehicle, such as would be the case for a head-on vehicle-to-vehicle collision, support component alignment issues may result in less than ideal use of crash features. If not given a proper reaction interface, passenger compartment intrusion values increase while energy absorption components go unused. By this methodology, in order to take full advantage of inherent passenger vehicle crash safety features, the reaction forces provided by the FUPD should replicate that of a wall. FUPD performance may therefore be analyzed through the comparison of impact forces in relation to vehicle collisions with a wall. FUPD robustness investigates this comparison over a range of initial crash conditions. Monitoring vehicle deformation supplies insight into the degree of underride experienced. Passenger vehicle deformation is measured from the vehicle's center of gravity throughout the duration of the collision. Graphing of the Impact Force vs. Deformation relationship creates the compatibility profile. The compatibility profile is selected for use when evaluating FUPD performance involving variation in design parameters at initial stages of development.

As outlined within the proposed strategy, initial understanding of performance improvement results in the narrowing of the search scope. After finding a number of improvements through iterative search, the compatibility profile method for performance evaluation becomes too broad of an analysis tool. The refinements made thus require a more sensitive method of evaluation. Occupant Compartment Intrusion is capable of filling this need. The IIHS maintains guidelines for the measurement of Occupant Compartment Intrusion for moderate overlap frontal collisions. Select locations from inside the passenger compartment surrounding the seated driver are monitored. Residual deformation values are measured. In total, eight points are evaluated, yielding an overview of the deformations experienced by the passenger vehicle safety cage [27]. In most instances, the residual values are calculated with respect to measurements made between the point of interest and a coordinate system near the rear seats [27]. Threshold ratings for performance evaluation are established on a per point basis.

10.2 Rear Underride Protection Device (RUPD)

Every large truck weighing 10,000 pounds or more is required to be equipped with a rear underride device if they were manufactured after January 24, 1998 [28]. These guards were based on the Federal Motor Vehicle Safety Standard (FMVSS) No. 223 "Rear Impact Guards", which specifies their requirements. They were designed to withstand certain forces without deflecting more than 125 mm. These forces are 50 kN at location P1, 50 kN at location P2 and 100 kN at location P3. Their positioning can be seen in Fig. 10.9 (Safety Assurance Office of Vehicle Safety Compliance).

The only form of underride protection devices on heavy trucks in North America are Rear Underride Protection devices (RUPDs). RUPDs are structural members

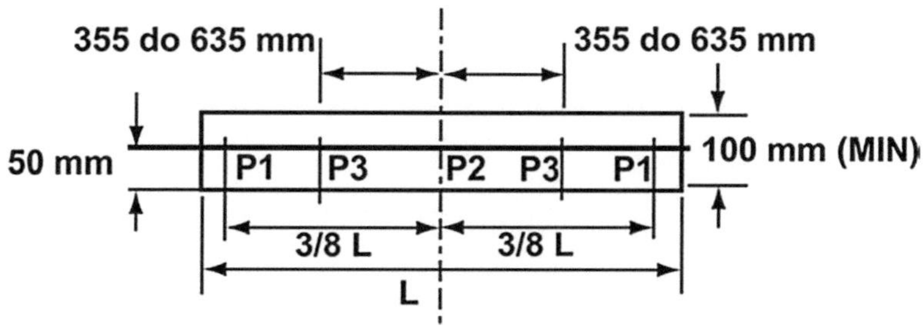

Fig. 10.9 Rear guard horizontal member and force positions (Safety assurance office of vehicle safety compliance)

Fig. 10.10 RUPDs on a manac trailer (new crash tests: underride guards on most big rigs leave passenger vehicle occupants at risk in certain crashes)

mounted on the rear end of the trailer, Fig. 10.9, for when passenger vehicles impact the trailer (2V1D). Geometric and performance testing methods are regulated by CMVSS 223 in Canada (Test Method 223 Rear Impact Guard, 2003), and FMVSS 223 in the United States (FMVSS 223, n.d.). Both regulations enhance the crashworthiness of the trailer and are in a good direction for providing underride protection to North America. However, the regulations have been under severe criticism after the Insurance Institute for Highway Safety physically tested various North American RUPDs with impacting passenger vehicles to prove the extreme inadequacies and failures of the regulation (Fig. 10.8) (New crash tests: Underride guards on most big rigs leave passenger vehicle occupants at risk in certain crashes) (Fig. 10.10).

The IIHS tested stationary trailers with various RUPDs being impacted by a passenger vehicle at 56 km/h with overlaps of 100%, 50% and 30%. This ignited pressure for the regulations to be revised by both countries (New crash tests: Underride guards on most big rigs leave passenger vehicle occupants at risk in certain crashes; NHTSA signals plan to address deaths in underride crashes; Rear underride guard mandate may extend to more trucks under NHTSA proposal).

Similar to ECE R93 quasi-static loading, RUPDs tests require three sequential points load testing along the structure, Fig. 10.11. Quasi-static point loads at P1 require a force of 50 kN, 50 kN at P2, and 100 kN at P3 with the maximum allowable deformation under the load of 125 mm. The Canadian regulation requires a secondary test involving the application of a full guard test of the RUPDs with a 350 kN quasi-static load (Fig. 10.12).

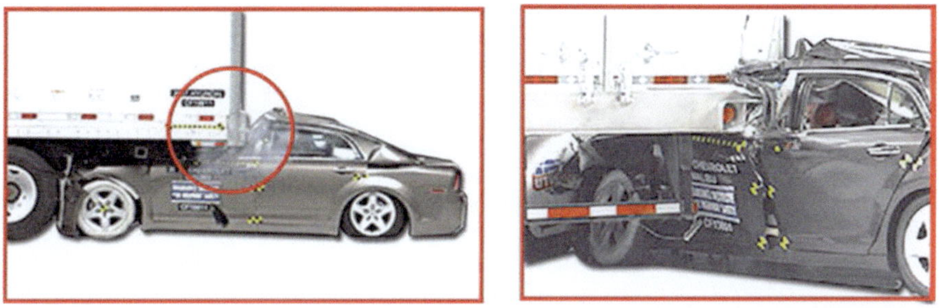

Fig. 10.11 Post-impact of passenger vehicles and the trailer rear end with a RUPD (new crash tests: underride guards on most big rigs leave passenger vehicle occupants at risk in certain crashes)

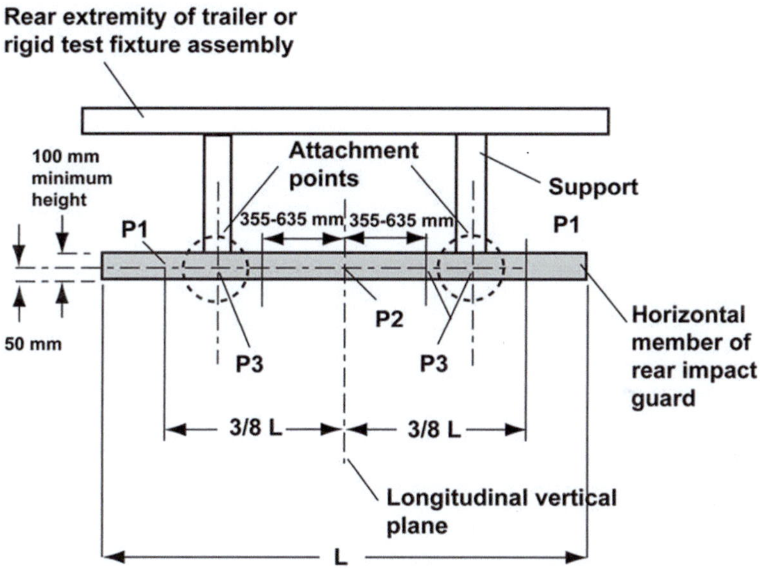

Fig. 10.12 Schematic of a secondary Canada Motor Vehicle Safety Standard (CMVSS 223) loading points for RUPDs

10.3 Side Underride Protection Device (SUPD)

10.3.1 Europe

On April 13, 1989, the council directive adopted a law concerning the use of lateral protection devices (LPD) for goods vehicles, trailers, and semi-trailers. This was adopted by Directive 89/297/EEC, which would then be called Regulation No. 73 [29] Addendum 72: Regulation No. 73 (UN/ECE, 2004). This regulation states that it is necessary to add side guards to heavy vehicles to reduce the risk of injury to

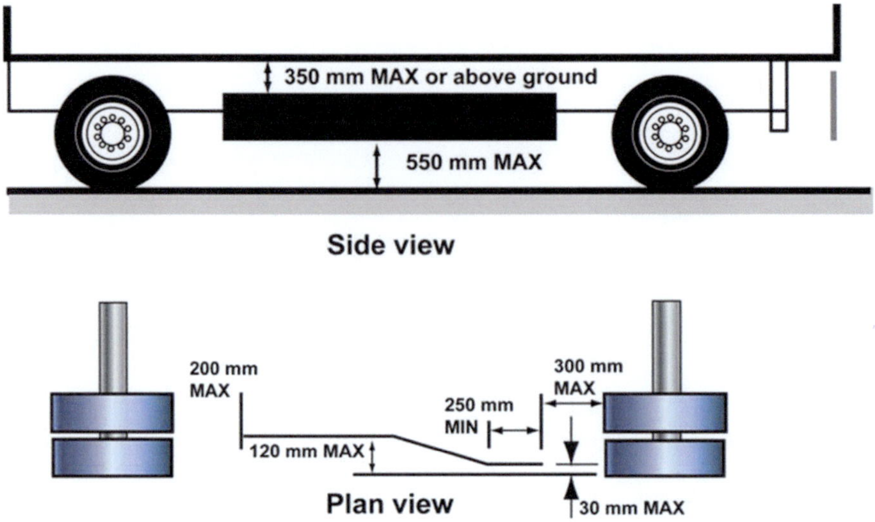

Fig. 10.13 Regulation no. 73 side lateral protection device requirements [12]

unprotected road users such as motorcyclists, pedestrians, and cyclists. The use of these guards would prevent those at risk from falling underneath the sides of large vehicles (The Council of the European Communities, 1989). This regulation applies to all vehicles in the N_2, N_3, O_3, and O_4 categories. The N_2 category represents all trucks with a gross vehicle weight between 3.5 and 12 tons and the N_3 category represents all trucks with a gross weight over 12 tons. The O_3 and O_4 categories represent all trailers with a gross vehicle weight between 3.5 to 10 tons and over 10 tons, respectively [30]. The law does not apply to tractors for semi-trailers, trailers designed for carrying long loads with long lengths and special purpose vehicles where it is not possible to fit the protection device (The Council of the European Communities, 1989). However, in 2008, Regulation No. 73 was changed to remove trailers with long loads of long lengths from the list [30]. Some of the technical requirements for such guards are shown in Fig. 10.13 [12].

Some dimensions of the LPD stipulate that the design must not add width to the vehicle and must not be more than 120 mm inward from the outermost surface. At the rear, the guard must not be more than 30 mm from the outer surface of the vehicle [29]. When testing the device, it must be able to withstand a force of 1 kN applied with a disk of 220 mm $\pm$ 10 mm in diameter. The device must react to the force with a maximum deformation of 30 mm over the rearmost 250 mm and a maximum deformation of 150 mm on the rest of the device.

10.3.2 Japan

Japan currently has regulations on pedestrian protection side guards, which are outlined in the Safety Regulations for Road Vehicle (Ministerial Ordinance) document and the Announcement (subordinate regulation). This Announcement states that ordinary-sized motor vehicles used to carry goods or vehicles with a gross vehicle weight greater than 8 tons must have pedestrian protection devices on both sides. These must comply with the requirements of strength and shape outlined in the Announcement. Their purpose is to prevent pedestrians, cyclists, and motorcyclists from being caught in the rear wheels of the larger vehicles. The requirements for this device are that the height must not exceed 450 mm from the ground and that the upper edge must not be higher than 650 mm from the ground [30].

10.3.3 Australia

The Australian government ruled that adopting side protection devices for heavy vehicles would not be beneficial. Although they adopted the European regulation that makes front underride protection devices mandatory, they determined that the cost of implementation of side guards was not justifiable. Their research showed that out of all the pedestrian underride accidents that happen each year, about 75% of the fatalities occur from a front impact, 10% from a rear impact and the rest from the side [30].

In November 2012, the Australian Trucking Association released an advisory procedure as a guide to operators about the use and application of side guards. The guards are to be applied to trailers on a voluntary basis. The regulation adopted is in compliance with Regulation 73 from Europe. Figure 10.14 from the procedure demonstrates some of the required dimensions for the devices [31].

It must be noted that all regulations mentioned above only specify devices that could be beneficial to unprotected road users such as cyclists, pedestrians, and motorcyclists. None of the laws discussed above outline devices that can be used for small passenger vehicles colliding with the side of heavy vehicles and tractor-trailers. If designed properly with optimization methods and new regulations, side guards can be implemented to protect pedestrians while being able to minimize the severity of damage caused by vehicle underrides.

10.3.4 Design Considerations

There is a very limited amount of research on the development of side underride protection devices for passenger car crashes. Some institutions have conducted private research to develop guards to reduce occupant injury experienced by passengers

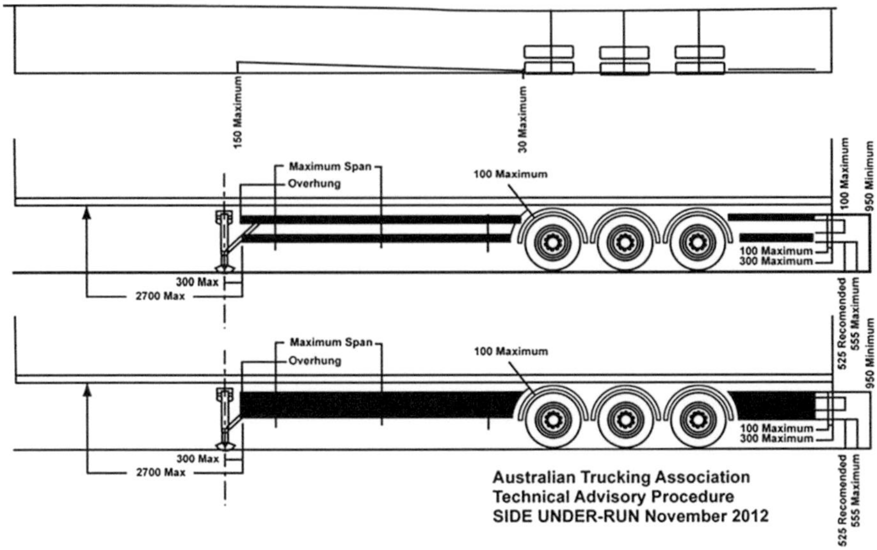

Fig. 10.14 Technical advisory procedures drawing for a side under-run device from the Australian trucking association [31]

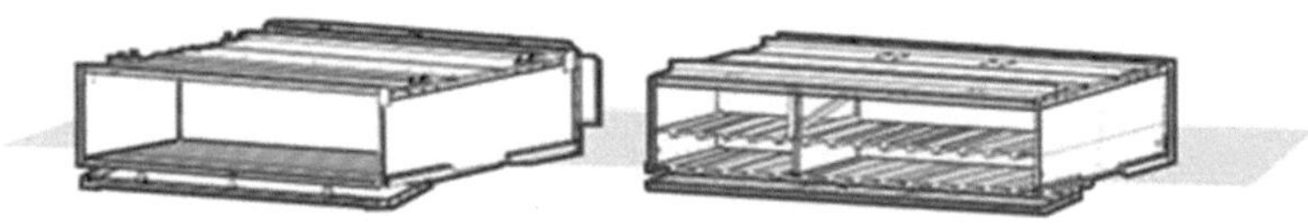

Fig. 10.15 Schematics of a conventional pallet box and reinforced pallet box [32]

of small vehicles. Some thesis work has also been done to create underride device designs. Others have studied the effect of large vehicle side underrides on smaller vehicles and their passengers.

10.3.4.1 Tested Designs

A project from APROSYS (Advanced Protection Systems) was started in 2004 to explore and develop a SUPD which would be integrated into a pallet box. This box allows for placement underneath the trailer, between the axles and the kingpin [32]. In the first test conducted, a Fiat Punto was driven at 65 km/h into a pallet box and in the second test, a Punto was driven into a reinforced pallet box, all while mounted on a test rig. The tests were also performed using an actual tractor-trailer. The trailer had a triple axle configuration, providing less area for side underride. It was further noted that a larger device would have to be installed on dual-axle trailers, which would result in more added weight (Fig. 10.15).

Fig. 10.16 Post-crash views of conventional and reinforced pallet box [32]

Figure 10.16 shows the conventional pallet box, which weighs 240 kg and the reinforced pallet box, which weighs 410 kg. Figure 10.17 shows the after effect of a vehicle colliding with the boxes.

For each crash, parameters such as the acceleration severity index (ASI), theoretical head impact velocity (THIV), and postcrash head deceleration (PHD) were calculated to evaluate and compare both designs. The results were very different. The data reported were the ASI for the conventional design was 1.2 g, the THIV was 50.95 km/h, and the PHD was 11.24 g. From the values, only the THIV was higher than its maximum acceptable value. The values for the reinforced pallet design were 2.29 g for the ASI, 70.55 km/h for the THIV, and 5.00 g for the PHD. Both the ASI and the THIV were higher than their limits. This, however, can be acceptable due to the lesser amount of vehicle deformation compared to the conventional design, which can be seen in Fig. 10.17. The report also outlines the deformations observed by the tests conducted. They concluded that the reinforced box did benefit passenger cars in the event of a crash [32].

Another study was conducted in a thesis paper submitted to the Wichita State University. The research mainly focused on rear underrides with a pliers guard. The guard was also tested on the side of the tractor to reduce side impact injury. The same principles and regulations that were applicable for the rear underride guards were used, since there were no current regulations for the side. The tests were conducted using three different speeds, 30, 40, and 50 mph, with a Ford Taurus weighing 1378 kg. Each speed was tested with three different configurations: without the guard, with the new pliers guard, and with the new pliers guard with horizontal cables [33].

The tests indicated that with the addition of the guards, there was a significant improvement in the deformation of the passenger compartment. There was also noticeable change between the collision with and without the guards; however, little change was shown between the results of the pliers guard and the pliers guard with horizontal cables. As the speed increased, so did the severity of the accidents. When comparing the accelerations at 50 mph, it was observed that without a guard, the Taurus experienced 39.9 g with the new pliers guard, while 48.1 g acceleration was observed and with the addition of horizontal bars, 44.5 g was recorded. The vehicles came to a quicker stop when the guard was installed, rather than passing underneath

Fig. 10.17 New pliers guard on side of large vehicle [33]

the truck, as happened when the guard was not present [33]. Table 10.1 outlines the findings of the tests.

Another paper was written concerning the scientific approach to side underrides with tractor-trailers [34]. This group reported that a main area of concern in obtaining information on side underride collisions lies in the reporting systems used by police. Further noted was a lack of appropriate coding and identification to properly report this type of accident. Some further observations made were small cars traveling at speeds over 25 mph completely passed under trailers; larger vehicles such as sedans that collide with trailers have impact on their roofs, which causes serious damage at speeds greater than 35 mph. They also noted that these larger sedans do not necessarily pass completely under the trailers. Larger, full-size vehicles were reported to underride trailers at speeds of 40 mph or higher. A series of crash tests created by the Midwest Institute of Safety were analyzed. These tests consisted of 32 configurations at various speeds from 7 to 37 mph and different angles of impact. The purpose was to develop an equation to determine the initial speed at which underride can occur. Two patterns were observed pertaining to deformation of the vehicle during side underride. The first was that the A pillar would deform by bending downward and backward when coming into contact with the trailer side edge. This causes the roof of the vehicle to crush rearwards and eventually flatten. The second pattern was when the A pillar did not catch the trailer side edge and instead, wedged the vehicle under the trailer [34].

Both concepts outlined in this section represent a solution to underride. However, there is much more work needed to design feasible guards for trailers of all sizes and configurations. The pallet box system proved to be beneficial to prevent cars from underriding. On the other hand, incorporating the system on a full 48-foot trailer can prove to be difficult due to the size and weight of such a device. The pliers guard also aided in stopping the car; however, this device was relatively small and was only tested using the principles for rear underride devices. For larger trailers or straight trucks, the results of tests conducted to validate this design could not be used. To advance the development of these devices, a regulation such as the one for the front and rear guards would have to be developed to test side guards. This regulation will have to consider all the possible configurations related to side underride crashes. The

Table 10.1 Comparison of the configurations at different speeds [33]

	30 mph	40 mph	50 mph
	48 km/h	65 km/h	80 km/h
No guard			
Displacement of tunnel at the end (mm)	2140	2590	3130
Velocity of tunnel at the end (mm/s)	4730	4920	5260
Maximum tunnel longitudinal acceleration (G)	–34	–39	–39.9
Maximum tunnel transverse acceleration (G)	22.2	25.1	28.9
Maximum tunnel velocity acceleration (G)	–32.8	–40.3	–44.8
New Pliers Guard			
Displacement of tunnel at the end (mm)	1460	1750	2130
Velocity of tunnel at the end (mm/s)	831	1870	2200
Maximum tunnel longitudinal acceleration (G)	–36.2	–42.3	–48.1
Maximum tunnel transverse acceleration (G)	–15.1	–13.8	–23.9
Maximum tunnel velocity acceleration (G)	–22.8	24	–35.2
New pliers guard with horizontal cables			
Displacement of tunnel at the end (mm)	1530	1830	2190
Velocity of tunnel at the end (mm/s)	735	1350	2030
Maximum tunnel longitudinal acceleration (G)	–37.6	–48.3	–44.5
Maximum tunnel transverse acceleration (G)	–15.2	–21.3	–23
Maximum tunnel velocity acceleration (G)	26.6	26.1	–34.6

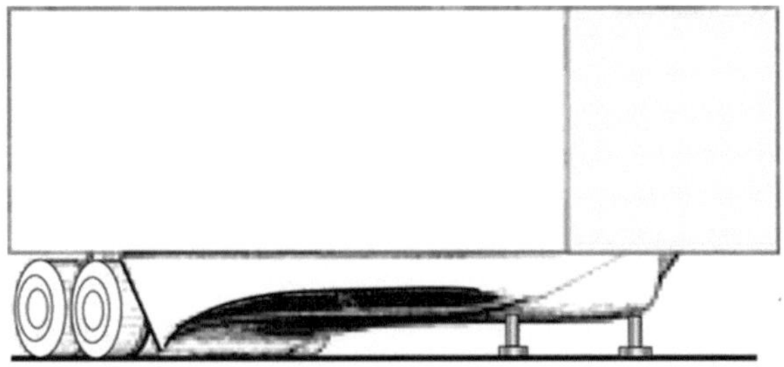

Fig. 10.18 Patent no.: US 7,780,224 B2 [35]

overall length of the device will be much greater than that of front and rear guards. Because of this, these devices require their own tests and regulations to be validated. Due to their size, optimization is required to produce the most efficient device while keeping the weight to a minimum.

10.3.4.2 Patent Overview

Some patents have been filed corresponding to possible side underride protection device designs. In 2008, a patent was filed regarding a crash attenuating underride guard. The proposed guards are described as aerodynamic fairings that can be used as underride guards as observed in Fig. 10.18. These are molded blocks placed underneath the trailers to reduce drag and prevent collision injury. The device itself consisted of an angled front section located in front of the trailer landing gear, allowing deflection of air and reduction of drag. The rear section is to be an angled section located in front of the rear wheels, allowing for air deflection away from the wheels and the rear of the trailer [35].

Another patent design utilizes a cable system as an underride guard for trailers. This system includes front and rear mounted brackets with cables extending over the length of the trailer which can be seen in Fig. 10.19. The patent claims that the front and rear mounting brackets must be located apart from each other, covering the length of each side of the trailer to prevent vehicle intrusion [36].

A third patent, which was published in 2008, claims the design of a side underride guard, meant for vehicles that have trailer portions with high ground clearance. The underride guard is to be attached to the trailer, along its side edge. An elongated member is to be used to stop vehicles from passing beneath the trailer. This member must be at an adequate distance from the bottom of the trailer to efficiently prevent cars from underriding. The elongated member is held in place by support members, consisting of both an upright and an angled beam, as can be seen in Fig. 10.20 [6].

The patents analyzed show possible designs for underride guards that can be feasible solutions to prevent cars from passing under trailers. The patent documentation explains the design and principles of the guards. They do not elaborate on the performance of the designs or give any detail on their ability to properly prevent underride.

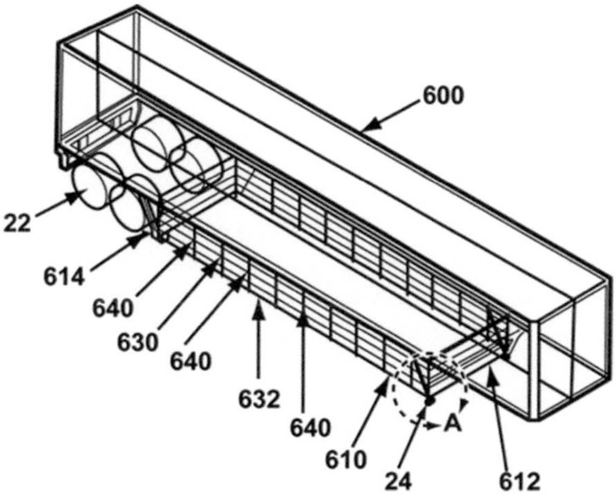

Fig. 10.19 Patent no.: US 8,162,384 B2 [36]

Fig. 10.20 Patent no.: US 2008/0116702 A1 [6]

These are design ideas with no publicly available documented testing, such as would allow for evaluation of their performance.

10.3.5 Aerodynamic Drag Reduction

Transport companies can see benefits from adding underride devices to their trailers by incorporating them with aerodynamic fairings. Research from Transport Canada on adapting aerodynamic fairings was conducted to investigate the benefits of these devices. Transport Canada has a Freight Sustainability Development Program to analyze and research objects that can reduce the fuel consumption of heavy vehicles [37]. It is stated that 50% more fuel is required to overcome drag when driving on highways. The addition of fairings can benefit tractor-trailers by reducing the turbulence, the drag and the crosswinds and creating smoother airflow. Two different kinds of fairing and side skirts were used for this research: the Freight Wing Belly Fairings and the Freight Wing "Low Rider" Fairings. Findings show that fuel consumption was reduced by 6.4%, which resulted in the saving of 339 liters of fuel per tractor per month. The results also determine that this amount of fuel would reduce greenhouse gases by 925 kg per trailer per month. Out of the 57 tractor-trailers used for testing, a total of 7,134 liters of fuel was saved, along with a 19,475 kg reduction in greenhouse gas emissions [37].

Another study consisted of testing and comparing four different types of drag reduction devices. The work concluded that the best design was a wedge type side skirt. This device starts at the front of the trailer as a point and extends along the length until it reaches each wheel, creating a long and narrow wedge-like shape. The alternative design was a simple side skirt design on each side of the trailer [38].

The National Research Council of Canada conducted wind tunnel testing of different drag reduction devices [39]. In this study, a Volvo VN660 was tested in a 9 m by 9 m wind tunnel with 28 ft and 40 ft trailers. The research was based on the estimation that trucks drive approximately 200,000 km a year and that 130,000 of those kilometers are clocked up cruising on highways at 100 km/h. Where the underbody drag components were concerned, the best overall design was the Freight Wing Belly Fairing with the low rider option. This device was tested on the 40 ft trailer. It was estimated that this device saved an average of 2,970 liters of fuel per year per transport. The runner up device was the Laydon Composites main and rear side skirts. This device had approximate savings of 2,355 liters of fuel per year [39]. By adding these devices to underride guards, manufacturers can benefit from the added fuel savings while making roads safer for small passenger cars and other road users.

10.3.5.1 Test Methods

The National Cooperative Highway Research Program (NCHRP) outlines in its report the recommended procedures for safety performance evaluation of highway features and a number of methods used to evaluate vehicle safety. This report outlines different types of evaluation criteria that can be used as guidelines for testing side underride devices. When test vehicles such as small cars are used for crash tests, the 700°C and 820°C class should be used. The 700°C class vehicle has a specified weight of 700 ± 25 kg with a 75 kg dummy. The 820°C weighs approximately 820 ± 25 kg, with a dummy weighing 75 kg. It must be noted that both of these vehicles were the top two most sold models for their given years and they were not more than 6 years old at the time of the test [40].

Calculations can be established post-accident to determine certain aspects of the collision. These include the THIV, the PHD, and the ASI. The THIV is used to determine the severity of the impact of a crash on a human [41]. The PHD calculates the acceleration at which the head travels during the crash [40]. Last, the ASI calculates the severity of a crash for an occupant sitting in a vehicle, with the x, y, and z axis accelerations over a moving time interval of ±50 ms [42].

The regulations and methods mentioned above can be used as benchmark configurations for testing side underride protection devices in North America.

10.4 Evaluation Example of Passenger Car Occupant Compartment Intrusion

The selection of appropriate evaluation criteria is paramount to the process of defining FUPD, RUPD, and SUPD design variable relationships. A number of quantitative metrics for testing roadside safety hardware subjected to passenger vehicle collisions are outlined as part of NCHRP 350. Occupant Impact Velocity (OIV), Occupant Compartment Deformation Index (OCDI), and Theoretical Head Impact Velocity (THIV) are just a few of the suggested evaluation models [1]. For initial selection of an UPD performance comparison method, one must outline the desired outcome. In other words, establish a definition related to an UPD performance increase. The common form of passenger vehicle regulation testing (as mentioned within the validation process) deals with a collision between the vehicle and either a rigid or deformable wall. The wall provides a large surface area with which the inherent crashworthiness features of the vehicle may react. While colliding head-on with a wall might be a rarity, these tests provide an optimistic view of passenger protection. Once the wall is replaced by a vehicle, such as would be the case for a head-on vehicle-to-vehicle collision, support component alignment issues may result in less than ideal use of crash features. If not given a proper reaction interface, passenger compartment intrusion values increase while energy absorption components go unused.

By this methodology, in order to take full advantage of inherent passenger vehicle crash safety features, the reaction forces provided by the UPD should replicate that of a wall. The UPD performance may therefore be analyzed through the comparison of impact forces in relation to vehicle collisions with a wall. Variations in UPD robustness investigate this comparison over a range of initial crash conditions. Monitoring vehicle deformation supplies insight into the degree of underride experienced. Passenger vehicle deformation is measured from the vehicle's center of gravity throughout the duration of the collision. Graphing the Impact Force vs. Deformation relationship creates the compatibility profile. The compatibility profile is selected for use when evaluating an UPD performance involving variation in design parameters at initial stages of development.

As outlined within the proposed strategy, initial understanding of performance improvement results in the narrowing of the search scope. After finding a number of improvements through iterative search, the compatibility profile method for performance evaluation becomes too broad of an analysis tool. The refinements made thus require a more sensitive method of evaluation. Occupant Compartment Intrusion is capable of filling this need. The IIHS maintains guidelines for the measurement of Occupant Compartment Intrusion for moderate overlap frontal collisions. Selected locations from inside the passenger compartment surrounding the seated driver are monitored. Residual deformation values are measured. In total, eight points are evaluated, yielding an overview of the deformations experienced by the passenger vehicle safety cage [27].

In most instances, the residual values are calculated with respect to measurements made between the point of interest and a coordinate system near the rear seats [27].

Table 10.2 Occupant compartment intrusion measurement points

1	Footrest
2	Left toepan
3	Center toepan
4	Right toepan
5	Brake pedal
6	Left instrument panel
7	Right instrument panel
8	Door or A-pillar (modified case)

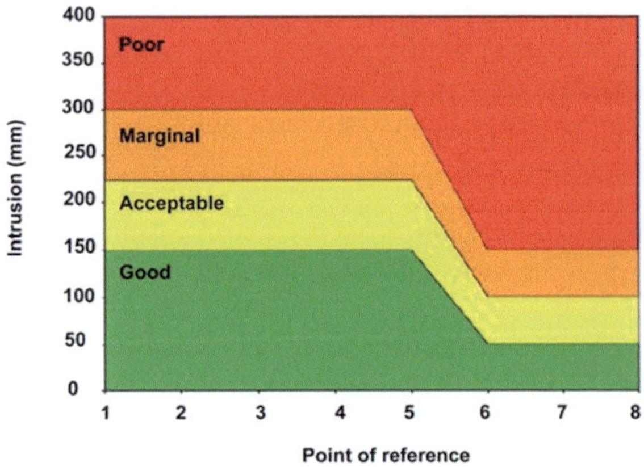

Fig. 10.21 IIHS occupant compartment intrusion thresholds [27]

Threshold ratings for performance evaluation are established on a per point basis. Table 10.2 outlines the points under scrutiny. A slight modification to the intrusion points of measurement is additionally set forth to better reflect the characteristics of an underride crash. In the cases denoted as "modified", measurement point 8 is taken from the A-Pillar instead of the door. The values from Table 10.2 are referenced within the adjacent figure shown in Fig. 10.21. The figure also classifies severity of intrusion based on measurement location with respect to known thresholds. LS-DYNA's IIHS card enables analysis of this data post-crash.

References

1. Malcolm HR (1997) The use of finite element analysis in roadside hardware design. Int J Crashworthiness 2(4):333–348
2. UNECE (1994) Regulation no. 93–front underrun protection
3. Albanese A (2008) Minister for infrastructure. In: Transport, regional development and locai government speech to the sydney institute, p 10
4. Sukeawa Y (2006) Japan's approach to fupd. In: Final workshop invitation of VC-COMPAT, Japan Automobile Research Institute

5. MacDonald T, El-Gindy M, Ghantae S, Critchley D, Ramachandra S (2013) A comprehensive review of front underride protection device design and evaluation. Int J Veh Syst Modell Test 8(3):282–294
6. Galipeau-Bélair P, El-Gindy M, Ghantae S, Critchley D, Ramachandra S (2013) A review of side underride statistics and protection device literature and designs. Int J Heavy Veh Syst 20(4):361–374
7. Todd M, Moustafa E-G, Srikanth G, Sarathy R, David C (2013) Rigid front underride protection device (fupd): compatibility and development via optimization. In: International design engineering technical conferences and computers and information in engineering conference, vol 55843. American Society of Mechanical Engineers, p V001T01A019
8. Blincoe LJ, Seay AG, Zaloshnja E, Miller TR, Romano EO, Luchter S, Spicer RS et al (2000) The economic impact of motor vehicle crashes. United States, National Highway Traffic Safety Administration, Technical report, p 2002
9. MacDonald T, El-Gindy M, Ghantae S, Critchley D, Ramachandra S (2014) Front underride protection device (fupd) development: design strategy with simultaneous loading. Int J Crashworthiness 19(2):196–207
10. Daniel B, John W, et al (2012) Survey of the status of truck safety: Brazil, China, Australia, and the United States
11. Dam R, de Coo PJA (2007) Report detailing final test procedure (s) and performance criteria for car to truck underrun. BASt, VC-Compact
12. John L, George R (2002) Review of truck safety: stage 1: frontal, side and rear underrun protection, vol 194. Monash University Accident Research Centre
13. Chislett W, Robinson TL (2010) Investigating the real-world effectiveness of introducing mandatory fitment of front underrun protection to heavy goods vehicles
14. Aleksandra K, Robert T (2008) Crash compatibility between heavy goods vehicles and passenger cars: structural interaction analysis and in-depth accident analysis. In: International Conference on Heavy Vehicles
15. Anderson J (2003) Inventory of current underrun devices. Cranfield Impact Centre, VC-Compat, Cranfield
16. Thomas W, Sven G, Michael F, Michael H, Ralf M (2011) Design of a tractor for optimised safety and fuel consumption. Technical report
17. Freund DM (2010) What a difference a quarter-century makes: Commercial motor vehicle and driver safety in the united states, 1984–2009. In: HVTT11: International Heavy Vehicle Symposium. Melbourne, Victoria, Australia, p 2010
18. Linus H, Björn B (2009) European truck aerodynamics–a comparison between conventional and coe truck aerodynamics and a look into future trends and possibilities. In: The Aerodynamics of Heavy Vehicles II: Trucks, Buses, and Trains. Springer, pp 469–477
19. Luo J, An H, Kamasamudram K, Currier N, Yezerets A, Watkins T, Allard L (2015) Impact of accelerated hydrothermal aging on structure and performance of cu-ssz-13 scr catalysts. SAE Int J Eng 8(3):1181–1186
20. Eric RT, Daniel LC, Sarah S, Anne TM (2017) Crash risk factors for interstate large trucks in north carolina. J. Saf. Res. 62:13–21
21. Alexander B, Michael K, Lars R, Ulrich B (2003) Passive safety of trucks in frontal and rear-end collisions with cars. In: Proceedings: International technical conference on the enhanced safety of vehicles, vol 2003. National Highway Traffic Safety Administration, p 10
22. TNO TRL, BASt C, Volvo UTAC, DAF GDV, Scania DC, CIM UPM (2007) Improvement of vehicle crash compatibility through the development of crash test procedures. In: TRL and TNO
23. Manuel C, Moustafa E-G, Chris F, Daniel M, Ali OA (2010) Truck front underride development: literature survey. Int J Heavy Veh Syst 17(1):18–34
24. Krusper A, Thomson R (2010) Energy-absorbing fupds and their interactions with fronts of passenger cars. Int J Crashworthiness 15(6):635–647
25. Bertrand L, Ivan MR (2012) Car-to-truck frontal crash compatibility. Master's thesis

26. Willem W (2005) Adaptive frontal structure design to achieve optimal deceleration pulses. In: 19th International Technical Conference on the Enhanced Safety of Vehicles Washington DC, USA, Paper, vol 05-0243. Citeseer
27. MacDonald T, EL-Gindy M, Srikanth G, Sarathy R, David C, (2014) Dual stage front underride protection devices (dsfupds): Collision interface and passenger compartment intrusion. Technical report, SAE Technical Paper
28. Kirk A (2010) The effectiveness of underride guards for heavy trailers. Technical report
29. Bulmer S (1994) Institutions and policy change in the european communities: the case of merger control1. Pub Admin 72(3):423–444
30. Patten JD, Tabra CV (2010) Side guards for trucks and trailers phase 1: Background investigation. Technical report
31. Australian Trucking Association et al (2012) Side under run protection. In: Industry Technical Council Advisory Procedure, Forrest
32. Aparicio A, Puttenstein H, Feist F (2009) Demonstration of truck side design improvements. In: Advanced Protection Systems (APROSYS)
33. Venkata Kiran KB (2006) Evaluation of energy absorbing pliers underride guards for rear and side of large trucks. PhD thesis
34. Trego A, Enz B, Head D, Oshida Y (2003) A scientific approach to tractor-trailer side underride analysis. Technical report, SAE Technical Paper
35. Mark AR (2010) Crash attenuating underride guard, August 24 2010. US Patent 7,780,224
36. Richard JG, Leonard WB, Rodney PE, Francis SS (2012) Side underride cable system for a trailer, April 24 2012. US Patent 8,162,384
37. Freight W (2011) Trial program on aerodynamic fairings. In: Transport Canada
38. Jason O, Kambiz S (2004) An experimental study of drag reduction devices for a trailer underbody and base. In: 34th AIAA Fluid Dynamics Conference and Exhibit, pp 2252
39. Jason L, Kevin RC (2006) Full-scale wind tunnel tests of production and prototype, second-generation aerodynamic drag-reducing devices for tractor-trailers. Technical report, SAE Technical Paper
40. Ross Jr HE, Sicking DL, Richard AZ, Jarvis DM (1993) Recommended procedures for the safety performance evaluation of highway features. Number 350
41. Laker IB, Payne AR (1992) International harmonization of testing and evaluation procedures for roadside safety features-theoretical head impact velocity concept. Transp Res Circular 396
42. Ross HE, Edward RP (1972) Criteria for guardrail need and location on embankments; volume i: Development of criteria. Texas Transp Inst, Texas A&M Univ Res Rept 140(4)